Engineering Economy

Tenth Edition

Engineering Economy

E. Paul Degarmo
Emeritus, University of California, Berkeley

William G. Sullivan
*Virginia Polytechnic Institute and State
    University*

James A. Bontadelli
The University of Tennessee, Knoxville

Elin M. Wicks
University of Missouri, Columbia

Prentice Hall
Upper Saddle River, New Jersey 07458

Library of Congress Cataloging-in-Publication Data

Engineering economy / E. Paul DeGarmo ... [et al.]—10th ed.
    p. cm.
   Rev. ed. of: Engineering economy / E. Paul DeGarmo, William G.
Sullivan, James A. Bontadelli. 9th ed. c1993.
  Includes bibliographical references and index.
  ISBN 0-13-382193-5
  1. Engineering economy. I. DeGarmo, E. Paul (Ernest Paul), 1907–.
  II. DeGarmo, E. Paul (Ernest Paul), 1907–. Engineering economy.
TA177.4.E539 1997
658.15′5—dc20   96-26501
          CIP

Editor-in-Chief: **Marcia Horton**
Acquisitions editor: **Alice Dworkin**
Managing editor: **Bayani Mendoza de Leon**
Production editor: **Irwin Zucker**
Art director: **Amy Rosen**
Assistant art director: **Rod Hernandez**

Creative director: **Paula Maylahn**
Cover design: **Rosemarie Votta**
Cover illustration: **Joseph Taylor**
Interior design: **Elm Street Publishing Services**
Manufacturing buyer: **Donna Sullivan**
Editorial assistant: **Phyllis Morgan**

Earlier editions entitled *Engineering Economy*, copyright © 1993, 1988, 1984, 1979, 1973, 1967, and 1960 by
Macmillan Publishing Company. Earlier editions entitled *Introduction to Engineering Economy*, copyright
© 1953 by Macmillan Publishing Company, and copyright © 1942 by E. P. DeGarmo and B. M. Woods.

The author and publisher of this book have used their best efforts in preparing this book. These efforts
include the development, research, and testing of the theories and programs to determine their effectiveness. The author and publisher make no warranty of any kind, expressed or implied, with regard to these
programs or the documentation contained in this book. The author and publisher shall not be liable in
any event for incidental or consequential damages in connection with, or arising out of, the furnishing,
performance, or use of these programs.

Printed in the United States of America

10 9 8 7 6 5 4 3 2 1

ISBN 0-13-382193-5

Prentice-Hall International (UK) Limited, London
Prentice-Hall of Australia Pty. Limited, Sydney
Prentice-Hall Canada Inc., Toronto
Prentice-Hall Hispanoamericana, S.A., Mexico
Prentice-Hall of India Private Limited, New Delhi
Prentice-Hall of Japan, Inc., Tokyo
Simon & Schuster Asia Pte. Ltd., Singapore
Editora Prentice-Hall do Brasil, Ltda., Rio de Janeiro

*This book is dedicated
to the ones we love.*

*E. Paul Degarmo
William G. Sullivan
James A. Bontadelli
Elin M. Wicks*

Contents

CHAPTER 3

Principles of Money–Time Relationships   63

PART II

Applications of Engineering Economy   139

CHAPTER 4

Applications of Money–Time Relationships   140

CHAPTER 5

Comparing Alternatives   188

CHAPTER 6

CHAPTER 7

CHAPTER 8

CHAPTER 9

CHAPTER 10

PART III
Special Topics in Engineering Economy   435

CHAPTER 11

CHAPTER 12

CHAPTER 13

CHAPTER 14

Probabilistic Risk Analysis   529

PART IV

Appendixes   575

APPENDIX

Preface

Engineering economy—what is it and why is it important? The initial reaction of many engineering students to these questions is "Money matters will be handled by someone else. It is not something I need to worry about." In reality, any engineering project must be not only physically realizable but also economically feasible. For example, a child's tricycle could be built with an aluminum frame or a composite frame. Some may argue that because the composite frame will be stronger and lighter, it is a better choice. However, there is not much of a market for thousand dollar tricycles. One might suggest that the above argument is ridiculously simplistic and that common sense would dictate choosing aluminum for the framing material. Although the scenario is an exaggeration, it reinforces the idea that the economic factors of a design weigh heavily in the design process, and that engineering economy is an integral part of that process, regardless of the engineering discipline.

In broad terms, for an engineering design to be successful, it must be technically sound and produce some benefits. These benefits must exceed the costs associated with the design in order for the design to enhance net value. The field of engineering economy is concerned with the systematic evaluation of the particular benefits and costs of projects involving engineering design and analysis. In other words, engineering economy quantifies the benefits and costs associated with engineering projects to determine if they make (or save) enough money to warrant their capital investments. Thus, engineering economy requires a coupling of technical analysis and economic feasibility to determine the best course of action for various engineering scenarios. As we shall see, engineering economy is as relevant to the design engineer who considers material selection as it is to the chief executive officer who approves capital expenditures for new ventures.

History of the Book

The original *Introduction to Engineering Economy*, authored by Woods and DeGarmo, appeared in 1942. This text has been in continuous classroom use for over 50 years, and over 400,000 students have learned about engineering economic analysis from its pages. The tenth edition of *Engineering Economy* has built upon the rich and time-tested teaching materials of earlier editions, and its publication makes it the second-oldest book on the market that deals exclusively with engineering economy.

Intended Use of the Book

This book has two primary purposes: (1) to provide students with a sound understanding of the principles, basic concepts, and methodology of engineering economy; and (2) to help them develop proficiency with these methods and with the process for making rational decisions regarding situations they are likely to encounter in professional practice. Consequently, *Engineering Economy* is intended to serve as a text for classroom instruction *and* as a basic reference for use by practicing engineers in all specialty areas (e.g., chemical, civil, electrical, industrial, and mechanical). The book is also useful to persons engaged in the management of technical activities.

As a textbook, the tenth edition is written principally for the first formal course in engineering economy. The contents of the book and the accompanying Instructor's Manual and Electronic Spreadsheets Supplement (both available from Prentice Hall) are organized for effective presentation and teaching of the subject matter. A three-credit-hour semester course should be able to cover the majority of topics in this edition, and there is sufficient depth and breadth to enable an instructor to arrange course content to suit individual needs. Moreover, because several advanced topics are included in this book, it can also be used for a second course in engineering economy.

Every chapter and appendix has been revised and updated to reflect current trends and issues. Also, numerous exercises that involve open-ended problem statements and iterative problem-solving skills are included throughout the book. A large portion of the 500+ end-of-chapter exercises are new, and many solved examples representing realistic problems that arise in various engineering disciplines are presented.

An engineering economy course may be classified, for Accreditation Board for Engineering and Technology (ABET) purposes, as part engineering science and part engineering design. It is generally advisable to develop and teach such a course at the upper division level, where the course incorporates the accumulated knowledge students have acquired in other areas of the curriculum also dealing with iterative problem solving, open-ended exercises/case studies, creativity in formulating and evaluating feasible solutions to problems, and consideration of realistic constraints (economic, esthetic, safety, etc.) in problem solving.

Instructional Features

The *Instructor's Manual* is designed as a comprehensive aid in teaching the text material. Full solutions of all problems at the end of each chapter are presented. Several *comprehensive examples (case studies)* have been included in the tenth edition. They provide the instructor with essential material for teaching both the first formal course and a second, more advanced course in engineering economy. These case studies demonstrate the integrated application of the principles, basic concepts, and methodologies that are needed by engineers in typical real-world situations. They also serve as a bridge from the classroom to professional practice.

Spreadsheet Supplement

A second supplement entitled *Electronic Spreadsheets for Engineering Economy Applications* is authored by James A. Alloway, Jr. Electronic spreadsheets are a mainstay in many undergraduate engineering economy courses; the spreadsheet supplement ensures that the tenth edition of *Engineering Economy* will maintain its leadership position by providing basic templates for all major topics in the text. In addition, it provides a concise summary of formulas and key concepts, which students will find invaluable for review and quick reference. To access the supplement, ftp to **ftp.prenhall.com**, log in as "anonymous" using your e-mail address as the password, and change to the directory: pub/ESM/industrial_engineering.s-047/sullivan/engineer.eco/alloway/spread.

The spreadsheet supplement marks the first time that a major engineering economy text has provided spreadsheet material at this level of detail via the Internet, and students and instructors will benefit in several ways. For instance, electronic publishing provides a means to increase the amount of material instructors can cover without increasing printing costs, and they do not have to make additional arrangements to provide access for their students. Enhancements and additions to the supplement will be available immediately. Students can "cut and paste" the chapter, notation, and formula summaries into their own study documents.

The greatest advantage is that it is no longer necessary to enter the spreadsheets by hand. The templates can be downloaded and opened directly in Lotus 1-2-3 Release 5 for Windows. Most other spreadsheet software packages (e.g., Excel) provide conversion utilities to convert these files into their respective native formats. Users can then modify the basic templates for the specific problem at hand. As a bonus, advanced templates have also been developed for such techniques as Monte Carlo simulation, three-factor simultaneous sensitivity analysis, and integer linear programming.

Acknowledgments

Many friends and associates have offered valuable and constructive ideas and concepts that have materialized into numerous improvements to the tenth edition.

To these individuals we express our most sincere appreciation. Also, to our students, who have unknowingly been the proving ground for much of this text, we offer our thanks. They have reacted, without compassion on some occasions, to make us aware of good and bad methods of presentation. It is our hope that they, and those who follow them, will find this book an invaluable guide to the study and practice of engineering economy.

We wish to thank the following individuals for their inputs to the tenth edition: Dr. Hampton R. Liggett of Northern Illinois University did an excellent job in updating Chapter 6; Dr. Dan Babcock, Professor Emeritus at the University of Missouri-Rolla provided invaluable contributions to virtually all chapters of the initial manuscript; Mr. Garry Coleman of Virginia Tech volunteered several end-of-chapter problems; Dr. Richard H. Bernhard of North Carolina State University made many helpful suggestions for improving our discussion of the benefit/cost ratio method in Chapter 6; Mr. James A. Brimson of ABM-I, Inc. was very constructive in his review of activity-based accounting (Appendix A); Ms. Tracy S. Turner, CPA, of Coulter & Justice, carefully examined the accuracy of income tax matter included in Chapter 7; Dr. James A. Alloway created a large number of the spreadsheets presented in this edition; and Dr. Samer Madanat of Purdue University was very encouraging in his review of an early manuscript.

Also for the many suggestions from other colleagues, students, and practicing engineers who have used the previous editions, we express our deep appreciation. We hope they will find this edition to be even more helpful.

E. P. D.

W. G. S.

J. A. B.

E. M. W.

About the Authors

William G. Sullivan

William G. Sullivan is professor of Industrial and Systems Engineering at Virginia Polytechnic Institute and State University. He is also a two-time recipient of the Eugene L. Grant Award for the best paper in *The Engineering Economist* (Volumes 29 and 36). Dr. Sullivan's current research interests include justification of advanced manufacturing technologies, the economic principles of engineering design, and activity-based costing applied to the design process. Dr. Sullivan is a registered professional engineer and a member of IIE, ASEE, IMA, and SME. He also serves as coeditor of the *International Journal of FAIM.* He is a fellow in the Institute of Industrial Engineers. His Ph.D. in Industrial and Systems Engineering was earned at the Georgia Institute of Technology.

James A. Bontadelli

James A. Bontadelli is a professor of Industrial Engineering at the University of Tennessee, Knoxville, and a former director of industrial engineering for the Tennessee Valley Authority. He obtained a Ph.D. in Industrial and Systems Engineering from The Ohio State University and is a registered professional engineer with more than thirty years of professional practice. He has been a co-author of *Engineering Economy* since the eighth edition.

Elin M. Wicks

Elin M. Wicks is an assistant professor of Industrial Engineering at the University of Missouri, Columbia. She obtained a Ph.D. in Industrial and Systems Engineering from the Virginia Polytechnic Institute and State University. Elin is a member of IIE and ACEE.

E. Paul DeGarmo

E. Paul DeGarmo is a professor emeritus of Industrial Engineering and Mechanical Engineering at the University of California, Berkeley. He was a co-author (with B. Woods) in 1942 of the first edition, entitled *Introduction to Engineering Economy*.

Engineering Economy

Background and Tools of Engineering Economy

[Engineering] is the art of doing that well with one dollar which any bungler can do with two.

—A. M. Wellington, *The Economic Theory of the Location of Railways*
(New York: John Wiley & Sons, 1887.)

C H A P T E R

1

Introduction to Engineering Economy

The objectives of Chapter 1 are to (1) introduce the subject of engineering economy, (2) discuss its critical role in engineering design and analysis, (3) discuss the basic principles of the subject, and (4) provide an overview of the book.

The following topics are discussed in this chapter:

The importance of this subject in engineering practice

Origins of engineering economy

The principles of engineering economy

Engineering economy and the design process

Accounting and engineering economy studies

Overview of the book

1.1 Introduction

The technological and social environments in which we live continue to change at a rapid rate. In recent decades, advances in science and engineering have made space travel possible, transformed our transportation systems, revolutionized the practice of medicine, and miniaturized electronic circuits so that a computer can be placed on a semiconductor chip. The list of such achievements seems almost endless. In your science and engineering courses you will learn about some of the physical laws that underlie these accomplishments.

The utilization of scientific and engineering knowledge for our benefit is achieved through the *design* of things we use, such as machines, structures, products, and services. However, these achievements don't occur without a price, monetary or otherwise. Therefore, the purpose of this book is to develop and illustrate the principles and methodology required to answer the basic economic question of any design: Do its benefits exceed its costs?

The Accreditation Board for Engineering and Technology states that engineering "is the profession in which a knowledge of the mathematical and natural sciences gained by study, experience, and practice is applied with judgment to develop ways to utilize, economically, the materials and forces of nature for the benefit of mankind."* In this definition the economic aspects of engineering are emphasized as well as the physical aspects. Clearly, it is essential that the economic part of engineering practice be accomplished well.

> *Engineering economy* is the discipline concerned with the economic aspects of engineering; it involves the systematic evaluation of the costs and benefits of proposed technical projects. The principles and methodology of engineering economy are an integral part of the daily management and operation of private-sector companies and corporations, regulated public utilities, government units or agencies, and nonprofit organizations. These principles are utilized to analyze alternative uses of financial resources, particularly in relation to the physical assets and the operation of an organization. Last, but certainly not least, engineering economy will prove to be invaluable to you in assessing the economic merits of alternative uses of your personal funds.

Therefore, engineering economy is the dollars-and-cents side of the decisions that engineers make as they work to position a firm to be profitable in a highly competitive marketplace. Inherent to these decisions are tradeoffs among different types of costs and the performance (response time, safety, weight, reliability, etc.) provided by the proposed design or problem solution. The mission of engineering economy is to balance these tradeoffs in the most economical manner. For instance, if an engineer at Ford Motor Company invents a new transmission lubricant that increases fuel mileage by 10% and extends the life of the transmission by 30,000 miles, how much can the company afford to spend to implement this invention? Engineering economy can provide an answer.

A few more of the myriad situations in which engineering economy plays a crucial role come to mind:

1. Choosing the best design for a high-efficiency gas furnace
2. Selecting the most suitable robot for a welding operation on an automotive assembly line
3. Making a recommendation about whether jet airplanes for an overnight delivery service should be purchased or leased
4. Considering the choice between reusable and disposable bottles for high-demand beverages

From these illustrations it should be obvious that engineering economy includes significant technical considerations. Thus, engineering economy involves technical analysis, with emphasis on the economic aspects, and has the objective of assisting

---

*Accreditation Board of Engineering and Technology, *Criteria for Accrediting Programs in Engineering in the United States*, New York, 1994.

decisions. This is true whether the decision maker is an engineer interactively an-alyzing alternatives at a computer-aided design workstation or the chief executive officer (CEO) considering a new project. *An engineer who is unprepared to excel at engineering economy is not properly equipped for his or her job.*

1.2 Origins of Engineering Economy

Cost considerations and comparisons are fundamental aspects of engineering prac-tice. This basic point was emphasized in Section 1.1. However, the development of engineering economy methodology, which is now used in nearly all engineering work, is relatively recent. This does not mean that, historically, costs were usually overlooked in engineering decisions. However, the perspective that ultimate econ-omy is a primary concern to the engineer and the availability of sound techniques to address this concern differentiate this aspect of modern engineering practice from that of the past.

A pioneer in the field was Arthur M. Wellington,* a civil engineer, who in the latter part of the nineteenth century specifically addressed the role of economic analysis in engineering projects. His particular area of interest was railroad building in the United States. This early work was followed by other contributions in which the emphasis was on techniques that depended primarily on financial and actuarial mathematics. In 1930, Eugene Grant published the first edition of his textbook.† This was a milestone in the development of engineering economy as we know it today. He placed emphasis on developing an economic point of view in engineering, and (as he stated in the preface) "this point of view involves a realization that quite as definite a body of principles governs the economic aspects of an engineering decision as governs its physical aspects." In 1942 Woods and DeGarmo wrote the first edition of this book, later titled *Engineering Economy.*

1.3 What Are the Principles of Engineering Economy?

The development, study, and application of any discipline must begin with a basic foundation. We define the foundation for engineering economy to be a set of principles, or fundamental concepts, that provide a comprehensive doctrine for developing the methodology.‡ These principles will be mastered by students

---

*A. M. Wellington, *The Economic Theory of Railway Location,* 2nd ed. (New York: John Wiley & Sons, 1887).

†E. L. Grant, *Principles of Engineering Economy* (New York: The Ronald Press Company, 1930).

‡The definition of the principles of engineering economy varies somewhat with different authors. Examples of other definitions may be found in the following works:
1. E. L. Grant, W. G. Ireson, and R. S. Leavenworth, *Principles of Engineering Economy,* 8th ed. (New York: John Wiley & Sons, 1989).
2. G. A. Fleischer, *Engineering Economy: Capital Allocation Theory* (Monterey, Calif: Brooks/Cole Engineering Division, 1984).
3. Report titled "Research Planning Conference for Developing a Research Framework for Engineer-ing Economics," Gerald J. Thuesen (editor), Georgia Institute of Technology, March 1986. The report was the result of National Science Foundation Grant MEA-8501237.

as they progress through this book. However, in engineering economic analysis, experience has shown that most errors can be traced to some violation or lack of adherence to the basic principles. We will define and discuss the foundation of the discipline in terms of seven principles.

P R I N C I P L E 1 —DEVELOP THE ALTERNATIVES:

The choice (decision) is among alternatives. The alternatives need to be identified and then defined for subsequent analysis.

A decision situation involves making a choice among two or more alternatives. Developing and defining the alternatives for detailed evaluation is important because of the resulting impact on the quality of the decision. Engineers and managers should place a high priority on this responsibility. Creativity and innovation are essential to the process.

One alternative that may be feasible in a decision situation is making no change to the current operation or set of conditions (i.e., doing nothing). If you judge this option feasible, make sure it is considered in the analysis. However, do not focus on the status quo to the detriment of innovative or necessary change.

P R I N C I P L E 2 —FOCUS ON THE DIFFERENCES:

Only the differences in expected future outcomes among the alternatives are relevant to their comparison and should be considered in the decision.

If all prospective outcomes of the feasible alternatives were exactly the same, there would be no basis or need for comparison. We would be indifferent among the alternatives and could make a decision using a random selection.

Obviously, only the differences in the future outcomes of the alternatives are important. Outcomes that are common to all alternatives can be disregarded in the comparison and decision. For example, if your feasible housing alternatives were two residences with the same purchase (or rental) price, price would be inconsequential to your final choice. Instead, the decision would depend on other factors such as location and annual operating and maintenance costs. This example illustrates, in a simple way, Principle 2, which emphasizes the basic purpose of an engineering economic analysis: to recommend a future course of action based on the differences among feasible alternatives.

P R I N C I P L E 3 —USE A CONSISTENT VIEWPOINT:

The prospective outcomes of the alternatives, economic and other, should be consistently developed from a defined viewpoint (perspective).

The perspective of the decision maker, which is often that of the owners of the firm, would normally be used. However, it is important that the viewpoint for the particular decision be first defined and then used consistently in the description, analysis, and comparison of the alternatives.

As an example, consider a public organization operating for the purpose of developing a river basin, including the generation and wholesale distribution of electricity from dams on the river system. A program is being planned to upgrade and increase the capacity of the power generators at two dam sites. What perspective should be used in defining the technical alternatives for the program? The "owners of the firm," in this example, means the segment of the public that will pay the cost of the program as well as benefit from it. Thus, the viewpoint of the "customer" should be adopted in this situation.

Now let us look at an example where the viewpoint may not be that of the owners of the firm. Suppose that the company in this example is a private firm, and the problem deals with providing a flexible benefits package for the employees. Also assume that the feasible alternatives for operating the plan all have the same future costs to the company. The alternatives, however, have differences from the perspective of the employees, and their satisfaction is an important decision criterion. The viewpoint for this analysis, comparison, and decision should be that of the employees of the company as a group, and the feasible alternatives should be defined from their perspective.

> **P R I N C I P L E 4 —USE A COMMON UNIT OF MEASURE:**
> Using a common unit of measurement to enumerate as many of the prospective outcomes as possible will make easier the analysis and comparison of the alternatives.

It is desirable to make as many prospective outcomes as possible *commensurable* (directly comparable). For economic consequences, a monetary unit such as dollars is the common measure. You should also try to translate other outcomes (which do not initially appear to be economic) into the monetary unit. This translation, of course, will not be feasible with some of the outcomes, but the additional effort toward this goal will enhance commensurability and make the subsequent analysis and comparison of alternatives easier.

What should you do with the outcomes that are not economic—that is, the expected consequences that cannot be translated (and estimated) using the monetary unit? First, if possible, quantify the expected future results using an appropriate unit of measurement for each outcome. If this is not feasible for one or more outcomes, describe these consequences explicitly so that the information is useful to the decision maker in the comparison of the alternatives.

> **P R I N C I P L E 5 —CONSIDER ALL RELEVANT CRITERIA:**
> Selection of a preferred alternative (decision making) requires the use of a criterion (or several criteria). The decision process should consider both the outcomes enumerated in the monetary unit and those expressed in some other unit of measurement or made explicit in a descriptive manner.

The decision maker will normally select the alternative that will best serve the long-term interests of the owners of the organization. In engineering economic analysis, the primary criterion relates to the long-term financial interests of the owners. This is based on the assumption that available capital will be allocated to provide maximum monetary return to the owners. Often, though, there are other organizational objectives you would like to achieve with your decision, and these should be considered and given weight in the selection of an alternative. These nonmonetary attributes and multiple objectives become the basis for additional criteria in the decision-making process.

> **P R I N C I P L E 6 —MAKE UNCERTAINTY EXPLICIT:**
> Uncertainty is inherent in projecting (or estimating) the future outcomes of the alternatives and should be recognized in their analysis and comparison.

The analysis of the alternatives involves projecting or estimating the future consequences associated with each of them. The magnitude and the impact of future outcomes of any course of action are uncertain. Even if the alternative involves no change from current operations, the probability is high that today's estimates of, for example, future cash receipts and expenses will not be what eventually occurs. Thus, dealing with uncertainty is an important aspect of engineering economic analysis and is the subject of Chapters 10 and 14.

> **P R I N C I P L E 7 —REVISIT YOUR DECISIONS:**
> Improved decision making results from an adaptive process; to the extent practicable, the initial projected outcomes of the selected alternative should be subsequently compared with actual results achieved.

A good decision-making process can result in a decision that has an undesirable outcome. Other decisions, even though relatively successful, will have results significantly different from the initial estimates of the consequences. Learning from and adapting based on our experience are essential; whether in the private or public sector of our economy, they are indicators of a good organization.

The evaluation of results versus the initial estimate of outcomes for the selected alternative is often considered impracticable or not worth the effort. Too often, no feedback to the decision-making process occurs. Organizational discipline is needed to ensure implemented decisions are routinely postevaluated and the results used to improve future analyses of alternatives and the quality of decision making. The percentage of important decisions in an organization that are not postevaluated should be small. For example, a common mistake made in the comparison of alternatives is the failure to examine adequately the impact of uncertainty in the estimates for selected factors on the decision. Only postevaluations will highlight this type of weakness in the engineering economy studies being done in an organization.

▼

1.4 Engineering Economy and the Design Process

> An engineering economy study is accomplished using a structured procedure and mathematical modeling techniques. The economic results are then used in a decision situation that involves two or more alternatives and normally includes other engineering knowledge and input.

A sound *engineering economic analysis procedure* incorporates the basic principles discussed in Section 1.3 and involves several steps. We represent the procedure, and will discuss it later in this section, in terms of the *seven steps* listed in the left-hand column of Table 1-1. There are several feedback loops (not shown) within the procedure. For example, within Step 1, information developed in evaluating the problem will be used as feedback to refine the problem definition. As another example, information from the analysis of alternatives may indicate the need to change one or more of them or to develop additional alternatives.

The seven-step procedure is also used to assist decision making within the engineering design process, shown as the right-hand column in Table 1-1. In this case, activities in the design process contribute information to related steps in the economic analysis procedure. The general relationship between the activities in the design process and the steps of the economic analysis procedure is indicated in Table 1-1.

Middendorf* states that "engineering design is an iterative, decision making activity whereby scientific and technological information is used to produce a

TABLE 1-1 The General Relationship Between the Economic Analysis Procedure and the Engineering Design Process	
Engineering Economic Analysis Procedure	**Engineering Design Process**
Step	*Activity*
1. Problem recognition, formulation, and evaluation.	1. Problem/need definition.
	2. Problem/need formulation and evaluation.
2. Development of the feasible alternatives.	3. Synthesis of possible solutions (alternatives).
3. Development of the cash flows for each alternative.	
4. Selection of a criterion (or criteria).	4. Analysis, optimization, and evaluation.
5. Analysis and comparison of the alternatives.	
6. Selection of the preferred alternative.	5. Specification of preferred alternative.
	6. Communication.
7. Performance monitoring and post-evaluation of results.	

*W. H. Middendorf, *Design of Devices and Systems* (New York: Marcel Dekker, Inc., 1986), p. 2.

system, device, or process which is different, in some degree, from what the designer knows to have been done before and which is meant to meet human needs." Also, we want to meet the human needs economically as emphasized in the definition of engineering in Section 1.1.

The engineering design process may be repeated in phases to accomplish a total design effort. For example, in the first phase, a full cycle of the process may be undertaken to select a conceptual or preliminary design alternative. Then, in the second phase, the activities are repeated to develop the preferred detailed design based on the selected preliminary design. The seven-step economic analysis procedure would be repeated as required to assist decision making in each phase of the total design effort.

1.4.1 Problem Definition

It is not adequate simply to think about a perplexing question or situation. Rather, a problem must be well understood and stated in an explicit form before the engineer proceeds with the rest of the analysis. The first step of the engineering economic analysis procedure (problem definition) is particularly important, since it provides the basis for the rest of the analysis.

The term *problem* is used here generically. It includes all decision situations for which an engineering economy analysis is required. Recognition of the problem is normally stimulated by internal or external organizational needs or requirements. An operating problem within a company (internal need) or a customer expectation about a product or service (external requirement) are examples.

Once recognized, formulation of the problem should be viewed from a *systems perspective*. That is, the boundary or extent of the situation needs to be carefully defined, thus establishing the elements of the problem and what constitutes its environment.

Evaluation of the problem includes refinement of needs and requirements, and information from the evaluation phase may change the original formulation of the problem. In fact, redefining the problem may be the most important part of the problem-solving process!

1.4.2 Search for Alternatives*

The two primary actions in Step 2 of the procedure are (1) searching for potential alternatives and (2) screening them to select a smaller group of feasible alternatives for detailed analysis and comparison in Step 5. The term *feasible* here means that each alternative selected for further analysis is judged, based on preliminary evaluation, to meet or exceed the specifications established for the situation.

---

*This is sometimes called *option development*. This important step is described in detail in A. B. Van Gundy, *Techniques of Structured Problem Solving*, 2nd ed. (New York: Van Nostrand Reinhold Co., 1988). For additional reading, see E. Lumsdaine and M. Lumsdaine, *Creative Problem Solving—An Introductory Course for Engineering Students* (New York: McGraw-Hill Book Co., 1990), and J. L. Adams, *Conceptual Blockbusting—A Guide to Better Ideas* (Reading, Mass: Addison-Wesley Publishing Co., 1986).

In the discussion of Principle 1 (Section 1.3), creativity and resourcefulness were emphasized as being absolutely essential to the development of feasible alternatives. The difference between good alternatives and great alternatives depends largely on an individual's or group's *problem-solving efficiency*. Such efficiency can be increased in the following ways:

1. Develop many redefinitions for the problem in Step 1.
2. Concentrate on redefining one problem at a time.
3. Avoid making judgments as new problem definitions are created.
4. Attempt to redefine a problem in terms that are dramatically different from the original Step 1 problem statement.
5. Make sure that the true problem is well researched and understood.

In searching for alternatives, several limitations invariably exist, including (1) lack of time and money, (2) preconceptions of what will and what will not work, and (3) lack of knowledge. Consequently, the analyst will be working with less than perfect problem solutions in the practice of engineering.

EXAMPLE 1-1

The management team of a small furniture-manufacturing company is under pressure to increase profitability in order to get a much-needed loan from the bank to purchase a more modern pattern-cutting machine. One proposed solution is to sell waste wood chips and shavings to a local charcoal manufacturer instead of using them to fuel space heaters for the company's office and factory areas.

(a) Formulate the company's problem. Next, reformulate the problem in a variety of creative ways.

(b) Develop at least one potential alternative for your reformulated problems in part (a). (Don't concern yourself with feasibility at this point.)

SOLUTION

(a) The company's problem appears to be that revenues are not sufficiently covering costs. Several reformulations can be posed, for example:

1. The problem is to increase revenues while reducing costs.
2. The problem is to maintain revenues while reducing costs.
3. The problem is an accounting system that provides distorted cost information.
4. The problem is that the new machine is really not needed (and hence there is no need for a bank loan).

(b) Based only on reformulation 1, an alternative is to sell wood chips and shavings as long as increased revenue exceeds extra expenses that may be required to heat the buildings. Another alternative is to discontinue the manufacture of specialty items and concentrate on standardized, high-volume products. Yet another alternative is to pool purchasing, accounting, engineering, and other white-collar support services with those of other small firms in the area.

1.4.3 Development of Investment Alternatives

"It takes money to make money," as the old saying goes. Did you know that in the United States the average firm spends over $100,000 in capital on each of its employees? So, to make money, each firm must invest capital to support its important human resource—but in what should an individual firm invest? There are usually hundreds of opportunities for a company to make money. Engineers are at the very heart of creating value by turning innovative and creative ideas into new and re-engineered commercial products and services. Most of these ideas require investment of money, and only a few of all feasible ideas can be developed, due to lack of time, knowledge, or resources.

Consequently, most investment alternatives created by good engineering ideas are drawn from a larger population of equally good problem solutions. But how can this larger set of equally good solutions be tapped into? Interestingly, studies have concluded that designers and problem solvers tend to pursue a few ideas that involve "patching and repairing" an old idea.* Truly new ideas are often excluded from consideration! This section outlines two approaches that have found wide acceptance in industry for developing sound investment alternatives by removing some of the barriers to creative thinking: (1) classical brainstorming and (2) the nominal group technique.

1.4.3.1 Classical Brainstorming Classical brainstorming is the most well-known and often-used technique for idea generation. It is based on the fundamental principles of *deferment of judgment* and that *quantity breeds quality*. There are four rules for successful brainstorming:

1. Criticism is ruled out.
2. Freewheeling is welcomed.
3. Quantity is wanted.
4. Combination and improvement are sought.

A. F. Osborn lays out a detailed procedure for successful brainstorming.† A classical brainstorming session has the following basic steps:

1. *Preparation*. The participants are selected, and a preliminary statement of the problem is circulated.
2. *Brainstorming*. A warm-up session with simple unrelated problems is conducted, the relevant problem and the four rules of brainstorming are presented, and ideas are generated and recorded using checklists and other techniques if necessary.
3. *Evaluation*. The ideas are evaluated relative to the problem.

Generally, a brainstorming group should consist of four to seven people, although some suggest larger groups.

---

*S. Finger and J. R. Dixon, "A Review of Research in Mechanical Engineering Design. Part I: Descriptive, Prescriptive, and Computer-Based Models of Design Processes," *Research in Engineering Design* (New York: Springer-Verlag, 1989).

†A. F. Osborn, *Applied Imagination,* 3rd ed., (New York: Charles Scribner's Sons, 1963).

1.4.3.2 Nominal Group Technique The nominal group technique (NGT), developed by Andre P. Delbecq and Andrew H. Van de Ven,* involves a structured group meeting designed to incorporate individual ideas and judgments into a group consensus. By correctly applying the NGT, it is possible for groups of people (preferably, five to ten) to generate investment alternatives or other ideas for improving the competitiveness of the firm. Indeed, the technique can be used to obtain group thinking (consensus) on a wide range of topics. For example, a question that might be given to the group is, "What are the most important problems/opportunities for improvement of ...?"

The technique, when properly applied, draws on the creativity of the individual participants, while reducing two undesirable effects of most group meetings: (a) the dominance of one or more participants and (b) the suppression of conflicting ideas. The basic format of an NGT session is as follows:

1. Individual silent generation of ideas
2. Individual round-robin feedback and recording of ideas
3. Group clarification of each idea
4. Individual voting and ranking to prioritize ideas
5. Discussion of group consensus results

The NGT session begins with an explanation of the procedure and a statement of question(s), preferably written by the facilitator.† The group members are then asked to prepare individual listings of alternatives, such as investment ideas or issues that they feel are crucial for the survival and health of the organization. This is known as the silent-generation phase and usually takes only a few minutes to "get the thoughts rolling." After this phase has been completed, the facilitator calls on each participant, in round-robin fashion, to present one idea from his or her list (or further thoughts as the round-robin session is proceeding). Each idea (or opportunity) is then identified in turn and recorded on a flip chart or board by the NGT facilitator, leaving ample space between ideas for comments or clarification. This process continues until all the opportunities have been recorded, clarified, and displayed for all to see. At this point a voting procedure is used to prioritize the ideas or opportunities. Finally, voting results lead to the development of group consensus on the topic being addressed.

1.4.4 Development of Prospective Outcomes

Step 3 of the engineering economic analysis procedure incorporates Principles 2, 3, and 4 from Section 1.3 and uses the basic *cash flow approach* employed in engineering economy. A cash flow occurs when money is transferred from one organization or individual to another. Thus, a cash flow represents the economic effects of an alternative in terms of money spent and received.

---

*A. Van de Ven and A. Delbecq, "The Effectiveness of Nominal, Delphi, and Interactive Group Decision Making Processes, " *Academy of Management Journal*, vol. 17, no. 4, December 1974, pp. 605–621.

†A good example of the NGT is given in D. S. Sink, "Using the Nominal Group Technique Effectively," *National Productivity Review*, Spring 1983, pp. 173–184.

Consider the concept of an organization having only one "window" to its external environment through which *all* monetary transactions occur—receipts of revenues and payments to suppliers, creditors, and employees. The key to developing the related cash flows for an alternative is estimating what would happen to the revenues and costs, as seen at this window, if the particular alternative were implemented. The *cash flow* for an alternative is the difference between all cash inflows (receipts or savings) and cash outflows (costs or expenses) during each time period.

In addition to the economic aspects of decision making, *nonmonetary factors* (*attributes*) often play a significant role in the final recommendation. Examples of objectives other than profit maximization or cost minimization that can be important to an organization include the following:

1. Meeting or exceeding customer expectations
2. Safety
3. Improving employee satisfaction
4. Maintaining production flexibility to meet changing demands
5. Meeting or exceeding all environmental requirements
6. Achieving good public relations; being an exemplary member of the community
7. Leveling cyclic fluctuations in production

1.4.5 Selection of Decision Criteria and Analysis of Alternatives

The selection of decision criteria (Step 4 of the analysis procedure) incorporates Principle 5. The decision maker will normally select the alternative that will best serve the long-term interests of the owners of the organization. Analysis of the economic aspects of an engineering decision situation (Step 5) is largely based on the cash flows developed for the feasible alternatives. The consideration of uncertainty in the estimates (Principle 6) may also be required, as well as a focus on the differences in cash flows between each pair of alternatives (Principle 2).

1.4.6 Selection of the Preferred Alternative

When the first five steps of the engineering economic analysis procedure have been done properly, the preferred alternative (Step 6) is simply a result of the total effort. Thus, the soundness of the technical-economic modeling and analysis techniques dictates the quality of the results obtained and the recommended course of action.

Step 6 is included in Activity 5 of the engineering design process (specification of the preferred alternative) when done as part of a design effort.

1.4.7 Performance Monitoring and Postevaluation of Results

This final step implements Principle 7 and is accomplished while and after the results achieved from the selected alternative are collected. Monitoring project performance during its operational phase improves the achievement of related goals and objectives and reduces the variability in desired results. Step 7 is also the follow-up step to a previous analysis, comparing actual results achieved with the previously estimated outcomes. The aim is to learn how to do better analyses, and the feedback from post-implementation evaluation is important to the continuing improvement of operations in any organization. Unfortunately, like Step 1, this final step is often not done consistently or well in engineering practice; therefore, it needs particular attention to ensure feedback for use in ongoing and subsequent studies.

EXAMPLE 1-2

Bad news—you have just wrecked your car! An automobile wholesaler offers you $2,000 for the car "as is." Also, your insurance company's claims adjuster estimates that there is $2,000 in damages to your car. Because you have collision insurance with a $1,000 deductibility provision, the insurance company mails you a check for $1,000. The odometer reading on your wrecked car is 58,000 miles.

You need another car immediately. What should you do?

SOLUTION

(a) *Develop your alternatives (Principle 1):*

1. Sell the wrecked car for $2,000 to the wholesaler and spend this money, the $1,000 insurance check, and all of your $7,000 life savings on a newer car. Total amount paid out of savings is $7,000, and the car will have 28,000 miles.
2. Spend the $1,000 insurance check and $1,000 of savings to fix the car. Total amount paid out of savings is $1,000, and the car will have 58,000 miles.
3. Spend the $1,000 insurance check and $1,000 of savings to fix the car, then sell the car for $4,500. Spend the $4,500 plus $5,500 of additional savings to buy the newer car. Total amount paid out of savings is $6,500, and the car will have 28,000 miles.
4. Give the car to a lesser mechanic, who will repair it for $1,100 ($1,000 insurance and $100 of your savings) but will take an additional month of repair time, and rent a car for that time at $400/month (paid out of savings). Total amount paid out of savings is $500, and the car will have 58,000 miles.
5. Same as 4, but then sell the car for $4,500 and use this money plus $5,500 of additional savings to buy the newer car. Total amount paid out of savings is $6,000, and the car will have 28,000 miles.

ASSUMPTIONS

1. The less reliable repair shop in Alternatives 4 and 5 will not take longer than the additional month to repair the car.
2. Each car will perform at the normal operating condition as it was originally intended and to the same total mileage before being sold or salvaged.
3. Interest earned on money remaining in savings is negligible.

(b) *Focus on the differences (Principle 2):*

1. Alternative 1 varies from all others because the car is not to be repaired at all but merely sold. This eliminates the benefit of the $500 increase in value of the

car when it is repaired and then sold. Also, this alternative leaves no money in the bank.

2. Alternative 2 varies from Alternative 1 because it allows the old car to be repaired. Alternative 2 differs from Alternatives 4 and 5 since it utilizes a more expensive ($500 more) and less risky repair facility. It also varies from Alternatives 3 and 5 because the car will be kept.

3. Alternative 3 gains an additional $500 by repairing the car and selling it to buy the same car as in Alternative 1.

4. Alternative 4 uses the same idea as Alternative 2 but involves a less expensive repair shop. This repair shop is more risky in the quality of its end product but will only cost $1,100 in repairs and $400 in an additional month's rental of a car.

5. Alternative 5 is the same as Alternative 4 but gains an additional $500 by selling the repaired car and purchasing a newer car as in Alternatives 1 and 3.

(c) *Use a consistent viewpoint (Principle 3):*
   The viewpoint taken here is that of the owner of the wrecked car.

(d) *Use a common unit of measure (Principle 4):*
   The dollar represents the value of the car to the owner. Hence, the dollar is used as the consistent value against which everything is to be measured. This reduces all decisions to a quantitative level, which can then be reviewed later with qualitative factors that may carry their own dollar value (e.g., how much is a reliable repair shop worth?).

(e) *Consider all relevant criteria (Principle 5):*

1. Alternative 1 is eliminated, because Alternative 3 gains the same end result and would also provide the car owner with $500 more cash. This is experienced with no change in the risk to the owner. Car value = $10,000, savings = 0.

2. Alternative 2 is a good alternative to consider, because it spends the least amount of cash, leaving $6,000 in the bank. Alternative 2 provides the same end result as Alternative 4 but costs $500 more to repair. Therefore, Alternative 2 is eliminated. Car value = $4,000, savings = $6,000.

3. Alternative 3 is eliminated, because Alternative 5 also repairs the car but at a lower out-of-savings cost ($500 difference), and both Alternatives 3 and 5 have the same end result of buying the newer car. Car value = $10,000, savings = $500.

4. Alternative 4 would be a good alternative, because it saves $500 by using a cheaper repair facility, provided that the risk of a poor repair job is judged to be small. Car value = $4,000, savings = $6,500.

5. Alternative 5 is the alternative *accepted*, because it repairs the car at a lower cost ($500 cheaper) but eliminates the risk of breakdown by selling it to someone else at an additional $500 gain. Car value = $10,000, savings = $1,000.

(f) *Make uncertainty explicit (Principle 6):*
   Among the uncertainties that can be found in this problem, the following are the most relevant to the decision. If the original car is repaired and kept, there is a possibility that it would have a higher frequency of breakdowns (based on personal experience). If a cheaper repair facility is used, the chance of a later

breakdown is even greater (based on personal experience). Also, the newer car purchased may be too expensive, based on the additional price paid (which is at least $6,000/30,000 miles = 20 cents per mile). Finally, the newer car may also have been in an accident and could have a worse repair history than the presently owned car.

(g) *Revisit your decisions (Principle 7):*

The newer car turned out after being "test driven" for 20,000 miles to be a real beauty. Mileage was great, and no repairs were needed. The systematic process of identifying and analyzing alternative solutions to this problem really paid off!

1.5 Accounting and Engineering Economy Studies

In Section 1.1 we emphasized that engineers and managers use the principles and methodology of engineering economy to assist decision making. Thus, engineering economy studies provide information on which current decisions pertaining to the *future* operation of an organization can be based.

After a decision to invest capital in a project has been made and the money has been invested, those who supply and manage the capital want to know the financial results. Therefore, accounting procedures are established so that financial events relating to the investment can be recorded and summarized and *financial performance* determined. At the same time, through the use of proper financial information, controls can be established and utilized to aid in guiding the operation toward the desired financial goals. *General accounting* and *cost accounting* are the procedures that provide these necessary services in a business organization. Thus, accounting data are primarily concerned with *past* and *current* financial events, even though such data are often used to make projections about the future.

Accounting procedures are similar to recording data in a scientific experiment. A recorder reads the pertinent gauges and meters and records all the essential data during the course of an experiment. From these data it is possible to determine the results of the experiment and to prepare a report. Similarly, an accountant records all significant financial events connected with an investment and the operation of an organization, and from these data he or she can determine what the results have been and can prepare financial reports. Just as an engineer, by understanding what is happening during the course of an experiment and making suitable corrections, can gain more information and better results from the experiment, managers must also rely on accounting reports to make corrective decisions in order to improve the current and future financial performance of the business.

General accounting is a source of much of the past financial data needed for estimating future financial conditions. Accounting is also a source of data for analyses of how well the results of a capital investment turned out compared to the results that were predicted in the engineering economic analysis.

Cost accounting, or management accounting, is a subset of accounting that is of particular importance because it is concerned principally with decision making and control in a firm. Consequently, it is the source of some of the cost data that are needed in engineering economy studies. Modern cost accounting may satisfy any or all of the following objectives:

1. To determine the cost of products or services
2. To provide a rational basis for pricing goods or services
3. To provide a means for controlling expenditures
4. To provide information on which operating decisions may be based and the results evaluated

Although the basic objectives of cost accounting are simple, the exact determination of costs usually is not. As a result, some of the procedures used are arbitrary conventions that make it possible to obtain reasonably accurate costs in some situations, but in many others the information is too aggregated and distorted to be relevant in managerial planning and control decisions.

Some of the inaccuracies of traditional cost accounting techniques have been remedied by a relatively recent methodology known as *activity-based accounting*. This methodology is aimed at producing more accurate and timely cost information primarily by (1) carefully tracing overhead to its causal activities and (2) assigning technology costs equitably over the entire product life cycle. Because overhead and technology account for as much as 60% of total product cost in many industries, improved cost reporting and control are made possible by being able to trace these two major cost components to the activities, and subsequently products, that truly create them.

An adequate understanding of the origins and meaning of accounting data is necessary to be able to interpret those data for use in engineering economy studies. Thus, a brief discussion of accounting, including activity-based accounting, is provided in Appendix A.

1.6 Overview of the Book

The contents of this book have been organized in three parts, with the chapters in a logical sequence for both teaching and applying the principles and methodology of engineering economy. The three parts of the book, and the chapters in each part, are

1. Part I: Background and Tools of Engineering Economy (Chapters 1–3)
2. Part II: Applications of Engineering Economy (Chapters 4–10)
3. Part III: Special Topics in Engineering Economy (Chapters 11–14)

In this first chapter, after a general discussion of the subject and its importance to the engineer, we presented the fundamental concepts of engineering economy in terms of seven basic principles, discussed the steps involved in an engineering economic analysis as well as their relationships to the principles, and then related the analysis procedure to the engineering design process. The interface between accounting and engineering economy was also discussed. Thus, in Chapter 1 the basic foundation of the subject has been established.

In Chapter 2, selected cost terms and other cost concepts important in engineering economy studies are presented. Particular emphasis is placed on the life cycle and its relative costs. The application of selected cost concepts is also discussed, including breakeven analysis, the economic principles of design, and present economy studies. Thus, in Chapter 2 the foundation of the subject is further extended to support the subsequent development and application of the methodology.

Chapter 3 concentrates on the concepts of money–time relationships, specifically the development of proper techniques to consider the time value of money in manipulating the future revenues and costs associated with an alternative (course of action). Then in Chapter 4 the methods commonly used in practice to analyze the economic consequences of an alternative are developed and demonstrated. These methods, and their proper use in the analysis and comparison of alternatives, are primary subjects of Chapter 5, which also includes a discussion of the appropriate time period for an analysis, as well as other topics related to the comparison of alternatives on a before-tax basis. Thus, Chapters 3, 4, and 5 together develop and discuss an essential part of the methodology needed for understanding the remainder of the book and for performing engineering economy studies on a before-tax basis. Chapter 6 presents the benefit–cost ratio method of project appraisal.

In Chapter 7, the additional techniques required to accomplish engineering economy studies on an after-tax basis are developed and discussed. In the private sector, most engineering economy studies are done on an after-tax basis. Therefore, this chapter adds to the basic part of the methodology developed in Chapters 3, 4, and 5. A part of Chapter 7 is concerned with depreciation under the modified accelerated cost recovery system authorized by the Tax Reform Act of 1986. However, techniques applicable to assets acquired prior to the effective date of the act are also included. Similarly, the remainder of Chapter 7 deals with performing after-tax analyses.

Chapter 8 considers the critical question of how to develop estimates of the future economic consequences associated with an alternative. The process associated with this step in an engineering economic analysis constitutes a crucial part of application and practice. This topic is placed in Chapter 8, instead of earlier in the book, so that the basic methodology of analyzing and comparing the estimated revenues and costs associated with alternatives on both a before-tax and an after-tax basis can be discussed in an integrated manner. Consequently, the development of these estimates of economic impact for an alternative is given concentrated attention in a chapter dedicated solely to this subject.

The effects of inflation (or deflation) and price changes are the topic of Chapter 9. This important subject needs to be understood by engineers in current practice. The concepts for handling price changes in an engineering economic analysis are discussed both comprehensively and pragmatically from an application viewpoint.

Concern over uncertainty is a reality in engineering practice. In Chapter 10 the impact of potential variation is considered between the estimated economic outcomes of an alternative and the results that may occur; that is, nonprobabilistic techniques for analyzing the consequences of uncertainty in future estimates of revenues and costs are developed and discussed.

Often, an organization must analyze whether existing assets should be continued in service or replaced with new assets to meet current and future operating needs. In Chapter 11 techniques for addressing this question are developed and discussed. Since the replacement of assets demands significant capital, decisions made in this area are important and require particular attention.

Replacement of assets is one source of demand for capital in an organization. In Chapter 12 the proper identification and analysis of all projects and other needs for capital within an organization, and the capital financing and capital allocation process to meet these needs, are addressed. This process is crucial to the welfare of an organization, since it affects most operating outcomes, whether in terms of current product quality and service effectiveness or long-term capability to compete in the world marketplace.

Privately owned, regulated public utilities are an important part of the U.S. economy. In Chapter 13, the unique characteristics of these firms, and the revenue requirements method of accomplishing engineering economy studies related to their operations, are discussed.

In Chapter 14, probabilistic techniques for analyzing the consequences of uncertainty in future cash flow estimates are discussed. Discrete and continuous probability concepts, as well as Monte Carlo simulation techniques, are included.

▼ 1.7 Summary

In this first chapter we have provided an introduction to the discipline of engineering economy, including a brief summary of its background and continuing development. The importance to an engineer of having a good working knowledge of the principles and methodology of the subject in today's workplace was emphasized, as well as the use of engineering economy in the design process. The engineer who is unprepared to handle the economic aspects of an engineering decision just as competently as its physical aspects is not properly equipped to perform his or her total job.

▼ 1.8 Problems

1-1. List 10 typical situations in the operation of an organization where an engineering economic analysis would significantly assist decision making. You may assume a specific type of organization (e.g., manufacturing firm, medical health center and hospital, transportation company, government agency) if it will assist in the development of your answer (state any assumptions). (1.1)*

1-2. Explain why the subject of engineering economy is important to the practicing engineer. (1.1–1.4)

1-3. Assume that your employer is a manufacturing firm that produces several different electronic consumer products. What are five nonmonetary factors (attributes) that may be important when a significant change is considered in the design of the current best-selling product? (1.3, 1.4)

1-4. Will the increased use of automation increase the importance of engineering economy studies? Why or why not?

1-5. Explain the meaning of the statement, "The choice (decision) is among alternatives." (1.3)

*The number(s) in parentheses at the end of a problem refer to the section(s) in that chapter most closely related to the problem.

1-6. Describe the outcomes that should be expected from a feasible alternative. What are the differences between potential alternatives and feasible alternatives? (1.4)

1-7. Define uncertainty. What are some of the basic causes of uncertainty in engineering economy studies? (1.3)

1-8. You have discussed with a coworker in the engineering department the importance of explicitly defining the viewpoint (perspective) from which future outcomes of a course of action being analyzed are to be developed. Explain what you mean by a viewpoint or perspective. (1.3)

1-9. Two years ago you were a member of the project team that analyzed whether your company should upgrade some building, equipment, and related facilities to support the expanding operation of the company. The project team analyzed three feasible alternatives, one of which makes no changes in facilities and the remaining two involve significant facility changes. Now you have been selected to lead a postevaluation team. Delineate your technical plan for comparing the estimated consequences (developed two years ago) of implementing the selected alternative with the results that have been achieved. (1.3, 1.4)

1-10. Describe how it might be feasible in an engineering economic analysis to consider the following different situations in terms of the monetary unit: (1.3)

a. A piece of equipment that is being considered as a replacement for an existing item has greater reliability; that is, the mean time between failures (MTBF) during operation of the new equipment has been increased 40% in comparison with the present item.

b. A company manufactures wrought iron patio furniture for the home market. Some changes in material and metal treatment, which involve increased manufacturing costs, are being considered to reduce the rusting problem significantly.

c. A large foundry operation has been in the same location in a metropolitan area for the past 35 years. Even though it is in compliance with current air pollution regulations, the continuing residential and commercial development of that area is causing an increasing expectation on the part of local residents for improved environmental control by the foundry. The company considers community relations to be important.

1-11. Explain the relationship between engineering economic analysis and engineering design. How does economic analysis assist decision making in the design process? (1.4)

1-12. For each of the seven steps in the engineering economic analysis process, describe the activities that normally would be accomplished in that part of the analysis. (1.4)

1-13. List a few advantages and disadvantages of (a) classical brainstorming and (b) the nominal group technique. (1.4)

1-14. *The Almost-Graduating Senior Problem.* Consider this situation faced by a first-semester senior in civil engineering who is exhausted from extensive job interviewing and penniless from excessive partying. Mary's impulse is to accept immediately a highly attractive job offer to work in her brother's successful manufacturing company. She would then be able to relax for a year or two, save some money, and then return to college to complete her senior year and graduate. Mary is cautious about this impulsive desire because it may lead to no college degree at all!

a. Develop at least two formulations for Mary's problem.

b. Identify feasible solutions for each problem formulation in part (a). *Be creative!*

Cost Concepts and the Economic Environment

The objectives of Chapter 2 are to (1) describe some of the basic cost terminology and concepts that are used throughout this book and (2) illustrate how they should be used in engineering economic analysis and decision making.

The following topics are discussed in this chapter:

Fixed, variable, and incremental costs

Recurring and nonrecurring costs

Direct, indirect, and overhead costs

Standard costs

Cash cost versus book cost

Sunk and opportunity costs

Life-cycle cost

The general economic environment

The relationship between price and demand

The total revenue function

Breakeven point relationships

Maximizing profit/minimizing cost

The average unit cost function

Cost-driven design optimization

Present economy studies

2.1 Introduction

Designing to meet economic needs and achieve competitive operations in private- and public-sector organizations depends on prudently balancing what is technically feasible and what is economically acceptable. Unfortunately, there is no shortcut method available to reach this balance between technical and economic feasibility. Thus, methods of engineering economic analysis should be used to provide results that will help to attain an acceptable balance.

The word *cost* (or *expense*) has meanings that vary in usage.* *Cost concepts* and other economic principles used in an engineering economy study depend on the situation and on the decision to be made. Consequently, the content of Chapter 2, which integrates cost concepts with principles of engineering economy, is important in the applications covered in subsequent chapters of the book.

2.2 Cost Terminology

When conducting engineering economy studies and communicating results, it is important to use consistent definitions for cost terms. Otherwise, common understanding is jeopardized. The definitions for selected terms provided in this section will help achieve that goal.

2.2.1 Fixed, Variable, and Incremental Costs

Fixed costs are those unaffected by changes in activity level over a feasible range of operations for the capacity or capability available. Typical fixed costs include insurance and taxes on facilities, general management and administrative salaries, license fees, and interest costs on borrowed capital.

Of course, any cost is subject to change, but fixed costs tend to remain constant over a specific range of operating conditions. When large changes in usage of resources occur, or when plant expansion or shutdown is involved, fixed costs will be affected.

Variable costs are those associated with an operation that vary in total with the quantity of output or other measures of activity level. If you were making an engineering economic analysis of a proposed change to an existing operation, the variable costs would be the primary part of the prospective differences between the present and changed operations as long as the range of activities is not significantly changed. For example, the costs of material and labor used in a product or service are variable costs—because they vary in total with the number of output units—even though the costs per unit stay the same.

An *incremental cost*, or *incremental revenue*, is the additional cost, or revenue, that results from increasing the output of a system by one (or more) units. Incremental

—————————
*For purposes of this book, the words *cost* and *expense* are used interchangeably.

cost is often associated with "go/no go" decisions that involve a limited change in output or activity level. For instance, the incremental cost per mile for driving an automobile may be $0.27, but this cost depends on considerations such as total mileage driven during the year (normal operating range), mileage expected for the next major trip, and the age of the automobile. Also, it is common to read of the "incremental cost of producing a barrel of oil" and the "incremental cost to the state for educating a student." As these examples indicate, the incremental cost (or revenue) is often quite difficult to determine in practice.

EXAMPLE 2-1

In connection with surfacing a new highway, a contractor has a choice of two sites on which to set up the asphalt mixing plant equipment. The contractor estimates that it will cost $1.15 per cubic yard per mile (yd^3-mile) to haul the asphalt paving material from the mixing plant to the job site. Factors relating to the two site alternatives are as follows (production costs at each site are the same):

Cost Factor	Site A	Site B
Average hauling distance	6 miles	4.3 miles
Monthly rental of site	$1,000	$5,000
Cost to set up and remove equipment	$15,000	$25,000
Hauling expense	$1.15/$yd^3$-mile	$1.15/$yd^3$-mile

Note: If Site B is selected, there will be an added charge of $96 per day for a flagman.

The job requires 50,000 cubic yards of mixed asphalt paving material. It is estimated that four months (17 weeks of five working days per week) will be required for the job. Compare the two sites in terms of their fixed, variable, and total costs. Which is the better site? For the selected site, how many cubic yards of paving material does the contractor have to deliver before starting to make a profit if paid $8.05 per cubic yard delivered to the job site?

SOLUTION

The fixed and variable costs for this job are indicated in the following table. Site rental and setup/removal costs (and the cost of the flagman at Site B) would be constant for the total job, but the hauling cost would vary in total amount with the distance and thus with the total output quantity of yd^3-miles.

Cost	Fixed	Variable	Site A		Site B	
Rent	✔			= $ 4,000		= $ 20,000
Setup/removal	✔			= 15,000		= 25,000
Flagman	✔			= 0	5(17)($96) =	8,160
Hauling		✔	6(50,000)($1.15) =	345,000	4.3(50,000)($1.15) =	247,250
			Total:	$364,000		$300,410

Thus Site B, which has the larger fixed costs, has the smaller total cost for the job.

The contractor will begin to make a profit at the point where total revenue equals total cost as a function of the cubic yards of asphalt pavement mix delivered. Based on Site B, we have:

$$4.3(\$1.15) = \$4.95 \text{ in variable cost per yd}^3 \text{ delivered}$$

$$\text{Total cost} = \text{total revenue}$$

$$\$53,160 + \$4.95x = \$8.05x$$

$$x = 17,149 \text{ yd}^3 \text{ delivered}$$

Therefore, by using Site B, the contractor will begin to make a profit on the job after delivering 17,149 cubic yards of material.

EXAMPLE 2-2

Four college students who live in the same geographical area intend to go home for Christmas vacation (a distance of 400 miles each way). One of the students has an automobile and agrees to take the other three if they will pay the cost of operating the automobile for the trip. When they return from the trip, the owner presents each of them with a bill for $102.40, stating that she has kept careful records of the cost of operating the car and that, based on an annual average of 15,000 miles, their cost per mile is $0.384. The three others feel that the charge is too high and ask to see the cost figures on which it is based. The owner shows them the following list:

Cost Element	Cost per Mile
Gasoline	$0.120
Oil and lubrication	0.021
Tires	0.027
Depreciation	0.150
Insurance and taxes	0.024
Repairs	0.030
Garage	0.012
Total	$0.384

The three riders, after reflecting on the situation, form the opinion that only the costs for gasoline, oil and lubrication, tires, and repairs are a function of mileage driven (variable costs) and thus could be caused by the trip. Because these four costs total only $0.198 per mile, and thus $158.40 for the 800-mile trip, the share for each student would be $158.40/3 = $52.80. Obviously, the opposing views are substantially different. Which, if either, is correct? What are the consequences of the two different viewpoints in this matter, and what should the decision-making criterion be?

SOLUTION

In this instance assume that the owner of the automobile agreed to accept $52.80 per person from the three riders, based on the variable costs that were purely incremental for the Christmas trip versus the owner's average annual mileage. That is, the $52.80 per person is the "with a trip" cost relative to the "without" alternative.

Now, what would the situation be if the three students, because of the low cost, returned and proposed another 800-mile trip the following weekend? And what if there were several more such trips on subsequent weekends? Quite clearly, what started out to be a small marginal (and temporary) change in operating conditions—from 15,000 miles per year to 15,800 miles—soon would become a normal operating condition of 18,000 or 20,000 miles per year. On this basis it would not be valid to compute the extra cost per mile as $0.198.

Because the normal operating range would change, the fixed costs would have to be considered. A more valid incremental cost would be obtained by computing the total annual cost if the car were driven, say, 18,000 miles, then subtracting the total cost for 15,000 miles of operation, and thereby determining the cost of the 3,000 additional miles of operation. From this difference the cost per mile for the additional mileage could be obtained. In this instance the total cost for 15,000 miles of driving per year was $15,000 \times \$0.384 = \$5,760$. If the cost of service— due to increased depreciation, repairs, and so forth—turned out to be $6,570 for 18,000 miles per year, evidently the cost of the additional 3,000 miles would be $810. Then the corresponding incremental cost per mile due to the increase in the operating range would be $0.27. Therefore, if several weekend trips were expected to become normal operation, the owner would be on more reasonable economic ground to quote an incremental cost of $0.27 per mile for even the first trip.

2.2.2 Recurring and Nonrecurring Costs

These two general cost terms are often used to describe various types of expenditures. *Recurring costs* are those that are repetitive and occur when an organization produces similar goods or services on a continuing basis. Variable costs are also recurring costs, because they repeat with each unit of output. But recurring costs are not limited to variable costs. A fixed cost that is paid on a repeatable basis is a recurring cost. For example, in an organization providing architectural and engineering services, office space rental—which is a fixed cost—is also a recurring cost.

Nonrecurring costs, then, are those that are not repetitive, even though the total expenditure may be cumulative over a relatively short period of time. Typically, nonrecurring costs involve developing or establishing a capability or capacity to operate. For example, the purchase cost for real estate upon which a plant will be built is a nonrecurring cost, as is the cost of constructing the plant itself.

2.2.3 Direct, Indirect, and Overhead Costs

These frequently encountered cost terms involve most of the cost elements that also fit into the previous overlapping categories of fixed and variable costs, and recurring and nonrecurring costs. *Direct costs* are those that can be reasonably measured and allocated to a specific output or work activity. The labor and material costs directly associated with a product, service, or construction activity are direct costs. For example, the materials needed to make a pair of scissors would be a direct cost.

Indirect costs are those that are difficult to attribute or allocate to a specific output or work activity. The term normally refers to types of costs that would involve too much effort to allocate directly to a specific output. In this usage they are costs allocated through a selected formula (such as, proportional to direct labor hours, direct labor dollars, or direct material dollars) to the outputs or work activities. For example, the costs of common tools, general supplies, and equipment maintenance in a plant are treated as indirect costs.

Overhead consists of plant operating costs that are not direct labor or direct material costs. In this book the terms *indirect costs, overhead,* and *burden* are used interchangeably. Examples of overhead include electricity, general repairs, property taxes, and supervision. Administrative and selling expenses are usually added to direct costs and overhead costs to arrive at a unit selling price for a product or service. (Appendix A provides a more detailed discussion of cost accounting principles.)

Various methods are used to allocate overhead costs among products, services, and activities. The most commonly used methods involve allocation in proportion to direct labor costs, direct labor hours, direct materials costs, the sum of direct labor and direct materials costs (referred to as *prime cost* in a manufacturing operation), or machine hours. In each of these methods it is necessary to know what the total overhead costs have been or are estimated to be for a time period (typically a year) to allocate them to the production (or service delivery) outputs. Also, total overhead costs are associated with a certain level of production. This is an important condition that should be remembered when dealing with unit cost data (see Section 2.5.2).

We can illustrate direct, indirect, and overhead costs using a typical project such as an addition to an existing plant. The work is planned, scheduled, and controlled—including cost control—by defined activities. Costs of labor and material for each activity are direct costs; that is, they are directly charged to each activity as they are used in accomplishing the work. Then there are other project costs associated with accomplishing the work that are very difficult to allocate directly to each construction activity.

2.2.4   Standard Costs

Standard costs are representative costs per unit of output that are established in advance of actual production or service delivery. They are developed from the direct labor hours, materials, and support functions (with their established costs per unit) planned for the production or delivery process. For example, a standard cost for manufacturing one unit of an automotive part such as a starter would be developed as follows:

Standard Cost Element	Sources of Data for Standard Costs
Direct labor	Process routing sheets, standard
+	times, standard labor rates
Direct material	Material quantities per unit, standard
+	unit material costs
Factory overhead costs	Total factory overhead costs allocated based on prime costs (direct labor plus direct material costs)

= Standard cost (per unit)

Standard costs play an important role in cost control and other management functions. Some representative uses are the following:

1. Estimating future manufacturing or service delivery costs.
2. Measuring operating performance by comparing actual cost per unit with the standard unit cost.
3. Preparing bids on products or services requested by customers.
4. Establishing the value of work-in-process and finished inventories.

2.2.5 Cash Cost Versus Book Cost

A cost that involves payment of cash is called a *cash cost* (and results in a cash flow) to distinguish it from one that does not involve a cash transaction and is reflected in the accounting system as a *noncash cost*. This noncash cost is often referred to as a *book cost*. Cash costs are estimated from the perspective established for the analysis (Principle 3, Section 1.3) and are the future expenses incurred for the alternatives being analyzed. Book costs are costs that do not involve cash payments, but rather represent the recovery of past expenditures over a fixed period of time. The most common example of book cost is the *depreciation* charged for the use of assets such as plant and equipment. In engineering economic analysis, only those costs that are cash flows or potential cash flows from the defined perspective for the analysis need to be considered. *Depreciation, for example, is not a cash flow* and is important in an analysis only because it affects income taxes, which are cash flows. We discuss the topics of depreciation and income taxes in Chapter 7.

2.2.6 Sunk Cost

A *sunk cost* is one that has occurred in the past and has no relevance to estimates of future costs and revenues related to an alternative course of action. Thus, a sunk cost is common to all alternatives, is not part of the future (prospective) cash flows, and can be disregarded in an engineering economic analysis. We need to be able to recognize sunk costs and then handle them properly in an analysis. Specifically, we need to be alert for the possible existence of sunk costs in any situation that involves a past expenditure that cannot be recovered, or capital that has already been invested and cannot be retrieved.

The concept of sunk cost is illustrated in the following simple example. Suppose that Joe College finds a motorcycle he likes and pays $40 as a down payment, which will be applied to the $1,300 purchase price but which must be forfeited if he decides not to take the cycle. Over the weekend, Joe finds another motorcycle he considers equally desirable for a purchase price of $1,230. For the purpose of deciding which cycle to purchase, the $40 is a sunk cost and thus would not enter into the decision except that it lowers the remaining cost of the first cycle. The decision then is between paying $1,260 ($1,300 − $40) for the first motorcycle versus $1,230 for the second motorcycle.

In summary, sunk costs result from past decisions and therefore are irrelevant in the analysis and comparison of alternatives that affect the future. Even though it is sometimes emotionally difficult to do, sunk costs should be ignored, except possibly to the extent that their existence assists you to anticipate better what will happen in the future.

EXAMPLE 2-3

A classic example of sunk cost involves the replacement of assets. Suppose that your firm is considering the replacement of a piece of equipment. It originally cost $50,000, is presently shown on the company records with a value of $20,000, and can be sold for an estimated $5,000. For purposes of replacement analysis, the $50,000 is a sunk cost. However, one view is that the sunk cost should be considered as the difference between the value shown in the company records and the present realizable selling price. According to this viewpoint, the sunk cost is $20,000 minus $5,000, or $15,000. Neither the $50,000 nor the $15,000, however, should be considered in an engineering economic analysis—except for the manner in which the $15,000 may affect income taxes, which will be discussed in Chapter 11.

2.2.7 Opportunity Cost

An *opportunity cost* is incurred because of the use of limited resources, such that the opportunity to use those resources to monetary advantage in an alternative use is foregone. Thus, it is the cost of the best rejected (i.e., foregone) opportunity and is often hidden or implied.

For example, suppose that a project involves the use of vacant warehouse space presently owned by a company. The cost for that space to the project should be the income or savings that possible alternative uses of the space may bring to the firm. In other words, the opportunity cost for the warehouse space should be the income derived from the best alternative use of the space. This may be more than or less than the average cost of that space obtained from the accounting records of the company.

Consider also a student who could earn $20,000 for working during a year but chooses instead to go to school for a year and spend $5,000 to do so. The opportunity cost of going to school for that year is $25,000: $5,000 cash outlay and $20,000 for income foregone. (This figure neglects the influence of income taxes and assumes that the student has no earning capability while in school.)

EXAMPLE 2-4

The concept of an opportunity cost is often encountered in analyzing the replacement of a piece of equipment or other capital asset. Let us reconsider Example 2-3, in which your firm considered the replacement of an existing piece of equipment that originally cost $50,000, is presently shown on the company records with a value of $20,000, but has a present market value of only $5,000. For purposes of an engineering economic analysis of whether to replace the equipment, the present investment in that equipment should be considered as $5,000, because, by keeping the equipment, the firm is giving up the *opportunity* to obtain $5,000 from its disposal. Thus, the $5,000 immediate selling price is really the investment cost of not replacing the equipment and is based on the opportunity cost concept.

2.2.8 Life-Cycle Cost

In engineering practice the term *life-cycle cost* is often encountered. This term refers to a summation of all the costs, both recurring and nonrecurring, related to a product, structure, system, or service during its life span. The *life cycle* is illustrated in Figure 2-1. Life cycle begins with identification of the economic need or want (the requirement) and ends with retirement and disposal activities. It is a time horizon that must be defined in the context of the specific situation—whether it is a highway bridge, a jet engine for commercial aircraft, or an automated flexible manufacturing cell for a factory. The end of the life cycle may be projected on a functional or an economic basis. For example, the amount of time that a structure or piece of equipment is able to perform economically may be shorter than that permitted by its physical capability. Changes in the design efficiency of a boiler illustrate this situation. The old boiler may be able to produce the steam required—but not economically enough for the intended use.

The life cycle may be divided into two general time periods: the acquisition phase and the operation phase. As shown in Figure 2-1, each of these phases is further subdivided into interrelated but different activity periods.

The acquisition phase begins with an analysis of the economic need or want—the analysis necessary to make explicit the requirement for the product, structure, system, or service. Then, with the requirement explicitly defined, the other activities in the acquisition phase can proceed in a logical sequence. The conceptual design activities translate the defined technical and operational requirements into a preferred preliminary design. Included in these activities are development of the feasible alternatives and engineering economic analyses to assist in selection of the preferred preliminary design. Also, advanced development and prototype-testing activities to support the preliminary design work occur during this period.

The next group of activities in the acquisition phase involves detailed design and planning for production or construction. This step is followed by the activities necessary to prepare, acquire, and make ready for operation the facilities and other resources needed for the production, delivery, or construction of the product, structure, system, or service involved. *Again, engineering economy studies are*

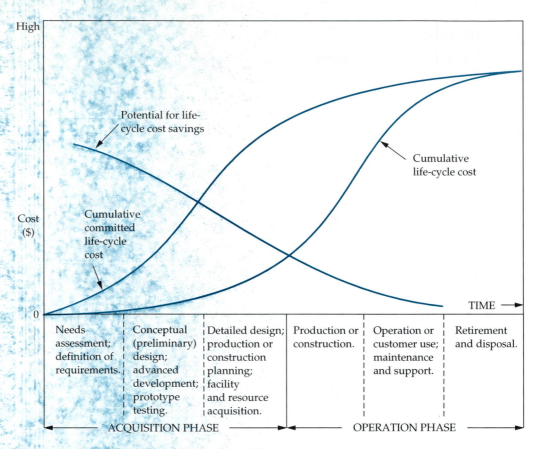

FIGURE 2-1 Phases of the Life Cycle and Their Relative Cost

an essential part of the design process to analyze and compare alternatives and to assist in determining the final detailed design.

In the operation phase the production, delivery, or construction of the end item(s) or service and their operation or customer use occur. This phase ends with retirement from active operation or use and, often, disposal of the physical assets involved. The priorities for engineering economy studies during the operation phase are (1) achieving efficient and effective support to operations, (2) determining whether (and when) replacement of assets should occur, and (3) projecting the timing of retirement and disposal activities.

Figure 2-1 shows relative cost profiles for the life cycle. The greatest potential for achieving life-cycle cost savings is early in the acquisition phase. How much of the life-cycle costs for a product (for example) can be saved is dependent on many factors. However, effective engineering design and economic analysis during this phase are critical in maximizing potential savings.

One aspect of cost-effective engineering design is the minimizing of the impact of design changes during the steps in the life cycle. In general, the cost of a design

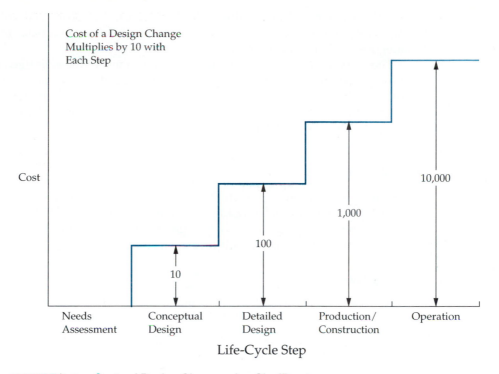

Cost of a Design Change Multiplies by 10 with Each Step

Cost

10,000

1,000

100

10

Needs Assessment | Conceptual Design | Detailed Design | Production/ Construction | Operation

Life-Cycle Step

FIGURE 2-2 Costs of Design Changes Are Significant Source: Reprinted with permission from *National Productivity Review*, Vol. 8, No. 3. Copyright 1989 by Executive Enterprises, Inc.; 22 West 21st Street; New York, NY 10010–6904. All rights reserved.

change increases by a multiple of approximately 10 with each step, as illustrated in Figure 2-2. Thus, there is a large savings incentive to have an excellent preliminary design on which to base the detailed design and to prevent any changes during the production/construction and operation stages of the life cycle.

The cumulative committed life-cycle cost curve increases rapidly during the acquisition phase. In general, approximately 80% of life-cycle costs are "locked in" at the end of this phase by the decisions made during requirements analysis and preliminary and detailed design. In contrast, as reflected by the cumulative life-cycle cost curve, only about 20% of actual costs occur during the acquisition phase, with about 80% being incurred during the operation phase.

Thus, one purpose of the life-cycle concept is to make explicit the interrelated effects of costs over the total life span for a product. An objective of the design process is to minimize the life-cycle cost—while meeting other performance requirements—by making the right tradeoffs between prospective costs during the acquisition phase and those during the operation phase.

The cost elements of the life cycle that need to be considered will vary with the situation. Because of their common use, however, several basic life-cycle cost categories will now be defined.

The *investment cost* is the capital required for most of the activities in the acquisition phase. In simple cases, such as acquiring specific equipment, an investment cost may be incurred as a single expenditure. On a large, complex construction project, however, a series of expenditures over an extended period could be incurred. This cost is also called a *capital investment*.

EXAMPLE 2–5

Consider a situation in which the equipment and related support for a new computer-aided design/computer-aided manufacturing (CAD/CAM) workstation are being acquired for your engineering department. The applicable cost elements and estimated expenditures are as follows:

Cost Element	Cost
Install a leased telephone line for communication	$ 1,100/month
Lease CAD/CAM software (includes installation and debugging)	550/month
Purchase hardware (CAD/CAM workstation)	20,000
Purchase a 19,200-baud modem	2,500
Purchase a high-speed printer	1,500
Purchase a four-color plotter	10,000
Shipping costs	500
Initial training (in house) to gain proficiency with CAD/CAM software	6,000

What is the investment cost of this CAD/CAM system?

SOLUTION

The investment cost in this example is the sum of all the cost elements except the two monthly lease expenditures—specifically, the sum of the initial costs for the CAD/CAM workstation, modem, printer, and plotter ($34,000); shipping cost ($500); and the initial training cost ($6,000). These cost elements result in a total investment cost of $40,500. The two cost elements that involve lease payments on a monthly basis (telephone line and CAD/CAM software) are part of the recurring costs in the operation phase.

The term *working capital* refers to the funds required for current assets (i.e., other than fixed assets such as equipment, facilities, etc.) that are needed for the start-up and subsequent support of operation activities. For example, products cannot be made or services delivered without having materials available in inventory. Functions such as maintenance cannot be supported without spare parts, tools, trained personnel, and other resources. Also, cash must be available to pay employee salaries and the other expenses of operation. The amount of working capital needed will vary with the project involved, and some or all of the investment in working capital is usually recovered at the end of a project's life.

Operation and maintenance cost includes many of the recurring annual expense items associated with the operation phase of the life cycle. The direct and indirect costs of operation associated with the five primary resource areas—people, machines, materials, energy, and information—are a major part of the costs in this category.

Disposal cost includes those nonrecurring costs of shutting down the operation and the retirement and disposal of assets at the end of the life cycle. Normally, costs associated with personnel, materials, transportation, and one-time special activities can be expected. These costs will be offset in some instances by receipts from the sale of assets with remaining market value. A classic example of a disposal cost is that associated with cleaning up a site where a chemical processing plant had been located.

2.3   The General Economic Environment

There are numerous general economic concepts that must be taken into account in engineering studies. In broad terms, economics deals with the interactions between people and wealth, and engineering is concerned with the cost-effective use of scientific knowledge to benefit humankind. This section introduces some of these basic economic concepts and indicates how they may be factors for consideration in engineering studies and managerial decisions.

2.3.1   Consumer and Producer Goods and Services

The goods and services that are produced and utilized may be divided conveniently into two classes. *Consumer goods and services* are those products or services that are directly used by people to satisfy their wants. Food, clothing, homes, cars, television sets, haircuts, opera, and medical services are examples. The producers of consumer goods and services must be aware of, and are subject to, the changing wants of the people to whom their products are sold. At the same time, the demand for such goods and services is directly related to people and in many cases, as will be discussed later, may be determined with considerable certainty.

Producer goods and services are used to produce consumer goods and services or other producer goods. Machine tools, factory buildings, buses, and farm machinery are examples. In the long run, producer goods serve to satisfy human wants, but only as a means to that end. Thus the amount of producer goods needed is determined indirectly by the amount of consumer goods or services that are demanded by people. However, because the relationship is much less direct than for consumer goods and services, the demand for and production of producer goods may greatly precede or lag behind the demand for the consumer goods that they will produce.

2.3.2 Measures of Economic Worth

> Goods and services are produced and desired because directly or indirectly they have *utility*—the power to satisfy human wants and needs. Thus they may be used or consumed directly, or they may be used to produce other goods or services that may, in turn, be used directly. Utility is most commonly measured in terms of *value*, expressed in some medium of exchange as the *price* that must be paid to obtain the particular item.

Much of our business activity, including engineering, focuses on increasing the utility (value) of materials and products by changing their form or location. Thus iron ore, worth only a few dollars per ton, significantly increases in value by being processed, combined with suitable alloying elements, and converted into razor blades. Similarly, snow, worth almost nothing when high in distant mountains, becomes quite valuable when it is delivered in melted form several hundred miles away to dry southern California.

2.3.3 Necessities, Luxuries, and Price Demand

Goods and services may be divided into two types—*necessities* and *luxuries*. Obviously, these terms are relative, because, for most goods and services, what one person considers a necessity may be considered a luxury by another. For example, a person living in one community may find that an automobile is a necessity to get to and from work. If the same person lived and worked in a different city, adequate public transportation might be available, and an automobile would be a luxury. For all goods and services, there is a relationship between the price that must be paid and the quantity that will be demanded or purchased. This general relationship is depicted in Figure 2-3. As the selling price per unit (p) is increased, there will be less demand (D) for the product, and as the selling price is decreased, the demand will increase. The relationship between price and demand can be expressed as a linear function:

$$p = a - bD \qquad \text{for } 0 \le D \le \frac{a}{b}, \text{ and } a > 0, b > 0 \qquad \textbf{(2-1)}$$

where a is the intercept on the price axis and $-b$ is the slope. Thus b is the amount by which demand increases for each unit decrease in p. Both a and b are constants. It follows, of course, that

$$D = \frac{a - p}{b} \qquad (b \ne 0) \qquad \textbf{(2-2)}$$

Although Figure 2-3 illustrates the general relationship between price and demand, this relationship would probably be different for necessities and luxuries. Consumers can readily forego the consumption of luxuries if the price is greatly increased, but they find it more difficult to reduce their consumption of true necessities. Also, they will use the money saved by not buying luxuries to pay the increased cost of necessities.

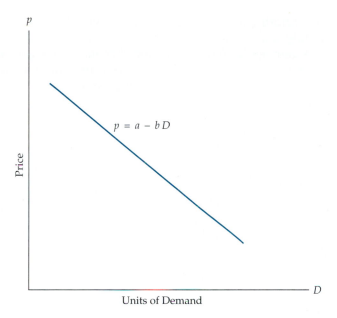

FIGURE 2-3 General Price–Demand Relationship (Note that price is considered to be the independent variable but is shown as the vertical axis. This convention is commonly used by economists.)

2.3.4 Competition

Because economic laws are general statements regarding the interaction of people and wealth, they are affected by the economic environment in which people and wealth exist. Most general economic principles are stated for situations in which *perfect competition* exists.

Perfect competition occurs in a situation in which any given product is supplied by a large number of vendors and there is no restriction on additional vendors entering the market. Under such conditions, there is assurance of complete freedom on the part of both buyer and seller. Perfect competition may never occur in actual practice, because of a multitude of factors that impose some degree of limitation upon the actions of buyers or sellers, or both. However, with conditions of perfect competition assumed, it is easier to formulate general economic laws.

The existing competitive situation is an important factor in most engineering economy studies. It will have a very real effect upon managerial decisions. Unless information is available to the contrary, it should be assumed that competitors do or will exist and that they produce a quality product or service, and the resulting effects should be taken into account.

Monopoly is at the opposite pole from perfect competition. A perfect monopoly exists when a unique product or service is only available from a single vendor and that vendor can prevent the entry of all others into the market. Under such conditions, the buyer is at the complete mercy of the vendor in terms of the availability and price of the product. Perfect monopolies rarely occur in practice, because (1) few products are so unique that substitutes cannot be used satisfactorily, and (2) governmental regulations prohibit monopolies if they are unduly restrictive.

A monopoly may be of great benefit to a producer in that it may permit control of the supply and the price to provide maximum profit. Although this might result in high prices for the product, higher *total* profits may be obtained from lower prices and wider distribution. Under some conditions a monopoly may avoid costly duplication of facilities and thus make possible lower prices for products and services. This situation is recognized by governing bodies in granting public utilities exclusive rights to render service in a given territory. Such a practice, for example, avoids having two electric power companies duplicate power-distribution lines and equipment in the same city. When vendors are granted such a monopolistic position, the governmental body also regulates the rates that the utility can charge for its services to ensure that the customers are not overcharged. Thus governmental regulation takes the place of competition in determining prices.

An *oligopoly* exists when there are so few suppliers of a product or service that action by one will almost inevitably result in similar action by the others. Thus if one of the only three gasoline stations in an isolated town raises the price of its product by 1 cent per gallon, the other two will probably do the same because they can do so and still retain their previous competitive positions.

2.3.5 The Total Revenue Function

The total revenue, TR, that will result from a business venture during a given period is the product of the selling price per unit, p, and the number of units sold, D. Thus

$$\text{TR} = \text{price} \times \text{demand} = p(D) \tag{2-3}$$

If the relationship between price and demand as given in Equation 2-1 is used,

$$\text{TR} = (a - bD)D = aD - bD^2 \qquad \text{for } 0 \le D \le \frac{a}{b} \text{ and } a > 0, b > 0 \tag{2-4}$$

The relationship between total revenue and demand for the condition expressed in Equation 2-4 may be represented by the curve shown in Figure 2-4. From calculus the demand, $\hat{D}$, that will produce maximum total revenue can be obtained by solving

$$\frac{d\text{TR}}{dD} = a - 2bD = 0 \tag{2-5}$$

Thus

$$\hat{D} = \frac{a}{2b} \tag{2-6}$$

To guarantee that $\hat{D}$ maximizes total revenue, check the second derivative to be sure it is negative:

$$\frac{d^2\text{TR}}{-dD^2} = -2b$$

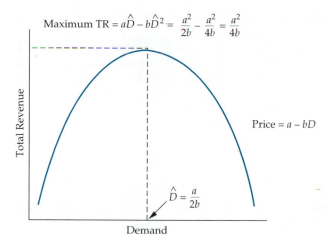

Total Revenue (vertical axis)

Maximum TR $= a\hat{D} - b\hat{D}^2 = \dfrac{a^2}{2b} - \dfrac{a^2}{4b} = \dfrac{a^2}{4b}$

Price $= a - bD$

$\hat{D} = \dfrac{a}{2b}$

Demand

FIGURE 2-4
Total Revenue Function as a Function of Demand

For example, if the equation for price is given by $50,000 - 200D$, the demand, $\hat{D}$, that maximizes total revenue is equal to $50,000/400 = 125$ units. It must be emphasized that, because of cost-volume relationships discussed in the following section, *most businesses would not obtain maximum profits by maximizing revenue.* Thus the cost-volume relationship must be considered and related to revenue.

At this point, attention is called to the derivative of the total revenue with respect to volume (demand), dTR/dD, which is called the *incremental* or *marginal revenue.*

2.3.6 Cost, Volume, and Breakeven Point Relationships

Fixed costs remain constant over a wide range of activities as long as the business does not permanently discontinue operations, but variable costs vary in total with the volume of output (Section 2.2.1). Thus, at any demand D, total cost is

$$C_T = C_F + C_V \tag{2-7}$$

where C_F and C_V denote fixed and variable costs, respectively. For the linear relationship assumed here,

$$C_V = (c_v)(D) \tag{2-8}$$

where c_v is the variable cost per unit. In this section we consider two scenarios for finding breakeven points. In the first scenario demand is a function of price. The second scenario assumes that price and demand are independent of each other.

Scenario 1 When total revenue, as depicted in Figure 2-4, and total cost, as given by Equations 2-7 and 2-8, are combined, the typical results as a function of demand are depicted in Figure 2-5. At *breakeven point* D'_1, total revenue is equal to total cost, and an increase in demand will result in a profit for the operation. Then at optimal demand, D^*, profit is maximized (Equation 2-10). At breakeven point D'_2, total revenue and total cost are again equal, but additional volume will result in an operating loss instead of a profit. Obviously, the conditions for

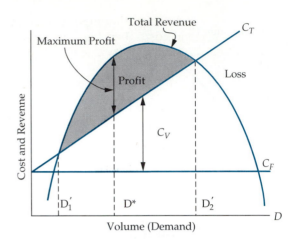

FIGURE 2-5

Combined Cost and Revenue
Functions, and Breakeven Points,
as Functions of Volume, and Their
Effect on Typical Profit

which breakeven and maximum profit occurs are our primary interest. First, at
any volume (demand), D,

$$\text{Profit (loss)} = \text{total revenue} - \text{total costs}$$

$$= (aD - bD^2) - (C_F + c_vD)$$

$$= -C_F + (a - c_v)D - bD^2 \qquad \text{for } 0 \leq D \leq \frac{a}{b} \qquad \textbf{(2-9)}$$

In order for a profit to occur, based on Equation 2-9, and to achieve the typical
results depicted in Figure 2-5, two conditions must be met:

1. $(a - c_v) > 0$; that is, the price per unit that will result in no demand has to be
   greater than the variable cost per unit (this avoids negative demand).
2. Total revenue (TR) must exceed total cost (C_T) for the period involved.

If these conditions are met, we can find the optimal demand at which maximum
profit will occur by taking the first derivative of Equation 2-9 with respect to D
and setting it equal to zero:

$$\frac{d(\text{profit})}{dD} = a - c_v - 2bD = 0$$

The optimal value of D that maximizes profit is

$$D^* = \frac{a - c_v}{2b} \qquad \textbf{(2-10)}$$

To ensure that we have *maximized* profit (rather than minimized it), the sign of the
second derivative must be negative. Checking this, we find that

$$\frac{d^2(\text{profit})}{dD^2} = -2b$$

which will be negative for $b > 0$ (as earlier specified). (Also recall that in cost minimization problems a positive-signed second derivative is necessary to guarantee
a minimum-valued optimal cost solution.)

An economic breakeven point for an operation occurs when total revenue equals total cost. Then for total revenue and total cost, as used in the development of Equations 2-9 and 2-10 and at any demand D,

$$\text{Total revenue} = \text{total cost} \quad \text{(breakeven point)}$$

$$aD - bD^2 = C_F + c_v D$$

$$-bD^2 + (a - c_v)D - C_F = 0 \tag{2-11}$$

Because Equation 2-11 is a quadratic equation with one unknown (D), we can solve for the breakeven points D_1' and D_2' (the roots of the equation).

$$D' = \frac{-(a - c_v) \pm [(a - c_v)^2 - 4(-b)(-C_F)]^{1/2}}{2(-b)} \tag{2-12}$$

With the conditions for a profit satisfied (Equation 2-9), the quantity in the brackets of the numerator (the discriminant) in Equation 2-12 will be greater than zero. This will ensure that D_1' and D_2' have real positive, and unequal, values.

EXAMPLE 2-6

A company produces an electronic timing switch that is used in consumer and commercial products made by several other manufacturing firms. The fixed cost (C_F) is $73,000 per month, and the variable cost (c_v) is $83 per unit. The selling price per unit is $p = \$180 - 0.02(D)$, based on Equation 2-1. For this situation (a) determine the optimal volume for this product and confirm that a profit occurs (instead of a loss) at this demand; and (b) find the volumes at which breakeven occurs; that is, what is the range of profitable demand?

SOLUTION

(a) $\quad D^* = \dfrac{a - c_v}{2b} = \dfrac{\$180 - 83}{2(0.02)} = 2{,}425$ units per month $\quad$ (Equation 2-10)

Is $(a - c_v) > 0$?

$$(\$180 - 83) = \$97, \quad \text{which is greater than 0.}$$

And is (total revenue $-$ total cost) > 0 for $D^* = 2{,}425$ units per month?

$$[\$180(2{,}425) - 0.02(2{,}425)^2] - [\$73{,}000 + 83(2{,}425)] = \$44{,}612$$

A demand of $D^* = 2{,}425$ units per month results in a maximum profit of $44,612 per month because the second derivative is negative (-0.04).

(b) $\qquad$ Total revenue $=$ total cost $\qquad$ (breakeven point)

$$-bD^2 + (a - c_v)D - C_F = 0 \qquad \text{(Equation 2-11)}$$

$$-0.02D^2 + (\$180 - \$83)D - \$73{,}000 = 0$$

$$-0.02D^2 + 97D - 73{,}000 = 0$$

And, from Equation 2-12,

$$D' = \frac{-97 \pm [(97)^2 - 4(-0.02)(-73,000)]^{0.5}}{2(-0.02)}$$

$$D'_1 = \frac{-97 + 59.74}{-0.04} = 932 \text{ units per month}$$

$$D'_2 = \frac{-97 - 59.74}{-0.04} = 3,918 \text{ units per month}$$

Thus, the range of profitable demand is 932 to 3,918 units per month.

Scenario 2 When the price per unit (p) for a product or service can be represented more simply as being independent of demand (versus being a linear function of demand, as assumed in Equation 2-1) and is greater than the variable cost per unit (c_v), a single breakeven point results. Then under the assumption that demand is immediately met, total revenue (TR) = $p(D)$. If the linear relationship for costs in Equations 2-7 and 2-8 is also used in the model, the typical situation is depicted in Figure 2-6.

EXAMPLE 2-7

An engineering consulting firm measures its output in a standard service hour unit, which is a function of the personnel grade levels in the professional staff. The variable cost (c_v) is $62 per standard service hour. The charge-out rate [i.e., selling price (p)] is $1.38(c_v) = \$85.56$ per hour. The maximum output of the firm is 160,000 hours per year, and its fixed cost (C_F) is $2,024,000 per year. For this

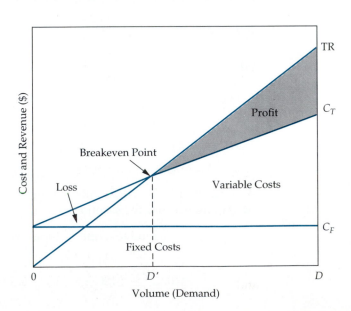

FIGURE 2-6

Typical Breakeven Chart with Price (p) a Constant

firm, (a) what is the breakeven point in standard service hours and in percentage of total capacity, and (b) what is the percentage reduction in the breakeven point (sensitivity) if fixed costs are reduced 10%; if variable cost per hour is reduced 10%; if both costs are reduced 10%; and if the selling price per unit is increased by 10%?

SOLUTION

(a)
$$\text{Total revenue} = \text{total cost} \qquad \text{(breakeven point)}$$
$$pD' = C_F + c_vD'$$
$$D' = \frac{C_F}{(p - c_v)} \tag{2-13}$$

and

$$D' = \frac{\$2,024,000}{(\$85.56 - \$62)} = 85,908 \text{ hours per year}$$

$$D' = \frac{85,908}{160,000} = 0.537, \quad \text{or } 53.7\% \text{ of capacity}$$

(b) 10 percent reduction in C_F:

$$D' = \frac{0.9(\$2,024,000)}{(\$85.56 - \$62)} = 77,318 \text{ hours per year}$$

and

$$\frac{85,908 - 77,318}{85,908} = 0.10, \quad \text{or a } 10\% \text{ reduction in } D'$$

10 percent reduction in c_v:

$$D' = \frac{\$2,024,000}{[\$85.56 - 0.9(\$62)]} = 68,011 \text{ hours per year}$$

and

$$\frac{85,908 - 68,011}{85,908} = 0.208, \quad \text{or a } 20.8\% \text{ reduction in } D'$$

10 percent reduction in both C_F and c_v:

$$D' = \frac{0.9(\$2,024,000)}{[\$85.56 - 0.9(\$62)]} = 61,210 \text{ hours per year}$$

and

$$\frac{85,908 - 61,210}{85,908} = 0.287, \quad \text{or a } 28.7\% \text{ reduction in } D'$$

10 percent increase in p:

$$D' = \frac{\$2,024,000}{[1.1(\$85.56) - \$62]} = 63,021 \text{ hours per year}$$

and

$$\frac{85,908 - 63,021}{85,908} = 0.266, \quad \text{or a } 26.6\% \text{ reduction in } D'$$

Thus, the breakeven point is more sensitive to a reduction in variable cost per hour than to the same percentage reduction in the fixed cost, but reduced costs in both areas should be sought. Furthermore, notice that the breakeven point in this example is highly sensitive to the selling price per unit, p. These results are summarized as follows:

Change in Factor Value(s)	Decrease in Breakeven Point
10% reduction in C_F	10.0%
10% reduction in c_v	20.8
10% reduction in C_F and in c_v	28.7
10% increase in p	26.6

The breakeven point for an operating situation can be determined in units of output, percentage utilization of capacity, or sales volume (demand). In Example 2-7(a), the breakeven point (D') was calculated in units of output (85,908 standard service hours) and then, using the total capacity figure (160,000 hours per year), it was also expressed as percentage utilization of capacity (53.7%). In terms of sales volume, the breakeven point in Example 2-7 is $85.56 (85,908) = $7,350,288.

Market competition often creates pressure to lower the breakeven point of an operation; the lower the breakeven point, the less likely that a loss will occur during market fluctuations. Also, if the selling price remains constant, a larger profit will be achieved at any level of operation above the reduced breakeven point.

2.3.7 Average Unit Cost Function

Most engineering projects and business operations are designed to operate more efficiently at a certain level of capacity utilization. Deviations from this level may affect the variable cost of operation and possibly the fixed cost, and will influence the *average unit cost* of the product or service. Any impacts will, in turn, affect the total profit.

The average unit cost (C_U) at any volume (demand D), within the capacity related to the fixed cost (C_F), is simply the total cost (C_T) at that volume divided by D.

$$C_U = \frac{C_T}{D} = \frac{C_F + C_V}{D} \tag{2-14}$$

When total variable costs are a linear function of demand, $C_V = c_v D$ (Equation 2-8), the average unit cost is

$$C_U = \frac{C_F}{D} + c_v \tag{2-15}$$

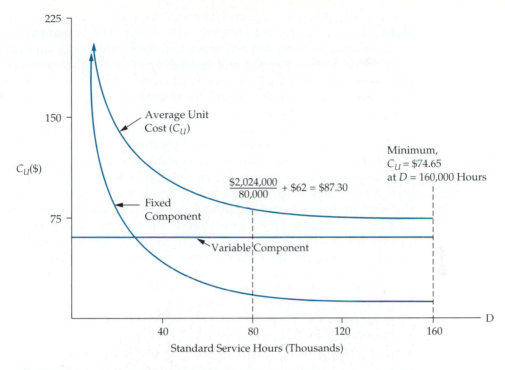

FIGURE 2-7 Average Unit Cost Function for Example 2-7

Applying Equation 2-15 to Example 2-7, the average unit cost is

$$C_U = \frac{\$2,024,000}{D} + \$62$$

and the result for the engineering consulting firm is shown in Figure 2-7. Thus, in the linear model situation, the minimum average cost occurs at the maximum output level that is feasible without affecting fixed costs.

2.4 Cost-Driven Design Optimization

As discussed in Section 2.2.8, engineers must maintain a *life-cycle* (i.e., "cradle to grave") viewpoint as they design products, processes, and services. Such a complete perspective ensures that engineers consider initial investment costs; operation and maintenance expenses and other annual expenses in later years; and environmental and social consequences over the life of their designs. In fact, a movement called *design for the environment* (DFE), or "green engineering," has prevention of waste, improved materials selection, and reuse/recycling of resources among its goals. Designing for energy conservation, for example, is a subset of green engineering. Another example is the design of an automobile bumper that can be easily recycled.

Minimum life-cycle cost, consistent with the consideration of other factors, is a goal achieved largely in the early stages of design. The attitude that the engineer can develop something that will work and then think about controlling the cost is fallacious, because, by the time most of the functional requirements are built into a design, many of the best opportunities for reducing costs have been missed. Engineers can accomplish a great deal toward the goal of minimizing life-cycle costs by simply keeping in mind its importance!

The importance of cost-effective design becomes clear when we examine Figure 2-8, which shows the proportion of production costs determined by various departments contrasted with annual percentages of operating costs incurred in these departments. In Figure 2-8 we see that 70% of design-related production costs are determined by 11% of a typical firm's annual operating costs for design. This fact underscores the importance of a design function in a company that considers all stages in a product's life cycle and strives to prevent costly changes in the production of a product or service. Think about how much money automobile manufacturers spend for recalls of cars that are already sold and in use. Clearly there are major costs involved in these late design modifications. This same idea was illustrated earlier in Figure 2-2, in which "downstream" design alterations were shown to cost inordinate amounts of money (and damage the reputation of the manufacturer).

Examples of cost minimization through effective design are plentiful in the practice of engineering. Consider the design of a heat exchanger in which tube material and configuration affect cost and dissipation of heat. The problems in this section designated as "cost-driven design optimization" are simple design models intended to illustrate the importance of cost in the design process. These problems show the procedure for determining an optimal design using cost concepts. We will consider discrete/continuous optimization problems that involve a single design variable, X. This variable is also called a *primary cost driver,* and knowledge of its behavior may allow a designer to account for a large portion of total cost behavior.

For cost-driven design optimization problems, the two main tasks are as follows:

1. Determine the optimal value for a certain alternative's design variable. For example, what velocity of an aircraft minimizes the total annual costs of owning and operating the aircraft?
2. Select the best alternative, each with its own unique value for the design variable. For example, what insulation thickness is best for a home in Virginia—R11, R19, R30, or R38?

In general, the cost models developed in these problems consist of three types of costs:

1. Fixed cost(s)
2. Cost(s) that vary *directly* with the design variable
3. Cost(s) that vary *indirectly* with the design variable

A simplified format of a cost model with one design variable is the following:

$$\text{Cost} = aX + \frac{b}{X} + k \tag{2-16}$$

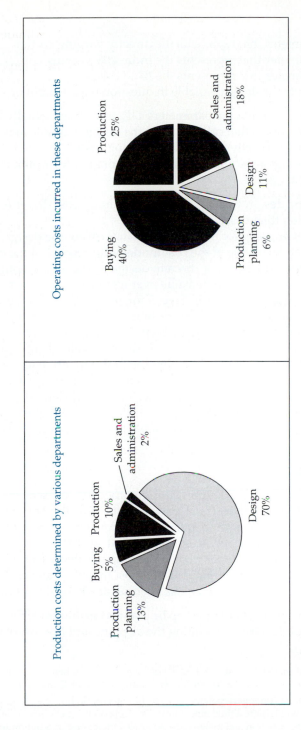

Production costs determined by various departments

Operating costs incurred in these departments

FIGURE 2-8 Production Costs and Operating Costs of Various Departments Source: Syan, C. S. and U. Menon, *Concurrent Engineering: Concepts, Implementation and Practice,* Chapman and Hall, London, 1994, p. 102.

where

a is a parameter that represents the directly varying cost(s)
b is a parameter that represents the indirectly varying cost(s)
k is a parameter that represents the fixed cost(s)
X represents the design variable in question (e.g., weight or velocity)

In a particular problem the parameters a, b, and k may actually represent the sum of a group of costs in that category, and the design variable may be raised to some power for either directly or indirectly varying costs.*

The following steps outline a general approach for optimizing a design with respect to cost.

1. Identify the design variable that is the primary cost driver (e.g., pipe diameter or insulation thickness).
2. Write an expression for the cost model in terms of the design variable.
3. Set the first derivative of the cost model with respect to the continuous design variable equal to zero. For discrete design variables, compute the value of the cost model for each discrete value over a selected range of potential values.
4. Solve the equation found in Step 3 for the optimum value of the continuous design variable.† For discrete design variables the optimum value has the minimum cost value found in Step 3. In Example 2-9 we will also use an incremental procedure for selecting the best-valued discrete design variable. This method is analogous to taking the first derivative for a continuous design variable and setting it equal to zero to determine an optimal value.
5. For continuous design variables use the second derivative of the cost model with respect to the design variable to determine whether the optimum value found in Step 4 corresponds to a global maximum or minimum.

EXAMPLE 2-8

The cost of operating a jet-powered commercial (passenger-carrying) airplane varies as the three-halves (3/2) power of its velocity; specifically, $C_0 = knv^{3/2}$, where n is the trip length in miles, k is a constant of proportionality, and v is velocity in miles per hour. It is known that at 400 miles per hour the *average* cost of operation is $300 per mile. The company that owns the aircraft wants to minimize the cost of operation, but that cost must be balanced against the cost of the passengers' time (C_c), which has been set at $300,000 per hour.

(a) At what velocity should the trip be planned to minimize the total cost, which is the sum of the cost of operating the airplane and the cost of passengers' time?

---

*A more general model is: Cost $= k + ax + b_1x^{e_1} + b_2x^{e_2} + \ldots$, where $e_1 = -1$ reflects costs that vary inversely with X, $e_2 = 2$ indicates costs that vary as the square of X, and so forth.

†If multiple optima (stationary points) are found in Step 4, finding the global optimum value of the design variable will require a little more effort. One approach is to systematically use each root in the second derivative equation and assign each point as a maximum or a minimum based on the sign of the second derivative. A second approach would be to use each root in the objective function and see which point best satisfies the cost function.

(b) How do you know that your answer for the problem in (a) minimizes the total cost?

SOLUTION

The equation for total cost (C_T) is:

$$C_T = C_0 + C_c = knv^{3/2} + (\$300{,}000 \text{ per hour})\left(\frac{n}{v}\right)$$

Now we solve for the value of k:

$$\frac{C_0}{n} = kv^{3/2}$$

$$\frac{\$300}{\text{mile}} = k\left(400\frac{\text{miles}}{\text{hour}}\right)^{3/2}$$

$$k = \frac{\$300/\text{mile}}{\left(400\frac{\text{miles}}{\text{hour}}\right)^{3/2}}$$

$$k = \frac{\$300/\text{mile}}{8000\left(\frac{\text{miles}^{3/2}}{\text{hours}^{3/2}}\right)}$$

$$k = \$0.0375\frac{\text{hours}^{3/2}}{\text{miles}^{5/2}}$$

Thus,

$$C_T = \left(\$0.0375\frac{\text{hours}^{3/2}}{\text{miles}^{5/2}}\right)(n \text{ miles})\left(v\frac{\text{miles}}{\text{hour}}\right)^{3/2} + \left(\frac{\$300{,}000}{\text{hour}}\right)\left(\frac{n \text{ miles}}{v\frac{\text{miles}}{\text{hour}}}\right)$$

$$C_T = \$0.0375nv^{3/2} + \$300{,}000\left(\frac{n}{v}\right)$$

Next, the first derivative is taken:

$$\frac{dC_T}{dv} = \frac{3}{2}(\$0.0375)nv^{1/2} - \frac{\$300{,}000n}{v^2} = 0$$

So,

$$0.05625v^{1/2} - \frac{300{,}000}{v^2} = 0$$

$$0.05625v^{5/2} - 300{,}000 = 0$$

$$v^{5/2} = \frac{300{,}000}{0.05625} = 5{,}333{,}333$$

$$v^* = (5{,}333{,}333)^{0.4} = 490.68 \text{ mph}$$

Finally, we check the second derivative to confirm a minimum cost solution:

$$\frac{d^2C_T}{dv^2} = \frac{0.028125}{v^{1/2}} + \frac{600,000}{v^3} \qquad \text{for } v > 0 \text{ and therefore } \frac{d^2C_T}{dv^2} > 0$$

The company concludes that $v = 490.68$ mph minimizes the total cost of this particular airplane's flight.

EXAMPLE 2-9

This example deals with a discrete optimization problem of determining the most economical amount of attic insulation for a large single-story home in Virginia. In general the heat lost through the roof of a single-story home is:

$$\begin{pmatrix} \text{Heat loss} \\ \text{in Btu} \\ \text{per hour} \end{pmatrix} = \begin{pmatrix} \Delta\text{Temperature} \\ \text{in °F} \end{pmatrix} \begin{pmatrix} \text{Area} \\ \text{in} \\ \text{ft}^2 \end{pmatrix} \begin{pmatrix} \text{Conductance in} \\ \text{Btu/hour} \\ \text{ft}^2\text{-°F} \end{pmatrix}$$

or

$$Q = (T_{\text{in}} - T_{\text{out}}) \cdot A \cdot U$$

In southwest Virginia the number of heating-days per year is approximately 230, and the annual heating degree-days equals 230 (65°F − 46°F) = 4,370 degree-days per year. Here 65°F is assumed to be the average inside temperature and 46°F is the average outside temperature each day.

Consider a 2,400 ft^2 single-family house in Blacksburg. The typical annual space-heating load for this size of a house is 100×10^6 Btu. That is, with no insulation in the attic we lose about 100×10^6 Btu per year.* Common sense dictates that the "no insulation" alternative is not attractive and is to be avoided.

With insulation in the attic, the amount of heat lost each year will be reduced. The value of energy savings that results from adding insulation and reducing heat loss is dependent on what type of residential heating furnace is installed. For this example we assume that an electrical resistance furnace is installed by the builder, and its efficiency is near 100%.

Now we're in a position to answer the question: what amount of insulation is most economical? An additional piece of data we need involves the cost of electricity, which is $0.074 per kWh. This can be converted to dollars per 10^6 Btu as follows (1 kWh = 3,413 Btu):

$$\frac{\text{kWh}}{3,413 \text{ Btu}} = \frac{293 \text{ kWh}}{10^6 \text{ Btu}}$$

$$\frac{293 \text{ kWh}}{10^6 \text{ Btu}} \left(\frac{\$0.074}{\text{kWh}} \right) \cong \$21.75/10^6 \text{ Btu}$$

The cost of several insulation alternatives and associated space heating loads for this house are given in the following table.

---

*100×10^6 Btu/yr $\cong \left(\dfrac{4,370 \text{ °F-days per year}}{1.00 \text{ efficiency}} \right)(2,400 \text{ ft}^2)(24 \text{ hr/day})\dfrac{0.397 \text{ Btu/hr}}{\text{ft}^2\text{-°F}}$, where 0.397 is the U-factor with no insulation.

	Amount of Insulation			
	R11	**R19**	**R30**	**R38**
Investment cost	$600	$900	$1,300	$1,600
Annual heating load (Btu/year)	74×10^6	69.8×10^6	67.2×10^6	66.2×10^6

In view of the data above, which amount of attic insulation is most economical? The life of the insulation is estimated to be 25 years.

SOLUTION

Set up a table to examine total life cycle costs.

	R11	**R19**	**R30**	**R38**
A. Investment cost	$600	$900	$1,300	$1,600
B. Cost of heat loss per year	$1,609.50	$1,518.15	$1,461.60	$1,439.85
C. Cost of heat loss over 25 years	$40,237.50	$37,953.75	$36,540	$35,996.25
D. Total life cycle costs (A + C)	$40,837.50	$38,853.75	$37,840	$37,596.25

Answer: To minimize total life cycle costs, select R38 insulation.

ANOTHER SOLUTION

Another approach for selecting the best alternative from a discrete set is to examine the incremental (Δ) differences among them (remember principle 2 on page 5?) when the alternatives are ranked from low investment cost to high investment cost. We illustrate this procedure here and return to it in Chapter 5.

We begin by examining total energy savings over 25 years for each added amount of insulation, less the added investment cost associated with each amount of insulation.

The following questions lead to the computation of the relevant tradeoffs involved:

1. What savings occur if we decide to insulate with R19 instead of R11?

$$R11 \rightarrow R19 \qquad \Delta \text{ Investment} = -\$300$$
$$\Delta \text{ Savings/year} = \$91.35$$
$$= [-1,518.15 - (-1,609.5)]$$
$$\Delta \text{ Savings over 25 years} = \$2,283.75$$

By putting in R19 rather than R11, the total net savings over 25 years is $1,983.75.

2. What total net savings are realized if we choose R30 instead of R19?

$$R19 \rightarrow R30 \qquad \Delta \text{ Investment} = -\$400$$
$$\Delta \text{ Savings/year} = \$56.55$$
$$= [-1,461.60 - (-1,518.15)]$$
$$\Delta \text{ Savings over 25 years} = \$1,413.75$$

The total net savings over 25 years is $1,013.75.

3. Finally, what net savings are achieved if we add the maximum amount of insulation (R38) rather than R30?

R30 → R38
Δ Investment = −$300
Δ Savings per year = $21.75
= [−1,439.85 − (−1,461.60)]
Δ Savings over 25 years = $543.75

The total net savings over 25 years is $243.75.

If we ignore the time value of money (to be discussed in Chapter 3) over the 25-year period and select the amount of attic insulation that gives us positive net savings, our best (most economical) choice would be R38.

CAUTION

This conclusion may change when we consider the time value of money (i.e., an interest rate greater than zero) in Chapter 3. In such a case it will not necessarily be true that adding more and more insulation is the optimal course of action.

2.5 Present Economy Studies

When the influence of time on money is not a significant consideration, cost analyses are usually called *present economy studies.* Typical situations involving present economy studies are as follows:

1. There is no initial investment of capital; only immediate operating costs and other factors are involved. For example, assume that you are employed by Company A and are making plans for a business trip. You can travel by commercial aircraft, which will require 3 hours of travel time and the rental of a car at your destination. The alternative is to travel by automobile, which will take 7 hours. Here the basic considerations are the immediate costs, the value of your time, and nonmonetary factors (e.g., fatigue).
2. There is an initial investment of capital, but after this investment cost the remaining life-cycle cost is estimated to be the same, or directly proportional to the initial investment. Thus, the alternative with the lowest investment cost will be the most economical. Consider the construction of an interstate highway bridge overpass. Whether a longitudinally reinforced concrete slab or a precast (prestressed) concrete design is used, the maintenance and other life-cycle costs of the two designs are usually assumed to be proportionally the same. (However, if one or more steel design alternatives were also being considered, a present economy study probably would not be appropriate. Maintenance and other related costs would be expected to vary among the alternatives, and the cost analysis should be based on the life cycle of the structure.)
3. The differences in revenues and costs among the alternatives all occur within a limited time period (1 year or less is a general guideline), or any future differences are estimated to remain proportional to those in the first time period. This is often the case when the decision is between alternative materials

in manufacturing. For example, if estimations show that using low-alloy/high-yield strength steel in a particular application gives better revenue and cost results than using low-carbon steel, this relative advantage would be expected to remain in future time periods.

In engineering practice, situations that give rise to present economy studies are quite common; several typical situations are discussed in the following sections. Recognizing these situations often saves considerable analysis effort.

2.5.1 Total Cost in Material Selection

In many cases economic selection among materials cannot be based solely on the costs of the materials. Frequently, a change in materials will affect the processing costs, and shipping costs may also be altered. A good example of this case is illustrated in Figure 2-9 for which annual demand is 1,000,000 units. The part shown was produced in considerable quantities on a high-speed turret lathe, using 1112 screw-machine steel costing $0.30 per pound. A study was conducted to determine whether it might be cheaper to use brass screw stock, costing $1.40 per pound. Because the weight of steel required per piece was 0.0353 pound and that of brass was 0.0384 pound, the material cost per piece was $0.0106 for steel and $0.0538 for brass. However, when the manufacturing and standards departments were consulted, it was found that, although 57.1 parts per hour were being produced using steel, the output would be 102.9 parts per hour if brass were used. Because the machine attendant was paid $7.50 per hour and the overhead cost for the turret lathe was $10.00 per hour, the total-cost comparison for the two materials was as follows:

	1112 Steel		Brass	
Material	$0.30 × 0.0353 =	$0.0106	$1.40 × 0.0384 =	$0.0538
Labor	$7.50/57.1 =	0.1313	$7.50/102.9 =	0.0729
Overhead[a]	$10.00/57.1 =	0.1751	$10.00/102.9 =	0.0972
Total cost per piece		$0.3170		$0.2239

Saving per piece by use of brass = $0.3170 − $0.2239 = $0.0931

[a] A given overhead rate applied to different alternatives without modification may be invalid for economic analyses even though it is useful in after-the-fact accounting allocations for whatever alternative is used.

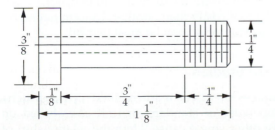

FIGURE 2-9
Small Screw Machine Product

Because a million parts were made each year, the saving of $93.10 per thousand was a substantial amount (a total of $93,100 for the year). It is also clear that costs other than the cost of material were of basic importance in the study.

The later history of this same product illustrates that shipping costs also must often be considered in selecting between materials. After the part had been made from brass stock for several years, it was found desirable to supply the domestic and foreign assembly plants of the company by using air freight for shipping. This led to a study of the possible use of a heat-treated aluminum alloy. This material cost $0.85 per pound and the cost of heat-treating each part, at an outside plant, was $0.018. Production studies indicated that the aluminum alloy could be machined at the same speeds as the brass stock.

The specific gravities of the brass and aluminum alloy are 8.7 and 2.75, respectively, and the raw and finished weights of the parts were as follows:

	Brass (lb)	Aluminum Alloy (lb)
Raw material	0.0384	(0.0384)(2.75/8.7) = 0.01213
Finished part	0.0150	(0.0150)(2.75/8.7) = 0.00474

Consequently, the comparative costs, including shipping at $3.00 per pound of finished part, were as follows:

	Brass	Aluminum Alloy
Material	$0.0538	$0.0103
Labor	0.0729	0.0729
Heat treatment	—	0.0180
Overhead	0.0972	0.0972
Shipping	0.0450	0.0142
Total cost per piece	$0.2689	$0.2126

A decision was made to use the aluminum alloy for the parts that were to be shipped by air to the subsidiary plants and for the parts that were consumed at the local plant. There was no advantage to using brass—even when shipping costs could be omitted.

Care should be taken in making economic selections between materials to ensure that any differences in yields or resulting scrap are taken into account. Commonly, alternative materials do not come in the same stock sizes, such as sheet sizes and bar lengths. This may considerably affect the yield obtained from a given weight of material. Similarly, the resulting scrap may differ for various materials. This factor can have serious economic implications when one of the materials is considerably more costly than another. Determination of these effects is an illustration of where experience may be most helpful.

2.5.2 Make-Versus-Purchase Studies

It is likely that more mistakes have been made in make-versus-purchase economic decisions due to improper use of standard or unit costs based on accounting records than due to any other single cause. Here the relationship of fixed costs, incremental costs, and unit costs is of the utmost importance. As mentioned previously in this chapter, all standard costs and unit costs are based on some level of activity, and a number of arbitrary cost allocations are used in their determination, such as in the allocation of indirect and other overhead costs. If operations are to be carried out at a different level, the unit costs are bound to be inaccurate. Because engineering economy studies often involve changes from existing conditions, the use of standard or unit costs may lead to considerable error. This fact is illustrated in Example 2-10.

EXAMPLE 2-10

A manufacturing plant consists of three departments: A, B, and C. Department A occupies 100 square meters in one corner of the plant. Product X is one of several products being produced in Department A. The daily production rate of X is 576 pieces. The cost accounting records show the following average daily production costs for Product X:

Direct labor	(1 operator working 4 hours per day at $22.50/hr, including fringe benefits, plus a part-time foreman at $30 per day)	$120.00
Material		86.40
Overhead	(at $0.82 per square meter of floor area)	82.00
	Total daily cost =	$288.40

The department foreman has recently learned about an outside company that sells Product X at $0.35 per piece. Accordingly, the foreman figured a daily cost of $0.35(576) = $201.60, resulting in a daily savings of $288.40 − $201.60 = $86.80. Therefore, a proposal was submitted to the plant manager for shutting down the production line of Product X and buying it from the outside company.

However, the plant manager decided not to accept the foreman's proposal based on Product X's unit cost after examining each cost component separately:

1. *Direct labor:* Because the foreman was supervising the manufacture of other products in Department A in addition to Product X, the only possible savings in labor would occur if the operator working 4 hours per day on Product X were not reassigned after this line is shut down. That is, a maximum savings of $90.00 per day would result.
2. *Materials:* The maximum savings on material will be $86.40. However, this figure could be lower if some of the material for Product X is obtained from scrap of another product.
3. *Overhead:* Because other products are made in Department A, no reduction in total floor space requirements will probably occur. Therefore, no reduction in overhead costs will result from discontinuing Product X. It has been estimated that there will be daily savings in overhead of about $3.00 due to a reduction in power costs and in insurance premiums.

Thus, the firm will save at most $90.00 in direct labor, $86.40 in materials, and $3.00 in overhead, which totals $179.40 per day. This daily savings estimate would not exceed the $201.60 to be paid to the outside company if Product X is purchased. For this reason, the plant manager rejected the proposal of the foreman and continued the manufacture of Product X.

In conclusion, Example 2-10 shows how an erroneous decision might be made by using the unit cost of Product X without detailed analysis. The fixed cost portion of Product X's unit cost, which is present even if the manufacture of Product X is discontinued, was not properly accounted for in the original analysis by the foreman.

2.5.3 Alternative Machine Speeds

Machines frequently can be operated at various speeds, resulting in different rates of product output. However, this usually results in different frequencies of machine downtime to permit servicing or maintaining the machine, such as resharpening or adjusting tooling. Such situations lead to present economy studies to determine the optimum or preferred operating speed. We first assume that there is an unlimited amount of work to be done.

A simple example of this type involves the planing of lumber. Lumber put through the planer increases in value by $0.10 per board foot. When the planer is operated at a cutting speed of 5,000 feet per minute, the blades have to be sharpened after 2 hours of operation, and lumber can be planed at the rate of 1,000 board feet per hour. When the machine is operated at 6,000 feet per minute, the blades have to be sharpened after $1\frac{1}{2}$ hours of operation, and the rate of planing is 1,200 board feet per hour. Each time the blades are changed, the machine has to be shut down for 15 minutes. The blades, unsharpened, cost $50 per set and can be sharpened 10 times before having to be discarded. Sharpening costs $10 per set. The crew that operates the planer changes and resets the blades. At what speed should the planer be operated?

Because the labor cost for the crew would be the same for either speed of operation, and because there was no discernible difference in wear upon the planer, these factors did not have to be included in the study.

In problems of this type, the operating time plus the delay time due to the necessity for tool changes constitute a cycle time that determines the output from the machine. The time required for a complete cycle determines the number of cycles that can be completed in a period of available time—for example, 1 day—and a certain portion of each complete cycle is productive. The actual productive time will be the product of the productive time per cycle and the number of cycles per day.

	Value per Day
At 5,000 feet per minute	
Cycle time = 2 hours + 0.25 hour = 2.25 hours	
Cycles per day = 8 ÷ 2.25 = 3.555	
Value added by planing = 3.555 × 2 × 1,000 × $0.10 =	$711.00*
Cost of resharpening blades = 3.555 × $10 = $35.55	
Cost of blades = 3.555 × $50/10 = 17.78	
Total cost	−53.33
Net increase in value per day	$657.67

At 6,000 feet per minute
    Cycle time $= 1.5$ hours $+ 0.25$ hour $= 1.75$ hours
    Cycles per day $= 8 \div 1.75 = 4.57$
    Value added by planing $= 4.57 \times 1.5 \times 1,200 \times \$0.10 =$ $\$822.60$*
    Cost of resharpening blades $= 4.57 \times \$10 = \$45.70$
    Cost of blades $= 4.57 \times \$50/10 = 22.85$
        Total cost -68.55
    Net increase in value per day $\$754.05$

---

*The units work out as follows: (cycles/day)(hours/cycle)(board-feet/hour)(dollar value/board-foot) = dollars/day

Thus, in this example, it is more economical to operate at the higher speed, in spite of the more frequent sharpening of blades that is required.

    It should be noted that this analysis assumes that the added production of lumber can be used. If, for example, the maximum production needed is equal to or less than that obtained by the slower machine speed ($1,000 \times 3.555$ cycles $\times$ 2 hours $= 7,110$ board feet per day), then the value added would be the same for each speed, and the decision then should be based on which speed minimizes total cost.

EXAMPLE 2-11

The problem presented in Section 2.5.3 assumes that every board-foot of lumber that is planed can be sold. If there is limited demand for the lumber, a correct choice of machining speeds can be made by *minimizing total cost*. Suppose now we want to know the better machining speed when only one job requiring 6,000 board-feet of planing is considered.

SOLUTION

For a fixed planing requirement of 6,000 board-feet, the value added by planing is $6,000(\$0.10) = \$600$ for either cutting speed. Hence, we want to minimize total cost.

    At 5,000 feet per minute cutting speed:

Cycle time $= 2.25$ hours
Production per cycle $= 2(1,000)$ board-feet
Number of cycles $= 6,000/2,000 = 3$, or 6.75 hours
Total cost $= 3(\$10/\text{cycle}) + 3(\$50/10) = \$45$

At 6,000 feet per minute cutting speed:

Cycle time $= 1.75$ hours
Production per cycle $= 1.5(1,200)$ board-feet
Number of cycles $= 6,000/1,800 = 3.33$, or 5.83 hours
Total cost $= 3.33(\$10/\text{cycle}) + 3.33(\$50/10) = \$50$

For a 6,000 board-foot job, select the slower cutting speed to minimize cost.

This type of study is also of great importance in connection with metal-cutting machine tool operations. Changes of cutting speeds can have a great effect on tool life. In addition, because the costs of machine tools and wage rates have increased, it is important that productivity be maintained at as high a level as possible. Under these conditions, it has frequently been found that increased cutting speeds give greater overall economy, even though the cutting-tool life is considerably less than was accepted practice in former years. This is particularly true if rapid means can be devised for changing tools when required. A method of this type called "single minute exchange of dies" (SMED) was developed by Japanese engineers.

2.6   Summary

In this chapter we have discussed selected cost terminology and concepts important in engineering economic analysis. A listing of important abbreviations and notation, by chapter, is provided in Appendix B. It is important that the meaning and use of various cost terms are understood in order to communicate effectively.

Several general economic concepts were discussed and illustrated. First, the ideas of consumer and producer goods and services, measures of economic growth, competition, and necessities and luxuries were covered. Then, some relationships among costs, price, and volume (demand) were discussed. Included were the concepts of optimal volume, breakeven points, and the average unit cost function. The economic principles of design optimization were also illustrated.

The use of present economy studies in noncomplex engineering decision making can provide satisfactory results and save considerable analysis effort. When an adequate engineering economic analysis can be accomplished by considering the various monetary consequences that occur in a short time period (usually 1 year or less), then a present economy study should be used.

2.7   References

BIERMAN, H., and S. SMIDT. *The Capital Budgeting Decision: Economic Analysis of Investment Projects,* 8th ed. New York: Macmillan Publishing Co., 1993.

BLECKE, CURTIS J. *Financial Analysis for Decision Making.* Englewood Cliffs, N.J.: Prentice-Hall, Inc., 1980.

SCHWEYER, HERBERT E. *Analytic Models for Managerial and Engineering Economics.* New York: Reinhold Publishing Corp., 1964.

2.8   Problems

2-1. A company in the process industry produces a chemical compound that is sold to manufacturers for use in the production of certain plastic products. The plant that produces the compound employs approximately 300 people. Develop a list of six different cost elements that would be *fixed* and a similar list of six cost elements that would be *variable.* (2.2)

2-2. Refer to Problem 2-1 and your answer to it. (2.2)

   a. Develop a table that shows the cost elements you defined and classified as fixed and variable. Indicate which of these costs are also *recurring, nonrecurring, direct,* or *indirect.*

b. Identify one additional cost element for each of the cost categories: recurring, nonrecurring, direct, and indirect.

2-3. Classify each of the following cost items as mostly fixed or variable: (2.2)

Raw materials
Direct labor
Depreciation
Supplies
Utilities
Property taxes
Administrative salaries
Payroll taxes
Insurance (building and equipment)
Clerical salaries
Sales commissions
Rent
Interest on borrowed money

2-4. In your own words, describe the life-cycle cost concept. Why is the potential for achieving life-cycle cost savings greatest in the acquisition phase of the life cycle? (2.2)

2-5. Explain why perfect competition is an ideal that is difficult to attain in the United States. List several business situations in which perfect competition is approached. (2.3)

2-6. Suppose we know that $p = 1,000 - D/5$, where p = price in dollars and D = annual demand. The *total* cost per year can be approximated by $\$1,000 + 2D^2$. (2.3)

a. Determine the value of D that maximizes profit.

b. Show in part (a) that profit has been maximized rather than minimized.

2-7. A company produces circuit boards used to update outdated computer equipment. The fixed cost is $42,000 per month and the variable cost is $53 per circuit board. The selling price per unit is $p = \$150 - 0.02D$. Maximum output of the plant is 4,000 units per month. (2.3)

a. Determine optimum demand for this product.

b. What is the maximum profit per month?

c. At what volumes does breakeven occur?

d. What is the company's range of profitable demand?

2-8. A company has established that the relationship between the sales price for one of its products and the quantity sold per month is approximately $D = 780 - 10p$ units (D is the demand or quantity sold per month, and p is the price in dollars). The fixed cost is $800 per month, and the variable cost is $30 per unit produced. What number of units, D^*, should be produced per month and sold to maximize net profit? What is the maximum profit per month related to the product? (2.3)

2-9. A company estimates that the relationship between unit price and demand per month for a potential new product is approximated by $p = \$100.00 - \$0.10D$. The company can produce the product by increasing fixed costs $17,500 per month, and the estimated variable cost is $40.00 per unit. What is the optimal demand, D^*, and based on this demand, should the company produce the new product? Why? (2.3)

a. Work out the complete solution by differential calculus, starting with the formula for profit or loss per month.

b. Solve graphically for an approximate answer.

2-10. A large wood products company is negotiating a contract to sell plywood overseas. The fixed cost that can be allocated to the production of plywood is one million dollars per month. The variable cost per thousand board feet is $131.50. The price charged will be determined by $p = \$700 - (0.05)D$ per 1,000 board feet (2.3)

a. For this situation determine the optimal monthly sales volume for this product and calculate the profit (or loss) at the optimal volume.

b. What is the range of profitable demand during a month?

2-11. A company produces and sells a consumer product, and thus far has been able to control the volume of the product by varying the selling price. The company is seeking to maximize its net profit. It has been concluded that the relationship between price and demand, per month, is approximately $D = 500 - 5p$, where p is the price per unit in dollars. The fixed cost is $1,000 per month, and the variable cost is $20 per unit. Obtain the answer, both mathematically and graphically, to the following questions: (2.3)

a. What is the optimal number of units that should be produced and sold per month?

	Anytown	Yourtown	(Pr. 2-13)
Average hauling distance	4 miles	3 miles	
Annual rental fee for solid waste site	$5,000	$100,000	
Hauling cost	$1.50/yd^3-mile	$1.50/yd^3-mile	

b. What is the maximum profit per month?

c. What are the breakeven sales quantities (range of profitable demand volume)?

2-12. A company estimates that as it increases its sales volume by decreasing the selling price of its product, revenue $= aD - bD^2$ (where D represents the units of demand per month, with $0 \le D \le a/b$). The fixed cost is $1,000 per month, and the variable cost is $4 per unit. If $a = \$6$ and $b = \$0.001$, determine the sales volume for maximum profit, and the maximum profit per month. (2.3)

2-13. A municipal solid waste site must be located outside Anytown or Yourtown. After sorting, some of the solid refuse will be transported to an electric power plant where it will be used as fuel. Data for the hauling of refuse from each town to the power plant is shown in the table above.

a. If the power plant will pay $8.00 per cubic yard of sorted solid waste delivered to the plant, where should the solid waste site be located? Use the town's viewpoint and assume that 200,000 cubic yards of refuse will be hauled to the plant for one year only. One site must be selected. (2.2)

b. Referring to the electric power plant above, the cost Y in dollars per hour to produce electricity is $Y = 12 + 0.2X + 0.27X^2$, where X is in megawatts. Revenue in dollars per hour from the sale of electricity is $16X - 0.2X^2$. Find the value of X that gives maximum profit. (2.3)

2-14. A plant has a capacity of 4,100 hydraulic pumps per month. The fixed cost is $504,000 per month. The variable cost is $166 per pump, and the sales price is $328 per pump (assume that sales equal output volume). What is the breakeven point in number of pumps per month? What percentage reduction will occur in the breakeven point if fixed costs are reduced by 18% and unit variable costs by 6%? (2.3)

2-15. Suppose that the ABC Corporation has a production (and sales) capacity of $1,000,000 per month. Its fixed costs—over a considerable range of volume—are $350,000 per month, and the variable costs are $0.50 per dollar of sales. (2.3)

a. What is the annual breakeven point volume (D')? Develop (graph) the breakeven chart.

b. What would be the effect on D' of decreasing the variable cost per unit by 25% if the fixed costs thereby increased by 10%?

c. What would be the effect on D' if the fixed costs were decreased by 10% and the variable cost per unit increased by the same percentage?

2-16. Refer to Problem 2-15. Graph the average unit cost function for this situation as originally given for (a). At what annual output within the present annual production (and sales) capacity of $12,000,000 does the minimum average cost per unit (C_u) occur? Why? (2.3)

2-17. The fixed cost related to the production of a product is $500,000 per year. Assume that the variable cost is $20,000 and the selling price is $30,000 for each percentage point of annual output capacity (which equals sales demand). Thus, the maximum sales per year are $3,000,000 (at 100% of output capacity), and we have: (2.3)

$C_F = \$500,000$ per year (Fixed cost)

$c_v = \$20,000/1\%$ of annual (Variable output capacity cost/unit)

$p = \$30,000/1\%$ of annual (Selling output capacity price/unit)

a. Determine the breakeven point for this situation.

b. Develop the mathematical expression for profit or loss in this situation as a function of demand, D.

2-18. A plant operation has fixed costs of $2,000,000 per year, and its output capacity is 100,000 electrical appliances per year. The variable cost is $40 per unit, and the product sells for $90 per unit.

a. Construct the economic breakeven chart.

b. Compare annual profit when the plant is operating at 90% of capacity with the plant operation at 100% capacity. Assume that the first 90% of capacity output is sold at $90 per unit, and the remaining 10% of production is sold at $70 per unit. (2.3)

2-19. The fixed cost for a steam line per meter of pipe is $350X + $50 per year. The cost for loss of heat from the pipe per meter is $4.8/X^{1/2} per year. Here X represents the thickness of insulation in meters, and X is a continuous design variable. (2.4)

a. What is the optimum thickness of the insulation?

b. How do you know that your answer in (a) minimizes total cost per year?

c. What is the basic tradeoff being made in this problem?

2-20. A farmer estimates that if he harvests his soybean crop now, he will obtain 1,000 bushels, which he can sell at $3.00 per bushel. However, he estimates that this crop will increase by an additional 1,200 bushels of soybeans for each week he delays harvesting, but the price will drop at a rate of 50 cents per bushel per week; in addition, it is likely that he will experience spoilage of approximately 200 bushels per week for each week he delays harvesting. When should he harvest his crop to obtain the largest net cash return, and how much will be received for his crop at that time? (2.4)

2-21. A recent engineering graduate was given the job of determining the best production rate for a new type of casting in a foundry. After experimenting with many combinations of hourly production rates and total production cost per hour, he summarized his findings in Table I below. The engineer then talked to the firm's marketing specialist, who provided these estimates of selling price per casting as a function of production output (see Table II below). (2.4)

a. What production rate would you recommend to maximize total profits?

b. How sensitive is the rate in (a) to changes in total production cost per hour?

2-22. The cost of operating a large ship (C_0) varies as the square of its velocity (v); specifically, $C_0 = knv^2$, where n is the trip length in miles and k is a constant of proportionality. It is known that at 12 miles/hour the *average* cost of operation is $100 per mile. The owner of the ship wants to minimize the cost of operation, but it must be balanced against the cost of the perishable cargo (C_c), which the customer has set at $1,500 per hour. At what velocity should the trip be planned to minimize the total cost (C_T), which is the sum of the cost of operating the ship and the cost of perishable cargo? (2.4)

2-23. Refer to Example 2-9. If the cost of electricity is $0.04 per kWh (instead of $0.074 per kWh), which thickness of insulation should be recommended? Recall that the life of the insulation was assumed to be 25 years. Comment on the basic tradeoff that is being made in this problem (relative to Example 2-9). (2.4)

2-24. A certain item can be readily purchased from a local vendor for $0.50 per unit. The shop foreman in your company has proposed manufacturing the item in an idle part of the production area. He has computed that labor, materials, and overhead per unit would be $0.15, $0.20, and $0.15, respectively. However, he contends that overhead should not be included in the manufactured cost, so that making the item is less expensive than purchasing it. Do you agree with the foreman's analysis? What other factors might have a bearing on this decision? (2.5)

2-25. Two machines are being considered for the production of a part. The material cost and selling price per part are $6 and $12, respectively. (Refer to the table at the top of the next page.) Assuming that all parts that are not rejected can be sold, which machine should be selected based on *expected daily profit*? The investment cost is the same for both machines and can be ignored. (2.5)

2-26. Either tool steel or carbon steel can be used for the set of tools on a certain lathe. It is neces-

Table I	Total cost/hour	$1,000	$2,600	$3,200	$3,900	$4,700	(Pr. 2-21)
	Castings produced/hour	100	200	300	400	500	
Table II	Selling price/casting	$20.00	$17.00	$16.00	$15.00	$14.50	
	Castings produced/hour	100	200	300	400	500	

	Machine A	Machine B	(Pr. 2-25)
Production rate	100 parts/hour	130 parts/hour	
Hours available for production	7 hours/day	6 hours/day	
Operator cost	$20/hour	$20/hour	
Parts rejected (negligible value)	3%	10%	

sary to sharpen the tools periodically. Relevant information for each is shown at the bottom of this page.

The cost of the lathe operator is $14.00 per hour, including the tool-changing time during which he is idle. The tool changer costs $20.00 per hour for just the time he is changing tools. Variable overhead costs for the lathe are $28.00 per hour, including tool-changing time. Which type of steel should be used to minimize overall costs? (2.5)

2-27. An automatic machine can be operated at three speeds, with the following results:

Speed	Output (pieces per hour)	Time Between Tool Grinds (hours)
A	400	15
B	480	12
C	540	10

A set of unsharpened tools costs $150 and can be ground 20 times. The cost of each grinding is $25. The time required to change and reset the tools is 1.5 hours, and such changes are made by a tool-setter who is paid $8.00 per hour. Overhead on the machine is charged at the rate of $3.75 per hour, including tool-changing time. Assume that all output produced can be used. At which speed should the machine be operated? The basic tradeoff in this problem is between the rate of output and tool usage. (2.5)

2-28. A company is analyzing a make-versus-purchase situation for a component used in several products, and the engineering department

has developed these data:

Option A: Purchase 10,000 items per year at a fixed price of $8.50 per item. The cost of placing the order is negligible according to the present cost accounting procedure.

Option B: Manufacture 10,000 items per year using available capacity in the factory. Cost estimates are direct materials = $5.00 per item and direct labor = $1.50 per item. Manufacturing overhead is allocated at 200% of direct labor (= $3.00 per item).

a. Based on these data, should the item be purchased or manufactured? (2.5)

b. If manufacturing overhead can be traced directly to this item—thus avoiding the 200% overhead rate—and it amounts to $2.15 per item, what choice should be recommended? (Traceable overhead is possible with an activity-based cost accounting procedure; is incremental to the manufacture of the part; and consists of such cost elements as employee training, material handling, quality control, supervision, and utilities.) Traceable overhead associated with purchasing this item (vendor certification, benchmarking, etc.) is $0.50 per item.

2-29. In the design of an automobile radiator, an engineer has a choice of using either a brass–copper alloy casting or a plastic molding. Either material provides the same service. However, the brass–copper alloy casting weighs 25 pounds, compared with 16 pounds for the plastic molding. Every pound of extra weight in the

	Carbon Steel	Tool Steel	(Pr. 2-26)
Output at optimum speed	100 pieces/hour	130 pieces/hour	
Time between tool grinds	3 hours	6 hours	
Time required to change tools	1 hour	1 hour	
Cost of unsharpened tools	$400.00	$1200.00	
Number of times tools can be ground	10	5	

automobile has been assigned a penalty of $4 to account for increased fuel consumption during the life cycle of the car. The brass–copper alloy casting costs $3.35 per pound, whereas the plastic molding costs $7.40 per pound. Machining costs per casting are $6.00 for the brass–copper alloy. Which material should the engineer select, and what is the difference in units costs? (2.5)

2-30. A construction company has a contract to build a sewer in a Central American country. It must dig a ditch 3 feet wide by 5 feet deep and 10,000 feet long. Local laborers are available at $1.50 per hour who could, *on the average,* remove one-half cubic yard per hour.

If a ditch-digging machine is rented in the United States and shipped to the job, it would be able to excavate at the rate of one-half cubic yards per minute. The rental on this machine would be $150 per day for all the time it was in use *or* in transit. Five days would be required for shipment each way. Freight costs would total $4,000. The operator for this machine would be brought from the United States and would have to be paid $150 per day for his services and equipment maintenance, including travel time. His round-trip airplane fare would cost $600, and one day of travel would be required each way. He would work an 8-hour day while on the job. If the machine is used, three local laborers also would be required. All workers and machine rental must be paid on a full 8-hour-day basis. What would you recommend? (2.5)

2-31. In some countries motorists are required to drive with their headlights on at all times. General Motors is beginning to equip their cars with daytime running lights. Most people would agree that driving with the headlights on at night is cost-effective with respect to extra fuel consumption and safety considerations. Given the following data and any additional assumptions you feel are necessary, analyze the cost effectiveness of driving with your headlights on during the day by answering the following questions. Cost-effective means that benefits outweigh (exceed) the costs. (2.5)

Cost of fuel = $1.15 per gallon.
Average distance traveled per year = 15,000 miles.
Average cost of an accident = $2,500.
Purchase price of headlights = $25.00 per set (2 headlights).
Average time car is in operation per year = 350 hours.
Average life of a headlight = 200 operating hours.
Average fuel consumption = 1 gallon per 30 miles.

a. What are the extra costs associated with driving with your headlights on during the day?
b. What are the benefits associated with driving with your headlights on during the day?
c. What additional assumptions (if any) do you need to complete your analysis?
d. It is cost-effective to drive with your headlights on during the day? Be sure to support your recommendation with the necessary calculations.

2-32. One method for developing a mine containing an estimated 100,000 tons of ore will result in the recovery of 62% of the available ore deposit and will cost $23 per ton of material removed. A second method of development will recover only 50% of the ore deposit, but it will cost only $15 per ton of material removed. Subsequent processing of the removed ore recovers 300 pounds of metal from each ton of processed ore and costs $40 per ton of ore processed. The recovered metal can be sold for $0.80 per pound. Which method for developing the mine should be used if your objective is to maximize total profit from the mine? (2.5)

2-33. Which of the following statements are true and which are false? (all sections)
a. Working capital is a variable cost.
b. The greatest potential for cost savings occurs in the operation phase of the life cycle.
c. If the capacity of an operation is significantly changed (e.g., a manufacturing plant), the fixed costs will also change.
d. The initial investment cost for a project is a nonrecurring cost.
e. Variable costs per output unit are a recurring cost.
f. A noncash cost is a cash flow.

75% of driving takes place between dawn and dusk.
2% of fuel consumption is due to accessories (radio, headlights, etc.).

g. Goods and services have utility because they have the power to satisfy human wants and needs.

h. The demand for necessities is more inelastic than the demand for luxuries.

i. Indirect costs can normally be allocated to a specific output or work activity.

j. Present economy studies are often done when the time value of money is not a significant factor in the situation.

k. Overhead costs normally include all costs that are not direct costs.

l. Optimal volume (demand) occurs when total costs equal total revenues.

m. Standard costs per unit of output are established in advance of actual production or service delivery.

n. A related sunk cost will normally affect the prospective cash flows associated with a situation.

o. The life cycle needs to be defined within the context of the specific situation.

p. The greatest commitment of costs occurs in the acquisition phase of the life cycle.

q. The average unit cost function is a linear function of demand.

2-34. *Brain Teaser*

The student chapter of the American Society of Mechanical Engineers is planning a 6-day trip to the national conference in Albany, New York. For transportation the group will rent a car from either the State Tech Motor Pool or a local car dealer. The Motor Pool charges $0.26 per mile, has no daily fee, and the motor pool pays for the gas. The car dealer charges $25 per day and $0.14 per mile, but the group must pay for the gas. The car's fuel rating is 20 miles per gallon, and the price of gas is estimated to be $1.20 per gallon. (2.3)

a. At what point, in miles, is the cost of both options equal?

b. The car dealer has offered a special student discount and will give the students 100 free miles per day. What is the new breakeven point?

c. Suppose now that the Motor Pool reduces its all-inclusive rate to $0.23 per mile and the car dealer increases its rate to $25 per day and $0.21 per mile. In this case, the car dealer wants to encourage student business, so he offers 1,000 free miles for the entire 6-day trip. He claims that if more than 882 miles are driven, students will come out ahead with one of his rental cars. If the students anticipate driving 1,600 miles (total), from whom should they rent a car? Is the car dealer's claim entirely correct?

3

*P*rinciples of Money–Time Relationships

*T*he objective of this chapter is to describe the return to capital in the form of interest (or profit) and to illustrate how basic equivalence calculations are made with respect to the time value of capital in engineering economy studies.

The following topics are discussed in this chapter:

Return to capital
Origins of interest
Simple interest
Compound interest
The concept of equivalence
Cash flow diagrams/tables
Interest formulas
Arithmetic sequences of cash flows
Geometric sequences of cash flows
Interest rates that vary with time
Nominal versus effective interest rates
Continuous compounding

3.1 Introduction

The term *capital* refers to wealth in the form of money or property that can be used to produce more wealth. The majority of engineering economy studies involve commitment of capital for extended periods of time, so the effect of time must be considered. In this regard, it is recognized that a dollar today is worth more than a dollar one or more years from now because of the interest (or profit) it can earn. Therefore, money has a *time value*.

3.2 Why Consider Return to Capital?

Capital in the form of money for the people, machines, materials, energy, and other things needed in the operation of an organization may be classified into two basic categories. *Equity capital* is that owned by individuals who have invested their money or property in a business project or venture in the hope of receiving a profit. *Debt capital*, often called *borrowed capital*, is obtained from lenders (e.g., through the sale of bonds) for investment. In return the lenders receive interest from the borrowers.

Normally, lenders do not receive any other benefits that may accrue from the investment of borrowed capital. They are not owners of the organization and do not participate as fully as the owners in the risks of the project or venture. Thus, lenders' fixed return on the capital loaned, in the form of interest, is more assured (has less risk) than the receipt of profit by the owners of equity capital. If the project or venture is successful, the return (profit) to the owners of equity capital can be substantially more than the interest received by lenders of capital. However, the owners could lose some or all of their money invested, whereas the lenders still could receive all the interest owed plus repayment of the money borrowed by the firm.

There are fundamental reasons why return to capital in the form of interest and profit is an essential ingredient of engineering economy studies. First, interest and profit pay the providers of capital for forgoing its use during the time the capital is being used. The fact that the supplier can realize a return on capital acts as an *incentive* to accumulate capital by savings, thus postponing immediate consumption in favor of creating wealth in the future. Second, interest and profit are payments for the *risk* the investor takes in permitting another person, or an organization, to use his or her capital.

In typical situations investors must decide whether the expected return on their capital is sufficient to justify buying into a proposed project or venture. If capital is invested in a project, investors would expect, as a minimum, to receive a return at least equal to the amount they have sacrificed by not using it in some other available opportunity of comparable risk. This interest or profit available from an alternative investment is the *opportunity cost* of using capital in the proposed undertaking. Thus, whether borrowed capital or equity capital is involved, there is a cost for the capital employed in the sense that the project and venture must provide a return sufficient to be financially attractive to suppliers of money or property.

In summary, whenever capital is required in engineering and other business projects and ventures, it is essential that proper consideration be given to its cost (i.e., time value). The remainder of this chapter deals with time value of money principles, which are vitally important to the proper evaluation of engineering projects that form the foundation of a firm's competitiveness, and hence to its very survival.

3.3 The Origins of Interest

Like taxes, interest has existed from earliest recorded human history. Records reveal its existence in Babylon in 2000 B.C.. In the earliest instances interest was paid in money for the use of grain or other commodities that were borrowed; it was also paid in the form of grain or other goods. Many existing interest practices stem from early customs in the borrowing and repayment of grain and other crops.

History also reveals that the idea of interest became so well established that a firm of international bankers existed in 575 B.C., with home offices in Babylon. The firm's income was derived from the high interest rates it charged for the use of its money for financing international trade.

Throughout early recorded history, typical annual rates of interest on loans of money were in the neighborhood of 6 to 25%, although legally sanctioned rates as high as 40% were permitted in some instances. The charging of exorbitant interest rates on loans was termed *usury*, and prohibition of usury is found in the Bible (see Exodus 22: 21–27).

During the Middle Ages, interest taking on loans of money was generally outlawed on scriptural grounds. In 1536 the Protestant theory of usury was established by John Calvin, and it refuted the notion that interest was unlawful. Consequently, interest taking again became viewed as an essential and legal part of doing business. Eventually, published interest tables became available to the public.

3.4 Simple Interest

When the total interest earned or charged is linearly proportional to the initial amount of the loan (principal), the interest rate, and the number of interest periods for which the principal is committed, the interest and interest rate are said to be *simple*. Simple interest is not used frequently in modern commercial practice.

When simple interest is applicable, the total interest, I, earned or paid may be computed in the formula

$$I = (P)(N)(i) \tag{3-1}$$

where P = principal amount lent or borrowed
    N = number of interest periods (e.g., years)
    i = interest rate per interest period.

The total amount repaid at the end of N interest periods is $P + I$. Thus, if \$1,000 is loaned for three years at a simple interest rate of 10% per year, the interest earned will be

$$I = \$1,000 \times 3 \times 0.10 = \$300$$

The total amount owed at the end of three years would be $\$1,000 + \$300 = \$1,300$. Notice that the cumulative amount of interest owed is a linear function of time until the interest is repaid (usually not until the end of period N).

▼
3.5 Compound Interest

Whenever the interest charge for any interest *period* (a year, for example) is based on the remaining principal amount plus any accumulated interest charges up to the *beginning* of that period, the interest is said to be *compound*. The effect of compounding of interest can be seen in the following table for $1,000 loaned for three periods at an interest rate of 10% compounded each period.

Period	(1) Amount Owed at Beginning of Period	(2) = (1) × 10% Interest Amount for Period	(3) = (1) + (2) Amount Owed at End of Period
1	$1,000	$100	$1,100
2	$1,100	$110	$1,210
3	$1,210	$121	$1,331

As you can see, a total of $1,331 would be due for repayment at the end of the third period. If the length of a period is one year, the $1,331 at the end of three periods (years) can be compared with the $1,300 given earlier for the same problem with simple interest. A graphical comparison of simple interest and compound interest is given in Figure 3-1. The difference is due to the effect of *compounding*, which is essentially the calculation of interest on previously earned interest. This difference would be much greater for larger amounts of money, higher interest rates, or greater numbers of years. Thus, simple interest does consider the time value of money but does not involve compounding of interest. Compound interest is much more common in practice than simple interest and is used throughout the remainder of this book.

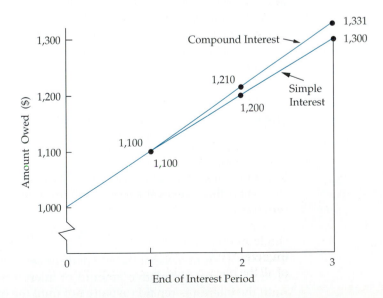

FIGURE 3-1

Illustration of Simple Versus Compound Interest

3.6 The Concept of Equivalence

Alternatives should be compared as far as possible when they produce similar results, serve the same purpose, or accomplish the same function. This is not always possible in some types of economy studies, as we shall see later, but now our attention is directed at answering the question: How can alternatives for providing the same service or accomplishing the same function be compared when interest is involved over extended periods of time? Thus, we should consider the comparison of alternative options, or proposals, by reducing them to an *equivalent basis* that is dependent on (1) the interest rate, (2) the amounts of money involved, (3) the timing of the monetary receipts and/or expenses, and (4) the manner in which the interest, or profit, on invested capital is paid and the initial capital recovered.

To better understand the mechanics of interest and to expand on the notion of economic equivalence, consider a situation in which we borrow $8,000 and agree to repay it in four years at an interest rate of 10% per year. There are many plans by which the principal of this loan (i.e., $8,000) and the interest on it can be repaid. For simplicity, we have selected four plans to demonstrate the idea of economic equivalence. Here *equivalence* means that all four plans are equally desirable to the borrower. In each the interest rate is 10% per year and the original amount borrowed is $8,000; thus differences among the plans rest with items (3) and (4) above. The four plans are shown in Table 3-1, and it will soon be apparent that all are *equivalent* at an interest rate of 10% per year.

In Plan 1, $2,000 of the loan principal is repaid at the end of each of years one through four. As a result, the interest we repay at the end of a particular year is affected by how much we still owe on the loan at the *beginning* of that year. Our end-of-year payment is just the sum of $2,000 and interest computed on the beginning-of-year amount owed.

Plan 2 indicates that none of the loan principal is repaid until the end of the fourth year. Our interest cost each year is $800, and it is repaid at the end of years one through four. Because interest does not accumulate in either Plan 1 or Plan 2, compounding of interest is not present. Notice that $3,200 in interest is paid in Plan 2, whereas only $2,000 is paid in Plan 1. We had the use of the $8,000 principal for four years in Plan 2 but, on average, had the use of much less than $8,000 in Plan 1.

Plan 3 requires that we repay equal end-of-year amounts of $2,524 each. Later in this chapter (Section 3.9) we will show how the $2,524 per year is computed. For our purposes here, the student should observe that the four end-of-year payments in Plan 3 completely repay the $8,000 loan principal with interest at 10% per year.

Finally, Plan 4 shows that no interest and no principal are repaid for the first three years of the loan period. Then at the end of the fourth year, the original loan principal plus the accumulated interest for the four years is repaid in a single lump-sum amount of $11,712.80 (rounded in Table 3-1 to $11,713). Plan 4 involves compound interest. The total amount of interest repaid in Plan 4 is highest of all the plans considered. Not only was the principal repayment in Plan 4 deferred

TABLE 3-1   Four Plans for Repayment of $8,000 in Four Years with Interest at 10% Per Year

(1) Year	(2) Amount Owed at Beginning of Year	(3) = 10% × (2) Interest Accrued for Year	(4) = (2) + (3) Total Money Owed at End of Year	(5) Principal Payment	(6) = (3) + (5) Total End-of-Year Payment (Cash Flow)
Plan 1: At End of Each Year Pay $2,000 Principal Plus Interest Due					
1	$ 8,000	$ 800	$8,800	$2,000	$ 2,800
2	6,000	600	6,600	2,000	2,600
3	4,000	400	4,400	2,000	2,400
4	2,000	200	2,200	2,000	2,200
	20,000 $-yr	$2,000 (total interest)		$8,000	$10,000 (total amount repaid)
Plan 2: Pay Interest Due at End of Each Year and Principal at End of Four Years					
1	$ 8,000	$ 800	$8,800	$ 0	$ 800
2	8,000	800	8,800	0	800
3	8,000	800	8,800	0	800
4	8,000	800	8,800	8,000	8,800
	32,000 $-yr	$3,200 (total interest)		$8,000	$11,200 (total amount repaid)
Plan 3: Pay in Four Equal End-of-Year Payments					
1	$ 8,000	$ 800	$8,800	$1,724	$ 2,524
2	6,276	628	6,904	1,896	2,524
3	4,380	438	4,818	2,086	2,524
4	2,294	230	2,524	2,294	2,524
	20,960 $-yr	$2,096 (total interest)		$8,000	$10,096 (total amount repaid)
Plan 4: Pay Principal and Interest in One Payment at End of Four Years					
1	$ 8,000	$ 800	$8,800	$ 0	$ 0
2	8,800	880	9,680	0	0
3	9,680	968	10,648	0	0
4	10,648	1,065	11,713	8,000	11,713
	37,130 $-yr	$3,713 (total interest)		8,000	$11,713 (total amount repaid)

until the end of year four, but we also deferred all interest payment until that time. If annual interest rates rise above 10% per year during the period of the loan, can you see that Plan 4 causes bankers to turn gray-haired rather quickly?

This brings us back to the notion of economic equivalence. If interest rates remain constant at 10% for the plans shown in Table 3-1, all four plans are equivalent. This assumes that one can freely borrow and lend at the 10% rate. Hence, we would be indifferent about whether the principal is repaid early in the loan's life (e.g., Plans 1 and 3) or repaid at the end of year four (e.g., Plans 2 and 4). *Economic equivalence is established, in general, when we are indifferent between a future payment, or series of future payments, and a present sum of money.*

To see *why* the four plans in Table 3-1 are equivalent at 10%, we could plot the amount owed at the beginning of each year (column 2) versus the year. The area under the resultant curve represents the *dollar-years* that the money is owed. For example, the dollar-years for Plan 1 equals 20,000, which is obtained from this graph:

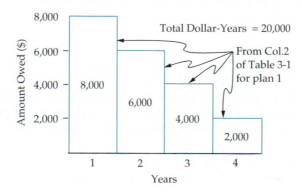

When total dollar-years are calculated for each plan and divided into total interest paid over the four years (the sum of column 3), the ratio is found to be constant:

Plan	Area Under Curve (Dollar-Years) (Sum of Col. 2 in Table 3-1)	Total Interest Paid (Sum of Col. 3 in Table 3-1)	Ratio of Total Interest to Dollar-Years
1	$20,000	$2,000	0.10
2	32,000	3,200	0.10
3	20,960	2,096	0.10
4	37,130	3,713	0.10

Because the ratio is constant at 0.10 for all plans, we can deduce that all repayment methods considered in Table 3-1 are equivalent, even though each involves a different total end-of-year payment in column 6. Dissimilar dollar-years of borrowing, by itself, does not necessarily mean that different loan repayment plans *are* or *are not* equivalent. In summary, equivalence is established when total interest paid, divided by dollar-years of borrowing, is a constant ratio among financing plans (i.e., alternatives).

One last important point to emphasize is that the loan repayment plans of Table 3-1 are equivalent only at an interest rate of 10%. If these plans are evaluated with methods presented later in this chapter at interest rates other than 10%, one plan can be identified that is superior to the other three. For instance, when $8,000 has been lent at 10% interest and subsequently the cost of borrowed money increases to 15%, the *lender* would prefer Plan 1 in order to recover his or her funds quickly so that they might be reinvested elsewhere at higher interest rates.

3.7 Notation and Cash Flow Diagrams/Tables

The following notation is utilized in formulas for compound interest calculations:

i = effective interest rate per interest period

N = number of compounding periods

P = present sum of money; the *equivalent* value of one or more cash flows at a reference point in time called the present

F = future sum of money; the *equivalent* value of one or more cash flows at a reference point in time called the future

A = end-of-period cash flows (or *equivalent* end-of-period values) in a uniform series continuing for a specified number of periods, starting at the end of the first period and continuing through the last period

The use of cash flow (time) diagrams and/or tables is strongly recommended for situations in which the analyst needs to clarify or visualize what is involved when flows of money occur at various times. In addition, viewpoint (remember Principle 3?) is an essential feature of cash flow diagrams.

The difference between total cash inflows (receipts) and cash outflows (expenditures) for a specified period of time (e.g., one year) is the net cash flow for the period. As discussed in Chapter 2, cash flows are important in engineering economy because they form the basis for evaluating alternatives. Indeed, the usefulness of a cash flow diagram for economic analysis problems is analogous to that of the free-body diagram for mechanics problems.

Figure 3-2 shows a cash flow diagram for Plan 4 of Table 3-1, and Figure 3-3 depicts the net cash flows of Plan 3. These two figures also illustrate the definition of the preceding symbols and their placement on a cash flow diagram. Notice that all cash flows have been placed at the end of the year to correspond with the convention used in Table 3-1. In addition, a viewpoint has been specified.

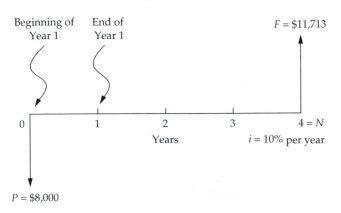

FIGURE 3-2
Cash Flow Diagram for Plan 4 of Table 3-1
(Lender's Viewpoint)

FIGURE 3-3
Cash Flow Diagram for Plan 3
of Table 3-1 (Lender's Viewpoint)

The cash flow diagram employs several conventions:

1. The horizontal line is a *time scale,* with progression of time moving from left to right. The period (e.g., year, quarter, month) labels can be applied to intervals of time rather than to points on the time scale. Note, for example, that the end of Period 2 is coincident with the beginning of Period 3. When the end-of-period cash flow convention is used, period numbers are placed at the end of each time interval as illustrated in Figures 3-2 and 3-3.
2. The arrows signify cash flows and are placed at the end of the period. If a distinction needs to be made, downward arrows represent expenses (negative cash flows or cash outflows) and upward arrows represent receipts (positive cash flows or cash inflows).
3. The cash flow diagram is dependent on the point of view. For example, the situations shown in Figures 3-2 and 3-3 were based on cash flow as seen by the lender. If the directions of all arrows had been reversed, the problem would have been diagrammed from the borrower's viewpoint.

EXAMPLE 3-1

Before evaluating the economic merits of a proposed investment, the XYZ Corporation insists that its engineers develop a cash flow diagram of the proposal. An investment of $10,000 can be made that will produce uniform annual revenue of $5,310 for five years and then have a market value of $2,000 at the end of year five. Annual expenses will be $3,000 at the end of each year for operating and maintaining the project. Draw a cash flow diagram for the five-year life of the project. Use the corporation's viewpoint.

SOLUTION
As shown in Figure 3-4, the initial investment of $10,000 and annual expenses of $3,000 are cash outflows, while annual revenues and the market (salvage) value are cash inflows.

Notice that the beginning of a given year is the end of the preceding year. For example, the beginning of year two is the end of year one.

Example 3-2 presents a situation in which cash flows are represented in tabular form to facilitate the analysis of plans/designs.

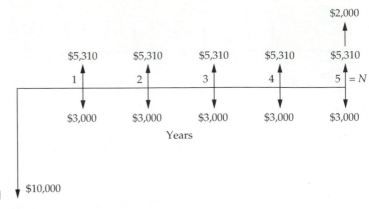

FIGURE 3-4
Cash Flow Diagram for Example 3-1

EXAMPLE 3-2

In a company's renovation of a small office building, two feasible alternatives for upgrading the heating, ventilation, and air conditioning (HVAC) system have been identified. Either Alternative A *or* Alternative B must be implemented. The costs are:

Alternative A *Rebuild (overhaul) the existing HVAC system*
- Equipment, labor, and materials to upgrade $18,000
- Annual cost of electricity   32,000
- Annual maintenance expenses   2,400

Alternative B *Install a new HVAC system that utilizes existing ductwork*
- Equipment, labor, and materials to install $60,000
- Annual cost of electricity   9,000
- Annual maintenance expenses   16,000
- Replacement of a major component 4 years hence . .   9,400

At the end of eight years, the estimated market value for Alternative A is $2,000, and for Alternative B it is $8,000. Assume that both alternatives will provide comparable service (comfort) over an eight-year time period, and assume that the major component replaced in Alternative B will have no market value at the end of year eight. (1) Use a cash flow table and end-of-year convention to tabulate the net cash flows for both alternatives. (2) Determine the annual net cash flow difference between the alternatives (B − A). (3) Compute the cumulative difference through the end of year eight. (The cumulative difference is the sum of differences, B − A, from year zero through year eight.)

SOLUTION

The cash flow table (company's viewpoint) for this example is shown in Table 3-2. Based on these results, several points can be made: (1) doing nothing is not an option—either A or B must be selected; (2) even though positive and negative cash flows are included in the table, on balance we are investigating two "cost-only" alternatives; (3) a decision between the two alternatives can be made just as easily

TABLE 3-2   Cash Flow Table for Example 3-2

End-of-Year	Alternative A Net Cash Flow	Alternative B Net Cash Flow	Difference (B − A)	Cumulative Difference
0 (now)	−$ 18,000	−$ 60,000	−$42,000	−$42,000
1	− 34,400	− 25,000	9,400	− 32,600
2	− 34,400	− 25,000	9,400	− 23,200
3	− 34,400	− 25,000	9,400	− 13,800
4	− 34,400	− 25,000 − 9,400	0	− 13,800
5	− 34,400	− 25,000	9,400	− 4,400
6	− 34,400	− 25,000	9,400	5,000
7	− 34,400	− 25,000	9,400	14,400
8	− 34,400 + 2,000	− 25,000 + 8,000	15,400	29,800
Total	−$291,200	−$261,400		

on the *difference* in cash flows (i.e., on the avoidable difference) as it can on the stand-alone net cash flows for Alternatives A and B; (4) Alternative B has cash flows identical to those of Alternative A *except for* the differences shown in the table, so if the avoidable difference can "pay its own way," Alternative B is the recommended choice; (5) cash flow changes caused by inflation or other suspected influences could have easily been inserted into the table and included in the analysis; and (6) it takes six *years* for the extra $42,000 investment in Alternative B to generate sufficient cumulative savings in annual expenses to justify the higher investment (this ignores the time value of money). So, which alternative is better? We'll be able to answer this question later when we consider the time value of money in order to recommend choices between alternatives.

It should be apparent that a cash flow table clarifies the timing of cash flows, the assumptions that are being made, and the data that are available. A cash flow table is often useful when the complexity of a situation makes it difficult to show all cash flow amounts on a diagram.

The remainder of Chapter 3 deals with the development and illustration of equivalence (time value of money) principles for assessing the economic attractiveness of investments such as those proposed in Examples 3-1 and 3-2.

Viewpoint: In most examples presented in this chapter, the company's (investor's) viewpoint will be taken.

3.8 Interest Formulas Relating Present and Future Equivalent Values of Single Cash Flows

Figure 3-5 shows a cash flow diagram involving a present single sum, P, and a future single sum, F, separated by N periods with interest at $i\%$ per period.

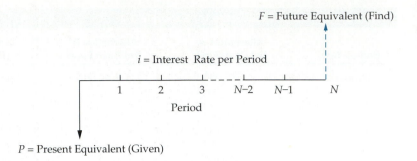

FIGURE 3-5
General Cash Flow Diagram
Relating Present Equivalent
and Future Equivalent of
Single Payments

Throughout this chapter a *dashed arrow,* such as that shown in Figure 3-5, indicates the quantity to be determined. Two formulas relating a given P and its unknown equivalent F are provided in Equations 3-2 and 3-3.

3.8.1 Finding *F* When Given *P*

If an amount of P dollars is invested at a point in time and $i\%$ is the interest (profit or growth) rate per period, the amount will grow to a future amount of $P + Pi = P(1+i)$ by the end of one period; by the end of two periods, the amount will grow to $P(1+i)(1+i) = P(1+i)^2$; by the end of three periods, the amount will grow to $P(1+i)^2(1+i) = P(1+i)^3$; and by the end of N periods the amount will grow to

$$F = P(1+i)^N \tag{3-2}$$

EXAMPLE 3-3

Suppose that you borrow $8,000 now, promising to repay the loan principal plus accumulated interest in four years at $i = 10\%$ per year. How much would you repay at the end of four years?

SOLUTION

Year	Amount Owed at Start of Year		Interest Owed for Each Year		Amount Owed at End of Year		Total End-of-Year Payment
1	P	$= \$ 8{,}000$	iP	$= \$ 800$	$P(1+i)$	$= \$ 8{,}800$	0
2	$P(1+i)$	$= \$ 8{,}800$	$iP(1+i)$	$= \$ 880$	$P(1+i)^2$	$= \$ 9{,}680$	0
3	$P(1+i)^2$	$= \$ 9{,}680$	$iP(1+i)^2$	$= \$ 968$	$P(1+i)^3$	$= \$10{,}648$	0
4	$P(1+i)^3$	$= \$10{,}648$	$iP(1+i)^3$	$= \$1{,}065$	$P(1+i)^4$	$= \$11{,}713$	$F = \$11{,}713$

In general, we see that $F = P(1+i)^N$, and the total amount to be repaid is $11,713. This further illustrates Plan 4 in Table 3-1 in terms of notation that we shall be using throughout this book.

The quantity $(1+i)^N$ in Equation 3-2 is commonly called the *single payment compound amount factor.* Numerical values for this factor are given in the second column from the left in the tables of Appendix C for a wide range of values of i

and N. In this book we shall use the functional symbol $(F/P, i\%, N)$ for $(1 + i)^N$. Hence Equation 3-2 can be expressed as

$$F = P(F/P, i\%, N) \tag{3-3}$$

where the factor in parentheses is read "find F given P at $i\%$ interest per period for N interest periods." Note that the sequence of F and P in F/P is the same as in the initial part of Equation 3-3, where the unknown quantity, F, is placed on the left-hand side of the equation. This sequencing of letters is true of all functional symbols used in this book and makes them easy to remember.

Another example of finding F when given P, together with a cash flow diagram and solution, is given in Table 3-3. Note in Table 3-3 that for each of the six common discrete compound interest circumstances covered, two problem statements are given—(a) *in borrowing–lending terminology* and (b) *in equivalence terminology*—but they both represent the same cash flow situation. Indeed, there are generally many ways in which a given cash flow situation can be expressed.

In general, a good way to interpret a relationship such as Equation 3-3 is that the calculated amount, F, at the point in time at which it occurs, *is equivalent to* (i.e., can be traded for) the known value, P, at the point in time at which it occurs, for the given interest or profit rate, i.

3.8.2 Finding *P* When Given *F*

From Equation 3-2, $F = P(1 + i)^N$. Solving this for P gives the relationship

$$P = F\left(\frac{1}{1 + i}\right)^N = F(1 + i)^{-N} \tag{3-4}$$

The quantity $(1 + i)^{-N}$ is called the *single payment present worth factor*. Numerical values for this factor are given in the third column of the tables in Appendix C for a wide range of values of i and N. We shall use the functional symbol $(P/F, i\%, N)$ for this factor. Hence

$$P = F(P/F, i\%, N) \tag{3-5}$$

EXAMPLE 3-4

An investor (owner) has an option to purchase a tract of land that will be worth $10,000 in six years. If the value of the land increases at 8% each year, how much should the investor be willing to pay now for this property?

SOLUTION

The purchase price can be determined from Equation 3-5 and Table C-11 in Appendix C as follows:

$$P = \$10,000(P/F, 8\%, 6)$$
$$P = \$10,000(0.6302)$$
$$= \$6,302$$

TABLE 3-3   Discrete Cash Flow Examples Illustrating Equivalence

Example Problems (all using an interest rate of $i = 10\%$ per year—see Table C-13 of Appendix C)

To Find:	Given:	(a) In Borrowing-Lending Terminology:	(b) In Equivalence Terminology:	Cash Flow Diagram[a]	Solution
For single cash flows:					
F	P	A firm borrows $1,000 for eight years. How much must it repay in a lump sum at the end of the eighth year?	What is the future equivalent at the end of eight years of $1,000 at the beginning of those eight years?	$P = \$1,000$ $N = 8$ 0 $F = ?$	$F = P(F/P, 10\%, 8)$ $= \$1,000(2.1436)$ $= \$2,143.60$
P	F	A firm wishes to have $2,143.60 eight years from now. What amount should be deposited now to provide for it?	What is the present equivalent of $2,143.60 received eight years from now?	$F = \$2,143.60$ 0 $N = 8$ $P = ?$	$P = F(P/F, 10\%, 8)$ $= \$2,143.60(0.4665)$ $= \$1,000.00$
For uniform series:					
F	A	If eight annual deposits of $187.45 each are placed in an account, how much money has accumulated immediately after the last deposit?	What amount at the end of the eighth year is equivalent to eight end-of-year payments of $187.45 each?	$F = ?$ 1 2 3 4 5 6 7 8 $A = \$187.45$	$F = A(F/A, 10\%, 8)$ $= \$187.45(11.4359)$ $= \$2,143.60$

P	A	How much should be deposited in a fund now to provide for eight end-of-year withdrawals of $187.45 each?	What is the present equivalent of eight end-of-year payments of $187.45 each?	$A = \$187.45$, years 1 2 3 4 5 6 7 8, $P = ?$	$P = A(P/A, 10\%, 8)$ $= \$187.45(5.3349)$ $= \$1,000.00$
A	F	What uniform annual amount should be deposited each year in order to accumulate $2,143.60 at the time of the eighth annual deposit?	What uniform payment at the end of eight successive years is equivalent to $2,143.60 at the end of the eighth year?	$F = \$2,143.60$, years 1 2 3 4 5 6 7 8, $A = ?$	$A = F(A/F, 10\%, 8)$ $= \$2,143.60(0.0874)$ $= \$187.45$
A	P	What is the size of eight equal annual payments to repay a loan of $1,000? The first payment is due one year after receiving the loan.	What uniform payment at the end of eight successive years is equivalent to $1,000 at the beginning of the first year?	$P = \$1,000$, years 1 2 3 4 5 6 7 8, $A = ?$	$A = P(A/P, 10\%, 8)$ $= \$1,000(0.18745)$ $= \$187.45$

^{a}The cash flow diagram represents the example as stated in borrowing-lending terminology.

Another example of this type of problem, together with a cash flow diagram and solution, is given in Table 3-3.

3.9 Interest Formulas Relating a Uniform Series (Annuity) to Its Present and Future Equivalent Values

Figure 3-6 shows a general cash flow diagram involving a series of uniform (equal) receipts, each of amount A, occurring at the end of each period for N periods with interest at $i\%$ per period. Such a uniform series is often called an *annuity*. It should be noted that the formulas and tables to be presented are derived such that A occurs at the end of each period, and thus:

1. P (present equivalent value) occurs one interest period before the first A (uniform amount).
2. F (future equivalent value) occurs at the same time as the last A, and N periods after P.
3. A (annual equivalent value) occurs at the end of periods 1 through N, inclusive.

The timing relationship for P, A, and F can be observed in Figure 3-6. Four formulas relating A to F and P will be developed.

3.9.1 Finding *F* When Given *A*

If a cash flow in the amount of A dollars occurs at the end of each period for N periods and $i\%$ is the interest (profit or growth) rate per period, the future equivalent value, F, at the end of the Nth period is obtained by summing the

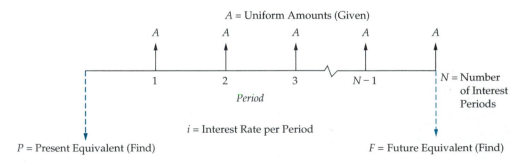

FIGURE 3-6   General Cash Flow Diagram Relating Uniform Series (Ordinary Annuity) to Its Present Equivalent and Future Equivalent Values

future equivalents of each of the cash flows. Thus,

$$F = A(F/P, i\%, N-1) + A(F/P, i\%, N-2) + A(F/P, i\%, N-3) + \cdots$$
$$+ A(F/P, i\%, 1) + A(F/P, i\%, 0)$$
$$= A[(1+i)^{N-1} + (1+i)^{N-2} + (1+i)^{N-3} + \cdots + (1+i)^1 + (1+i)^0]$$

The bracketed terms comprise a geometric sequence having a common ratio of $(1+i)^{-1}$. Recall that the sum of the first N terms (S_N) of a geometric sequence is

$$S_N = \frac{a_1 - ba_N}{1-b} \quad (b \neq 1)$$

where a_1 is the first term in the sequence, a_N is the last term, and b is the common ratio. If we let $b = (1+i)^{-1}, a_1 = (1+i)^{N-1}$, and $a_N = (1+i)^0$, then

$$F = A \left[\frac{(1+i)^{N-1} - \dfrac{1}{(1+i)}}{1 - \dfrac{1}{(1+i)}} \right]$$

which reduces to

$$F = A \left[\frac{(1+i)^N - 1}{i} \right] \tag{3-6}$$

The quantity $\{[(1+i)^N - 1]/i\}$ is called the *uniform series compound amount factor*. Numerical values for this factor are given in the fourth column of the tables in Appendix C for a wide range of values of i and N. We shall use the functional symbol $(F/A, i\%, N)$ for this factor. Hence Equation 3-6 can be expressed as

$$F = A(F/A, i\%, N) \tag{3-7}$$

Examples of this type of problem are provided here and in Table 3-3.

EXAMPLE 3-5

(a) Suppose you make 15 equal annual deposits of $1,000 each into a bank account paying 5% interest per year. The first deposit will be made one year from today. How much money can be withdrawn from this bank account immediately after the 15th deposit?

SOLUTION
The value of A is $1,000, N equals 15 years, and $i = 5\%$ per year. Immediately after the 15th payment, the future equivalent amount is

$$F = \$1,000(F/A, 5\%, 15)$$
$$= \$1,000(21.5786)$$
$$= \$21,578.60$$

Notice in the cash flow diagram below that the value of F is coincident with the last payment of $1,000.

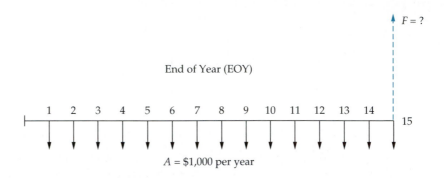

(b) To illustrate further the amazing effects of compound interest, we consider the credibility of this statement: "If you are 20 years of age and save $1.00 each day for the rest of your life, you can become a millionaire." Let's assume you live to age 80, and the annual interest rate is 10% ($i = 10\%$). Under these specific conditions, we compute the future compound amount (F) to be

$$F = \$365/\text{yr.} \ (F/A, 10\%, 60 \text{ years})$$
$$= \$365 \ (3{,}034.81)$$
$$= \$1{,}107{,}706$$

Thus, the statement is true for the assumptions given! The moral is to *start saving early* and let the "magic" of compounding work on your behalf!

3.9.2 Finding *P* When Given *A*

From Equation 3-2, $F = P(1+i)^N$. Substituting for F in Equation 3-6, one determines that

$$P(1 + i)^N = A \left[\frac{(1 + i)^N - 1}{i} \right]$$

Dividing both sides by $(1 + i)^N$,

$$P = A \left[\frac{(1 + i)^N - 1}{i(1 + i)^N} \right] \tag{3-8}$$

Thus, Equation 3-8 is the relation for finding the present equivalent value (as of the beginning of the first period) of a uniform series of end-of-period cash flows of amount A for N periods. The quantity in brackets is called the *uniform series present worth factor*. Numerical values for this factor are given in the fifth column of the tables in Appendix C for a wide range of values of i and N. We shall use the functional symbol $(P/A, i\%, N)$ for this factor. Hence

$$P = A(P/A, i\%, N) \tag{3-9}$$

EXAMPLE 3-6

If a certain machine undergoes a major overhaul now, its output can be increased by 20%—which translates into additional cash flow of $20,000 at the end of each year for five years. If $i = 15\%$ per year, how much can we afford to invest to overhaul this machine?

SOLUTION

The increase in cash flow is $20,000 per year, and it continues for five years at 15% annual interest. The upper limit on what we can afford to spend is

$$P = \$20,000(P/A, 15\%, 5)$$
$$= \$20,000(3.3522)$$
$$= \$67,044$$

EXAMPLE 3-7

Suppose that your rich uncle has $1,000,000 that he wishes to distribute to his heirs at the rate of $100,000 per year. If the $1,000,000 is deposited in a bank account that earns 6% interest per year, how many years will it take to completely deplete the account? How long will it take if the account earns 8% interest per year instead of 6%?

SOLUTION

From Table C-9, solve for N in the following equation: $\$1,000,000 = \$100,000(P/A, 6\%, N)$; $N = 15.7$ years. When the interest rate is increased to 8% per year, it will take 20.9 years to bring the account balance to zero, which is found by solving this equation: $10 = (P/A, 8\%, N)$.

3.9.3 Finding *A* When Given *F*

Taking Equation 3-6 and solving for A, one finds that

$$A = F\left[\frac{i}{(1 + i)^N - 1}\right] \tag{3-10}$$

Thus, Equation 3-10 is the relation for finding the amount, A, of a uniform series of cash flows occurring at the end of N interest periods that would be equivalent to (have the same value as) its future equivalent value occurring at the end of the last period. The quantity in brackets is called the *sinking fund factor*. Numerical values for this factor are given in the sixth column of the tables in Appendix C for a wide range of values of i and N. We shall use the functional symbol $(A/F, i\%, N)$ for this factor. Hence

$$A = F(A/F, i\%, N) \tag{3-11}$$

EXAMPLE 3-8

An enterprising student is planning to have personal savings totaling $1,000,000 when she retires at age 65. She is now 20 years old. If the annual interest rate will average 7% over the next 45 years on her savings account, what equal end-of-year amount must she save to accomplish her goal?

SOLUTION

The future amount, F, is $1,000,000. The equal annual amount this student must place in a *sinking fund* that grows to $1,000,000 in 45 years at 7% annual interest (see Table C-10) is

$$A = \$1,000,000 \, (A/F, 7\%, 45)$$

$$= \$1,000,000(0.0035)$$

$$= \$3,500$$

Another example of this type of problem, together with a cash flow diagram and solution, is given in Table 3-3.

3.9.4 Finding *A* When Given *P*

Taking Equation 3-8 and solving for A, one finds that

$$A = P\left[\frac{i(1 + i)^N}{(1 + i)^N - 1}\right] \tag{3-12}$$

Thus, Equation 3-12 is the relation for finding the amount, A, of a uniform series of cash flows occurring at the end of each of N interest periods that would be equivalent to, or could be traded for, the present equivalent P, occurring at the beginning of the first period. The quantity in brackets is called the *capital recovery factor*.* Numerical values for this factor are given in the seventh column of the tables in Appendix C for a wide range of values of i and N. We shall use the functional symbol $(A/P, i\%, N)$ for this factor. Hence

$$A = P(A/P, i\%, N) \tag{3-13}$$

An example that utilizes the equivalence between a present lump-sum loan amount and a series of equal uniform annual payments starting at the end of year

---

*The capital recovery factor is more conveniently expressed as $i/[1 - (1 + i)^{-N}]$ for computation with a hand-held calculator.

one and continuing through year four is provided in Table 3-1 as Plan 3. Equation 3-13 yields the equivalent value of A that repays the $8,000 loan plus 10% interest per year over four years:

$$A = \$8,000(A/P, 10\%, 4) = \$8,000(0.3155) = \$2,524$$

The entries in columns three and five of Plan 3 in Table 3-1 can now be better understood. Interest owed at the end of year one equals $8,000(0.10), and therefore the principal repaid out of the total end-of-year payment of $2,524 is the difference, $1,724. At the beginning of year two, the amount of principal owed is $8,000 − $1,724 = $6,276. Interest owed at the end of year two is $6,276(0.10) ≅ $628, and the principal repaid at that time is $2,524 − $628 = $1,896. The remaining entries in Plan 3 are obtained by performing these calculations for years three and four.

A graphical summary of Plan 3 is given in Figure 3-7. Here it can be seen that 10% interest is being paid on the beginning-of-year amount owed and that year-end payments of $2,524, consisting of interest and principal, bring the amount owed to $0 at the end of the fourth year. (The exact value of A is $2,523.77 and produces an exact value of $0 at the end of four years.) It is important to note that all the uniform series interest factors in Table 3-3 involve the same concept as the one illustrated in Figure 3-7.

Another example of a problem where we desire to compute an equivalent value for A, from a given value of P and a known interest rate and number of compounding periods, is given in Table 3-3.

For an annual interest rate of 10%, the reader should now be convinced from Table 3-3 that $1,000 at the beginning of year one is equivalent to the $187.45 at the end of years one through eight, which is then equivalent to $2,143.60 at the end of year eight.

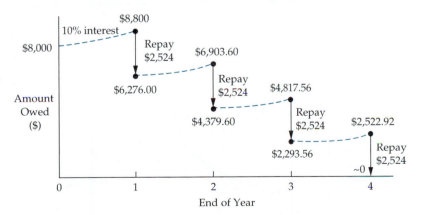

FIGURE 3-7   Relationship of Cash Flows for Plan 3 of Table 3-1 to Repayment of the $8,000 Loan Principal

3.9.5 Interest Factor Relationships: Summary

This section is summarized by presenting equations and graphs of the relationships between an annuity and its present and future equivalent values.

Equations:

$$(A/P, i\%, N) = \frac{1}{(P/A, i\%, N)} \tag{3-14}$$

$$(A/F, i\%, N) = \frac{1}{(F/A, i\%, N)} \tag{3-15}$$

$$(F/A, i\%, N) = (P/A, i\%, N)(F/P, i\%, N) \tag{3-16}$$

$$(P/A, i\%, N) = \sum_{k=1}^{N} (P/F, i\%, k) \tag{3-17}$$

$$(F/A, i\%, N) = \sum_{k=1}^{N} (F/P, i\%, N-k) \tag{3-18}$$

$$(A/F, i\%, N) = (A/P, i\%, N) - i \tag{3-19}$$

Graphs:

(For a fixed value of N, these graphs help to visualize the preceding equations):

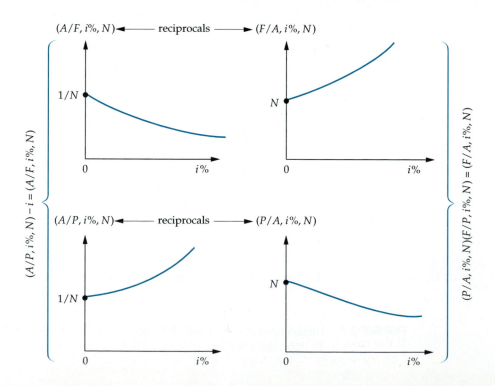

TABLE 3-4   Discrete Compounding Interest Factors and Symbols[a]

To Find:	Given:	Factor by Which to Multiply "Given"[a]	Factor Name	Factor Functional Symbol[b]
For single cash flows:				
F	P	$(1 + i)^N$	Single payment compound amount	$(F/P, i\%, N)$
P	F	$\dfrac{1}{(1 + i)^N}$	Single payment present worth	$(P/F, i\%, N)$
For uniform series (annuities):				
F	A	$\dfrac{(1 + i)^N - 1}{i}$	Uniform series compound amount	$(F/A, i\%, N)$
P	A	$\dfrac{(1 + i)^N - 1}{i(1 + i)^N}$	Uniform series present worth	$(P/A, i\%, N)$
A	F	$\dfrac{i}{(1 + i)^N - 1}$	Sinking fund	$(A/F, i\%, N)$
A	P	$\dfrac{i(1 + i)^N}{(1 + i)^N - 1}$	Capital recovery	$(A/P, i\%, N)$

[a]i, effective interest rate per interest period; N, number of interest periods; A, uniform series amount (occurs at the end of each interest period); F, future equivalent; P, present equivalent.
[b]The functional symbol system is used throughout this book.

▼ 3.10 Interest Formulas for Discrete Compounding and Discrete Cash Flows

Table 3-4 above provides a summary of the six most common discrete compound interest factors, utilizing notation of the preceding sections. The formulas are for *discrete compounding*, which means that the interest is compounded at the end of each finite-length period, such as a month or a year. Furthermore, the formulas also assume discrete (i.e., lump-sum) cash flows spaced at the end of equal time intervals on a cash flow diagram. Discrete compound interest factors are given in Appendix C, where the assumption is made that i remains constant during the N compounding periods.

▼ 3.11 Deferred Annuities (Uniform Series)

All annuities (uniform series) discussed to this point involve the first cash flow being made at the end of the first period, and they are called *ordinary annuities*. If the cash flow does not begin until some later date, the annuity is known as a *deferred annuity*. If the annuity is deferred J periods ($J < N$), the situation is as

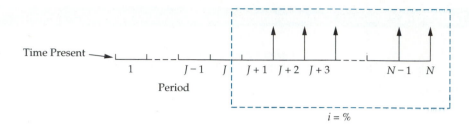

FIGURE 3-8   General Cash Flow Representation of a Deferred Annuity (Uniform Series)

portrayed in Figure 3-8, in which the entire framed ordinary annuity has been moved forward from "time present," or "time 0," by J periods. Remember that in an annuity deferred for J periods the first payment is made at the end of period $(J + 1)$, assuming that all periods involved are equal in length.

The present equivalent at the end of period J of an annuity with cash flows of amount A is, from Equation 3-9, $A(P/A, i\%, N - J)$. The present equivalent of the single amount $A(P/A, i\%, N - J)$ as of time 0 will then be

$$A(P/A, i\%, N - J)(P/F, i\%, J)$$

EXAMPLE 3-9

To illustrate the preceding discussion, suppose that a father, on the day his son is born, wishes to determine what lump amount would have to be paid into an account bearing interest of 12% per year to provide withdrawals of $2,000 on each of the son's 18th, 19th, 20th, and 21st birthdays.

SOLUTION

The problem is represented in Figure 3-9. One should first recognize that an ordinary annuity of four withdrawals of $2,000 each is involved, and that the present equivalent of this annuity occurs at the 17th birthday when a $(P/A, i\%, N - J)$ factor is utilized. In this problem, $N = 21$ and $J = 17$. It is often helpful to use a *subscript* with P or F to denote the respective point in time. Hence

$$P_{17} = A(P/A, 12\%, 4) = \$2,000(3.0373) = \$6,074.60$$

Note the dashed arrow in Figure 3-9, denoting P_{17}. Now that P_{17} is known, the next step is to calculate P_0. With respect to P_0, P_{17} is a future equivalent, and hence it could also be denoted F_{17}. Money at a given point in time, such as the end of

FIGURE 3-9
Cash Flow Diagram of the Deferred Annuity
Problem in Example 3-9

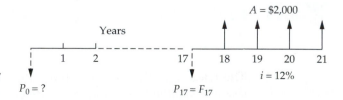

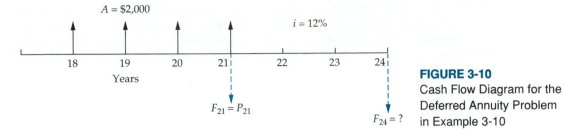

FIGURE 3-10
Cash Flow Diagram for the Deferred Annuity Problem in Example 3-10

period 17, is the same regardless of whether it is called a present equivalent or a future equivalent. Hence

$$P_0 = F_{17}(P/F, 12\%, 17) = \$6{,}074.60(0.1456) = \$884.46$$

which is the amount that the father would have to deposit on the day his son is born.

EXAMPLE 3-10

As an addition to the problem in Example 3-9, suppose that the father wishes to determine the equivalent worth of the four \$2,000 withdrawals as of the son's 24th birthday. This could mean that four amounts were never withdrawn or possibly that the son took them and immediately redeposited them in an account also earning interest at 12% per year. Using our subscript system, we wish to calculate F_{24} as shown in Figure 3-10 above.

SOLUTION

One way to work this is to calculate

$$F_{21} = A(F/A, 12\%, 4) = \$2{,}000(4.7793) = \$9{,}558.60$$

To determine F_{24}, F_{21} can now be denoted P_{21}, and

$$F_{24} = P_{21}(F/P, 12\%, 3) = \$9{,}558.60(1.4049) = \$13{,}428.88$$

Another, quicker way to work the problem is to recognize that $P_{17} = \$6{,}074.60$ and $P_0 = \$884.46$ are each equivalent to the four \$2,000 withdrawals. Hence one can find F_{24} directly, given P_{17} or P_0. Using P_0, we obtain

$$F_{24} = P_0(F/P, 12\%, 24) = \$884.46(15.1786) = \$13{,}424.86$$

which closely approximates the previous answer. The two numbers differ by \$4.02, which can be attributed to round-off error in the interest factors.

3.12 Equivalence Calculations Involving Multiple Interest Formulas

The reader should now be comfortable with equivalence problems that involve discrete compounding of interest and discrete cash flows. All compounding of interest takes place once per time period (e.g., a year), and to this point cash flows also occur once per time period. This section provides two examples involving two or more equivalence calculations to solve for an unknown quantity. End-of-year cash flow convention is used. Again, the interest rate is constant over the N time periods.

EXAMPLE 3-11

Figure 3-11 depicts an example problem with a series of year-end cash flows extending over eight years. The amounts are $100 for the first year, $200 for the second year, $500 for the third year, and $400 for each year from the fourth through the eighth. These could represent something like the expected maintenance expenditures for a certain piece of equipment or payments into a fund. Note that the payments are shown at the end of each year, which is a standard assumption (convention) for this book and for economic analyses in general unless one has information to the contrary. It is desired to find the (a) present equivalent expenditure, P_0; (b) future equivalent expenditure, F_8; and (c) annual equivalent expenditure, A, of these cash flows if the annual interest rate is 20%.

SOLUTION

(a) To find the equivalent P_0, one needs to sum the equivalent values of all payments as of the beginning of the first year (time zero). The required movements of money through time are shown graphically in Figure 3-11a.

$$
\begin{aligned}
P_0 = \ & F_1(P/F, 20\%, 1) & = \ & \$100(0.8333) & = \$ \ & 83.33 \\
& + F_2(P/F, 20\%, 2) & & + \$200(0.6944) & & + 138.88 \\
& + F_3(P/F, 20\%, 3) & & + \$500(0.5787) & & + 289.35 \\
& + A(P/A, 20\%, 5) \times (P/F, \ 20\%, 3) & & + \$400(2.9900) \times (0.5787) & & + 692.26 \\
\hline
& & & & & \$1{,}203.82
\end{aligned}
$$

(b) To find the equivalent F_8, one can sum the equivalent values of all payments as of the end of the eighth year (time eight). Figure 3-11b indicates these movements of money through time. However, since the equivalent P_0 is already known to be $1,203.82, one can calculate directly

$$F_8 = P_0(F/P, 20\%, 8) = \$1{,}203.82(4.2998) = \$5{,}176.19$$

(c) The equivalent A of the irregular cash flows can be calculated directly from either P_0 or F_8 as follows:

$$A = P_0(A/P, 20\%, 8) = \$1{,}203.82(0.2606) = \$313.73$$

or

$$A = F_8(A/F, 20\%, 8) = \$5{,}176.19(0.0606) = \$313.73$$

The computation of A from P_0 and F_8 is shown in Figure 3-11c. Thus, one finds that the irregular series of payments shown in Figure 3-11 is equivalent to $1,203.82 at time zero, $5,176.19 at time eight, or a uniform series of $313.73 at the end of each of the eight years.

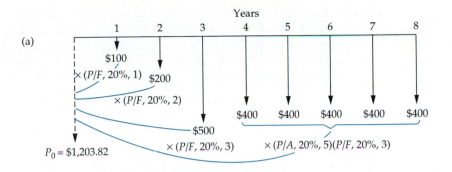

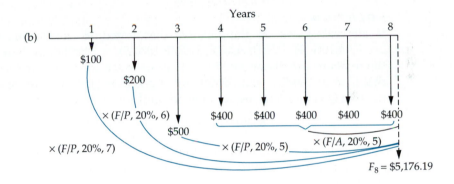

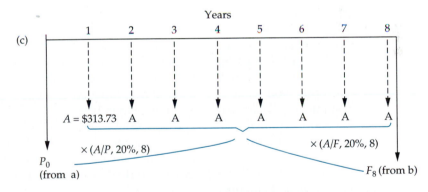

FIGURE 3-11   Example 3-11 for Calculating the Equivalent *P*, *F*, and *A* Values

EXAMPLE 3-12

Transform the cash flows on the left-hand side of Figure 3-12 to their equivalent cash flows on the right-hand side. That is, take the left-hand quantities as givens and determine the unknown value of *Q* in terms of *H* in Figure 3-12. The interest rate is 10% per year. (Notice that "⇐⇒" means "equivalent to.")

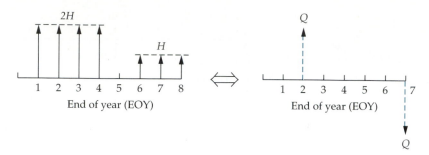

FIGURE 3-12   Cash Flow Diagrams for Example 3-12

SOLUTION

If all cash flows on the left are discounted to year zero, we have $P_0 = 2H(P/A,10\%,4) + H(P/A,10\%,3)(P/F,10\%,5) = 7.8839H$. When cash flows on the right are also discounted to year zero, we can solve for Q in terms of H. [Notice that Q at the end of year (EOY) two is positive, Q at EOY seven is negative, and the two Q values must be equal in amount.]

$$7.8839H = Q(P/F,10\%,2) - Q(P/F,10\%,7)$$

or

$$Q = 25.172H$$

EXAMPLE 3-13

Suppose you start a savings plan in which you save $500 each year for 15 years. You make your first payment at age 22 and then leave the accumulated sum in the savings plan (and make NO more annual payments) until you reach age 65, at which time you withdraw the total accumulated amount. The average annual interest rate you'll earn on this savings plan is 10%.

A friend of yours (exactly your age) from Minnesota State University waits 10 years to start her savings plan (i.e., she is age 32). She decides to save $2,000 each year in an account earning interest at the rate of 10% per year. She will make these annual payments until she is 65 years old, at which time she will withdraw the total accumulated amount.

How old will you be when your friend's *accumulated* savings amount (including interest) exceeds yours? State any assumptions you think are necessary.

SOLUTION

Creating cash flow diagrams for Example 3-13 is an important first step in solving for the unknown number of years, N, until the future equivalent values of both savings plans are equal. The two diagrams are shown in Figure 3-13. The future equivalent (F) of your plan is $500(F/A,10\%,15)(F/P,10\%,N-36)$, and that of your friend is $F' = \$2,000(F/A,10\%,N-31)$. It is clear that N, the age at which $F = F'$, is greater than 32. Assuming that the interest rate remains constant

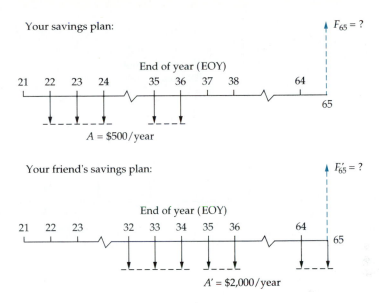

FIGURE 3-13   Cash Flow Diagrams for Example 3-13

at 10% per year, the value of N can be determined by trial and error:

N	Your Plan's F	Friend's F'
36	$15,886	$12,210
38	$19,222	$18,974
39	$21,145	$22,872
40	$23,259	$27,159

By the time you reach age 39, your friend's accumulated savings will exceed yours. (If you had deposited $1,000 instead of $500, you would be over 76 years old when your friend's plan surpassed yours. Moral: start saving early!)

3.13 Interest Formulas Relating a Uniform Gradient of Cash Flows to Its Annual and Present Equivalents

Some problems involve receipts or expenses that are projected to increase or decrease by a uniform *amount* each period, thus constituting an arithmetic sequence of cash flows. For example, because of leasing a certain type of equipment, maintenance and repair savings relative to purchasing the equipment may increase by a roughly constant amount each period. This situation can be modeled with a uniform gradient.

Figure 3-14 is a cash flow diagram of a sequence of end-of-period cash flows increasing by a constant amount, G, in each period. The G is known as the *uniform gradient amount*. Note that the timing of cash flows on which the derived formulas

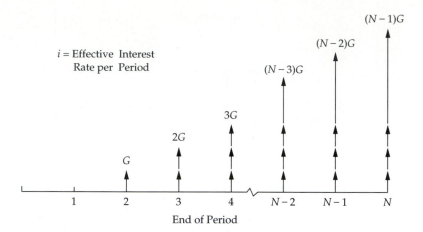

FIGURE 3-14

Cash Flow Diagram for a Uniform Gradient Increasing by G Dollars per Period

and tabled values are based is as follows:

End of Period	Cash Flows
1	0
2	G
3	$2G$
.	.
.	.
.	.
$N - 1$	$(N - 2)G$
N	$(N - 1)G$

Notice that the first cash flow occurs at the end of period two.

3.13.1   Finding F When Given G

The future equivalent, F, of the arithmetic sequence of cash flows shown in Figure 3-14 is

$$F = G(F/A, i\%, N - 1) + G(F/A, i\%, N - 2) + \cdots$$
$$+ G(F/A, i\%, 2) + G(F/A, i\%, 1)$$

or

$$F = G\left[\frac{(1 + i)^{N-1} - 1}{i} + \frac{(1 + i)^{N-2} - 1}{i} + \cdots\right.$$
$$\left. + \frac{(1 + i)^2 - 1}{i} + \frac{(1 + i)^1 - 1}{i}\right]$$

$$= \frac{G}{i}[(1 + i)^{N-1} + (1 + i)^{N-2} + \cdots$$
$$+ (1 + i)^2 + (1 + i)^1 + 1] - \frac{NG}{i}$$

$$= \frac{G}{i}\left[\sum_{k=0}^{N-1}(1+i)^k\right] - \frac{NG}{i}$$

$$= \frac{G}{i}(F/A, i\%, N) - \frac{NG}{i} \tag{3-20}$$

Instead of working with future equivalent values, it is usually more practical to deal with annual and present equivalents in Figure 3-14.

3.13.2   Finding *A* When Given *G*

From Equation 3-20, it is easy to develop an expression for *A* as follows:

$$A = F(A/F, i, N)$$

$$= \left[\frac{G}{i}(F/A, i, N) - \frac{NG}{i}\right](A/F, i, N)$$

$$= \frac{G}{i} - \frac{NG}{i}(A/F, i, N)$$

$$= \frac{G}{i} - \frac{NG}{i}\left[\frac{i}{(1+i)^N - 1}\right]$$

$$= G\left[\frac{1}{i} - \frac{N}{(1+i)^N - 1}\right] \tag{3-21}$$

The term in braces in Equation 3-21 is called the *gradient to uniform series conversion factor*. Numerical values for this factor are given on the right side of Appendix C for a range of *i* and *N* values. We shall use the functional symbol $(A/G, i\%, N)$ for this factor. Thus

$$A = G(A/G, i\%, N) \tag{3-22}$$

3.13.3   Finding *P* When Given *G*

We may now utilize Equation 3-21 to establish the equivalence between *P* and *G*:

$$P = A(P/A, i\%, N)$$

$$= G\left[\frac{1}{i} - \frac{N}{(1+i)^N - 1}\right]\left[\frac{(1+i)^N - 1}{i(1+i)^N}\right]$$

$$= G\left[\frac{(1+i)^N - 1 - Ni}{i^2(1+i)^N}\right]$$

$$= G\left\{\frac{1}{i}\left[\frac{(1+i)^N - 1}{i(1+i)^N} - \frac{N}{(1+i)^N}\right]\right\} \tag{3-23}$$

The term in braces in Equation 3-23 is called the *gradient to present equivalent conversion factor*. It can also be expressed as $(1/i)[(P/A, i\%, N) - N(P/F, i\%, N)]$.

Numerical values for this factor are given in column 8 of Appendix C for a wide assortment of i and N values. We shall use the functional symbol $(P/G, i\%, N)$ for this factor. Hence

$$P = G(P/G, i\%, N) \qquad \text{(3-24)}$$

3.13.4 Computations Using G

Be sure to notice that the direct use of gradient conversion factors applies when there is no cash flow at the end of period one, as in Example 3-14. There may be an A amount at the end of period one, but it is treated separately, as illustrated in Examples 3-15 and 3-16. A major advantage of using gradient conversion factors (i.e., computational time savings) is realized when N becomes large.

EXAMPLE 3-14

As an example of the straightforward use of the gradient conversion factors, suppose that certain end-of-year cash flows are expected to be $1,000 for the *second* year, $2,000 for the third year, and $3,000 for the fourth year, and that if interest is 15% per year, it is desired to find the (a) present equivalent value at the beginning of the first year, and (b) uniform annual equivalent value at the end of each of the four years.

SOLUTION

Observe that this schedule of cash flows fits the model of the arithmetic gradient formulas with $G = \$1,000$ and $N = 4$ (see Figure 3-14). Note that there is no cash flow at the end of the first period.

(a) The present equivalent can be calculated as

$$P_0 = G(P/G, 15\%, 4) = \$1,000(3.79) = \$3,790$$

(b) The annual equivalent can be calculated from Equation 3-22 as

$$A = G(A/G, 15\%, 4) = \$1,000(1.3263) = \$1,326.30$$

Of course, once P_0 is known, the value of A can be calculated as

$$A = P_0(A/P, 15\%, 4) = \$3,790(0.3503) = \$1,326.30$$

EXAMPLE 3-15

As a further example of the use of arithmetic gradient formulas, suppose that one has cash flows as follows:

End of Year	Cash Flows ($)
1	−5,000
2	−6,000
3	−7,000
4	−8,000

and that one wishes to calculate their present equivalent at $i = 15\%$ per year using arithmetic gradient interest formulas.

SOLUTION

The schedule of cash flows is depicted in the top diagram of Figure 3-15. The bottom two diagrams of Figure 3-15 show how the original schedule can be broken into two separate sets of cash flows, a uniform series of $5,000 *payments*, plus an arithmetic gradient *payment* of $1,000 that fits the general gradient model for which factors are tabled. The summed present equivalents of these two separate sets of payments equal the present equivalent of the original problem. Thus, using the symbols shown in Figure 3-15, we have

$$P_{0T} = P_{0A} + P_{0G}$$
$$= -A(P/A, 15\%, 4) - G(P/G, 15\%, 4)$$
$$= -\$5,000(2.8550) - \$1,000(3.79) = -\$14,275 - 3,790 = -\$18,065$$

The annual equivalent of the original cash flows could be calculated with the aid of Equation 3-22 as follows:

$$A_T = A + A_G$$
$$= -\$5,000 - \$1,000(A/G, 15\%, 4) = -\$6,326.30$$

A_T is equivalent to P_{0T} because $-\$6,326.30(P/A, 15\%, 4) = -\$18,061$, which is the same value obtained previously (subject to round-off error).

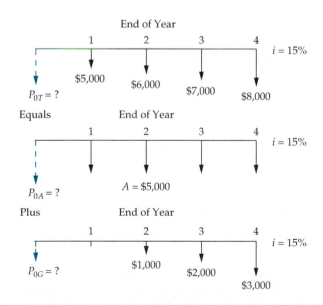

FIGURE 3-15 Example 3-15 Involving an Increasing Arithmetic Gradient

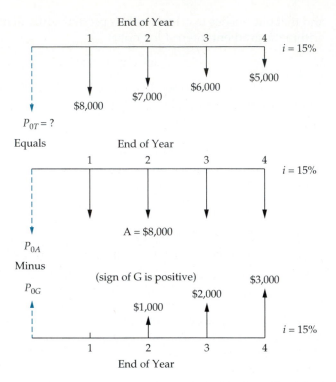

FIGURE 3-16   Example 3-16 Involving a Decreasing Arithmetic Gradient

EXAMPLE 3-16

For another example of the use of arithmetic gradient formulas, suppose that one has cash flows that are timed in exact reverse of the situation depicted in Example 3-15. The top diagram of Figure 3-16 shows the following sequence of cash flows:

End of Year	Cash Flows ($)
1	−8,000
2	−7,000
3	−6,000
4	−5,000

Calculate the present equivalent at $i = 15\%$ per year using arithmetic gradient interest factors.

SOLUTION

The bottom two diagrams of Figure 3-16 show how these cash flows can be broken into two separate sets of cash flows. It must be remembered that the arithmetic gradient formulas and tables provided are for increasing gradients only. Hence one must subtract an *increasing* gradient of payments that *did not* occur. Thus

$$P_{0T} = P_{0A} - P_{0G}$$
$$= -A(P/A, 15\%, 4) + G(P/G, 15\%, 4)$$
$$= -\$8{,}000(2.8550) + \$1{,}000(3.79)$$
$$= -\$22{,}840 + \$3{,}790 = -\$19{,}050$$

Again, the annual equivalent of the original decreasing series of cash flows can be calculated by the same rationale:

$$A = A - A_G$$
$$= -\$8{,}000 + \$1{,}000(A/G, 15\%, 4)$$
$$= -\$6{,}673.70$$

Note from Examples 3-15 and 3-16 that the present equivalent of $-\$18{,}065$ for an increasing arithmetic gradient series of payments is different from the present equivalent of $-\$19{,}050$ for an arithmetic gradient of payments of identical amounts *but reversed timing*. This difference would be even greater for higher interest rates and gradient amounts and exemplifies the marked effect of timing of cash flows on equivalent values. It is also helpful to observe that the sign of G corresponds to the general slope of the cash flows over time. For instance, in Figure 3-15 the slope of the cash flows is negative (G is negative), whereas in Figure 3-16 the slope is positive (G is positive).

3.14 Interest Formulas Relating a Geometric Sequence of Cash Flows to Its Present and Annual Equivalents

Some economic equivalence problems involve projected cash flow patterns that are changing at an average *rate*, $\bar{f}$, each period. A fixed amount of a commodity that inflates in price at a constant rate each year is a typical situation that can be modeled with a geometric sequence of cash flows. The resultant end-of-period cash flow pattern is referred to as a *geometric gradient series* and has the general appearance shown in Figure 3-17. Notice that *the initial cash flow in this series, A_1, occurs at the end of period 1* and that $A_k = (A_{k-1})(1 + \bar{f}), 2 \leq k \leq N$. The Nth term in this geometric sequence is $A_N = A_1(1 + \bar{f})^{N-1}$, and the common ratio throughout the sequence is $(A_k - A_{k-1})/A_{k-1} = \bar{f}$. Be sure to notice that $\bar{f}$ can be positive *or* negative.

Each term in Figure 3-17 could be discounted, or compounded, at interest rate i per period to obtain a value of P or F, respectively. However, this becomes quite tedious for large N, so it is convenient to have a single equation instead.

To develop a compact expression for P at interest rate i per period for the cash flows of Figure 3-17, consider the following summation:

$$P = \sum_{k=1}^{N} A_k(1 + i)^{-k} = \sum_{k=1}^{N} A_1(1 + \bar{f})^{k-1}(1 + i)^{-k}$$

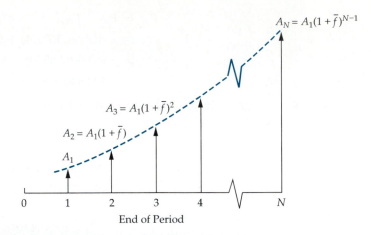

FIGURE 3-17

Cash Flow Diagram for a Geometric Sequence of Cash Flows Increasing at a Constant Rate of $\bar{f}$ per Period

or

$$P = \frac{A_1}{1+\bar{f}} \sum_{k=1}^{N} \left(\frac{1+\bar{f}}{1+i}\right)^k \tag{3-25}$$

When $i \neq \bar{f}$, we can simplify Equation 3-25 by defining a "convenience rate," i_{CR}, as follows:

$$i_{CR} = \frac{1+i}{1+\bar{f}} - 1 \tag{3-26}$$

The convenience rate can also be written as $i_{CR} = (i - \bar{f})/(1 + \bar{f})$. In the situation where $i \neq \bar{f}$, Equation 3-25 can thus be rewritten as

$$
\begin{aligned}
P &= \frac{A_1}{1+\bar{f}} \sum_{k=1}^{N} \left(\frac{1+i}{1+\bar{f}}\right)^{-k} \\
&= \frac{A_1}{1+\bar{f}} \sum_{k=1}^{N} (1 + i_{CR})^{-k} \\
&= \frac{A_1}{1+\bar{f}} (P/A, i_{CR}\%, N)^* \tag{3-27}
\end{aligned}
$$

Equation 3-27 makes use of the fact that

$$(P/A, i_{CR}\%, N) = \sum_{k=1}^{N} (1 + i_{CR})^{-k} = \sum_{k=1}^{N} (P/F, i_{CR}\%, k)$$

When $i = \bar{f}$ and $i_{CR} = 0$, Equation 3-27 reduces to

$$P = \frac{A_1}{1+\bar{f}} (P/A, 0\%, N) = \frac{NA_1}{1+\bar{f}} \tag{3-28}$$

---

*When $\bar{f}$ exceeds i, i_{CR} is negative and the above summation is valid only when N is finite.

The interested reader can verify Equation 3-28 by applying L'Hôpital's Rule to the $(P/A, i_{CR}\%, N)$ factor in Equation 3-27 and taking the limit as $i_{CR} \rightarrow 0$.

Values of i_{CR} used in connection with Equation 3-27 are typically not included in the tables in Appendix C. Because i_{CR} is usually a noninteger interest rate, resorting to the definition of a $(P/A, i_{CR}\%, N)$ factor (see Table 3-4) and substituting terms into it is a satisfactory way to obtain values of these interest factors.

The end-of-period uniform annual equivalent, A, of a geometric gradient series can be determined from Equation 3-27 (or Equation 3-28) as follows:

$$A = P(A/P, i\%, N) \qquad \text{(3-29)}$$

The year zero "base" of this annuity, which increases at a constant rate of $\bar{f}\%$ per period, is A_0 and equals

$$A_0 = P(A/P, i_{CR}\%, N) \qquad \text{(3-30)}$$

The difference between A and A_0 can be seen in Figure 3-18. Finally, the future equivalent of this geometric gradient series is simply

$$F = P(F/P, i\%, N) \qquad \text{(3-31)}$$

Additional discussion of geometric sequences of cash flows is provided in Chapter 9 (Section 9.3), which deals with inflation and price changes.

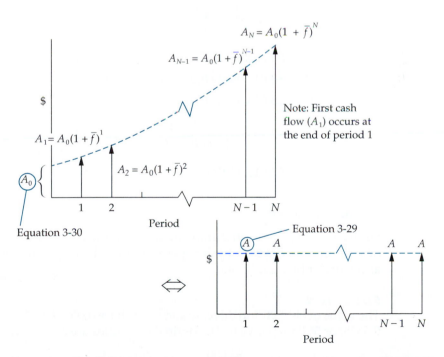

FIGURE 3-18   Graphical Interpretation of A and A_0 Terms in a Geometric Gradient Series when $\bar{f} > 0$

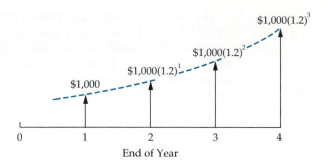

FIGURE 3-19
Cash Flow Diagram for Example 3-17

End of Year

EXAMPLE 3-17

Consider the end-of-year geometric sequence of cash flows in Figure 3-19 and determine the P, A, A_0, and F equivalent values. The rate of increase is 20% per year after the first year, and the interest rate is 25% per year.

SOLUTION

$$P = \frac{\$1,000}{1.2}\left(P/A, \frac{25\% - 20\%}{1.20}, 4\right) = \$833.33(P/A, 4.167\%, 4)$$

$$= \$833.33\left[\frac{(1.04167)^4 - 1}{0.04167(1.04167)^4}\right]$$

$$= \$833.33(3.6157) = \$3,013.08$$

$$A = \$3,013.08(A/P, 25\%, 4) = \$1,275.86$$

$$A_0 = \$3,013.08(A/P, 4.167\%, 4)$$

$$= \$3,103.08\left[\frac{0.04167(1.04167)^4}{(1.04167)^4 - 1}\right] = \$833.34$$

$$F = \$3,013.08(F/P, 25\%, 4) = \$7,356.15$$

EXAMPLE 3-18

Suppose that the geometric gradient in Example 3-17 begins with $1,000 at the end of year one and *decreases* by 20% per year after the first year. Determine P, A, A_0, and F under this condition.

SOLUTION
The value of $\bar{f}$ is -20% in this case and $i_{CR} = [(1+i)/(1+\bar{f})]-1 = (1.25/0.80)-1 = 0.5625$, or 56.25% per year. The desired quantities are as follows:

$$P = \frac{\$1,000}{0.8}(P/A, 56.25\%, 4) = \$1,250(1.4795)$$

$$= \$1,849.38$$

$$A = \$1{,}849.38(A/P, 25\%, 4) = \$783.03$$
$$A_0 = \$1{,}849.38(A/P, 56.25\%, 4) = \$1{,}250.00$$
$$F = \$1{,}849.38(F/P, 25\%, 4) = \$4{,}515.08$$

3.15 Interest Rates That Vary with Time

When the interest rate on a loan can vary with, for example, the Federal Reserve Board's discount rate, it is necessary to take this into account when determining the future equivalent value of the loan. It is becoming common to see interest-rate "escalation riders" on some types of loans. Example 3-19 demonstrates how this situation is treated.

EXAMPLE 3-19

A person has made an arrangement to borrow $1,000 now and another $1,000 two years hence. The entire obligation is to be repaid at the end of four years. If the projected interest rates in years one, two, three, and four are 10%, 12%, 12%, and 14%, respectively, how much will be repaid as a lump-sum amount at the end of four years?

SOLUTION

This problem can be solved by compounding the amount owed at the beginning of each year by the interest rate that applies to each individual year and repeating this process over the four years to obtain the total future equivalent value:

$$F_1 = \$1{,}000(F/P, 10\%, 1) = \$1{,}100$$
$$F_2 = \$1{,}100(F/P, 12\%, 1) = \$1{,}232$$
$$F_3 = (\$1{,}232 + \$1{,}000)(F/P, 12\%, 1) = \$2{,}500$$
$$F_4 = \$2{,}500(F/P, 14\%, 1) = \$2{,}850$$

To obtain the present equivalent of a series of future cash flows subject to varying interest rates, a procedure similar to the preceding one would be utilized with a sequence of $(P/F, i_k\%, k)$ factors. In general, the present equivalent value of a cash flow occurring at the end of period N can be computed with Equation 3-32, where i_k is the interest rate for the kth period (the symbol $\prod$ means "the product of"):

$$P = \frac{F_N}{\prod_{k=1}^{N}(1 + i_k)} \tag{3-32}$$

For instance, if $F_4 = \$1{,}000$ and $i_1 = 10\%, i_2 = 12\%, i_3 = 13\%$, and $i_4 = 10\%$

$$P = \$1{,}000(P/F, 10\%, 1)(P/F, 12\%, 1)(P/F, 13\%, 1)(P/F, 10\%, 1)$$
$$= \$1{,}000[(0.9091)(0.8929)(0.8850)(0.9091)] = \$653$$

3.16 Nominal and Effective Interest Rates

Very often the interest period, or time between successive compounding, is less than one year. It has become customary to quote interest rates on an annual basis, followed by the compounding period if different from one year in length. For example, if the interest rate is 6% per interest period and the interest period is six months, it is customary to speak of this rate as "12% compounded semiannually." Here the annual rate of interest is known as the *nominal rate*, 12% in this case. A nominal interest rate is represented by r. But the actual annual rate on the principal is not 12% but something greater, because compounding occurs twice during the year.

Consequently, the frequency at which a nominal interest rate is compounded each year can have a pronounced effect on the dollar amount of total interest earned. For instance, consider a principal amount of $1,000 to be invested for three years at a nominal rate of 12% compounded semiannually. The interest earned during the first six months would be $1,000 \times (0.12/2) = \$60$.

Total principal and interest at the beginning of the second six-month period is

$$P + Pi = \$1,000 + \$60 = \$1,060$$

The interest earned during the second six months would be

$$\$1,060 \times (0.12/2) = \$63.60$$

Then total interest earned during the year is

$$\$60.00 + \$63.60 = \$123.60$$

Finally, the *effective* annual interest rate for the entire year is

$$\frac{\$123.60}{\$1,000} \times 100 = 12.36\%$$

If this process is repeated for years two and three, the *accumulated* (compounded) *amount of interest* can be plotted as in Figure 3-20. Suppose that the same $1,000 had been invested at 12% compounded *monthly*, which is 1% per month. The accumulated interest over three years that results from monthly compounding is shown in Figure 3-21.

The actual or exact rate of interest earned on the principal during one year is known as the *effective rate*. It should be noted that effective interest rates are always expressed on an annual basis unless specifically stated otherwise. In this text the effective interest rate per year is customarily designated by i and the nominal interest rate per year by r. In engineering economy studies in which compounding is annual, $i = r$. The relationship between effective interest, i, and nominal interest, r, is

$$
\begin{aligned}
i &= (1 + r/M)^M - 1 \\
&= (F/P, r/M, M) - 1
\end{aligned}
$$

(3-33)

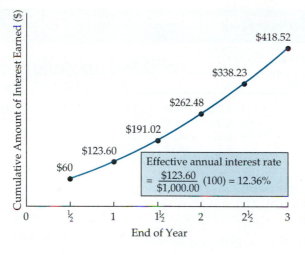

FIGURE 3-20

$1,000 Compounded at a Semiannual Frequency (r = 12%)

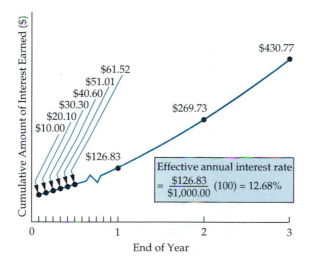

FIGURE 3-21

$1,000 Compounded at a Monthly Frequency (r = 12%)

where M is the number of compounding periods per year. It is now clear from Equation 3-33 why $i > r$ when $M > 1$.

The effective rate of interest is useful for describing the compounding effect of interest earned on interest in during one year. Table 3-5 shows effective rates for various nominal rates and compounding periods.

Interestingly, the federal truth in lending law now requires a statement regarding the annual percentage rate (APR) being charged in contracts involving borrowed money. The APR is a nominal interest rate and *does not* account for compounding that may occur, or be appropriate, during a year. Before this legislation was passed by Congress in 1969, creditors had no obligation to explain how interest charges were determined or what the true cost of money on a loan was.

TABLE 3-5   Effective Interest Rates for Various Nominal Rates and Compounding Frequencies

Compounding Frequency	Number of Compounding Periods per Year, M	Effective Rate (%) for Nominal Rate of					
		6%	8%	10%	12%	15%	24%
Annually	1	6.00	8.00	10.00	12.00	15.00	24.00
Semiannually	2	6.09	8.16	10.25	12.36	15.56	25.44
Quarterly	4	6.14	8.24	10.38	12.55	15.87	26.25
Bimonthly	6	6.15	8.27	10.43	12.62	15.97	26.53
Monthly	12	6.17	8.30	10.47	12.68	16.08	26.82
Daily	365	6.18	8.33	10.52	12.75	16.18	27.11

As a result, borrowers were generally unable to compute their APR and compare different financing plans.

EXAMPLE 3-20

A credit card company charges an interest rate of 1.375% per month on the unpaid balance of all accounts. The annual interest rate, they claim, is 12(1.375%) = 16.5%. What is the effective rate of interest per year being charged by the company?

SOLUTION

Interest tables in Appendix C are based on time periods that may be annual, quarterly, monthly, and so on. Because we have no 1.375% tables (or 16.5% tables), Equation 3-33 must be used to compute the effective rate of interest in this example:

$$i = \left(1 + \frac{0.165}{12}\right)^{12} - 1$$

$$= 0.1781, \text{ or } 17.81\%/\text{year}$$

Note that $r = 12(1.375\%) = 16.5\%$, which is the APR. It is true that $r = M(r/M)$, as seen in Example 3-20, where r/M is the interest rate per period.

▼ 3.17 Interest Problems with Compounding More Often Than Once per Year

3.17.1 Single Amounts

If a nominal interest rate is quoted and the number of compounding periods per year and number of years are known, any problem involving future, annual, or present equivalent values can be calculated by straightforward use of Equations 3-3 and 3-33, respectively.

EXAMPLE 3-21

Suppose that a $100 lump-sum amount is invested for 10 years at a nominal interest rate of 6% compounded quarterly. How much is it worth at the end of the tenth year?

SOLUTION

There are four compounding periods per year, or a total of $4 \times 10 = 40$ periods. The interest rate per interest period is $6\%/4 = 1.5\%$. When the values are used in Equation 3-3, one finds that

$$F = P(F/P, 1.5\%, 40) = \$100.00(1.015)^{40} = \$100.00(1.814) = \$181.40$$

Alternatively, the effective interest rate from Equation 3-33 is 6.14%. Therefore, $F = P(F/P, 6.14\%, 10) = \$100.00(1.0614)^{10} = \$181.40$.

3.17.2 Uniform Series and Gradient Series

When there is more than one compounded interest period per year, the formulas and tables for uniform series and gradient series can be used *as long as* there is a cash flow at the end of each interest period, as shown in Figures 3-6 and 3-14 for a uniform annual series and a uniform gradient series, respectively.

EXAMPLE 3-22

Suppose that one has a bank loan for $10,000, which is to be repaid in equal *end-of-month* installments for five years with a nominal interest rate of 12% compounded monthly. What is the amount of each payment?

SOLUTION

The number of installment payments is $5 \times 12 = 60$, and the interest rate per month is $12\%/12 = 1\%$. When these values are used in Equation 3-13, one finds that

$$A = P(A/P, 1\%, 60) = \$10,000(0.0222) = \$222$$

Notice that there is a cash flow at the end of each month (interest period), including month 60, in this example.

EXAMPLE 3-23

Certain operating savings are expected to be 0 at the end of the first six months, to be $1,000 at the end of the second six months, and to increase by $1,000 at the end of each six-month period thereafter for a total of four years. It is desired to find the equivalent uniform amount, A, at the end of each of the eight 6-month periods if the nominal interest rate is 20% compounded semiannually.

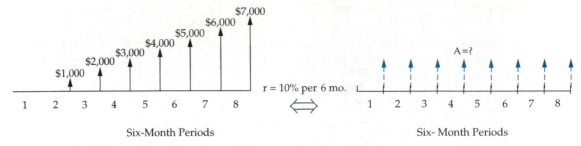

FIGURE 3-22   Arithmetic Gradient with Compounding More Often Than Once per Year in Example 3-23

SOLUTION

A cash flow diagram is given in Figure 3-22, and the solution is

$$A = G(A/G, 10\%, 8) = \$1,000(3.0045) = \$3,004.50$$

The symbol "$\Leftrightarrow$" in Figure 3-22 indicates that the left-hand cash flow diagram is *equivalent to* the right-hand cash flow diagram when the correct value of A has been determined. In Example 3-23, the interest rate per six-month period is 10%, and cash flows occur every six months.

▼ 3.18   Interest Problems with Cash Flows Less Often Than Compounding Periods

In general, if i is the effective interest rate per interest period and there is a uniform cash flow, X, at the *end* of each Kth interest period ($K > 1$), then the equivalent amount, A, at the end of each interest period is

$$A = X(A/F, i\%, K) \tag{3-34}$$

By similar reasoning, if i is the effective interest rate per interest period and there is a uniform cash flow, X, at the *beginning* of each Kth interest period, then the equivalent amount, A, at the end of each interest period is

$$A = X(A/P, i\%, K) \tag{3-35}$$

EXAMPLE 3-24

In cash flow diagram (a), write an equation to convert the three amounts, $X, into their annual equivalent value over N years ($N = 3Z$) when the interest rate is $i\%$ per year. For cash flow diagram (b), determine the annual equivalent amount over 18 years when $i = 10\%$ per year.

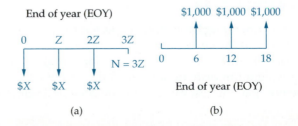

SOLUTION

(a) By using Equation 3-35, the value of A corresponding to the three payments of $X is

$$A = \$X(A/P, i\%, Z)$$

(b) Equation 3-34 enables one to calculate the value of A which extends from end-of-year 1 through end-of-year 18:

$$A = \$1,000(A/F, 10\%, 6) = \$129.60$$

EXAMPLE 3-25

Suppose that there exists a series of 10 end-of-year receipts of $1,000 each, and it is desired to compute their equivalent worth as of the *end* of the tenth year if the nominal interest rate is 12% compounded quarterly. The cash flows are depicted in Figure 3-23.

SOLUTION

Interest is 12%/4 = 3% per quarter, but the uniform series cash flows do not occur at the end of each quarter. In such cases, one can make special adaptations to fit the interest formulas to the tables provided. To solve this type of problem, (1) compute an equivalent cash flow for the time interval that corresponds to the stated compounding frequency, *or* (2) determine an effective interest rate for the interval of time separating the cash flows.

The *first* adaptation procedure is to take the number of compounding periods over which a cash flow occurs and convert the cash flow into its equivalent uniform end-of-period series. The upper cash flow diagrams in Figure 3-24 show this approach applied to the first year (four interest periods) in the example of Figure 3-23. The uniform end-of-quarter amount, equivalent to $1,000 at the end of the year with interest at 3% per quarter, can be calculated by using Equation 3-34:

$$A = F(A/F, 3\%, 4) = \$1,000(0.2390) = \$239$$

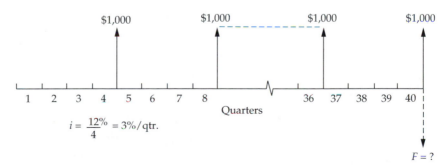

FIGURE 3-23   Uniform Series with Cash Flows Less Often Than Compounding Periods in Example 3-25

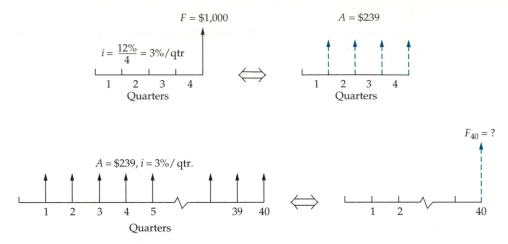

FIGURE 3-24   First Adaptation to Solve Example 3-25

Thus, $239 at the end of each quarter is equivalent to $1,000 at the end of each year. This is true not only for the first year but also for each of the 10 years under consideration. Hence the original series of 10 end-of-year cash flows of $1,000 each can be converted to a problem involving 40 end-of-quarter amounts of $239 each, as shown in the lower cash flow diagrams of Figure 3-24.

The future equivalent at the end of the 10th year (40th quarter) may then be computed as

$$F_{40} = A(F/A, 3\%, 40) = \$239(75.4012) = \$18,021$$

The *second* procedure for handling cash flows occurring less often than compounding periods is to find the exact interest rate for each time period *separating* cash flows and then to straightforwardly apply the interest formulas and tables for the exact interest rate. For Example 3-25, interest is 3% per quarter and payments occur each year. Hence, the interest rate to be found is the exact rate each year, or the *effective rate* per year. The effective rate per year that corresponds to 3% per quarter (12% nominal) can be found from Equation 3-33.

$$\left(1 + \frac{0.12}{4}\right)^4 - 1 = (F/P, 3\%, 4) - 1 = 0.1255$$

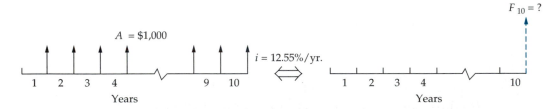

FIGURE 3-25   Second Adaptation to Solve Example 3-25

Hence, the original problem in Figure 3-23 can now be expressed as shown in Figure 3-25. The future equivalent of this series can then be found as

$$F_{10} = A(F/A, 12.55\%, 10) = \$1,000(F/A, 12.55\%, 10) = \$18,022$$

Because interest factors are not commonly tabled for $i = 12.55\%$, one must compute the $(F/A, 12.55\%, 10)$ factor by substituting $i = 0.1255$ and $N = 10$ into its algebraic equivalent, $[(1 + i)^N - 1]/i$ (see Table 3-4).

The second procedure just illustrated is probably the more popular way to deal with problems in which cash flows occur every K ($K > 1$) compounding periods. By using the second procedure, the basic question becomes: "How do we find an effective interest rate for the fixed time interval (K compounding periods) separating the cash flows?" We now formalize this procedure by using a more general version of Equation 3-33 to determine an effective interest rate per K compounding periods:

$$i(\text{per } K \text{ compounding periods}) = (1 + r/M)^K - 1 \qquad \textbf{(3-36)}$$

where K = number of compounding periods per fixed time interval separating cash flows

         r = nominal interest rate per year

    M = number of compounding periods per year

EXAMPLE 3-26

Determine the present equivalent, P, of this cash flow diagram:

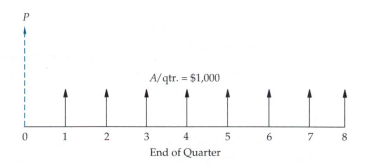

The nominal interest rate is 15% *compounded monthly*. Cash flows occur every three months (once per quarter).

SOLUTION

By using Equation 3-36, we determine the effective interest rate per quarter: $i/\text{qtr.} = (1 + \frac{0.15}{12})^3 - 1 = (1.0125)^3 - 1 = 0.038$, or 3.8%. Then $P = \$1,000 \times (P/A, 3.8\%, 8) = \$6,788.70$. Note that the second procedure can also be easily used with nonuniform cash flows (e.g., gradients).

▼ 3.19 Interest Formulas for Continuous Compounding and Discrete Cash Flows

In most business transactions and economy studies, interest is compounded at the end of discrete periods of time and, as has been discussed previously, cash flows are assumed to occur in discrete amounts at the end of such periods. *This practice will be used throughout the remaining chapters of this book.* However, it is evident that in most enterprises cash is flowing in and out in an almost continuous stream. Because cash, whenever it's available, can usually be used profitably, this situation creates opportunities for very frequent compounding of the interest earned. So that this condition can be dealt with (modeled) when continuously compounded interest rates are available, the concepts of continuous compounding and continuous cash flow are sometimes used in economy studies. Actually, the effects of these procedures compared to those of discrete compounding are rather small in most cases.

Continuous compounding assumes that cash flows occur at discrete intervals (e.g., once per year) but that compounding is continuous throughout the interval. For example, with a nominal rate of interest per year of r, if the interest is compounded M times per year, one unit of principal will amount to $[1 + (r/M)]^M$ at the end of one year. Letting $M/r = p$, the foregoing expression becomes

$$\left[1 + \frac{1}{p}\right]^{rp} = \left[\left(1 + \frac{1}{p}\right)^p\right]^r \qquad \text{(3-37)}$$

Because

$$\lim_{p \to \infty}\left(1 + \frac{1}{p}\right)^p = e^1 = 2.71828\ldots,$$

Equation 3-37 can be written as e^r. Consequently, the *continuously compounded compound amount factor (single cash flow)* at $r\%$ nominal interest for N years is e^{rN}. Using our functional notation, we express this as

$$(F/P, \underline{r}\%, N) = e^{rN} \qquad \text{(3-38)}$$

Note that the symbol $\underline{r}$ is directly comparable to that used for discrete compounding and discrete cash flows ($i\%$) except that $\underline{r}\%$ is used to denote the nominal rate *and* the use of continuous compounding.

Since e^{rN} for continuous compounding corresponds to $(1 + i)^N$ for discrete compounding, e^r is equal to $(1 + i)$. Hence, we may correctly conclude that

$$i = e^r - 1 \qquad \text{(3-39)}$$

By using this relationship, the corresponding values of $(P/F), (F/A)$, and (P/A) for continuous compounding may be obtained from Equations 3-4, 3-6, and 3-8, respectively, by substituting $e^r - 1$ for i in these equations. Thus, for continuous compounding and discrete cash flows,

$$(P/F, \underline{r}\%, N) = \frac{1}{e^{rN}} = e^{-rN} \tag{3-40}$$

$$(F/A, \underline{r}\%, N) = \frac{e^{rN} - 1}{e^r - 1} \tag{3-41}$$

$$(P/A, \underline{r}\%, N) = \frac{1 - e^{-rN}}{e^r - 1} = \frac{e^{rN} - 1}{e^{rN}(e^r - 1)} \tag{3-42}$$

Values for $(A/P, \underline{r}\%, N)$, and $(A/F, \underline{r}\%, N)$ may be derived through their inverse relationships to $(P/A, \underline{r}\%, N)$ and $(F/A, \underline{r}\%, N)$, respectively. Numerous continuous compounding, discrete cash flow interest factors and their uses are summarized in Table 3-6.

Because continuous compounding is used infrequently in this text, detailed values for $(A/F, \underline{r}\%, N)$ and $(A/P, \underline{r}\%, N)$ are not given in Appendix D. However, the tables in Appendix D do provide values of $(F/P, \underline{r}\%, N)$, $(P/F, \underline{r}\%, N)$, $(F/A, \underline{r}\%, N)$, and $(P/A, \underline{r}\%, N)$ for a limited number of interest rates.

TABLE 3-6 Continuous Compounding and Discrete Cash Flows: Interest Factors and Symbols[a]

To Find:	Given:	Factor by Which to Multiply "Given"	Factor Name	Factor Functional Symbol
For single cash flows:				
F	P	e^{rN}	Continuous compounding compound amount (single cash flow)	$(F/P, \underline{r}\%, N)$
P	F	e^{-rN}	Continuous compounding present equivalent (single cash flow)	$(P/F, \underline{r}\%, N)$
For uniform series (annuities):				
F	A	$\dfrac{e^{rN} - 1}{e^r - 1}$	Continuous compounding compound amount (uniform series)	$(F/A, \underline{r}\%, N)$
P	A	$\dfrac{e^{rN} - 1}{e^{rN}(e^r - 1)}$	Continuous compounding present equivalent (uniform series)	$(P/A, \underline{r}\%, N)$
A	F	$\dfrac{e^r - 1}{e^{rN} - 1}$	Continuous compounding sinking fund	$(A/F, \underline{r}\%, N)$
A	P	$\dfrac{e^{rN}(e^r - 1)}{e^{rN} - 1}$	Continuous compounding capital recovery	$(A/P, \underline{r}\%, N)$

[a] $\underline{r}$, nominal annual interest rate, compounded continuously; N, number of periods (years); A, annual equivalent amount (occurs at the end of each year); F, future equivalent; P, present equivalent.

Note that tables of interest and annuity factors for continuous compounding are tabulated in terms of nominal annual rates of interest.

EXAMPLE 3-27

Suppose that one has a present loan of $1,000 and desires to determine what equivalent uniform end-of-year payments, A, could be obtained from it for 10 years if the nominal interest rate is 20% compounded continuously ($M = \infty$).

SOLUTION

Here we utilize this formulation:

$$A = P(A/P, \underline{r}\%, N)$$

Since the (A/P) factor is not tabled for continuous compounding, we substitute its inverse (P/A), which is tabled in Appendix D. Thus

$$A = P \times \frac{1}{(P/A, \underline{20}\%, 10)} = \$1,000 \times \frac{1}{3.9054} = \$256$$

Note that the answer to the same problem, with discrete annual compounding ($M = 1$), is

$$A = P(A/P, 20\%, 10)$$
$$= \$1,000(0.2385) = \$239$$

EXAMPLE 3-28

An individual needs $12,000 immediately as a down payment on a new home. Suppose that he can borrow this money from his insurance company. He must repay the loan in equal payments every six months over the next eight years. The nominal interest rate being charged is 7% compounded continuously. What is the amount of each payment?

SOLUTION

The nominal interest rate per six months is 3.5%. Thus, A each six months is $12,000(A/P, \underline{r} = 3.5\%, 16)$. By substituting terms in Equation 3-42 and then using its inverse, we determine the value of A per six months to be $997, that is

$$A = \$12,000 \left[\frac{1}{(P/A, \underline{r} = 3.5\%, 16)} \right] = \frac{\$12,000}{12.038} = \$997$$

3.20 Interest Formulas for Continuous Compounding and Continuous Cash Flows

Continuous flow of funds means a series of cash flows occurring at infinitesimally short intervals of time; this corresponds to an annuity having an infinite number of short periods. This model could apply to companies having receipts and expenses that occur frequently during each working day. In such cases the interest is normally compounded continuously. If the nominal interest rate per year is r and there are p payments per year, which amount to a total of one unit per year, then, by using Equation 3-8, the present equivalent at the beginning of the year (for one year) is

$$P = \frac{1}{p}\left\{ \frac{[1 + (r/p)]^p - 1}{r/p[1 + (r/p)]^p} \right\} = \frac{[1 + (r/p)]^p - 1}{r[1 + (r/p)]^p} \tag{3-43}$$

The limit of $[1 + (r/p)]^p$ as p approaches infinity is e^r. Calling the present equivalent of one unit per year, flowing continuously and with continuous compounding of interest, the *continuous compounding present equivalent factor (continuous, uniform cash flow over one period)*, one finds

$$(P/\overline{A}, r\%, 1) = \frac{e^r - 1}{re^r} \tag{3-44}$$

where $\overline{A}$ is the amount flowing uniformly and continuously over one year (here $1).

For $\overline{A}$ flowing each year over N years, as depicted in Figure 3-26,

$$(P/\overline{A}, r\%, N) = \frac{e^{rN} - 1}{re^{rN}} \tag{3-45}$$

which is the *continuous compounding present equivalent factor (continuous, uniform cash flows)*.

Equation 3-44 can also be written

$$(P/\overline{A}, r\%, 1) = e^{-r}\left[\frac{e^r - 1}{r} \right] = (P/F, r\%, 1)\left[\frac{e^r - 1}{r} \right]$$

Because the present equivalent of $1 per year, flowing continuously with continuous compounding of interest, is $(P/F, r\%, 1)(e^r - 1)/r$, it follows that $(e^r - 1)/r$ must also be the compound amount of $1 per year, flowing continuously with continuous compounding of interest. Consequently, the *continuous compounding*

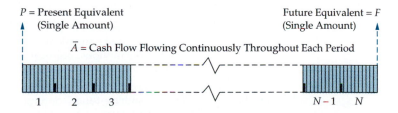

P = Present Equivalent
(Single Amount)

Future Equivalent = F
(Single Amount)

$\overline{A}$ = Cash Flow Flowing Continuously Throughout Each Period

1   2   3 $N - 1$ N

r = Nominal Interest Rate
Compounded Continuously

FIGURE 3-26 General Cash Flow Diagram for Continuous Compounding, Continuous Cash Flows

compound amount factor (continuous, uniform cash flow over one year) is

$$(F/\overline{A}, \underline{r}\%, 1) = \frac{e^r - 1}{r} \tag{3-46}$$

For N years,

$$(F/\overline{A}, \underline{r}\%, N) = \frac{e^{rN} - 1}{r} \tag{3-47}$$

Equation 3-47 can also be developed by integration in this manner:

$$F = \overline{A} \int_0^N e^{rt}\, dt = \overline{A}\left(\frac{1}{r}\right) \int_0^N re^{rt}\, dt$$

or

$$F = \frac{\overline{A}}{r}(e^{rt}) \Big|_0^N = \overline{A}\left[\frac{e^{rN} - 1}{r}\right]$$

This is the *continuous compounding compound amount factor (continuous uniform cash flows for N years).*

Values of $(P/\overline{A}, \underline{r}\%, N)$ and $(F/\overline{A}, \underline{r}\%, N)$ are given in the tables in Appendix D for various interest rates. Values for $(\overline{A}/P, \underline{r}\%, \underline{N})$ and $(\overline{A}/F, \underline{r}\%, N)$ can be readily obtained through their inverse relationship to $(P/\overline{A}, \underline{r}\%, N)$ and $(F/\overline{A}, \overline{r}\%, N)$, respectively. A summary of these factors and their use is given in Table 3-7.

TABLE 3-7   **Continuous Compounding Continuous Uniform Cash Flows: Interest Factors and Symbols[a]**

To Find:	Given:	Factor by Which to Multiply "Given"[a]	Factor Name	Factor Functional Symbol
F	$\overline{A}$	$\dfrac{e^{rN} - 1}{r}$	Continuous compounding compound amount (continuous, uniform cash flows)	$(F/\overline{A}, \underline{r}\%, N)$
P	$\overline{A}$	$\dfrac{e^{rN} - 1}{re^{rN}}$	Continuous compounding present equivalent (continuous, uniform cash flows)	$(P/\overline{A}, \underline{r}\%, N)$
$\overline{A}$	F	$\dfrac{r}{e^{rN} - 1}$	Continuous compounding sinking-fund (continuous, uniform cash flows)	$(\overline{A}/F, \underline{r}\%, N)$
$\overline{A}$	P	$\dfrac{re^{rN}}{e^{rN} - 1}$	Continuous compounding capital recovery (continuous, uniform cash flows)	$(\overline{A}/P, \underline{r}\%, N)$

[a]r, nominal annual interest rate, compounded continuously; N, number of periods (years); $\overline{A}$, amount of money flowing continuously and uniformly during each period; F, future equivalent; P, present equivalent.

EXAMPLE 3-29

What will be the future equivalent amount at the end of five years of a uniform, continuous cash flow, at the rate of $500 per year for five years, with interest compounded continuously at the nominal annual rate of 8%?

SOLUTION

$$F = \overline{A}(F/\overline{A}, 8\%, 5) = \$500 \times 6.1478 = \$3,074$$

Note that if this cash flow had been in year-end amounts of $500 with discrete annual compounding of $i = 8\%$, the future equivalent amount would have been

$$F = A(F/A, 8\%, 5) = \$500 \times 5.8666 = \$2,933$$

If the year-end payments had occurred with 8% nominal interest compounded continuously, the future equivalent would then have been

$$F = A(F/A, 8\%, 5) = \$500 \times 5.9052 = \$2,953$$

It is clear that for a given A amount and continuous compounding of a given nominal interest rate, continuous funds flow produces the largest-valued future equivalent amount.

EXAMPLE 3-30

What is the future equivalent of $10,000 per year that flows continuously for 8.5 years if the nominal interest is 10% compounded continuously?

SOLUTION

There are seventeen six-month periods in 8.5 years, and the r per six months is 5%. The $\overline{A}$ every six months is $5,000, so $F = \$5,000(F/\overline{A}, 5\%, 17) = \$133,964.50$. This formulation is utilized to enable us to find an interest factor having an integer-valued N. The same answer could have been obtained by resorting to the definition of the $(F/\overline{A}, r\%, N)$ factor given in Table 3-7 with $N = 8.5$ years:

$$F = \$10,000 \left[\frac{e^{0.10(8.5)} - 1}{0.10} \right]$$

$$= \$133,964.50$$

3.21 Additional Solved Problems

This section contains many solved problems that further illustrate the economic equivalence concepts of Chapter 3.

PROBLEM 1

Given the following information, determine the value of each "?" in the following table.

Loan Principal = $10,000
Interest Rate = 8% per year
Duration of Loan = 3 years

End of Year k	Interest Paid	Principal Repayment
1	$800	?
2	$553.60	$3,326.40
3	?	?

SOLUTION

A uniform annual payment scheme is implied by the table entries. Thus, the total annual payment = $10,000 $(A/P, 8\%, 3)$ = $3,880. At the end of year one, the principal repayment will be $3,880 − $800 = $3,080. At the beginning of year three, the remaining principal to be repaid is $10,000 − $3,080 − $3,326.40 = $3,593.60. Therefore, the interest paid during year three is about 0.08($3,593.60) = $286.40. (Some rounding is present in this problem because of four significant digits in the interest tables.)

PROBLEM 2

Suppose the 8% interest rate in Problem 1 is a nominal interest rate. If compounding occurs monthly, what is the effective annual interest rate?

SOLUTION
Use Equation 3-33 to find

$$i = \left(1 + \frac{0.08}{12}\right)^{12} - 1$$

$$= 0.083 \ (8.3\% \text{ effective annual interest rate}).$$

PROBLEM 3

Compare the interest earned by $9,000 for five years at 8% simple interest per year with the interest earned by the same amount for five years at 8% compounded annually. Explain why a difference occurs.

SOLUTION

Simple interest:
$I = (P)(N)(i) = \$9,000(0.08)(5) = \underline{\$3,600}$
Total $= \$9,000 + \$3,600 = \$12,600$

Compound interest:
$F = P(F/P, 8\%, 5) = \$9,000(1.4693) = \$13,223.70$
Total interest $= \$13,223.70 - \$9,000 = \underline{\$4,223.70}$

There is a difference in the amount of interest earned because compounding allows interest from previous years to earn interest, whereas simple interest does not.

PROBLEM 4

With the minimum number of interest factors, find the value of X below so that the two cash flow diagrams are equivalent when the interest rate is 10% per year. *Set-up only.*

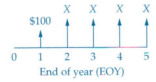

SOLUTION

Use EOY one as the reference point and three interest factors:

$$\$1,000 + \$800(P/A, 10\%, 4) - \$200(P/G, 10\%, 4) = 100 + X(P/A, 10\%, 4)$$

$$X = \frac{\$900 + \$800(P/A, 10\%, 4) - \$200(P/G, 10\%, 4)}{(P/A, 10\%, 4)}$$

PROBLEM 5

Set up an expression for the value of Z on the left-hand cash flow diagram that establishes equivalence with the right-hand cash flow diagram. The nominal interest rate is 12% compounded quarterly.

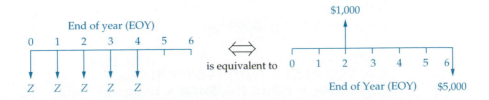

SOLUTION

$$i = (1 + 0.12/4)^4 - 1 \quad \text{(Equation 3-33)}$$

$$\approx 0.1255 \ (12.55\%)$$

$$-Z - Z(P/A, 12.55\%, 4) = \$1{,}000(P/F, 12.55\%, 2) - \$5{,}000(P/F, 12.55\%, 6)$$

$$Z = \frac{\$1{,}000(P/F, 12.55\%, 2) - \$5{,}000(P/F, 12.55\%, 6)}{[-1 - (P/A, 12.55\%, 4)]}$$

PROBLEM 6

A student decides to make semi-annual payments of $500 each into a bank account that pays an APR (nominal interest) of 8% compounded weekly. How much money will this student have accumulated in this bank account at the end of 20 years? Assume only one (the final) withdrawal is made.

SOLUTION

Using Equation 3-36, i per six months (26 weeks) is equal to

$$\left(1 + \frac{0.04}{26}\right)^{26} - 1 = 0.0408 \ (4.08\%)$$

Then F at the end of year 20 will be: $F = \$500(F/A, 4.08\%, 40)$ or

$$F = \$500\left[\frac{(1.0408)^{40} - 1}{0.0408}\right] = \$500\left[\frac{4.9510 - 1}{0.0408}\right] = \$48{,}419$$

PROBLEM 7

Consider an end-of-year (EOY) geometric gradient, which lasts for eight years, whose initial value at EOY one is $5,000, and $\bar{f} = 6.04\%$ per year thereafter. Find the equivalent uniform gradient amount over the same time period if the initial value of the uniform gradient at the end of year one is $4,000. Complete the following questions in determining the value of the gradient amount, G. The nominal interest rate is 8% compounded semi-annually.
(a) What is i_{CR}?

$$i = \left(1 + \frac{0.08}{2}\right)^2 - 1 = 0.0816$$

$$= 8.16\%$$

$$i_{CR} = \frac{1 + 0.0816}{1 + 0.0604} - 1 = 0.02 \ (2\%)$$

(b) What is P_0 for the geometric gradient?

$$P_0 = \frac{\$5{,}000}{1 + 0.0604}(P/A, 2\%, 8)$$

$$= \$34{,}541$$

(c) What is P_0' of the uniform (arithmetic) gradient?

$$P_0' = \$4,000(P/A, 8.16\%, 8) + G(P/G, 8.16\%, 8)$$

(d) What is the value of G?

Set $P_0 = P_0'$ and solve for G. Answer: $G = \$662.53$

PROBLEM 8

An individual makes five annual deposits of $2,000 in a savings account that pays interest at a rate of 4% per year. One year after making the last deposit, the interest rate changes to 6% per year. Five years after the last deposit the accumulated money is withdrawn from the account. How much is withdrawn?

SOLUTION

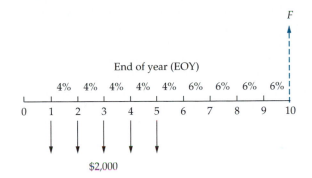

$$F = \$2,000(F/A, 4\%, 5)(F/P, 4\%, 1)(F/P, 6\%, 4) = \$14,223$$

PROBLEM 9

Some future amount, F, is equivalent to $2,000 being received every 6 months over the next 12 years. The nominal interest rate is 20% compounded continuously. What is the value of F?

SOLUTION

$$F = \frac{\$2,000}{6 \text{ mo.}}(F/A, \underline{10}\% \text{ 6 mo.}, \text{24 six-month periods})$$

$$= \$190,607.40$$

PROBLEM 10

What is the value of P that is equivalent to $\overline{A} = \$800/\text{yr}$ ($800 flowing continuously each year) for 11.2 years? The nominal rate of interest is 10%, continuously compounded.

SOLUTION

$$P = \frac{\$800}{\text{yr.}}(P/\overline{A}, \underline{10}\%/\text{yr.}, 11.2 \text{ years})$$

$$= \$800 \left(\frac{e^{0.10(11.2)} - 1}{0.10e^{0.10(11.2)}}\right) = \$800 \left(\frac{e^{1.12} - 1}{0.10e^{1.12}}\right)$$

$$\cong \$5,390$$

3.22 Spreadsheet Applications

Appendices C and D tabulate the most common interest factors for a variety of interest rates and a number of compounding intervals. However, frequently we use an interest rate that does not have a corresponding table in the appendices. In this situation we must resort to using the equations that define the interest factors. This process can be facilitated by using a spreadsheet.

Figure 3-27 shows a spreadsheet (and corresponding cell formulas) that can be used to generate the discrete compounding factor values for a given interest rate (i) and number of compounding periods (N). A similar spreadsheet for generating the factor values in the case of continuous compounding is shown in Figure 3-28.

Figure 3-29 is a spreadsheet model that can compute effective interest rates. Given a nominal interest rate (r) and the number of compounding periods per year

	A	B
1	i	30.0%
2	N	7
3		
4		
5	*(F/P, i%, N) =*	6.2749
6	*(P/F, i%, N) =*	0.1594
7	*(F/A, i%, N) =*	17.5828
8	*(P/A, i%, N) =*	2.8021
9	*(A/F, i%, N) =*	0.0569
10	*(A/P, i%, N) =*	0.3569
11		
12	*(P/G, i%, N) =*	5.6218
13	*(A/G, i%, N) =*	2.0063
14	*(F/G, i%, N) =*	35.2761

Cell	Name
B1	i
B2	N

Cell	Contents
B5	$(1+i)\wedge N$
B6	$1/(1+i)\wedge N$
B7	$((1+i)\wedge N-1)/i$
B8	$((1+i)\wedge N-1)/(i*(1+i)\wedge N)$
B9	$i/((1+i)\wedge N-1)$
B10	$i*(1+i)\wedge N/((1+i)\wedge N-1)$
B12	$(((1+i)\wedge N-1)/(i*(1+i)\wedge N)-N/(1+i)\wedge N)/i$
B13	$(1/i)-N/((1+i)\wedge N-1)$
B14	$((1+i)\wedge N-1)/(i\wedge2-N)/i$

FIGURE 3-27   Spreadsheet for Generating Interest Factor Values with Discrete Compounding

	A	B
1	r	15.0%
2	N	5
3		
4		
5	(F/P, r%, N) =	2.1170
6	(P/F, r%, N) =	0.4724
7	(F/A, r%, N) =	6.9021
8	(P/A, r%, N) =	3.2603
9	(A/F, r%, N) =	0.1449
10	(A/P, r%, N) =	0.3067

Cell	Name
B1	r
B2	N

Cell	Formula
B5	=EXP(r*N)
B6	=EXP(-r*N)
B7	=(EXP(r*N)-1)/(EXP(r)-1)
B8	=(EXP(r*N)-1)/(EXP(r*N)*(EXP(r)-1))
B9	=(EXP(r)-1)/(EXP(r*N)-1)
B10	=(EXP(r*N)*(EXP(r)-1))/(EXP(r*N)-1)

FIGURE 3-28 Spreadsheet for Generating Interest Factor Values with Continuous Compounding

	A	B	C	D	E
1	Nominal interest rate, r				12%
2	Compounding periods per year, M				12
3	Number of compounding periods				
4	per fixed time interval				
5	separating cash flows, K				4
6					
7	Effective interest rates:				
8	i (annual)				12.68%
9	i (per K compounding periods)				4.06%

Cell	Name
E1	r
E2	M
E5	K

Cell	Contents
E8	=((1 + r/M)^M)-1
E9	=((1 + r/M)^K)-1

FIGURE 3-29 Spreadsheet for Computing Effective Interest Rates

(M), the spreadsheet computes the effective annual interest rate. If cash flows occur less frequently than compounding periods—for example, monthly compounding and quarterly cash flows—the spreadsheet can compute the effective interest rate for the time interval separating cash flows.

3.23 Summary

Chapter 3 has presented the fundamental time value of money relationships that are used throughout the remainder of this book. Considerable emphasis has been placed on the notion of economic equivalence, whether the relevant cash flows and interest rates are discrete or continuous. Students should feel comfortable with the material in this chapter before embarking on their journey through subsequent chapters. Important abbreviations and notation in Chapter 3 are listed in Appendix B, which will serve as a handy reference in your use of this book.

3.24 References

Au, T., and T. P. Au. *Engineering Economics for Capital Investment Analysis.* Boston: Allyn and Bacon, 1983.

Bussey, L. E., and T. G. Eschenbach. *The Economic Analysis of Industrial Projects.* Englewood Cliffs, N.J.: Prentice-Hall Inc., 1992.

Fleischer, G. A. *Introduction to Engineering Economy.* Monterey, Calif: Brooks/Cole, 1994.

Thuesen, G. J., and W. J. Fabrycky. *Engineering Economy,* 8th ed. Englewood Cliffs, N.J.: Prentice-Hall, Inc., 1993.

White, J. A., M. H. Agee, and K. E. Case. *Principles of Engineering Economic Analysis,* 3rd ed. New York: John Wiley, 1989.

3.25 Problems

The numbers in parentheses () that follow these problems indicate the section from which each problem is taken. (Refer to Appendix G of this book for answers to even-numbered problems.)

3-1. What lump-sum amount of interest will be paid on a $10,000 loan that was made on August 1, 1996, and repaid on November 1, 1999, with ordinary simple interest at 10% per year? (3.4)

3-2. Draw a cash flow diagram for $10,500 being loaned out at an interest rate of 12% per year over a period of six years. How much simple interest would be repaid as a lump-sum amount at the end of the sixth year? (3.4, 3.7)

3-3. What is the future equivalent of $1,000 invested at 6% simple interest per year for $2\frac{1}{2}$ years? (3.4)
  a. $1,157   **b.** $1,188   **c.** $1,184
  d. $1,175   **e.** $1,150

3-4. How much interest is *payable each year* on a loan of $2,000 if the interest rate is 10% per year when half of the loan principal will be repaid as a lump sum at the end of four years and the *other half* will be repaid in one lump-sum amount at the end of eight years? How much interest will be paid over the eight-year period? (3.6)

3-5. In Problem 3-4, if the interest had not been paid each year but had been added to the outstanding principal plus accumulated interest, how much interest would be due to the lender as a lump

sum at the end of the eighth year? How much extra interest is being paid here (as compared to Problem 3-4) and what is the reason for the difference? (3.6)

3-6.
  a. Suppose that in Plan 1 of Table 3-1, $4,000 in principal is to be repaid at the end of years two and four only. How much total interest would have been paid by the end of year four? (3.6)
  b. Rework Plan 3 of Table 3-1 when an annual interest rate of 8% is being charged on the loan. How much *principal* is now being repaid in the third year's total end-of-year payment? How much total interest has been paid by the end of the fourth year? (3.6, 3.9)

3-7.
  a. Based on the information, determine the value of each "?" in the table below, (3.6)

$$\text{Loan Principal} = \$10,000$$
$$\text{Interest Rate} = 6\%/\text{yr.}$$
$$\text{Duration of Loan} = 3 \text{ yrs.}$$

EOY k	Interest Paid	Principal Repayment
1	$600	?
2	$411.54	$3,329.46
3	?	?

b. What is the amount of principal owed at the *beginning* of year three?

c. Why is the total interest paid in (a) different from $10,000(1.06)^3 - 10,000 \approx \$1,910$ that would be repaid according to plan 4 in Table 3-1?

3-8. Compute the value of P_0 for each of the two alternatives discussed in Example 3-2. The annual interest rate is 10%. Which alternative should be selected, assuming that doing nothing is not an option? (3.9, 3.12)

3-9. What amount would need to be paid each January 1 into a savings account if at the end of 15 years (15 payments) you desired $10,000? Annual interest is 7%. (Note: The last payment will coincide with the time of the $10,000 balance.) (3.9)

3-10. A future amount, F, is equivalent to $1,500 now when six years separate the amounts and the annual interest rate is 12%. What is the value of F? (3.8)

3-11. A present obligation of $20,000 is to be repaid in equal uniform annual amounts, each of which includes repayment of the debt (principal) and interest on the debt, over a period of five years. If the interest rate is 10% per year, what is the amount of the annual repayment? (3.9)

3-12. Suppose that the $20,000 in Problem 3-11 is to be repaid at a rate of $4,000 per year plus the interest that is owed based on the beginning-of-year unpaid principal. Compute the total amount of interest repaid in this situation and compare it with that of Problem 3-11. Why are the two amounts different? (3.6)

3-13. A person wishes to accumulate $2,500 over a period of 15 years so that a cash payment can be made for a new roof on a summer cottage. To have this amount when it is needed, annual payments will be made into a savings account that earns 8% interest per year. How much must each annual payment be? Draw a cash flow diagram. (3.7, 3.9)

3-14. You have just learned that ABC Corporation has an investment opportunity that costs $35,000 and eight years later pays a lump-sum amount of $100,000. The cash flow diagram is shown at the top of the next column.

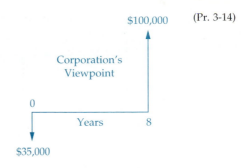

(Pr. 3-14)

$100,000

Corporation's Viewpoint

0

Years   8

$35,000

What interest rate per year would be earned on this investment? Calculate your answer to the nearest one-tenth of 1%. (3.8)

3-15. It is estimated that a copper mine will produce 10,000 tons of ore during the coming year. Production is expected to increase by 5% per year thereafter in each of the following six years. Profit per ton will be $14 for years one through seven.

a. Draw a cash flow diagram for this copper mine operation from the company's viewpoint. (3.7)

b. If the company can earn 10% per year on its capital, what is the future equivalent of the copper mine's cash flows at the end of year seven? (3.8 or 3.14)

3-16. Mrs. Green has just purchased a new car for $12,000. She makes a down payment of 30% of the negotiated price and then makes payments of $303.68 per month thereafter for 36 months. Furthermore, she believes the car can be sold for $3,500 at the end of three years. Draw a cash flow diagram of this situation from Mrs. Green's viewpoint. (3.7)

3-17. If $25,000 is deposited now into a savings account that earns 8% per year, what uniform annual amount could be withdrawn at the end of each year for ten years so that nothing would be left in the account after the tenth withdrawal? (3.9)

3-18. It is estimated that a certain piece of equipment can save $6,000 per year in labor and materials costs. The equipment has an expected life of five years and no salvage value. If the company must earn a 15% annual return on such investments, how much could be justified now for the purchase of this piece of equipment?

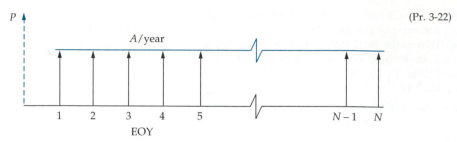

(Pr. 3-22)

Draw a cash flow diagram from the company's viewpoint. (3.7, 3.9)

3-19. Suppose that installation of Low-Loss thermal windows in your area is expected to save $150 a year on your home heating bill for the next 18 years. If you can earn 8% per year on other investments, how much could you afford to spend now for these windows? (3.9)

3-20. A proposed product modification to avoid production difficulties will require an immediate expenditure of $14,000 to modify certain dies. What annual savings must be realized to recover this expenditure in five years with interest at 10% per year? (3.9)

3-21. You can buy a machine for $100,000 that will produce a net income, after operating expenses, of $10,000 per year. If you plan to keep the machine for four years, what must the market (resale) value be at the end of four years to justify the investment? You must make a 12% annual return on your investment. (3.9)

3-22. Consider the cash flow diagram above and answer (a)–(d). (3.9)
- **a.** If $P = \$1,000$, $A = \$200$, and $i\% = 12\%$ per year, then $N = ?$
- **b.** If $P = \$1,000$, $A = \$200$, and $N = 10$ years, then $i = ?$

c. If $A = \$200$, $i\% = 12\%$ per year, and $N = 5$ years, then $P = ?$
d. If $P = \$1,000$, $i\% = 12\%$ per year, and $N = 5$ years, then $A = ?$

3-23. Use the rule of 72 to determine how long it takes to accumulate $2,000 in a savings account when $P = \$1,000$ and $i = 7\%$ per year. (3.8)

Rule of 72: The time required to double the value of a lump-sum investment that is allowed to compound is approximately

$$72 \div \text{annual interest rate (as a \%)}$$

3-24.
- **a.** Show that the following relationship is true: $(A/P, i, N) = i/[1 - (P/F, i, N)]$. (3.10)
- **b.** Show that $(A/G, 0\%, N) = (N - 1)/2$. (3.14)
- **c.** If an annuity, A, starts at the end of year one and continues each year thereafter forever, what is its P_0 equivalent value when $i = 12\%$ per year? (3.9)

3-25. Using the figure below, find the equivalent values of cash flows I, II, III, IV, and V to a single $1,000 cash flow at the end of 1995 when the interest rate is 5% per year. (3.8)

(Pr. 3-25)

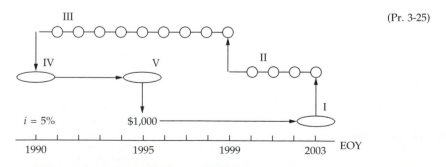

3-26. Suppose that $10,000 is borrowed now at 15% interest per year. A partial repayment of $3,000 is made four years from now. The amount that will remain to be paid then is most nearly: (3.8)
a. $7,000   **b.** $8,050   **c.** $8,500
d. $13,000   **e.** $14,490

3-27. How much should be deposited each year for 12 years if you wish to withdraw $309 each year for five years, beginning at the end of the 15th year? Let i = 8% per year. (3.11)

3-28. Suppose that you have $10,000 cash today and can invest it at 8% compound interest each year. How many years will it take you to become a millionaire? (3.8)

3-29. Equal end-of-year payments of $263.80 each are being made on a $1,000 loan at 10% effective interest per year. (3.6, 3.9)
a. How many payments are required to repay the entire loan?
b. Immediately after the third payment, what lump-sum amount would completely pay off the loan?

3-30. Maintenance costs for a small bridge with an expected 60-year life are estimated to be $1,000 each year for the first five years, followed by a $10,000 expenditure in the 15th year and a $10,000 expenditure in year thirty. If i = 10% per year, what is the equivalent uniform annual cost over the entire 60-year-period? (3.12)

3-31. In 1971 first-class postage for a one-ounce envelope was $0.08. In 1996, a first-class stamp for the same envelope cost $0.32. What compounded *annual* increase in the cost of first-class postage was experienced during the 25 years? (3.8)

3-32. You purchase special equipment that reduces defects by $10,000 per year on an item. This item is sold on contract for the next five years. After the contract expires, the special equipment will save approximately $2,000 per year for five years. You assume that the machine has no market value at the end of ten years. How much can you afford to pay for this equipment now if you require a 20% annual return on your investment? All cash flows are end-of-year amounts. (3.12)

3-33. John Q wants his estate to be worth $200,000 at the end of 10 years. His net worth is now zero. He can accumulate the desired $200,000 by depositing $14,480 at the end of each year for the

next 10 years. At what interest rate per year must his deposits be invested? (3.9)

3-34. What lump sum of money must be deposited into a bank account at the present time so that $500 per month can be withdrawn for five years, with the first withdrawal scheduled for six years from today? The interest rate is 1% per month. (*Hint*: Monthly withdrawals begin at the end of month 72.) (3.11)

3-35. Solve for the value of Z in the figure below so that the top cash flow diagram is equivalent to the bottom one. Let i = 12% per year. (3.12)

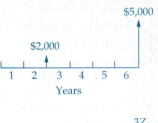

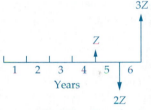

3-36. An individual is borrowing $100,000 at 8% interest compounded annually. The loan is to be repaid in equal annual payments over 30 years. However, just after the eighth payment is made, the lender allows the borrower to triple the annual payment. The borrower agrees to this increased payment. If the lender is still charging 8% per year, compounded annually, on the unpaid balance of the loan, what is the balance still owed just after the twelfth payment is made? (3.12)

3-37. A woman arranges to repay a $1,000 bank loan in 10 equal payments at a 10% effective annual interest rate. Immediately after her third payment she borrows another $500, also at 10% per year. When she borrows the $500, she talks the banker into letting her repay the remaining debt of the first loan and the entire amount of the second loan in 12 equal annual payments. The first of these 12 payments would be made

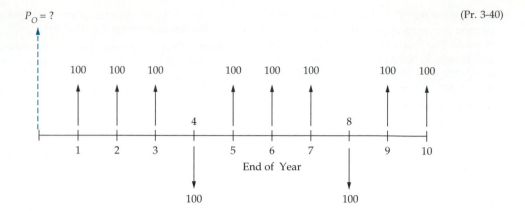

(Pr. 3-40)

one year after she receives the $500. Compute the amount of each of the 12 payments. (3.12)

3-38. A loan of $4,000 is to be repaid over a period of eight years. During the first four years, exactly half of the *loan principal* is to be repaid (along with accumulated compound interest) by a uniform series of payments of A_1 dollars per year. The other half of the loan principal is to be repaid over four years with accumulated interest by a uniform series of payments of A_2 dollars per year. If $i = 9\%$ per year, what are A_1 and A_2? (3.12)

3-39. On January 1, 1996 a person's savings account was worth $200,000. Every month thereafter this person makes a cash contribution of $676 to the account. If the fund is expected to be worth $400,000 on January 1, 2001, what annual rate of interest is being earned on this fund? (3.17)

3-40. Determine the present equivalent value at time 0 in the cash flow diagram above when $i = 10\%$ per year. Try to minimize the number of interest factors you use. (3.12)

3-41. Transform the cash flows on the left-hand side of the top diagram below to their equivalent amount, F, shown on the right-hand side. The interest rate is 10% per year. (3.12)

3-42. Determine the value of W on the right-hand side of the bottom diagram below that makes the two cash flow diagrams equivalent when $i = 12\%$ per year. (3.12)

3-43. Determine the value of Z on the left-hand side of the cash flow diagram at the top of page 127 that establishes equivalence with the right-hand side. The interest rate is 12% per year. (3.12)

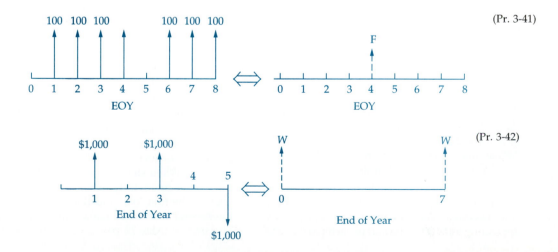

(Pr. 3-41)

(Pr. 3-42)

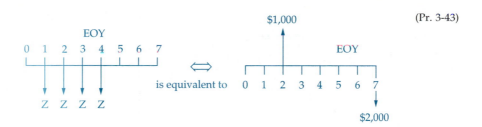

(Pr. 3-43)

3-44. Determine the value of "A" (uniform annual amount) that is equivalent to the following cash flow pattern. The interest rate is 10% per year. (3.12)

End of Year	Amount
0	$ 800
1	1,000
2	1,000
3	1,100
4	1,200
5	1,300
6	1,400
7	1,500
8	1,600
9	1,700
10	1,800

3-45. A certain fluidized-bed combustion vessel has an investment cost of $100,000, a life of 10 years, and negligible market (resale) value. Annual costs of materials, maintenance, and electric power for the vessel are expected to total $8,000. A major relining of the combustion vessel will occur during the fifth year at a cost of $20,000; during this year, the vessel will *not* be in service. If the interest rate is 15% per year, what is the lump-sum equivalent cost of this project at the present time? (3.12)

3-46. Suppose that $400 is deposited each year into a bank account that pays interest annually (*i* = 10%). If 12 payments are made into the account, how much would be accumulated in this fund by the end of the twelfth year? The first payment occurs at time zero (now). (3.9)

3-47. An expenditure of $20,000 is made to modify a material-handling system in a small job shop. This modification will result in first-year savings of $2,000, second-year savings of $4,000, and savings of $5,000 per year thereafter. How many years must the system last if a 20% return on investment is required? The system is tailor-made for this job shop and has no market (salvage) value at any time. (3.12)

3-48. Determine the present equivalent and annual equivalent value of the cash flow pattern shown below when *i* = 8% per year. (3.13)

3-49. Find the uniform annual amount that is equivalent to a uniform gradient series in which the first year's payment is $500, the second year's payment is $600, the third year's payment is $700, and so on, and there is a total of 15 payments. The annual interest rate is 8%. (3.13)

3-50. Suppose that annual income from a rental property is expected to start at $1,200 per year and decrease at a uniform amount of $40 each year for the 15-year expected life of the property. The investment cost is $8,000 and *i* is 9% per year. Is this a good investment? Assume that the investment occurs at time zero (now) and the annual income is first received at the end of year one. (3.13)

3-51. For a repayment schedule that starts at the end of year three at $Z and proceeds for years four through ten at $2Z, $3Z ..., what is the value of Z if the principal of this loan is $10,000 and the interest rate is 10% per year? (3.13)

3-52. If $10,000 now is equivalent to 4Z at the end of year two, 3Z at the end of year three, 2Z at

End of Year	0	1	2	3	4	5	6	7	(Pr. 3-48)
Amount ($)	−1,500	+500	+500	+500	+400	+300	+200	+100	

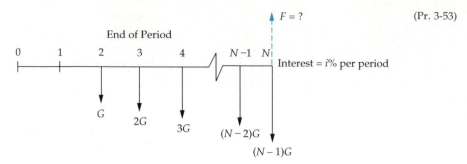

(Pr. 3-53)

the end of year four, and Z at the end of year five, what is the value of Z when $i = 8\%$ per year? (3.13)

3-53. Refer to the cash flow diagram shown above and solve for the unknown quantity in parts (a) through (d) that makes the equivalent value of cash outflows equal to the equivalent value of the cash inflow, F. (3.13)

a. If $F = \$10,000$, $G = \$600$, and $N = 6$, then $i = ?$

b. If $F = \$10,000$, $G = \$600$, and $i = 5\%$ per period, then $N = ?$

c. If $G = \$1,000$, $N = 12$, and $i = 10\%$ per period, then $F = ?$

d. If $F = \$8,000$, $N = 6$, and $i = 10\%$ per period, then $G = ?$

3-54. Solve for P_0 in the cash flow diagram shown below by using only two interest factors. The interest rate is 11% per year. (3.13)

3-55. At the top of page 129, what is the value of K on the left-hand cash flow diagram that is

equivalent to the right-hand cash flow diagram? Let $i = 15\%$ per year. (3.13)

3-56. For the second cash flow diagram on page 129, complete this equivalence equation: (3.13) $P_0 = \$100(P/A, 10\%, 4) +$ _____ (This can be completed with one more term.)

3-57. Calculate the future equivalent at the end of 1999, at 8% per year, of the following series of cash flows. (3.13)

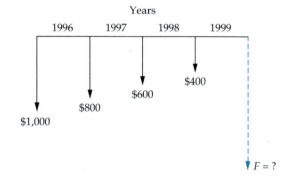

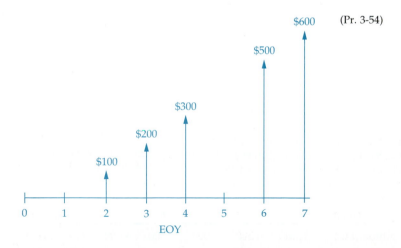

(Pr. 3-54)

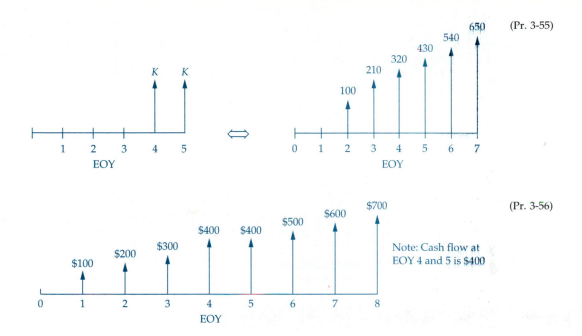

(Pr. 3-55)

(Pr. 3-56)

Note: Cash flow at
EOY 4 and 5 is $400

3-58. Convert the following cash flow pattern to a uniform series of end-of-year costs over a seven-year period. Let $i = 12\%$ per year. (3.12)

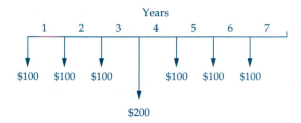

3-59. Suppose that the parents of a young child decide to make annual deposits into a savings account, with the first deposit being made on the child's fifth birthday and the last deposit being made on the fifteenth birthday. Then starting on the child's eighteenth birthday, the withdrawals shown below will be made. If the effective annual interest rate is 8% during this period of time, what are the annual deposits in years five through fifteen? (3.13)

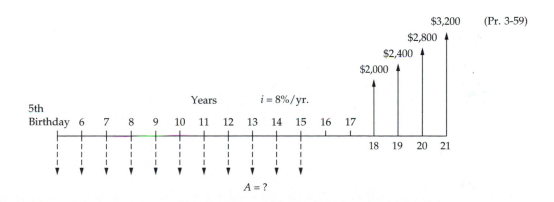

(Pr. 3-59)

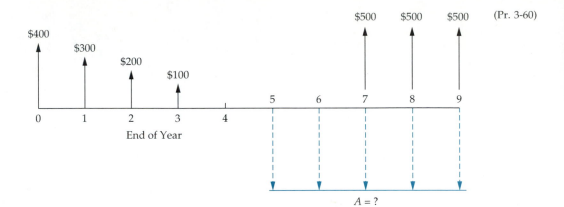

(Pr. 3-60)

3-60. Find the value of the unknown quantity in the cash flow diagram above to establish equivalence of cash inflows and outflows. Let $i = 12\%$ per year. Use a uniform gradient factor in your solution. (3.13)

3-61. The heat loss through the exterior walls of a certain poultry processing plant is estimated to cost the owner $3,000 next year. A salesman from Superfiber Insulation, Inc., has told you, the plant engineer, that he can reduce the heat loss by 80% with the installation of $15,000 worth of Superfiber now. If the cost of heat loss rises by $200 per year (gradient) after the next year and the owner plans to keep the present building for 15 more years, what would you recommend if the interest rate is 12% per year? (3.13)

3-62. Find the equivalent value of Q in the cash flow diagram in the next column. (3.13)

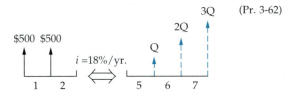

(Pr. 3-62)

3-63. What value of N comes closest to making the left-hand cash flow diagram below equivalent to the one on the right? Let $i = 15\%$ per year. (3.13)

3-64. Find the value of B on the left-hand diagram (top of page 131) that makes the two cash flow diagrams equivalent at $i = 10\%$ per year. (3.13)

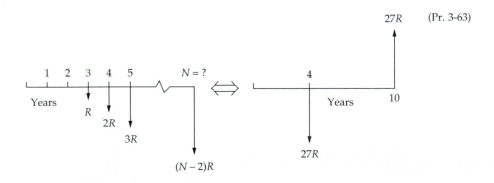

(Pr. 3-63)

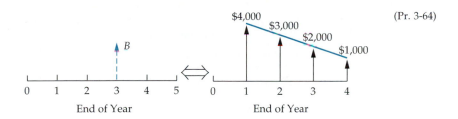

(Pr. 3-64)

3-65. You are the manager of a large crude oil refinery. As part of the refining process, a certain heat exchanger (operated at high temperatures and with abrasive material flowing through it) must be replaced every year. The replacement and downtime cost in the first year is $175,000. It is expected to increase due to inflation at a rate of 8% per year for five years, at which time this particular heat exchanger will no longer be needed. If the company's cost of capital is 18% per year, how much could you afford to spend for a higher-quality heat exchanger so that these annual replacement and downtime costs could be eliminated? (3.14)

3-66. A geometric gradient that increases at $\bar{f} = 6\%$ per year for 15 years is shown in the following diagram. The annual interest rate is 15%. What is the present equivalent value of this gradient? (3.14)

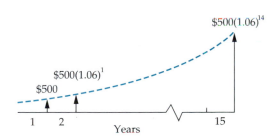

3-67. In a geometric sequence of annual cash flows starting at end of year *zero*, the value of A_0 is $1,304.35 (which *is* a cash flow). The value of the last term in the series, A_{10}, is $5,276.82. What is the equivalent value of A for years one through ten? Let $i = 20\%$ per year. (3.14)

3-68. An electronic device is available that will reduce this year's labor costs by $10,000. The

equipment is expected to last for eight years. If labor costs increase at an average rate of 7% per year and the interest rate is 12% per year, answer these questions:
a. What is the maximum amount that we could justify spending for the device?
b. What is the uniform annual equivalent value (A) of labor costs over the eight-year period?
c. What annual year zero amount (A_0) that inflates at 7% per year is equivalent to the answer in part (a)? (3.14)

3-69. Determine the present equivalent (at time zero) of the following geometric sequence of cash flows. Let $i = 15.5\%$ per year and $\bar{f} = 10\%$. (3.14)

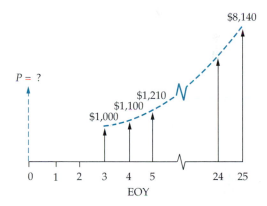

3-70. Rework problem 3-69 when the cash flow at the end of year three is $8,140, and cash flows at the end of years 4 through 25 *decrease* by 10% per year (i.e., $\bar{f} = -10\%$ per year). (3.14)

3-71. For the cash flow diagram at the top of page 132, solve for X such that the cash receipt in year zero is equivalent to the cash outflows in years one through six. (3.14)

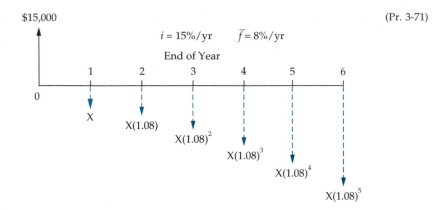

(Pr. 3-71)

3-72. An end-of-year (EOY) geometric gradient lasts for 10 years, whose initial value at EOY three is $5,000 and $\bar{f} = 6.04\%$ per year thereafter. Find the equivalent uniform gradient amount (G) over the same time period (beginning in year 1 and ending in year 12) if the initial value of the uniform gradient at EOY one is $4,000. Answer the following questions in determining the value of the gradient amount, G. The interest rate is 8% nominal, compounded semiannually. (3.13, 3.14)
   a. What is i_{CR}?
   b. What is P_0 for the geometric gradient?
   c. What is P_0 of the uniform (arithmetic) gradient?
   d. What is the value of G?

3-73. *Set up an expression* for the unknown quantity, Z, in the cash flow diagram at the top of page 133. (3.13, 3.14)

3-74. An individual makes six annual deposits of $2,000 in a savings account that pays interest at a rate of 4% compounded annually. Two years after making the last deposit, the interest rate changes to 7% compounded annually. Ten years after the last deposit the accumulated money is withdrawn from the account. How much is withdrawn? (3.15)

3-75. Compute the effective annual interest rate in each of these situations: (3.16)
   a. 10% nominal interest, compounded semiannually.
   b. 10% nominal interest compounded quarterly.
   c. 10% nominal interest compounded weekly.

3-76. Sixty monthly deposits are made into an account paying 6% nominal interest compounded monthly. If the objective of these deposits is to accumulate $100,000 by the end of the fifth year, what is the amount of each deposit? (3.17)
   a. $1,930   **b.** $1,478   **c.** $1,667
   d. $1,430   **e.** $1,695

3-77.
   a. What extra semiannual expenditure for five years would be justified for the maintenance of a machine in order to avoid an overhaul costing $3,000 at the end of five years? Assume nominal interest at 10%, compounded semiannually. (3.17)
   b. What is the annual equivalent value of $125,000 now when 12% nominal interest per year is compounded monthly? Let $N = 15$ years. (3.17)

3-78.
   a. What equal monthly payments will repay an original loan of $1,000 in six months at a nominal rate of 6% compounded monthly? What is the effective annual interest rate? (3.17)
   b. For part (a), what is the effective quarterly interest rate? (3.18)

3-79. Determine the current amount of money that must be invested at 12% nominal interest, compounded monthly, to provide an annuity of $10,000 (per year) for six years, starting ten years from now. The interest rate remains constant over this entire period of time. (3.17)

3-80. Find the present equivalent value of the following series of payments: $50 at the end of each month for 125 months at a nominal rate of 15% compounded monthly. (3.17)

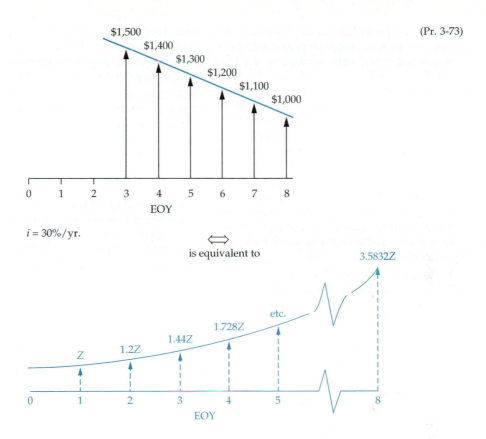

(Pr. 3-73)

3-81. Determine the present equivalent value for $1,000 paid every three months over a period of seven years in each of these situations: (3.18)

  a. The nominal interest rate is 12%, compounded annually.

  b. The nominal interest rate is 12%, compounded quarterly.

  c. The nominal interest rate is 12%, compounded weekly.

3-82. Suppose that you have just borrowed $7,500 at 10% nominal interest compounded quarterly. What is the total lump-sum, compounded amount to be paid by you at the end of a 10-year loan period? (3.17)

3-83. How many deposits of $100 each must you make at the end of each month if you desire to accumulate $3,350 for a new home entertainment center? Your savings account pays 9% nominal interest, compounded monthly. (3.17)

3-84. You have used your credit card to purchase automobile tires for $340. Unable to make payments for seven months, you then write a letter of apology and enclose a check to pay your bill in full. The credit card company's nominal interest rate is 16.5% compounded monthly. For what amount should you write the check? (3.17)

3-85. How long does it take a given amount of money to triple if the money is invested at a nominal rate of 12%, compounded monthly? (3.17)

3-86. Find the single lump-sum equivalent at end of year four of a geometric gradient series which starts at $10,000 at end of year two, with cash flows increasing by 6.3% per year for the following nine years. The nominal interest rate is 12% per year compounded monthly. (3.14, 3.18)

3-87.

  a. A certain savings and loan association advertises that it pays 8% nominal interest, compounded quarterly. What is the *effective* interest rate per annum? If you deposit $5,000 now and plan to withdraw it in three years, how much would your account be worth at that time? (3.17)

b. If instead you decide to deposit $800 every year for three years, how much could be withdrawn at the end of the third year? Suppose that, instead, you deposit $400 every six months for three years. What would the accumulated amount be? (3.18)

3-88. The effective annual interest rate, i, has been determined to be 26.82% (based on monthly compounding). Calculate how much can be spent now to avoid future computer software maintenance expenses of $1,000 *per quarter* for the next five years. (3.18)

3-89. If the nominal interest rate is 10% and compounding is semiannual, what is the present equivalent value of the receipts in the following diagram? (3.13, 3.17)

Years

3-90. What is the monthly payment on a loan of $15,000 for five years at a nominal rate of interest of 9% compounded monthly? (3.17)

a. $214   **b.** $250   **c.** $312
d. $324   **e.** $381

3-91. The effective annual interest rate, i, is stated to be 19.2%. What is the nominal interest rate per year, r, if continuous compounding is being used? (3.19)

3-92. Find the value of A that is equivalent to the uniform gradient shown in the diagram at the bottom of this page if the nominal interest rate is 12% compounded monthly. (3.13, 3.18)

3-93. Suppose that you have a money market certificate earning an annual rate of interest, which varies over time as follows:

Year k	1	2	3	4	5
i_k	14%	12%	10%	10%	12%

If you invest $5,000 in this certificate at the beginning of year one and do not add or withdraw any money for five years, what is the value of the certificate at the end of the fifth year? (3.15)

3-94. Determine the present equivalent value of the cash flow diagram at the top of page 135 when the annual interest rate, i_k, varies as indicated. (3.15)

3-95. What is the value of F_4 in the following cash flow diagram? (3.15)

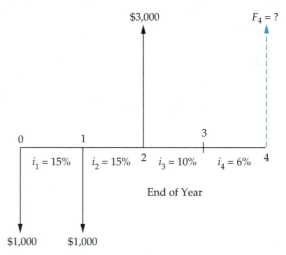

End of Year

(Pr. 3-92)

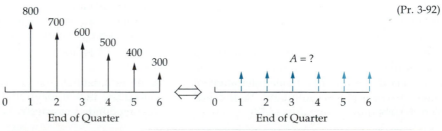

End of Quarter

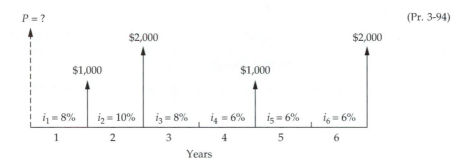

P = ? (Pr. 3-94)

3-96. Indicate whether the following statements are true (T) or False (F). (all sections)
   a. **T F** Interest is money paid for the use of equity capital.
   b. **T F** $(A/F, i\%, N) = (A/P, i\%, N) + i$
   c. **T F** Simple interest ignores the time value of money principle.
   d. **T F** Cash flow diagrams are analogous to free body diagrams for mechanics problems.
   e. **T F** $1,791 ten years from now is equivalent to $900 now if the interest rate equals 8% per year.
   f. **T F** It is always true that $i > r$ when $M \geq 2$.
   g. **T F** Suppose that a lump-sum of $1,000 is invested at $r = 15\%$ for eight years. The future equivalent is greater for daily compounding than it is for continuous compounding.
   h. **T F** For a fixed amount, F dollars, that is received at EOY N, the "A equivalent" increases as the interest rate increases.
   i. **T F** For a specified value of F at EOY N, P at time zero will be larger for $r = 10\%$ per year than it will be for $r = 10\%$ per year, compounded monthly.

3-97. If a nominal interest rate of 8% is compounded continuously, determine the unknown quantity in each of the following situations. (3.19)
   a. What uniform end-of-year amount for 10 years is equivalent to $8,000 at the end of year 10?
   b. What is the present equivalent value of $1,000 per year for 12 years?
   c. What is the future equivalent at the end of the sixth year of $243 payments made every six months during the six years? The first pay-

ment occurs six months from the present and the last occurs at the end of the sixth year.
   d. Find the equivalent lump-sum amount at the end of year nine when $P_0 = \$1,000$ and a nominal interest rate of 8% is compounded continuously.

3-98. Find the value of the unknown quantity Z in the following diagram such that the equivalent cash outflow equals the equivalent cash inflows when $r = 10\%$ compounded continuously. (3.19)

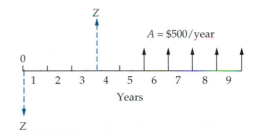

3-99. A man deposited $10,000 in a savings account when his son was born. The nominal interest rate was 5% per year, compounded continuously. On the son's eighteenth birthday, the accumulated sum is withdrawn from the account. How much will this accumulated amount be? (3.19)

3-100. Find the value of P in the cash flow diagram on the top of page 136. (3.19)

3-101. Your rich uncle has just offered to make you wealthy! For every dollar you save in an insured, continuously compounded, bank account during the next 10 years, he will give you a dollar to match it. Because your modest income permits you to save $3,000 per year for each of the next 10 years, your uncle will be willing to give you

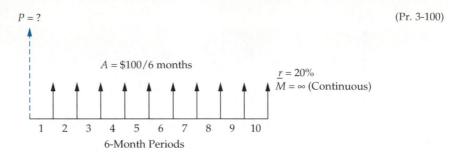

(Pr. 3-100)

$30,000 at the end of the tenth year. If you desire a *total* of $70,000 ten years from now, what annual interest rate would you have to earn on your insured bank account to make your goal possible? (3.19)

3-102. A person needs $12,000 immediately as a down payment on a new home. Suppose that she can borrow this money from her company credit union. She will be required to repay the loan in equal payments *made every six months* over the next 12 years. The annual interest rate being charged is 10% compounded continuously. What is the amount of each payment? (3.19)

3-103.

a. What is the present equivalent of a uniform series of annual payments of $3,500 each for five years if the interest rate, compounded continuously, is 10%?

b. The amount of $7,000 is invested in a certificate of deposit (CD) and will be worth $16,000 in nine years. What is the continuously compounded nominal (annual) interest rate for this CD? (3.19)

3-104.

a. Many persons prepare for retirement by making monthly contributions to a savings pro-

gram. Suppose that $200 is set aside each month and invested in a savings account that pays 12% interest per year, compounded continuously. Determine the accumulated savings in this account at the end of 30 years.

b. In part (a), suppose that an annuity will be withdrawn from savings that have been accumulated at the end of year 30. The annuity will extend from the end of year 31 to the end of year 40. What is the value of this annuity if the interest rate and compounding frequency in part (a) do not change? Refer to the diagram below. (3.19)

3-105.

a. What is the future equivalent of a continuous funds flow amounting to $10,500 per year when $r = 15\%, M = \infty$, and $N = 12$ years?

b. If the nominal interest rate is 10% per year, continuously compounded, what is the future equivalent of $10,000 per year flowing continuously for 8.5 years? See the cash flow diagram at the top of page 137.

c. Let $\overline{A} = \$7,859$ per year with $r = 15\%, M = \infty$. How many years will it take to have $1 million in this account? (3.20)

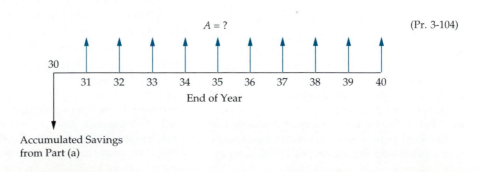

(Pr. 3-104)

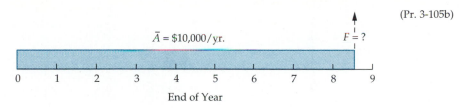

(Pr. 3-105b)

$\bar{A} = \$10,000/\text{yr.}$ $F = ?$

0 1 2 3 4 5 6 7 8 9

End of Year

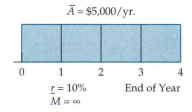

$\bar{A} = \$5,000/\text{yr.}$

0 1 2 3 4

$r = 10\%$ End of Year
$M = \infty$

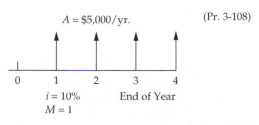

$A = \$5,000/\text{yr.}$ (Pr. 3-108)

0 1 2 3 4

$i = 10\%$ End of Year
$M = 1$

3-106. For how many years must an investment of $63,000 provide a continuous flow of funds at the rate of $16,000 per year so that a nominal interest rate of 10%, continuously compounded, will be earned? (3.20)

3-107. What is the present equivalent of the following continuous funds flow situations?
  a. $1,000,000 per year for four years at 10% compounded continuously.
  b. $6,000 per year for 10 years at 8% compounded annually.
  c. $500 per quarter for 6.75 years at 12% compounded continuously. (3.20)

3-108. What is the difference in present equivalents for the two cash flow diagrams shown in the second figure above? (3.20)

3-109. Mark each statement true (T) or false (F), and fill in the blanks in part f.
  a. **T F** The nominal interest rate will always be less than the effective interest rate when $r = 10\%$ and $M = \infty$.

  b. **T F** A certain loan involves monthly repayments of $185 over a 24-month period. If $r = 12\%$ per year, more than half of the principal is still owed on this loan after the *tenth* monthly payment is made.

  c. **T F** $1,791 ten years from now is equivalent to $900 now if the nominal interest rate is 8% compounded semiannually.

  d. **T F** If i (expressed as a decimal) is added to the series capital-recovery factor, the series sinking-fund factor will be obtained.

  e. **T F** The $(P/A, i\%, N)$ factor equals $N \cdot (P/F, i\%, 1)$.

  f. Fill in the missing interest factor:
    i. $(P/A, i\%, N)(\underline{\hspace{1.5cm}}) = (F/A, i\%, N)$
    ii. $(A/G, i\%, N)(P/A, i\%, N) = (\underline{\hspace{1.5cm}})$

Applications of Engineering Economy

The final test of any system is, does it pay?

—Frederick W. Taylor, testimony before the Special Committee of the U.S. House of Representatives (January 25, 1912)

Applications of Money–Time Relationships

*T*he two primary objectives of this chapter are (1) to illustrate several basic methods for making engineering economy studies considering the time value of money, and (2) to describe briefly the underlying assumptions and interrelationships among these methods.

The following topics are discussed in this chapter:

Determining the minimum attractive rate of return

The present worth method

The future worth method

The annual worth method

The internal rate of return method

The external rate of return method

The payback (payout) period method

Investment balance diagrams

Comprehensive industrial examples

4.1 Introduction

All engineering economy studies of capital projects should consider the return that a given project will or should produce. A basic question this book addresses is whether a proposed capital investment and its associated expenditures can be recovered by revenue (or savings) over time *in addition to* a return on the capital that is sufficiently attractive in view of the risks involved and the potential alternative uses. The interest and money–time relationships discussed in Chapter 3 emerge as essential ingredients in answering this question, and they are applied to many different types of problems in this chapter.

Because patterns of capital investment, revenue (or savings) cash flows, and disbursement cash flows can be quite different in various projects, there is no single method for performing engineering economic analyses that is ideal for all cases. Consequently, several methods are commonly used.

In this chapter we concentrate on the correct use of five methods for evaluating the economic profitability of a single proposed problem solution (i.e., *alternative*). Later, in Chapter 5, multiple alternatives are evaluated. The five methods described in Chapter 4 are present worth (PW), annual worth (AW), future worth (FW), internal rate of return (IRR), and external rate of return (ERR). The first three methods convert cash flows resulting from a proposed solution into their equivalent worth at some point (or points) in time by using an interest rate known as the *minimum attractive rate of return (MARR)*. The concept of a MARR, as well as the determination of its value, is discussed in the following section. The IRR and ERR methods produce annual rates of profit, or returns, resulting from an investment, and are then compared against the MARR.

A sixth method, the payback period, is also discussed briefly in this chapter. The payback period is a measure of the *speed* with which an investment is recovered by the cash inflows it produces. This measure, in its most common form, ignores time value of money principles. For this reason, the payback method is often used to supplement information produced by the five primary methods featured in this chapter.

Unless otherwise specified, the end-of-period cash flow convention and discrete compounding of interest are used throughout this and subsequent chapters. A planning horizon, or study (analysis) period, of N compounding periods (usually years) is used to evaluate prospective investments throughout the remainder of the book.

4.2 Determining the Minimum Attractive Rate of Return

The minimum attractive rate of return (MARR) is usually a policy issue resolved by the top management of an organization in view of numerous considerations. Among these considerations are the following:

1. The amount of money available for investment, and the source and cost of these funds (i.e., equity funds or borrowed funds).
2. The number of good projects available for investment and their purpose (i.e., whether they sustain present operations and are *essential,* or expand on present operations and are *elective*).
3. The amount of perceived risk associated with investment opportunities available to the firm and the estimated cost of administering projects over short planning horizons versus long planning horizons.

4. The type of organization involved (i.e., government, public utility, or competitive industry).

In theory the MARR, which is sometimes called the *hurdle rate,* should be chosen to maximize the economic well-being of an organization, subject to the types of considerations just listed. How an individual firm accomplishes this in practice is far from clear-cut and is frequently the subject of discussion. One popular approach to establishing a MARR involves the *opportunity cost* viewpoint described in Chapter 2, and it results from the phenomenon of *capital rationing.* For purposes of this chapter, capital rationing exists when management decides to limit the total amount of capital invested. This situation may arise when the amount of available capital is insufficient to sponsor all worthy investment opportunities.

A simple example of capital rationing is given in Figure 4-1, where the cumulative investment requirements of seven acceptable projects are plotted against the prospective annual rate of profit of each. Figure 4-1 shows a limit of $6 million on available capital. In view of this limitation the last funded project would be E, with a prospective rate of profit of 19% per year, and the best *rejected* project is F. In this case the MARR by the opportunity cost principle would be 16% per year. By *not* being able to invest in project F, the firm would presumably be forfeiting the chance to realize a 16% annual return. As the amount of investment capital and opportunities available change over time, the firm's MARR will also change.

Superimposed on Figure 4-1 is the approximate cost of obtaining the $6 million, illustrating that project E is acceptable only as long as its annual rate of profit exceeds the cost of raising the last $1 million. As shown in Figure 4-1, the cost of capital will tend to increase gradually as larger sums of money are acquired through increased borrowing (debt) and/or new issuances of common stock (equity). One last observation in connection with Figure 4-1 is that the perceived risk associated

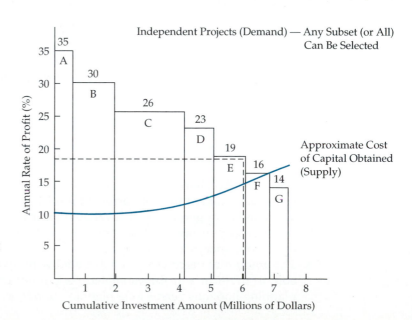

FIGURE 4-1

Determination of the MARR Based on the Opportunity Cost Viewpoint. A popular measure of annual rate of profit is "internal rate of return." (Discussed later in this chapter.)

with financing and undertaking the seven projects has been determined by top management to be acceptable.

EXAMPLE 4-1

Consider the following schedule, which shows prospective annual rates of profit for a company's portfolio of capital investment projects (this is the *demand* for capital):

Expected Annual Rate of Profit	Investment Requirements (Thousands of Dollars)	Cumulative Investment
40% and over	$ 2,200	$ 2,200
30–39.9%	3,400	5,600
20–29.9%	6,800	12,400
10–19.9%	14,200	26,600
Below 10%	22,800	49,400

If the supply of capital obtained from internal and external sources has a cost of 15% per year for the first $5,000,000 invested and then increases 1% for every $5,000,000 thereafter, what is this company's MARR when using an opportunity cost viewpoint?

SOLUTION

Cumulative capital demand versus supply can be plotted against prospective annual rate of profit, as shown in Figure 4-2. The point of intersection is approximately 18% per year, which represents a realistic estimate of this company's MARR when using the opportunity cost viewpoint.

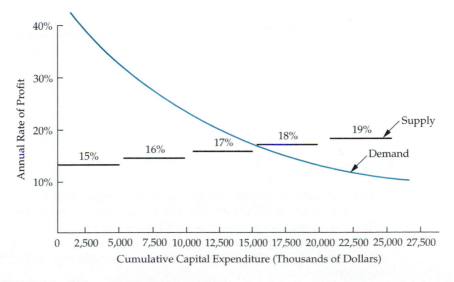

FIGURE 4-2 Solution Graph for Example 4-1

▼
4.3 The Present Worth Method

The present worth (PW) method is based on the concept of equivalent worth of all cash flows relative to some base or beginning point in time called the present. That is, all cash inflows and outflows are discounted to the present point in time at an interest rate that is generally the MARR.

The PW of an investment alternative is a measure of how much money an individual or a firm could afford to pay for the investment in excess of its cost. Or, stated differently, a positive PW for an investment project is a dollar amount of profit over the minimum amount required by investors. It is assumed that cash generated by the alternative is available for other uses that earn interest at a rate equal to the MARR.

To find the PW as a function of $i\%$ (per interest period) of a series of cash inflows and outflows, it is necessary to discount future amounts to the present by using the interest rate over the appropriate study period (years, for example) in the following manner:

$$PW(i\%) = F_0(1 + i)^0 + F_1(1 + i)^{-1} + F_2(1 + i)^{-2} + \cdots$$
$$+ F_k(1 + i)^{-k} + \cdots + F_N(1 + i)^{-N}$$
$$= \sum_{k=0}^{N} F_k(1 + i)^{-k} \tag{4-1}$$

where i = effective interest rate, or MARR, per compounding period
   k = index for each compounding period ($0 \le k \le N$)
   F_k = future cash flow at the end of period k
   N = number of compounding periods in the planning horizon (i.e., study period)

The relationship given in Equation 4-1 is based on the assumption of a *constant interest rate* throughout the life of a particular project. If the interest rate is assumed to change, the PW must be computed in two or more steps, as was illustrated in Chapter 3.

The higher the interest rate and the further into the future a cash flow occurs, the lower its PW is. This is shown graphically in Figure 4-3. As long as the PW (i.e., present equivalent of cash inflows minus cash outflows) is greater than or equal to zero, the project is economically justified; otherwise, it is not acceptable.

EXAMPLE 4-2

An investment of $10,000 can be made in a project that will produce a uniform annual revenue of $5,310 for five years and then have a market (salvage) value of $2,000. Annual expenses will be $3,000 each year. The company is willing to accept any project that will earn 10% per year or more, before income taxes, on all invested capital. Show whether this is a desirable investment by using the PW method.

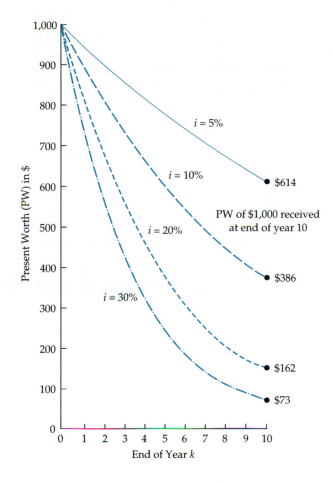

FIGURE 4-3

PW of $1,000 Received at the End of Year k at an Interest Rate of i% per year

SOLUTION

	PW	
	Cash Outflows	**Cash Inflows**
Annual revenue: $5,310(P/A, 10\%, 5)$		$20,125
Market (salvage) value: $2,000(P/F, 10\%, 5)$		1,245
Investment	−$10,000	
Annual expenses: $3,000(P/A, 10\%, 5)$	− 11,370	
Total	−$21,370	$21,370
Total PW		$ 0

Because total PW(10%) = $0, the project is shown to be marginally acceptable.

EXAMPLE 4-3

A piece of new equipment has been proposed by engineers to increase the productivity of a certain manual welding operation. The investment cost is $25,000,

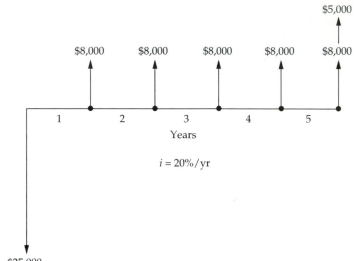

FIGURE 4-4
Cash Flow Diagram for Example 4-3 $25,000

and the equipment will have a market value of $5,000 at the end of a study period of five years. Increased productivity attributable to the equipment will amount to $8,000 per year after extra operating costs have been subtracted from the revenue generated by the additional production. A cash flow diagram for this investment opportunity is given in Figure 4-4. If the firm's MARR (before income taxes) is 20% per year, is this proposal a sound one? Use the PW method.

SOLUTION

$$PW = PW \text{ of cash inflows} - PW \text{ of cash outflows}$$

or

$$PW(20\%) = \$8,000(P/A, 20\%, 5) + \$5,000(P/F, 20\%, 5) - \$25,000$$
$$= \$934.29$$

Because PW(20%) > 0, this equipment is economically justified.

Based on Example 4-3, Table 4-1 can be used to plot the cumulative PW of cash flows through year k. The graphs of cumulative PW shown in Figure 4-5 at $i = 20\%$ and $i = 0\%$ are plotted from columns (C) and (D) of Table 4-1, respectively.

The MARR in Example 4-3 (and in other examples throughout this chapter) is to be interpreted as an effective interest rate (i). Here $i = 20\%$ per year. Cash flows are discrete, end-of-year amounts. If *continuous compounding* had been specified for a nominal interest rate (r) of 20% per year, the PW would have been calculated by using the interest factors presented in Appendix D:

TABLE 4-1 Cumulative PW Calculations for Example 4-3

End of Year k	(A) Net Cash Flow	(B) Present Worth of Cash Flow at $i = 20\%/\text{yr.}$	(C) Cumulative PW at $i = 20\%/\text{yr.}$ through Year k	(D) Cumulative PW at $i = 0\%/\text{yr.}$ through Year k
0	−$25,000	−$25,000	−$25,000	−$25,000
1	8,000	6,667	− 18,333	− 17,000
2	8,000	5,556	− 12,777	− 9,000
3	8,000	4,630	− 8,147	− 1,000
4	8,000	3,858	− 4,289	7,000
5	13,000	5,223	+ 934	20,000

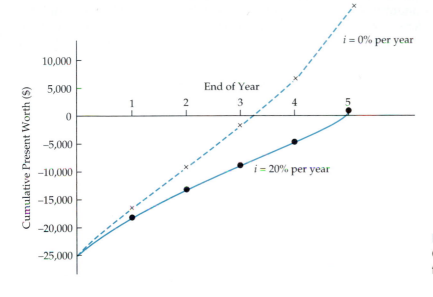

FIGURE 4-5
Graph of Cumulative PW for Example 4-3

$$\text{PW }(\underline{r} = 20\%) = -\$25{,}000 + \$8{,}000 \, (P/A, \underline{r} = 20\%, 5)$$
$$+ \$5{,}000 \, (P/F, \underline{r} = 20\%, 5)$$
$$= -\$25{,}000 + \$8{,}000 \, (2.8551) + \$5{,}000 \, (0.3679)$$
$$= -\$319.60$$

Consequently, with continuous compounding, the equipment would not be economically justifiable. The reason is that the higher effective annual interest rate ($e^{0.20} - 1 = 0.2214$) reduces the PW of future positive cash flows but does not affect the PW of the capital invested at the beginning of Year 1.

4.3.1 Bond Value

A bond provides an excellent example of commercial value being the PW of the future net cash flows that are expected to be received through ownership of an interest-bearing certificate. Thus, the value of a bond, at any time, is the PW of future cash receipts. For a bond, let

Z = face, or par, value
C = redemption or disposal price (usually equal to Z)
r = bond rate (nominal interest) per interest period
N = number of periods before redemption
i = bond *yield* rate per period
V_N = value (price) of the bond N interest periods prior to redemption— this is a PW measure of merit.

The owner of a bond is paid two types of payments by the borrower. The first consists of the series of periodic interest payments he or she will receive until the bond is retired. There will be N such payments, each amounting to rZ. These constitute an annuity of N payments. In addition, when the bond is retired or sold, the bondholder will receive a single payment equal in amount to C. The PW of the bond is the sum of present worths of these two types of payments at the bond's yield rate ($i\%$):

$$V_N = C(P/F, i\%, N) + rZ(P/A, i\%, N) \tag{4-2}$$

EXAMPLE 4-4

Find the current price (PW) of a 10-year bond paying 6% per year (payable semi-annually) that is redeemable at par value, if bought by a purchaser to *yield 10% per year*. The face value of the bond is $1,000.

$$N = 10 \times 2 = 20 \text{ periods}$$
$$r = 6\%/2 = 3\% \text{ per period}$$
$$i = [(1.10)^{1/2} - 1]100 \approx 4.9\% \text{ per semi-annual period}$$
$$C = Z = \$1,000$$

SOLUTION
Using Equation 4-2, we obtain

$$V_N = \$1,000(P/F, 4.9\%, 20) + \$1,000(0.03)(P/A, 4.9\%, 20)$$
$$= \$384.10 + \$377.06 = \$761.16$$

EXAMPLE 4-5

A bond with a face value of $5,000 pays interest of 8% per year. This bond will be redeemed at par value at the end of its 20-year life, and the first interest payment is due one year from now.

(a) How much should be paid now for this bond in order to receive a yield of 10% per year on the investment?

(b) If this bond is purchased now for $4,600, what annual yield would the buyer receive?

SOLUTION

(a) By using Equation 4-2, the value of V_N can be determined:

$$V_N = \$5,000(P/F, 10\%, 20) + \$5,000(0.08)(P/A, 10\%, 20)$$
$$= \$743.00 + \$3,405.44 = \$4,148.44$$

(b) Here we are given $V_N = \$4,600$ and we must find the value of $i\%$ in Equation 4-2:

$$\$4,600 = \$5,000(P/F, i'\%, 20) + \$5,000(0.08)(P/A, i'\%, 20)$$

To solve for $i'\%$, we can resort to an iterative trial-and-error procedure (e.g., try 8.5%, 9.0%, etc.), to determine that $i'\% = 8.9\%$ per year.

EXAMPLE 4-6

A certain U.S. Treasury bond that matures in eight years has a face value of $10,000. This means that the bondholder will receive $10,000 cash when the bond's maturity date is reached. The bond stipulates a fixed nominal interest rate of 8% per year, but interest payments are made to the bondholder every three months, therefore each payment amounts to 2% of the face value.

A prospective buyer of this bond would like to earn 10% nominal interest per year on his or her investment because interest rates in the economy have risen since the bond was issued. How much should this buyer be willing to pay for the bond?

SOLUTION

To establish the value of this bond in view of stated conditions, the PW of future cash flows during the next eight years (the study period) must be evaluated. Interest payments are quarterly. Because the prospective buyer desires to obtain 10% *nominal interest per year* on the investment, the PW is computed at $i = 10\%/4 = 2.5\%$ per quarter for the remaining $8(4) = 32$ quarters of the bond's life:

$$V_N = \$10,000(0.02)\,(P/A, 2.5\%, 32) + \$10,000(P/F, 2.5\%, 32)$$
$$= \$4,369.84 + \$4,537.71$$
$$= \$8,907.55$$

Thus, the buyer should pay no more than $8,907.55 when 10% nominal interest per year is desired.

4.4 The Future Worth Method

Because a primary objective of all time value of money methods is to maximize the future wealth of the owners of a firm, the economic information provided by the future worth (FW) method is very useful in capital investment decision situations. The future worth is based on the equivalent worth of all cash inflows and outflows at the end of the planning horizon (study period) at an interest rate that is generally the MARR. Also, the FW of a project is equivalent to its PW; that is, $FW = PW(F/P, i\%, N)$. If $FW \geq 0$ for a project, it would be economically justified.

Equation 4-3 summarizes the general calculations necessary to determine a project's future worth:

$$FW(i\%) = F_0(1+i)^N + F_1(1+i)^{N-1} + \cdots + F_N(1+i)^0$$

$$= \sum_{k=0}^{N} F_k(1+i)^{N-k} \tag{4-3}$$

EXAMPLE 4-7

Evaluate the FW of the potential improvement project described in Example 4-3. Show the relationship between FW and PW for this example.

SOLUTION

$$\begin{aligned}
FW\,(20\%) =\ & -\$25{,}000(F/P, 20\%, 5) \\
& + \$8{,}000(F/P, 20\%, 4) + \$8{,}000(F/P, 20\%, 3) \\
& + \$8{,}000(F/P, 20\%, 2) + \$8{,}000(F/P, 20\%, 1) + \$13{,}000 \\
=\ & \$2{,}324.80
\end{aligned}$$

Again, the project is shown to be a good investment. The FW is a multiple of the equivalent PW value:

$$PW\,(20\%) = \$2{,}324.80(P/F, 20\%, 5) = \$934.29$$

To this point the PW and FW methods have used a known and constant MARR over the study period. Each method produces a measure of merit expressed in dollars and is equivalent to the other. The difference in economic information provided is relative to the point in time used (i.e., the present for the PW versus the future, or end of the study period, for the FW).

4.5 The Annual Worth Method

The annual worth (AW) of a project is an equal annual series of dollar amounts, for a stated study period, that is *equivalent* to the cash inflows and outflows at an interest rate that is generally the MARR. Hence, the AW of a project is annual equivalent revenues or savings ($\underline{R}$) minus annual equivalent expenses ($\underline{E}$), less its annual equivalent capital recovery (CR) amount, which is defined in Equation 4-5. An annual equivalent value of $\underline{R}$, $\underline{E}$, and CR is computed for the study period, N, which is usually in years. In equation form the AW, which is a function of $i\%$, is

$$AW(i\%) = \underline{R} - \underline{E} - CR(i\%) \tag{4-4}$$

Also, we need to notice that the AW of a project is equivalent to its PW and FW. That is, AW = PW $(A/P, i\%, N)$, and AW = FW $(A/F, i\%, N)$. Hence, it can be easily computed for a project from these other equivalent values.

As long as the AW is greater than or equal to zero, the project is economically attractive; otherwise, it is not. An AW of zero means that an annual return exactly equal to the MARR has been earned.

TABLE 4-2   Calculation of Equivalent Annual CR Amount

Year	Value of Investment at Beginning of Year[a]	Uniform Loss in Value	Interest on Beginning-of-Year Investment at $i = 10\%$	CR Amount Each Year	PW of CR Amount at $i = 10\%$
1	$10,000	$1,600	$1,000	$2,600	$2,600(P/F, 10\%, 1) = $2,364
2	8,400	1,600	840	2,440	$2,440(P/F, 10\%, 2) = $2,016
3	6,800	1,600	680	2,280	$2,280(P/F, 10\%, 3) = $1,713
4	5,200	1,600	520	2,120	$2,120(P/F, 10\%, 4) = $1,448
5	3,600	1,600	360	1,960	$1,960(P/F, 10\%, 5) = \underline{$1,217}
					$8,758

$$CR = \$8,758(A/P, 10\%, 5) = \$2,310$$

[a]This is also referred to later as the *beginning-of-year unrecovered investment*.

The capital recovery (CR) amount for a project is the equivalent uniform annual *cost* of the capital invested. It is an annual amount that covers the following two items:

1. Loss in value of the asset
2. Interest on invested capital (i.e., at the MARR)

As an example, consider a machine or other asset that will cost $10,000, last five years, and have a salvage (market) value of $2,000. Further, the MARR is 10% per year.

It can be shown that no matter which method of calculating an asset's loss in value over time is used, the equivalent annual CR amount is the same. For example, if uniform loss in value is assumed, the equivalent annual CR amount is calculated to be $2,310, as shown in Table 4-2.

There are several convenient formulas by which CR amount (cost) may be calculated to obtain the result in Table 4-2. Probably the easiest formula to understand involves finding the annual equivalent of the initial capital investment and then subtracting the annual equivalent of the salvage value. Thus

$$CR(i\%) = I(A/P, i\%, N) - S(A/F, i\%, N) \tag{4-5}$$

where I = initial investment for the project*
S = salvage (market) value at the end of the study period
N = project study period

When Equation 4-5 is applied to the example in Table 4-2, the CR amount is

$$CR(10\%) = \$10,000(A/P, 10\%, 5) - \$2,000(A/F, 10\%, 5)$$
$$= \$10,000(0.2638) - 2,000(0.1638) = \$2,310$$

Another way to calculate the CR amount is to add an annual sinking fund amount (or deposit) to the interest on the original investment. Thus

---

*In some cases the investment will be spread over several periods. In such situations, I is the PW of all investment amounts.

$$CR\,(i\%) = (I - S)\,(A/F, i\%, N) + I\,(i\%) \tag{4-6}$$

When Equation 4-6 is applied to the example in Table 4-2, the CR amount is

$$CR\,(10\%) = (\$10{,}000 - \$2{,}000)\,(A/F, 10\%, 5) + \$10{,}000\,(10\%)$$
$$= \$8{,}000(0.1638) + \$10{,}000(0.10) = \$2{,}310$$

Yet another way to calculate the CR amount is to add the equivalent annual cost of the uniform loss in value of the investment to the interest on the salvage value.

$$CR\,(i\%) = (I - S)\,(A/P, i\%, N) + S\,(i\%) \tag{4-7}$$

Applied to the example used previously,

$$CR\,(10\%) = (\$10{,}000 - \$2{,}000)\,(A/P, 10\%, 5) + \$2{,}000\,(10\%)$$
$$= \$8{,}000(0.2638) + \$2{,}000(0.10) = \$2{,}310$$

EXAMPLE 4-8

By using the AW method and Equation 4-4, determine whether the equipment described in Example 4-3 should be recommended.

SOLUTION

The AW method applied to Example 4-3 yields the following:

$$\text{AW}(20\%) = \overbrace{\$8{,}000}^{R\,-\,E} - \overbrace{[\$25{,}000\,(A/P, 20\%, 5) - \$5{,}000\,(A/F, 20\%, 5)]}^{\text{CR amount (Eq. 4-5)}}$$
$$= \$8{,}000 - (\$8{,}359.50 - \$671.90)$$
$$= \$312.40$$

Because its AW (20%) is positive, the equipment more than pays for itself over a period of five years while earning a 20% return per year on the unrecovered investment. In fact, the annual equivalent "surplus" is $312.40, which means that the equipment provided more than a 20% return on beginning-of-year unrecovered investment. This piece of equipment should be recommended as an attractive investment opportunity. Also, we can confirm that the AW (20%) in Example 4-8 is equivalent to PW (20%) = $934.29 in Example 4-3 and FW (20%) = $2,324.80 in Example 4-7. That is, AW = $934.29 $(A/P, 20\%, 5)$ = $312.40 and also AW = $2,324.80 $(A/F, 20\%, 5)$ = $312.40.

EXAMPLE 4-9

An investment company is considering building a 25-unit apartment complex in a growing town. Because of the long-term growth potential of the town, it is felt that the company could average 90% of full occupancy for the complex each year. If the following items are reasonably accurate estimates, what is the minimum monthly rent that should be charged if a 12% MARR (per year) is desired? Use the AW method.

Land investment cost	$50,000
Building investment cost	$225,000
Study period, N	20 years
Rent per unit per month	?
Upkeep expense per unit per month	$35
Property taxes and insurance per year	10% of *total* initial investment

SOLUTION

The procedure for solving this problem is first to determine the equivalent AW of all costs at the MARR of 12% per year. To earn exactly 12% on this project, the annual rental income, adjusted for 90% occupancy, must equal the AW of costs:

$$\text{Initial investment cost} = \$50,000 + \$225,000 = \$275,000$$

$$\text{Taxes and insurance/year} = 0.1(\$275,000) = \$27,500$$

$$\text{Upkeep/year} = \$35(12 \times 25)(0.9) = \$9,450$$

$$\text{CR cost/year (Equation 4-5)} = \$275,000(A/P, 12\%, 20) - \$50,000(A/F, 12\%, 20)$$
$$= \$36,123$$

(We assume that investment in land is recovered at the end of year 20 and that annual upkeep is directly proportional to the occupancy rate.)

$$\text{Equivalent AW (of costs)} = -\$27,500 - \$9,450 - \$36,123 = -\$73,073$$

Therefore, the minimum *annual* rental required equals $73,073 and with annual compounding ($M = 1$) the monthly rental amount, $\hat{R}$, is

$$\hat{R} = \frac{\$73,073}{(12 \times 25)(0.9)} = \$270.64$$

Many decision makers prefer the AW method because it is relatively easy to interpret when one is accustomed to working with annual income statements and cash flow summaries.

4.6 The Internal Rate of Return Method

The internal rate of return (IRR) method is the most widely used rate of return method for performing engineering economic analyses. It is sometimes called by several other names, such as *investor's method, discounted cash flow method,* and *profitability index.*

This method solves for the interest rate that equates the equivalent worth of an alternative's cash inflows (receipts or savings) to the equivalent worth of cash outflows (expenditures, including investment costs). Equivalent worth may be computed with any of the three methods discussed earlier. The resultant interest rate is termed the *internal rate of return (IRR).*

For a single alternative, the IRR is not positive unless (1) both receipts and expenses are present in the cash flow pattern and (2) the sum of receipts exceeds the sum of all cash outflows. Be sure to check both of these conditions in order to avoid the unnecessary work involved with finding that the IRR is *negative.*

(Visual inspection of the total net cash flow will determine whether the IRR is zero or less.)

By using a PW formulation, the IRR is the $i'\%$* at which

$$\sum_{k=0}^{N} R_k(P/F, i'\%, k) = \sum_{k=0}^{N} E_k(P/F, i'\%, k) \tag{4-8}$$

where R_k = net revenues or savings for the kth year
         E_k = net expenditures including any investment costs for the kth year
         N = project life (or study period)

Once i' has been calculated, it is compared with the MARR to assess whether the alternative in question is acceptable. *If $i' \geq$ MARR, the alternative is acceptable; otherwise, it is not.*

A popular variation of Equation 4-8 for computing the IRR for an alternative is to determine the i' at which its *net* PW is zero. In equation form, the IRR is the value of i' at which

$$PW = \sum_{k=0}^{N} R_k(P/F, i'\%, k) - \sum_{k=0}^{N} E_k(P/F, i'\%, k) = 0 \tag{4-9}$$

For an alternative with a single investment cost at the present time ($k = 0$) followed by a series of positive cash inflows over N, a graph of PW versus the interest rate typically has the general convex form shown in Figure 4-6. The point at which PW = 0 in Figure 4-6 defines $i'\%$, which is the project's IRR.

The value of $i'\%$ can also be determined as the interest rate at which FW = 0 or AW = 0. For example, by setting net FW equal to zero, Equation 4-10 would result:

$$FW = \sum_{k=0}^{N} R_k(F/P, i'\%, N-k) - \sum_{k=0}^{N} E_k(F/P, i'\%, N-k) = 0 \tag{4-10}$$

Another way to interpret the IRR is through an *investment balance diagram.* Figure 4-7 shows how much of the original investment in an alternative is still to be recovered as a function of time. The downward arrows in Figure 4-7 represent annual returns, ($R_k - E_k$) for $1 \leq k \leq N$, against the unrecovered investment and the dashed lines indicate the opportunity cost of interest, or profit, on the beginning-of-year investment balance. *The IRR is the value of i' in Figure 4-7 that causes the unrecovered investment balance to exactly equal 0 at the end of the study period (year N)* and thus represents the *internal* earning rate of a project. It is important to notice that $i'\%$ is calculated on the *beginning-of-year unrecovered* investment through the life of a project rather than on the total initial investment. Additional examples of investment balance diagrams are given in Section 4.9.

The method of solving Equations 4-8 through 4-10 normally involves trial-and-error calculations until the $i'\%$ is converged upon or can be interpolated. Example 4-10 presents a typical solution using the common convention of "+" signs for cash inflows and "−" signs for cash outflows.

---

* i' is often used in place of i to mean the interest rate that is to be determined.

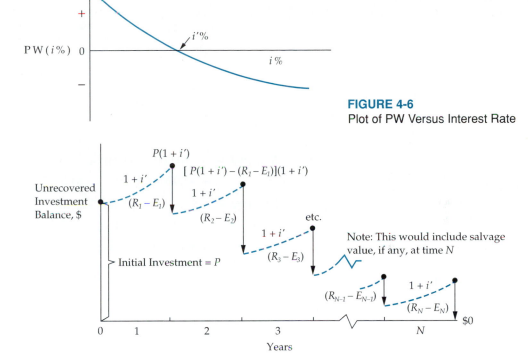

FIGURE 4-6
Plot of PW Versus Interest Rate

FIGURE 4-7   Investment Balance Diagram Showing IRR

EXAMPLE 4-10 (RESTATEMENT OF EXAMPLE 4-2)

A capital investment of $10,000 can be made in a project that will produce a uniform annual revenue of $5,310 for five years and then have a salvage value of $2,000. Annual expenses will be $3,000. The company is willing to accept any project that will earn at least 10% per year, before income taxes, on all invested capital. Determine whether it is acceptable by using the IRR method.

SOLUTION

In this example we immediately see that the sum of positive cash flows ($13,550) exceeds the sum of negative cash flows ($10,000). Thus, it is likely that a positive-valued $i'\%$ can be determined. By writing an equation for the PW of the project's total net cash flow and setting it equal to zero, we can compute the IRR.

$$PW = 0 = -\$10,000 + (\$5,310 - \$3,000)(P/A, i'\%, 5)$$
$$+ \$2,000(P/F, i'\%, 5); \quad i'\% = ?$$

If we did not already know the answer from Example 4-2 ($i' = 10\%$), we would probably try a relatively low i', such as 5%, and a relatively high i', such as 15%. Linear interpolation will be used to solve for i', and the procedure illustrated in Figure 4-8 should not exceed a range of 10%.

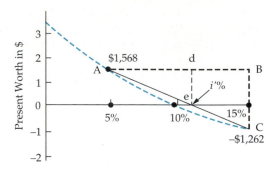

FIGURE 4-8
Use of Linear Interpolation to Find the
Approximation of IRR for Example 4-10

$$\text{At } i' = 5\%: PW = -\$10{,}000 + \$2{,}310(4.3295)$$
$$+ \$2{,}000(0.7835) = +\$1{,}568$$

$$\text{At } i' = 15\%: PW = -\$10{,}000 + \$2{,}310(3.3522)$$
$$+ \$2{,}000(0.4972) = -\$1{,}262$$

Because we have both a positive and a negative PW, the answer is bracketed. The dashed curve in Figure 4-8 is what we are linearly approximating. The answer, $i'\%$, can be determined by using the similar triangles dashed in Figure 4-8.

$$\frac{\text{line AB}}{\text{line BC}} = \frac{\text{line Ad}}{\text{line de}}$$

$$\frac{15\% - 5\%}{\$1{,}568 - (-\$1{,}262)} = \frac{i'\% - 5\%}{\$1{,}568 - \$0}$$

or

$$i'\% = 5\% + \frac{\$1{,}568}{\$1{,}568 - (-\$1{,}262)}(15\% - 5\%)$$
$$= 5\% + 5.5\% = 10.5\%$$

This approximate solution illustrates the trial-and-error process, together with linear interpolation. The error in this answer is due to nonlinearity of the PW function and would be less if the range of interest rates used in the interpolation had been smaller.

From the result of Example 4-2, we already know that the project is minimally acceptable and that $i' = $ MARR $= 10\%$ per year. We can confirm this by substituting $i = 10\%$ in the PW equation as follows:

$$PW\,(10\%) = -\$10{,}000 + (\$5{,}310 - \$3{,}000)\,(P/A, 10\%, 5)$$
$$+ \$2{,}000\,(P/F, 10\%, 5) = 0$$

EXAMPLE 4-11 (RESTATEMENT OF EXAMPLE 4-3)

A piece of new equipment has been proposed by engineers to increase the productivity of a certain manual welding operation. The investment cost is $25,000, and the equipment will have a salvage value of $5,000 at the end of its expected life of five years. Increased productivity attributable to the equipment will amount to $8,000 per year after extra operating costs have been subtracted from the value of the additional production. A cash flow diagram for this equipment was given in Figure 4-4. Evaluate the IRR of the proposed equipment. Is the investment a good one? Recall that the MARR is 20%.

SOLUTION

By using Equation 4-9, the following expression is obtained:

$$\$8,000\ (P/A, i'\%, 5) + \$5,000\ (P/F, i'\%, 5) - \$25,000 = 0; \quad i' = ?$$

To solve the above equation by trial and error, use Table 4-3. The PW computations in Table 4-3 are illustrated in Figure 4-9.

By inspection, the value of $i'\%$ where PW = 0 is about 22%. For most applications an $i'\%$ value of 22% is accurate enough because our major concern is whether $i'\%$ equals or exceeds the MARR. A more precise value of i' can be determined by directly solving the above equation with repeated trial-and-error calculations ($i' = 21.577\%$). Clearly, this equipment is economically attractive because 21.577% > 20%.

A final point needs to be made for Example 4-11. The investment balance diagram is provided in Figure 4-10, and the reader should notice that $i' = 21.577\%$

TABLE 4-3   Computation of Selected PW (i) in Example 4-11

i'	PW($i'\%$)
0.00	$8,000(5) + $5,000(1) - $25,000 = $20,000
0.10	8,000(3.7908) + 5,000(0.6209) - 25,000 = 8,430.90
0.20	8,000(2.9906) + 5,000(0.4019) - 25,000 = 934.30
0.25	8,000(2.6893) + 5,000(0.3277) - 25,000 = -1,847.10
0.30	8,000(2.436) + 5,000(0.2693) - 25,000 = -4,165.50

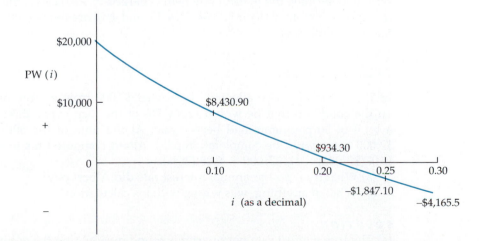

FIGURE 4-9   PW Plotted Against i for Example 4-11

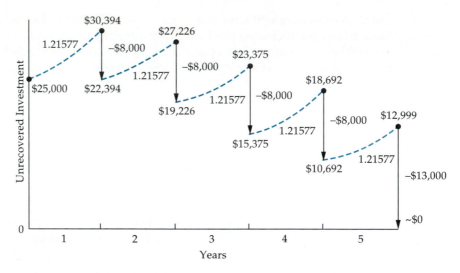

FIGURE 4-10   Investment Balance Diagram for Example 4-11

is a rate of return calculated on the beginning-of-year unrecovered investment. The IRR is *not* an average return each year based on the total investment of $25,000.

A rather common application of the IRR method is in so-called *installment financing* types of problems. These problems are associated with financing arrangements for purchasing merchandise "on time." An interest, or finance, charge is typically based on the total amount of money borrowed and is often paid by the borrower at the beginning of the loan instead of on the unpaid loan balance, as illustrated by Figure 4-10. This finance charge typically involves *early repayment of interest* and leads to an actual interest rate that often greatly exceeds the stated interest rate. To determine the interest rate being charged in such cases, the IRR method is frequently employed. Examples 4-12, 4-13, and 4-14 are representative installment financing problems.

EXAMPLE 4-12

In 1915 Albert Epstein allegedly borrowed $7,000 from a large New York bank on the condition that he would repay 7% of the loan every three months, until a total of 50 payments had been made. At the time of the 50th payment, the $7,000 loan would be completely repaid. Albert computed his interest rate to be $[0.07(\$7,000) \times 4]/\$7,000 = 0.28$ (28%).

(a) What true *effective* annual interest rate did Albert pay?
(b) What, if anything, was wrong with his calculation?

SOLUTION
(a) The true interest rate per quarter is found by equating the equivalent value of the amount borrowed to the equivalent value of the amounts repaid. Equating the

AW amounts per quarter, we find:

$$\$7,000 \, (A/P, i'\%/\text{qtr.}, 50 \text{ qtrs.}) = 0.07 \, (\$7,000)\text{per quarter}$$
$$(A/P, i'\%, 50) = 0.07$$

Linearly interpolating to find $i'\%$/qtr. by using similar triangles is the next step:

$$(A/P, 6\%, 50) = 0.0634$$
$$(A/P, 7\%, 50) = 0.0725$$

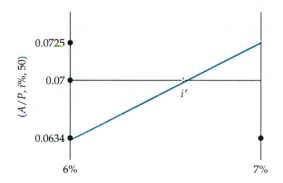

$$\frac{7\% - i'\%}{7\% - 6\%} = \frac{0.0725 - 0.07}{0.0725 - 0.0634}$$

$$i'\% = 7\% - 1\% \left(\frac{0.0025}{0.0091} \right)$$

$$\text{or } i'\% \simeq 6.73\% \text{ per quarter}$$

Now we can compute the effective $i'\%$ per year that Albert was paying:

$$i'\% = [(1.0673)^4 - 1]100$$

$$\simeq 30\% \text{ per year}$$

(b) Even though Albert's answer of 28% is close to the true value of 30%, his calculation is insensitive to how long his payments were made. For instance, he would get 28% for an answer when 20, 50, or 70 quarterly payments of $490 were made! For 20 quarterly payments the true effective interest rate is 14.5% per year, and for 70 quarterly payments it is 31% per year. As more payments are made, the true annual effective interest rate being charged by the bank will increase, but Albert's method would not reveal by how much.

EXAMPLE 4-13

The Fly-by-Night finance company advertises a "bargain 6% plan" for financing the purchase of automobiles. To the amount of the loan being financed, 6% is added for each year money is owed. This total is then divided by the number of months over which the payments are to be made, and the result is the amount of the monthly payments. For example, a woman purchases a $10,000 automobile under this plan and makes an initial cash payment of $2,500. She wishes to pay the balance in 24 monthly payments.

Purchase price	=	$10,000
− Initial payment	=	2,500
= Balance due, (P_0)	=	7,500
+ 6% finance charge $= 0.06 \times 2$ years $\times \$7,500$	=	900
= Total to be paid	=	8,400
∴ Monthly payments = $8,400/24	= $	350

What effective annual rate of interest does she actually pay?

SOLUTION

Because there are to be 24 payments of $350 each, made at the end of each month, these constitute an annuity (A) at some unknown rate of interest, $i'\%$, that should be computed only upon the unpaid balance. In this example, the amount owed on the automobile (i.e., the unpaid balance) is $7,500. Letting $P_0 = \$7,500$, the following equivalence expression is utilized to compute the unknown monthly interest rate:

$$P_0 = A(P/A, i'\%, N)$$
$$\$7,500 = \$350(P/A, i'\%, 24)$$
$$(P/A, i'\%, 24) = \frac{\$7,500}{\$350} = 21.43$$

Examining the interest tables for P/A factors at $N = 24$ that come closest to 21.43, one finds that $(P/A, 3/4\%, 24) = 21.8891$ and $(P/A, 1\%, 24) = 21.2434$.
  Linear interpolation gives

$$i'\% = 1\% - \frac{21.43 - 21.2434}{21.8891 - 21.2434}(1\% - 3/4\%) = 0.93\%$$

Since payments are monthly, 0.93% is the interest rate being charged per month. The nominal rate paid on the borrowed money is 0.93%(12) = 11.16% compounded monthly. This corresponds to an effective annual interest rate of $[(1 + 0.0093)^{12} - 1] \times 100 \cong 12\%$. What appeared at first to be a real bargain turns out to involve effective annual interest at twice the stated rate.

EXAMPLE 4-14

A small company needs to borrow $160,000. The local (and only) banker makes this statement: "We can loan you $160,000 at a very favorable rate of 12% per year for a five-year loan. However, to secure this loan you must agree to establish a checking account (with no interest) in which the *minimum* average balance is $32,000. In addition, your interest payments are due at the end of each year and the principal will be repaid in a lump-sum amount at the end of year five." What is the true effective annual interest rate being charged?

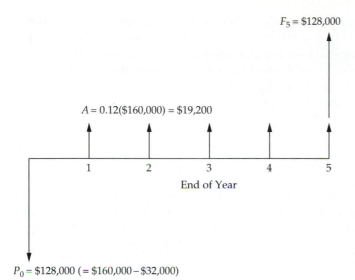

$F_5 = \$128,000$

$A = 0.12(\$160,000) = \$19,200$

1 2 3 4 5

End of Year

$P_0 = \$128,000\ (= \$160,000 - \$32,000)$

FIGURE 4-11
Cash Flow Diagram for Example 4-14

SOLUTION

The cash flow diagram from the banker's viewpoint appears in Figure 4-11. When solving for an unknown interest rate, it is good practice to draw a *cash flow diagram* prior to writing an equivalence relationship. The interest rate (IRR) that establishes equivalence of positive and negative cash flows can now easily be computed:

$$P_0 = F_5(P/F, i'\%, 5) + A(P/A, i'\%, 5)$$
$$\$128,000 = \$128,000(P/F, i'\%, 5) + \$19,200(P/A, i'\%, 5)$$

If we try $i' = 15\%$, we discover that $\$128,000 = \$128,000$. Therefore, the true effective interest rate is 15% per year.

4.6.1 Difficulties Associated with the IRR Method

The PW, AW, and FW methods assume that net receipts less expenses (positive recovered funds) each time period are reinvested at the MARR during the study period, N. However, the IRR method is based on the assumption that recovered funds, if not consumed in each time period, are reinvested at $i'\%$ rather than at the MARR. This latter assumption may not mirror reality in some problems, thus making IRR an unacceptable method for analyzing engineering alternatives.*

Other difficulties with the IRR method include its computational intractability and the occurrence of multiple IRRs in some types of problems. A procedure for dealing with seldom-experienced multiple rates of return is discussed and demonstrated in Appendix 4-A. Generally speaking, multiple rates are not meaningful for

---

*See H. Bierman and S. Smidt, *The Capital Budgeting Decision: Economic Analysis of Investment Projects* (New York: Macmillan Publishing Company, 1984). The term *internal* rate of return means that the value of this measure depends only on the cash flows from an investment and not on any assumptions about reinvestment rates: "One does not need to know the reinvestment rates to compute the internal rate of return. However, one may need to know the reinvestment rates to compare alternatives" (p. 60).

decision-making purposes, and another method of evaluation (e.g., PW) should be utilized.

Another possible drawback to the IRR method is that it must be carefully applied and interpreted in the analysis of two or more alternatives when only one of them is to be selected (i.e., mutually exclusive alternatives). This is discussed further in Chapter 5. The key advantage of the method is its widespread acceptance by industry, where various types of rates of return and ratios are routinely used in making project selections.

4.7 The External Rate of Return Method

The reinvestment assumption of the IRR method noted previously may not be valid in an engineering economy study. For instance, if a firm's MARR is 20% per year and the IRR for a project is 42.4%, it may not be possible for the firm to reinvest net cash proceeds from the project at much more than 20%. This situation, coupled with the computational demands and possible multiple interest rates associated with the IRR method, has given rise to other rate of return methods that can remedy some of these weaknesses.

One such method is the external rate of return (ERR) method. It directly takes into account the interest rate (ϵ) external to a project at which net cash flows generated (or required) by the project over its life can be reinvested (or borrowed). If this external reinvestment rate, which is usually the firm's MARR, happens to equal the project's IRR, then the ERR method produces results identical to those of the IRR method.

In general, three steps are used in the calculating procedure. First, all net cash outflows are discounted to time 0 (the present) at ϵ% per compounding period. Second, all net cash inflows are compounded to period N at ϵ%. Third, the external rate of return, which is the interest rate that establishes equivalence between the two quantities, is determined. The *absolute value* of the present equivalent worth of the net cash outflows at ϵ% (first step) is used in this last step. In equation form, the ERR is the i'% at which

$$\sum_{k=0}^{N} E_k(P/F, \epsilon\%, k)(F/P, i'\%, N) = \sum_{k=0}^{N} R_k(F/P, \epsilon\%, N - k) \qquad \textbf{(4-11)}$$

where R_k = excess of receipts over expenses in period k
E_k = excess of expenditures over receipts in period k
N = project life or number of periods for the study
ϵ = external reinvestment rate per period

Graphically, we have the following (the numbers relate to the three steps):

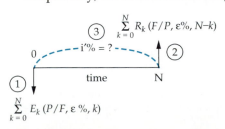

A project is acceptable when $i'\%$ of the ERR method is greater than or equal to the firm's MARR.

The external rate of return method has two basic advantages over the IRR method:

1. It can usually be solved for directly rather than by trial and error.
2. It is not subject to the possibility of multiple rates of return. (*Note:* The multiple rate of return problem with the IRR method is discussed in Appendix 4-A.)

EXAMPLE 4-15

Referring to Example 4-11, suppose that ϵ = MARR = 20% per year. What is the alternative's external rate of return, and is the alternative acceptable?

SOLUTION

By utilizing Equation 4-11, we have this relationship to solve for i':

$$\$25,000(F/P,i'\%,5) = \$8,000(F/A,20\%,5) + \$5,000$$

$$(F/P,i'\%,5) = \frac{\$64,532.80}{\$25,000} = 2.5813$$

$$i' = 20.88\%$$

Because i' > MARR, the alternative is barely justified.

EXAMPLE 4-16

When ϵ = 15% and MARR = 20%, determine whether the project whose total cash flow diagram appears below is acceptable. Notice in this example that the use of an $\epsilon\%$ different from the MARR is illustrated. This might occur if, for some reason, part or all of the funds related to a project are "handled" outside the firm's normal capital structure.

SOLUTION

$$E_0 = \$10,000 \ (k = 0)$$
$$E_1 = \$5,000 \ (k = 1)$$
$$R_k = \$5,000 \text{ for } k = 2,3,\ldots,6$$
$$|-\$10,000 - \$5,000(P/F,15\%,1)|(F/P,i'\%,6)$$
$$= \$5,000(F/A,15\%,5); \ i'\% = 15.3\%$$

The $i'\%$ is less than MARR = 20%, therefore, this project would be unacceptable according to the ERR method.

4.8 The Payback (Payout) Period Method

All methods presented thus far reflect the *profitability* of a proposed alternative for a study period of N. The payback method, which is often called the *simple payout method,* mainly indicates a project's *liquidity* rather than its profitability. Historically, the payback method has been used as a measure of a project's riskiness, since liquidity deals with how fast an investment can be recovered. A low-valued payback period is considered desirable. Quite simply, the payback method calculates the number of years required for cash inflows to just equal cash outflows. Hence the simple payback period is the *smallest* value of θ ($\theta \leq N$) for which this relationship is satisfied under our normal end-of-year cash flow convention. For a project where all capital investment occurs at time 0, we have:

$$\sum_{k=1}^{\theta}(R_k - E_k) - I \geq 0 \tag{4-12}$$

The *simple* payback period, θ, ignores the time value of money and all cash flows that occur after θ. If this method is applied to the investment project in Example 4-10, the number of years for payout can be calculated to be

$$\sum_{k=1}^{\theta=4.33}(\$5,310 - \$3,000)_k - \$10,000 > 0$$

Under end-of-year convention, θ would be rounded up to five years in this example. When $\theta = N$ (the last time period in the planning horizon), the salvage value is included in the determination of a payback period. As can be seen from Equation 4-12, the payback period does not indicate anything about project desirability except the speed with which the investment will be recovered. The payback period can produce misleading results, and it is recommended as supplemental information only in conjunction with one or more of the five methods previously discussed.

It is informative to plot undiscounted cash flows against time in order to see the simple payback period, θ. For Example 4-3, this is shown in Figure 4-5 as the 0% plot.

Sometimes the *discounted* payback period, $\theta'(\theta' \leq N)$, is calculated so that the time value of money is considered:

$$\sum_{k=1}^{\theta'}(R_k - E_k)(P/F, i\%, k) - I \geq 0 \tag{4-13}$$

where $i\%$ is the minimum attractive rate of return, I is the capital investment made at the present time ($k = 0$), and θ' is the smallest value that satisfies

Equation 4-13. Because $\theta' \geq \theta$, we can reduce the number of iterations required by Equation 4-13 to determine θ'.

This variation of the simple payback period produces the *breakeven life* of a project in view of the time value of money. However, neither payback period calculation includes cash flows occurring after θ (or θ'). This means that they do not take into consideration the economic life of physical assets. Thus, these methods will be misleading if one alternative that has a longer (less desirable) payout period than another produces a higher rate of return (or PW) on the invested capital.

Using the payout period to make investment decisions should generally be avoided except as a measure of how quickly invested capital will be recovered, which is an indicator of project risk. For instance, the breakeven life of the investment described in Table 4-1 is five years (in column C), which corresponds to the discounted payback period at $i = 20\%$. Payback periods of three years or less are often desired in U.S. industry, so under this criterion the equipment in Example 4-3 would be regarded as too risky even though its profitability is acceptable.

EXAMPLE 4-17

Considering the alternative presented in Example 4-11 that has end-of-year cash flows, calculate the simple payback period and the discounted payback period. In view of a minimum acceptable payback period of five years, is this alternative attractive?

SOLUTION
Payback period:

$$\left[\sum_{k=1}^{\theta=3.13} (\$8,000)_k \right] - \$25,000 \approx 0$$

Thus $\theta =$ four years (rounded up from 3.13 years) and the alternative is acceptable. Notice that the $5,000 salvage value was not considered in Equation 4-12 to determine the simple payback period.

Discounted payback period:

$$\left[\sum_{k=1}^{\theta'=5} (\$8,000)_k (P/F, 20\%, k) \right] + \$5,000(P/F, 20\%, 5) - \$25,000 > 0$$

or

$$\$8,000(P/A, 20\%, 5) + \$5,000(P/F, 20\%, 5) - \$25,000 > 0$$

at the project's expected life of $N = 5$ years. The alternative is minimally acceptable with the discounted payback method since $\theta' = 5$ years. In this case, the salvage value was considered in determining the value of θ'. Observe that when $k = 4$, Equation 4-13 is not satisfied.

▼ 4.9 Investment Balance Diagrams

The simple payback and discounted payback period methods tell us how long it takes cash inflows from a project to accumulate to equal (or exceed) the project's cash outflows. The cash flows are not discounted in the case of simple payback calculations, and they are discounted at the MARR to determine the discounted payback period. Refer to Figure 4-5 for a graphical summary of θ and θ' in Example 4-3.

Another useful method for describing how much money is tied up in a project and how the recovery of funds behaves over its estimated life is the *investment balance diagram.* The mechanics of this method were illustrated for a particular project in Figure 4-7 (where i' will be specified to be the MARR and negative

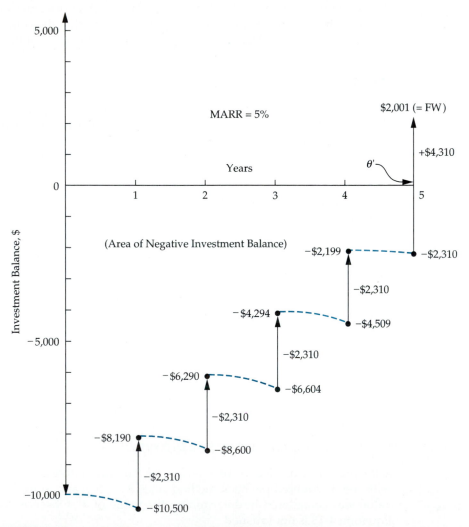

FIGURE 4-12   Investment Balance Diagram for Example 4-10

amounts were drawn above the line). Suppose we return to Example 4-10 and develop an investment balance diagram for this project when the MARR = 5% per year. This diagram, with positive amounts above the time axis, is shown in Figure 4-12 (page 166). It provides us with several pieces of information: the discounted

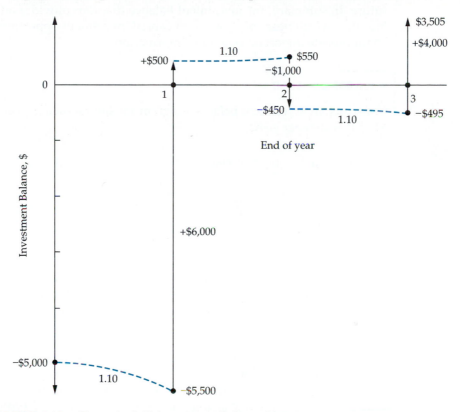

FIGURE 4-13   Example 4-18 Investment Balance Diagram

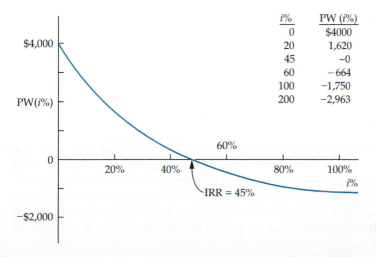

i%	PW (*i*%)
0	$4000
20	1,620
45	~0
60	−664
100	−1,750
200	−2,963

FIGURE 4-14

Example 4-18 Plot of PW versus *i*%

payback period (θ') is five years, the FW is $2,001, and the project has a negative investment balance until the end of the fifth year. An investor in this venture is "at risk" until the last year of the study period. This is not a comfortable situation when one fears losing money on capital investments that have an uncertain future. In summary, the investment balance diagram provides additional insight into the "worthiness" of a proposed capital investment opportunity and helps to communicate important economic information.

EXAMPLE 4-18

Construct an investment balance diagram for the following cash flow table. The MARR is 10% per year.

End of Year	Net Cash Flow
0	−$5,000
1	6,000
2	− 1,000
3	4,000

SOLUTION

The investment balance diagram is shown in Figure 4-13. You can see that money is committed to the project in years one and three and that the initial investment cost is completely recovered by the end of year one. The exposure to loss is much less in Figure 4-13 (page 167) than it was in Figure 4-12. In fact, FW (10%) = $3,505 and IRR = 45%, which confirms our good feeling about this capital investment. Furthermore, *the IRR is unique*, as demonstrated by the plot of PW(*i*%) versus *i*% in Figure 4-14 (page 167) for this example. Unique in this case means that the PW(*i*%) curve only intersects with the *i*% axis at one point. Thus, the IRR is unique even though there are three sign changes in the project's cash flow profile.

4.10 An Example of a Proposed Capital Investment to Reduce Costs

Many engineering projects are aimed at reducing costs rather than enhancing revenues. Example 4-19 illustrates a cost reduction opportunity.

EXAMPLE 4-19

A jewelry manufacturer is contemplating the installation of a system that will recover a larger proportion of the fine particles of gold and platinum that result from the various manufacturing operations. At the present time a little over $45,000 worth of these metals is being lost per year, and it is anticipated that, because of the growth of the company, this amount will increase by $5,000 each year for the next 10 years (the study period). The proposed system, involving a network of exhaust ducts and separators, will recover approximately two-thirds of the gold and platinum that would otherwise be lost. The complete installation would cost $140,000. The best estimates for the operating costs of the system, obtained from operations of similar systems, are $10,000 per year for operating expense, $1,800 per year for maintenance and repairs, and 2% of the first cost annually for property taxes and insurance. The company would require the capital investment to be recovered with interest in 10 years. The MARR of the company, before taxes, is 15%. Should the recovery system be installed?

SOLUTION

Such an investment is made to reduce some of the annual expenses, in this case the cost of the material used. Thus, the saving (income) to be obtained by making the capital investment is almost entirely within the control of the investors. The company knows exactly what expenses have been. If the efficiency of the proposed equipment is known, the only factors that should affect the saving are the variation of production, operation, and maintenance expenses of the proposed equipment. In most cases of this type, these items are known or may be predicted quite accurately. The company would have a good idea of how its volume might vary. Operation and maintenance expenses can usually be estimated accurately, especially if historical data are available on the proposed equipment.

Using the data given, we find that the PW and IRR calculations would be as shown in Table 4-4. In deciding whether or not the calculated internal rate of return of 16.8% is sufficient to justify the investment, each factor that might contribute to the risk must be examined so that a measure of the total risk may be obtained. In this case it appears that the factors are quite well controlled or known. There is little reason to believe that much more risk would be involved than is present in all the normal operations of the company. Thus, the company can use its own experience as a basis of comparison. If the company is sound and its business is quite stable, a return of 16.8% coupled with a simple payback period of six years should be satisfactory.

When capital is invested in a going concern in order to reduce costs, the risk is usually easier to assess and is often much less than when entirely new enterprises are involved. As a result, the rate of return required for such investments is often lower. A formal treatment of risk is given in Chapter 14.

4.11 An Example of a Large Industrial Investment Opportunity

The following example illustrates the before-tax cash flow analysis of a typical industrial plant expansion problem. It involves many types of cash flows that are summarized in a large cash flow table. Particular attention is directed to variable, fixed, and sunk costs that are present in addition to expenditures that are *capitalized* (e.g., investments) versus those such as cash outlays for materials that are *expensed*.

TABLE 4-4 Tabular Determination of PW and IRR[a] for the Proposed System in Example 4-19

End of Year, k	Investment	Recovery	Costs	Net Cash Flow	(P/F, 15%, k)	Present Worth at i = 15%	(P/F, 20%, k)	Present Worth at i = 20%
0	−$140,000	—	—	−$140,000	1.000	−$140,000	1.0000	−$140,000
1	—	$33,333	−$14,600	18,733	0.8696	16,290	0.8333	15,610
2	—	36,667	− 14,600	22,067	0.7561	16,685	0.6944	15,323
3	—	40,000	− 14,600	25,400	0.6575	16,701	0.5787	14,699
4	—	43,333	− 14,600	28,733	0.5718	16,430	0.4823	13,858
5	—	46,667	− 14,600	32,067	0.4972	15,944	0.4019	12,888
6	—	50,000	− 14,600	35,400	0.4323	15,303	0.3349	11,855
7	—	53,333	− 14,600	38,733	0.3759	14,560	0.2791	10,810
8	—	56,667	− 14,600	42,067	0.3269	13,752	0.2326	9,785
9	—	60,000	− 14,600	45,400	0.2843	12,907	0.1938	8,799
10	—	63,333	− 14,600	48,733	0.2472	12,047	0.1615	7,870
Total				$197,333		$ 10,619		−$ 18,503

[a]IRR $\cong 15\% + [\$10,619/(10,619 + 18,503)] \times (20\% - 15\%) \cong 16.8\%$.

EXAMPLE 4-20

I. PROBLEM STATEMENT   Product X-21, a food preservative, is manufactured in a facility with a nominal capacity of 180,000 pounds per year. The ultimate capacity of the facility can be increased, at some loss in cost efficiency, by process modifications and the use of extensive amounts of overtime, to 195,000 pounds per year. With the present marketing strategy, sales are expected to level off at about 190,000 pounds per year in 1997 and to remain at that volume. Therefore, the existing capacity will be sufficient under the present marketing strategy.

However, an opportunity has become available to enter a new market for an X-21 product modified slightly to give it better warm weather stability. A German firm has a trade secret on the modifying process but is willing to enter into a nondisclosure agreement for a one-time fee of $75,000. A proposal has been submitted to management to expand the existing X-21 plant to permit the company to enter this new market. *Management has requested an economic analysis of this proposal and a recommendation.* Your assignment is to provide this recommendation.

II. GENERAL BACKGROUND INFORMATION   **A.** *Demand schedule:* The demand schedule for X-21 is as follows:

Year	Total Demand (lb)	Incremental Sales from This Project (lb)
1996	175,000	0
1997	190,000	0
1998	220,000	30,000
1999	250,000	60,000
2000–2012	260,000	70,000

Both the modified and regular grades of X-21 will be sold at the same price of $39.50 per pound. All demand over 190,000 pounds per year is attributable to the new market.

B. *Existing manufacturing facility:* The existing facility consists of two production lines, each with an ultimate capacity of 97,500 pounds per year. A new line of the same capacity can be added in an existing building, which because of federal restrictions and its unique layout, cannot be used for any other purpose. That is, no cost need be allocated for the building space. The new line can be used for both the regular X-21 and the modified X-21.

C. *Expansion plan:* The total installed cost of the new line is estimated at $430,000, assuming project operation by January 1, 1998. Of this total, engineering expenditures of $40,000 would be required in 1996 and all other capital costs would be incurred in 1997.

D. *Project life:* The company considers all projects of this type and degree of risk to have a 15-year life. Thus, commercial operation of the new production line would begin in early 1998 and terminate at the end of year 2012.

E. *Inflation:* The analysis will ignore the effects of inflation (in Chapter 9, this topic is considered).

F. *Federal and state income taxes:* The analysis will ignore all income taxes (income taxes are considered in Chapter 7).

G. *Interest rate and financing:* The before-tax MARR for projects with this degree of risk is 20% per year (effective). (The financing decision for large projects of this type is discussed in Chapter 12.)

H. *Other profitable company operations:* These will offset any short-term negative cash flows.

I. *Expenditures:* Those occurring throughout a year will be assumed for cash flow purposes to occur at the end of the year.

III. CASH FLOW DATA

A. *Capitalizable project expenditures:* A total of $430,000 will be required to install the new production line. In 1996, $40,000 is needed, and the remainder of $390,000 would be expected in 1997. (Capitalizable expenditures are those expenditures that must be depreciated using the methods discussed in Chapter 7.)

B. *Project expenditures chargeable to operations:* Certain expenditures will result from the dismantlement and rearrangement of special equipment that will be required to fit the new line into the existing building. Estimated before-tax cost for this work is $48,600.

C. *Related projects and work orders:* Various types of auxiliary equipment will have to be replaced at five year intervals. Therefore, equipment replacement expenditures of $120,000 are planned for 2002 and 2007.

D. *Net change in working capital:* All firms require capital for day-to-day operations, including funds to support prepaid expenses, inventories, and bank balances. Such *working capital* has associated with it an opportunity cost that in many cases is a substantial item in an engineering economy study. Working capital caused by a project is normally treated as an asset whose first cost and salvage value are equal. Thus, the annual opportunity cost (i) of working capital in year k, WC_k, is $i(WC_k)$.

The change in working capital should be roughly proportional to the change in sales. Working capital in 1995 was approximately $200,000 when sales were 164,000 pounds. The following *net changes in working capital* are computed using this proportionality:

Year	Net Change from Previous Year
1996	$13,400
1997	18,300
1998	36,600
1999	36,600
2000	12,200
2001–2012	0

Only changes in 1998 and later are pertinent to the analysis.

E. *Prepaid know-how:* A single payment of $75,000 to the German firm is required before startup. This payment will be made at the end of 1997.

F. *Pretax cash flows from commercial operation (incremental due to the new line):*

1. *Fixed costs:* Fixed costs for supervisory salaries, general plant overhead, insurance, and property taxes will increase by $750,000 per year if this project is implemented. This increase will remain constant over the 15-year life of the project.

2. *Variable costs:* These costs consist primarily of raw materials costs, direct labor, utilities, and the variable portion of maintenance costs. Variable costs are expected to be $9.18 per pound in 1998. As operating experience is gained, yields should increase, resulting in variable costs of $8.83 per pound in 1999 and $7.56 per pound in 2000 and thereafter.

3. *General and administrative (G and A) costs:* G and A costs are budgeted for $900,000 in 1998 when the product is first introduced. After 1998, G and A costs are budgeted for $1,100,000 per year. This amounts to about 40% of sales revenue after sales have leveled out.

The following table summarizes the annual before-tax cash flows in the years 1998–2012 attributable to this proposed project.

Year	(1) Fixed Cost (thousands)	(2) Variable Cost (thousands)	(3) G & A Costs (thousands)	Sales Revenues (thousands)	Sales Revenue Less Costs in Columns 1–3 (thousands)
1998	$750	$275.4	$ 900	$1,185	−$740.4
1999	750	529.8	1,100	2,370	− 9.8
2000–2012	750	529.2	1,100	2,765	385.8

G. *Introductory costs:* Introductory costs will be relatively small. The only significant introductory cost is training personnel in modifications to the process and technical assistance during the startup. The total introductory cost in 1997 is estimated at $8,500.

H. *Sales Revenues:* During 1998, sales revenues are expected to be $39.50/pound × 30,000 pounds/year = $1,185,000. They level out to $39.50 × 70,000 pounds/year = $2,765,000 from 2000 through 2012.

IV. ANALYSIS RESULTS AND RECOMMENDATIONS

A cash flow worksheet for this example is given in Table 4-5 (page 174). From the before-tax cash flows in column (9), the project's IRR can be calculated based on Equation 4-9.

$$PW = 0 = -\$40,000 - \$522,100(P/F, i'\%, 1) - \$777,000(P/F, i'\%, 2) - \cdots$$
$$+ \$471,200(P/F, i'\%, 16)$$

By trial and error, $i' = 18.3\% < $ MARR.

The before-tax MARR is 20% and the PW at the end of 1996 is −$87,137. Thus, the IRR and PW measures of profitability indicate that the company should *not* enter the new market for a modified X-21 product. Even if these criteria had signaled a favorable project (e.g., IRR ≥ 20%), a detailed analysis of uncertainty should be performed and nonmonetary considerations analyzed before a decision is made.

TABLE 4-5   Cash Flow Worksheet for Example 4-20

End of Year	(1) Capitalizable Project Expenditures	(2) Expenditures Chargeable to Operations	(3) Related Projects and Work Orders	(4) Net Change in Working Capital	(5) Prepaid Know-How
1996	−$ 40,000				
1997	− 390,000	−$48,600			−$75,000
1998				−$36,600	
1999				− 36,600	
2000				− 12,200	
2001					
2002			−$120,000		
2003					
2004					
2005					
2006					
2007			− 120,000		
2008					
2009					
2010					
2011					
2012				+ 85,400[a]	

[a]This is return of working capital at the end of the project.   (*Continued*)

TABLE 4-5   Cash Flow Worksheet for Example 4.20   Cont'd

End of Year	(6) Total Investment Cash Flow: Cols. 1 + 2 + 3 + 4 + 5	(7) Before-Tax Cash Flows During Commercial Operation	(8) Introductory Costs	(9) Before-Tax Cash Flows for the Entire Project: Cols. 6 + 7 + 8
1996	−$ 40,000			−$ 40,000
1997	− 513,600		−$8,500	− 522,100
1998	− 36,600	−$740,400		− 777,000
1999	− 36,600	− 9,800		− 46,400
2000	− 12,200	385,800		373,600
2001		385,800		385,800
2002	− 120,000	385,800		265,800
2003		385,800		385,800
2004		385,800		385,800
2005		385,800		385,800
2006		385,800		385,800
2007	− 120,000	385,800		265,800
2008		385,800		385,800
2009		385,800		385,800
2010		385,800		385,800
2011		385,800		385,800
2012	+ 85,400[a]	385,800		471,200

[a]This is return of working capital at the end of the project.

4.12 Spreadsheet Applications

In this chapter, several measures of merit were introduced for evaluating engineering projects. Most spreadsheet packages include financial functions that can be used to simplify the computation of these measures. These functions and their parameters are described below.

Function	Description
NPV(i, $range$)	Returns the net present worth of the cash flows in $range$ using i as the interest rate one period prior to the first cash flow in $range$.
PMT(i, n, NPV())	Returns the value of the uniform end-of-period payments made on a loan with interest rate i, repayment period n, and principal amount NPV().
FV(i, n, PMT())	Returns the future worth (at end of period n) of n uniform payments of PMT() dollars when the interest rate is i.
IRR($range$, $guess$)	Returns the internal rate of return of the cash flows in $range$, where $guess$ is an initial estimate of the IRR. The MARR is usually a good guess.
MIRR($range$, i, ϵ)	Returns the external rate of return of the cash flows in $range$ where i is the interest rate charged on cash outflows and ϵ is the reinvestment rate for cash inflows.

The financial functions are based on the following assumptions, which agree with those presented in this book:

1. The per period interest rate, i, remains constant.
2. There is exactly one period between cash flows.
3. The period length remains constant.
4. The end-of-period cash flow convention is utilized.
5. The first cash flow in a range at the end of the first period.

The NPV( ) function is the most useful of the equivalent value financial functions; however, care should be taken to observe the assumptions of this function. The function is designed to compute the net present worth of a series of cash flows. According to assumption (5) above, the timing of net present worth returned is one interest period before the first cash flow. Thus, if you include the investment amount at $t = 0$ in the range of cash flows, the net present worth returned by the NPV() function is related to $t = -1$. One way to address this issue is to include cash flows for periods 1 through N in P_range, and then add to this value the capital investment amount. This is the approach taken in this book.

The equivalent worth and rate of return measures of merit are obtained with the following combinations of functions.

$$PW = NPV(MARR, P\_range) + Capital\ Investment$$

$$AW = -PMT(MARR, n, NPV(MARR, P\_range) + Capital\ Investment)$$

$$FW = FV(MARR, n, PMT(MARR, n, NPV(MARR, P\_range) + Capital\ Investment)$$

	A	B	C	D	E	F
1	Economic Measures of Merit:					
2						
3	MARR		20%			
4	Reinvestment Rate (ε)		20%			
5						
6	End of	Net	Cumulative		Cumulative	
7	Period	Cash Flow	PW (0%)		PW(MARR)	
8						
9	0	$ (25,000)	$ (25,000)		$ (25,000)	
10	1	$ 8,000	$ (17,000)		$ (18,333)	
11	2	$ 8,000	$ (9,000)		$ (12,778)	
12	3	$ 8,000	$ (1,000)		$ (8,148)	
13	4	$ 8,000	$ 7,000	*	$ (4,290)	
14	5	$ 13,000	$ 20,000		$ 934	**
15						
16	Present Worth		$ 934.28			
17	Annual Worth		$ 312.41			
18	Future Worth		$ 2,324.80			
19						
20	Internal Rate of Return		21.58%			
21	External Rate of Return		20.88%			
22						
23	Note:	*	denotes the simple payback period, and			
24		**	denotes the discounted payback period			

FIGURE 4-15   Spreadsheet for Computing Economic Measures of Merit for Example 4-11

$$IRR = IRR(MARR, range)$$

$$ERR = MIRR(range, i, \epsilon)$$

The payback period of a project can also be easily computed using a spreadsheet. By computing the cumulative present worth with $i = 0\%$ and $i = MARR$, it is easy to identify the simple and discounted payback periods, respectively.

Figure 4-15 shows a spreadsheet that calculates all of the economic measures of merit discussed in this chapter for the proposed project discussed in Example 4-11. The formulas in the highlighted cells are given in the following table.

Cell	Contents
C13	= B13 + C12
D13	= IF(AND(C13 >= 0, C12 < 0), "*", "")
E13	= E9 + NPV(C3, B10 : B13)
F13	= IF(AND(E13 >= 0, E12 < 0), " * *", "")
C16	= NPV(C3, B10 : B14) + B9
C17	= PMT(C3, 5, −(NPV(C3, B10 : B14) + B9))
C18	= FV(C3, 5, PMT(C3, 5, (NPV(C3, B10 : B14) + B9)))
C20	= IRR(B9 : B14, C3)
C21	= MIRR(B9 : B14, C3, C4)

4.13 Summary

Throughout this chapter we have examined five basic methods for evaluating the financial *profitability* of a single project: present worth, annual worth, future worth, internal rate of return, and external rate of return. Two supplemental methods for assessing a project's *liquidity* were also presented: the simple payback period and the discounted payback period. Computational procedures, assumptions, and acceptance criteria for all methods were discussed and illustrated with examples. Appendix B provides a listing of new abbreviations and notations that have been introduced in this chapter.

4.14 References

CANADA, J. R., W. G. SULLIVAN, and J. A. WHITE. *Capital Investment Decision Analysis for Engineering and Management*. 2nd ed. Englewood Cliffs, N.J.: Prentice Hall, Inc., 1996.

GRANT, E. L., W. G. IRESON, and R. S. LEAVENWORTH. *Principles of Engineering Economy*, 8th ed. New York: John Wiley & Sons, 1989.

MORRIS, W. T. *Engineering Economic Analysis*. Reston, Va.: Reston Publishing Co., 1976.

THUESEN, G. J., and W. J. FABRYCKY. *Engineering Economy*, 8th ed. Englewood Cliffs, N.J.: Prentice Hall, Inc., 1993.

WHITE, J. A., M. H. AGEE, and K. E. CASE. *Principles of Engineering Economic Analysis*, 3rd ed. New York: John Wiley & Sons, 1989.

4.15 Problems

Unless stated otherwise, discrete compounding of interest and end-of-period cash flows should be assumed in all problem exercises in the remainder of the book. All MARRs are "per year." The number in parentheses at the end of a problem refers to the chapter section(s) most closely related to the problem.

4-1. In your own words, what is the meaning of
   a. present worth? (4.3)
   b. capital recovery amount? (4.5)
   c. minimum attractive rate of return? (4.2)

4-2. You are faced with a decision on an investment proposal. Specifically, the estimated additional income from the investment is $180,000 per year; the investment cost is $640,000; and the estimated annual expenses are $44,000, which begin decreasing by $4,000 per year starting at the end of the third year. Assume an 8-year analysis period, no salvage value, and MARR = 15% per year. (4.3, 4.6)

a. What is the PW of this proposal?
b. What is the IRR of this proposal?

4-3.

a. Evaluate machine XYZ on the basis of the PW method when the MARR is 12% per year. Pertinent cost data are as follows: (4.3)

	Machine XYZ
Investment cost	$13,000
Useful life	15 years
Salvage value	$ 3,000
Annual operating expenses	100
Overhaul cost—end of fifth year	200
Overhaul cost—end of tenth year	550

b. Determine the capital recovery amount of machine XYZ by all three formulas presented in the text. (4.5)

4-4.

a. Determine the PW, FW, and AW of the following engineering project when the MARR is 15% per year. (4.3, 4.5)

	Proposal A
Investment cost	$10,000
Expected life	5 years
Market (salvage) value[a]	−$ 1,000
Annual receipts	8,000
Annual expenses	4,000

[a]A negative salvage value means that there is a net cost to dispose of an asset.

b. Determine the IRR of the project. Is it acceptable? (4.6)
c. What is the ERR for this project? Assume ϵ = 15% per year. (4.7)

4-5. Uncle Wilbur's trout ranch is now for sale for $40,000. Annual property taxes, maintenance, supplies, and so on are estimated to continue to be $3,000 per year. Revenues from the ranch are expected to be $10,000 next year and then to decline by $500 per year thereafter through the tenth year. If you bought the ranch, you would plan to keep it for only five years and at that time to sell it for the value of the land, which is $15,000. If your desired annual rate of return is 12%, should you become a trout rancher? Use the PW method. (4.3)

4-6. A company is considering constructing a plant to manufacture a proposed new product. The land costs $300,000, the building costs $600,000, the equipment costs $250,000, and $100,000 additional working capital is required. It is expected that the product will result in sales of $750,000 per year for 10 years, at which time the land can be sold for $400,000, the building for $350,000, and the equipment for $50,000. All of the working capital would be recovered. The annual expenses for labor, materials, and all other items are estimated to total $475,000. If the company requires an MARR of 25% per year on projects of comparable risk, determine if it should invest in the new product line. Use the PW method. (4.3)

4-7.

a. Draw a cash flow diagram for the bond that was described in Example 4-4.
b. If the bond in Example 4-4 is purchased to yield 5% per 6-month period (rather than i = 10% per year), the current purchase price would be how much? (4.3)

4-8. How much can be paid for a $5,000, 10% bond, with interest paid semiannually, if the bond matures 12 years hence? Assume that the purchaser will be satisfied with 8% nominal interest compounded semiannually. (4.3)

4-9. A 20-year bond with a face value of $5,000 is offered for sale at $3,800. The nominal rate of interest on the bond is 7%, paid semiannually. This bond is now 8 years old (i.e., the owner has received 16 semiannual interest payments). If the bond is purchased for $3,800, what effective annual rate of interest would be realized on this investment opportunity? (4.3)

4-10.

a. A company has issued 10-year bonds, with a face value of $1,000,000, in $1,000 units. Interest at 8% is paid quarterly. If an investor desires to earn 12% nominal interest (compounded quarterly) on $10,000 worth of these bonds, what would the selling price have to be?
b. If the company plans to redeem these bonds in total at the end of 10 years and establishes a sinking fund that earns 8%, compounded semiannually, for this purpose, what is the *annual* cost of interest and redemption? (4.3)

4-11. You bought a $1,000 bond at par (face value) that paid nominal interest at the rate of 10%, payable semiannually, and held it for 10 years. You then sold it at a price that resulted in a yield of 8% nominal interest compounded semiannually on your capital. What was the selling price? (4.3)

4-12. A small company bought BMI bonds at their face value on January 1, 1991. These bonds pay interest of 7.25% every six months (14.5% per year). The face value of the bonds is $100,000, and they mature on December 31, 2006. On January 1, 2001, these bonds were sold for $110,000. What interest rate (per six months) was earned by the company on the BMI bonds? (4.3)

4-13. Susie Queue has a $100,000 mortgage on her deluxe townhouse in urban Philadelphia. She makes monthly payments on a 10% nominal interest rate (compounded monthly) loan and has a 30-year mortgage. Home mortgages are presently available at a 7% nominal interest rate on a 30-year loan. Susie has lived in the townhouse for only two years, and she is considering refinancing her mortgage at a 7% nominal interest rate. The mortgage company informs her that the one-time cost to refinance the present mortgage is $4,500.

How many months must Susie continue to live in her townhouse to make the decision to refinance a good one? Her MARR is the return she can earn on a 30-month certificate of deposit that pays 1/2% per month (6% nominal interest). (4.3, 4.5)

4-14. On January 1, 1997, your brother bought a used car for $8,200, and he agreed to make a down payment of $1,500 and repay the balance in 36 equal payments, with the first payment due February 1. The nominal interest rate is 13.8% compounded monthly. During the summer your brother made enough money so that he decided to repay the entire balance due on the car as of September 1. How much did he repay on September 1? (4.3)

4-15

a. A car can be leased for $400 per month for 36 months and then purchased for $10,000 when the last payment is made. If $i = 1\%$ per month, the purchase price today should be how much? (4.3)

b. If the purchase price turned out to be $22,000, what would this tell you about the dealer? (4.3)

4-16. List the advantages and disadvantages of each of these five basic methods for performing engineering economy studies (PW, AW, FW, IRR, ERR).

4-17. The Anirup Food Processing Company is presently using an outdated method for filling 25-pound sacks of dry dog food. To compensate for weighing inaccuracies inherent to this packaging method, the process engineer at the plant has estimated that each sack is overfilled by 1/8 pound on the average. A better method of packaging is now available that would eliminate overfilling (and underfilling). The production quota for the plant is 300,000 sacks per year for the next six years, and a pound of dog food costs this plant $0.15 to produce. The present system has no salvage value and will last another four years, and the new method has an estimated life of four years with a salvage value equal to 10% of its investment cost, I. The present packaging operation expense is $2,100 per year more to maintain than the new method. If the MARR is 12% per year for this company, what amount, I, could be justified for the purchase of the new packaging method? (4.3)

4-18. Fill in the table at the bottom of this page when $P = \$10,000$, $S = \$2,000$ (at the end of four years), and $i = 15\%/\text{yr}$. Complete the table and show that the equivalent uniform CR amount equals $3,102.12. (4.5)

4-19.

a. A certain service can be performed satisfactorily by process R, which has a capital in-

Year	Investment at Beginning of Year	Opportunity Cost of Interest ($i = 15\%$)	Loss in Value of Asset During Year	Capital Recovery Amount	(Pr. 4-18)
1	$10,000		$3,000		
2			$2,000		
3			$2,000		
4					

Year	Investment at Beginning of Yr.	Opportunity Cost (5% per Yr.)	Loss in Value of Asset During Yr.	Capital Recovery Amount	(Pr. 4-21)
1	$1,000	$50	$(a)	$250	
2	(b)	(c)	200	240	
3	600	30	200	230	
4	(d)	20	(e)	(f)	

vestment cost of $8,000, an estimated life of 10 years, no salvage value, and annual net receipts (revenues − expenses) of $2,400. Assuming a MARR of 18% before income taxes, find the FW and AW of this process and specify whether you would recommend it. (4.4, 4.5)

b. A compressor that costs $2,500 to purchase has a 5-year useful life and a salvage value of $1,000 after five years. At a nominal interest rate of 12%, compounded quarterly, what is the *annual* CR amount of the compressor? (4.5)

4-20. You purchased a building five years ago for $100,000. Its annual maintenance expense has been $5,000 per year. At the end of three years, you spent $9,000 on roof repairs. At the end of five years (now), you sell the building for $120,000. During the period of ownership, you rented the building for $10,000 per year paid at the *beginning* of each year. Use the AW method to evaluate this investment when your MARR is 12% per year. (4.5)

4-21. Given that the purchase price of a machine is $1,000, and its market value at the end of year four is $300, complete the table above (values a through f) using an opportunity cost of 5% per year. Compute the equivalent uniform capital recovery amount, based on information from the completed table. (4.5)

4-22. Based on the following cash flow diagram, answer the following questions (4.3, 4.5, 4.8)

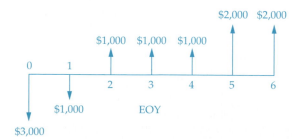

a. As $i \rightarrow \infty$, the PW equals _____.
b. The discounted payback period (θ') is _____ years. Let MARR = 12% per year.
c. If the cash flow at the end of year six had been −$2,000 instead of +$2,000, AW (0%) = _____.

4-23. A manufacturing firm has considerable excess capacity in its plant and is seeking ways to utilize it. The firm has been invited to submit a bid to become a subcontractor on a product that is not competitive with the one it produces but that, with the addition of $75,000 in new equipment, could readily be produced in its plant. The contract would be for five years at an annual output of 20,000 units.

In analyzing probable costs, direct labor is estimated at $1.00 per unit and new materials at $0.75 per unit. In addition, it is discovered that in each new unit one pound of scrap material can be used from the present operation, which is now selling for $0.30 per pound of scrap. The firm has been charging overhead at 150% of prime cost, but it is believed that for this new operation the incremental overhead, above maintenance, taxes, and insurance on the new equipment, would not exceed 60% of the direct labor cost. The firm estimates that the maintenance expenses on this equipment would not exceed $2,000 per year, and annual taxes and insurance would average 5% of the first cost. (*Note:* Prime cost = direct labor + direct materials cost).

While the firm can see no clear use for the equipment beyond the five years of the proposed contract, the owner believes it could be sold for $3,000 at that time. He estimates that the project will require $15,000 in working capital, and he wants to earn at least a 20% before-tax annual rate of return on all capital utilized. (4.3, 4.5)
a. What unit price should be bid?

b. Suppose that the purchaser of the product wants to sell it at a price that will result in a profit of 20% of the selling price. What should be the selling price?

4-24. To purchase a used automobile, you borrow $8,000 from Loan Shark Enterprises. They tell you the interest rate being charged is 1% per month for 35 months. They also charge you $200 for a credit investigation, so you leave with $7,800 in your pocket. The monthly payment they calculated for you is

$$\frac{8,000(0.01)(35) + \$8,000}{35} = \$308.57/\text{month}$$

If you agree to these terms and sign their contract, what is the actual APR (annual percentage rate) that you are paying? (4.6)

4-25. Suppose that you borrow $1,000 from the Easy Credit Company with the agreement to repay it over a 5-year period. Their stated interest rate is 9% per year. They show you the following items in determining the monthly payment: (4.6)

Principal $1,000
Total interest: 0.09 (5 years) ($1,000) $ 450

They ask you to pay 20% of the *interest* immediately, so you leave with $1,000 − $90 = $910 in your pocket. Your monthly payment is calculated as follows:

$$\frac{\$1,000 + \$450}{60} = \$24.17/\text{month}$$

a. Draw a cash flow diagram of this transaction.
b. Determine the effective *annual* interest rate.

4-26. Joe Roe is considering establishing a company to produce impellers for water pumps. An investment of $150,000 will be required for the plant and equipment, and $15,000 will be required for working capital. It is expected that the property will last for 15 years, at which time only the working capital part of the investment cost can be recovered. It is estimated that sales will be $200,000 per year and that operating expenses will be as follows:

Materials $40,000 per year
Labor $70,000 per year
Overhead $10,000 + 10% of sales
                   per year
Selling expense $5,000 per year

Joe has a regular job paying $40,000 per year, but he will keep that job even if he establishes this company. If Joe expects to earn at least 15% per year on his capital, should this investment be made? Use the IRR method. (4.6)

4-27. A meat-packing company is considering producing a new product, the entire domestic supply of which is now manufactured by three large companies. A new plant would be required that would cost $200,000 and be built on land that now is owned by the company but not used. The plant would have an expected life of 20 years. Annual costs for labor would be $60,000 and for materials $110,000. These would provide for an annual output of 600,000 pounds. This would constitute 25% of the present domestic consumption and would be 80% of the plant capacity. Advertising and other overhead expenses would amount to $50,000 per year. Annual taxes and insurance would total 3% of the value of the plant. The product is a very stable one and at present is selling for 55 cents per pound. Over a period of years the price has varied by ±15%.

a. Would you recommend that the company go ahead on this project? Capital is available and earns not less than 15% per year.
b. What is the minimum selling price for the product that would justify the investment? (4.5)

4-28. A machine that is not equipped with a brake "coasts" 30 seconds after the power is turned off upon completion of each workpiece, thus preventing removal of the work from the machine. The time per piece, exclusive of this stopping time, is two minutes. The machine is used to produce 40,000 pieces per year. The operator receives $16.50 per hour, and the machine overhead rate is $4.00 per hour. How much could the company afford to pay for a brake that would reduce the stopping time to three seconds if it had a life of five years? Assume zero market value, a MARR of 15% per year, and repairs and maintenance on the brake totaling no more than $250 per year. (4.3)

4-29. Your roommate borrowed money from a banker on the condition that he pay 7% of the loan every three months, until a total of 35 payments were made. Then the loan would be considered repaid. What effective *annual* interest rate did your roommate pay? Solve for the interest rate to the nearest 1/10 percent (use linear interpolation). (4.6)

4-30. Your boss has just presented you with the summary (below) of projected costs and annual receipts for a new product line. He asks you to calculate the before-tax IRR for this investment opportunity. What would you present to your boss, and how would you explain the results of your analysis? (It is widely known that the boss likes to see graphs of present worth versus interest rate for this type of problem.) The company's MARR is 10% per year. (4.6)

End of Year	Net Cash Flow
0	−$450,000
1	− 42,500
2	+ 92,800
3	+ 386,000
4	+ 614,600
5	−$202,200

4-31. Rework Problem 4-30 with the ERR method. Should the answers to the two problems be the same? Why? Let ϵ = MARR. (4.7)

4-32. In order to enter the market to produce a new toy for children, a manufacturer will have to make an immediate capital investment of $60,000 and additional investments of $5,000 at the end of one year and $3,000 more at the end of two years. Competing toys are now being produced by two large manufacturers. From a fairly extensive study of the market, it is believed that sufficient sales can be achieved to produce year-end, before-tax net cash flows as follows (investment costs not included):

Year	Cash Flow	Year	Cash Flow
1	−$10,000	6	$26,000
2	5,000	7	26,000
3	5,000	8	20,000
4	20,000	9	15,000
5	21,000	10	7,000

In addition, while it is believed that after 10 years the demand for the toy will no longer be sufficient to justify production, it is estimated that the physical assets would have a market value of about $8,000. If capital is worth not

less than 12% per year before taxes, would you recommend undertaking the project? Make a recommendation with each of these methods: (a) FW and (b) discounted payback with a maximum acceptable value of five years. (4.4, 4.8)

4-33. Find the IRR in each of these situations: (4.6)

a.

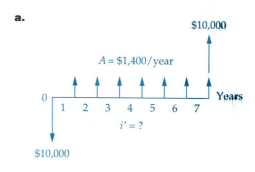

b.

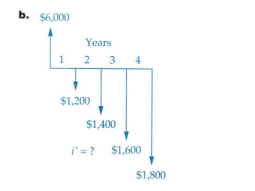

c. You purchased a used car for $4,200. After you make a $1,000 down payment on the car, the salesperson looks in her *Interest Calculations Made Simple* handbook and announces: "The monthly payments will be $160 for the next 24 months and the first payment is due one month from now." (Draw a cash flow diagram.)

4-34. Rework part (a) of Problem 4-33 by using the ERR method when ϵ = 8% per year. (4.7)

4-35. Plot the PW of part (a) of Problem 4-33 as a function of the interest rate. The MARR is equal to 8% per year. (4.3)

4-36. Draw an investment balance diagram for part (a) of Problem 4-33 using i = IRR (determined in that problem). (4.9)

4-37. A zero-coupon certificate involves payment of a fixed sum of money now with future lump-sum withdrawal of an accumulated amount. Earned interest is not paid out periodically, but instead compounds to become the major component of the accumulated amount paid when the zero-coupon certificate matures. Consider a certain zero-coupon certificate that was issued on March 25, 1993, and matures on January 30, 2010. A person who purchases a certificate for $13,500 will receive a check for $54,000 when the certificate matures. What is the annual interest rate (yield) that will be earned on this certificate? (4.3)

4-38. An individual approaches the Ajax Loan Agency for $1,000 to be repaid in 24 monthly installments. The agency advertises an interest rate of 1.5% per month. They proceed to calculate a monthly payment in the following manner:

Amount requested	$1,000
Credit investigation	25
Credit risk insurance	5
Total	$1,030

Interest: ($1030)(24)(0.015) = $371

Total owed: $1,030 + $371 = $1,401

Payment: $\dfrac{\$1,401}{24} = \$58.50$

What effective annual interest rate is the individual paying? (4.6)

4-39. A small company purchased now for $23,000 will lose $1,200 each year the first four years. An additional $8,000 invested in the company during the fourth year will result in a profit of $5,500 each year from the fifth year through the fifteenth year. At the end of 15 years the company can be sold for $33,000.
a. Determine the IRR. (4.6)
b. Calculate the FW if MARR = 12%. (4.4)
c. Calculate the ERR when $\epsilon = 12\%$. (4.7)

4-40.
a. Monthly amounts of $200 each are deposited into an account that earns 12% nominal interest, compounded quarterly. After 48 deposits of $200 each, what is the FW of the account? State your assumptions. (4.4)

b. A "Christmas Plan" requires deposits of $10 per week for 52 weeks each year. The stated nominal interest rate is 20% compounded weekly. What is the PW of the plan (beginning of week one)? (4.3)

4-41. In a certain foreign country, a man who wanted to borrow $10,000 for one year was informed that he would have to pay only $2,000 in interest (i.e., a 20% interest rate). The lender stated that the total owed to him, $12,000, would be repaid at the rate of $1,000 per month for 12 months. What was the true effective annual interest rate being charged in this transaction? Why is it greater than the apparent rate of 20%? (4.6)

4-42. Determine the effective interest rate per year being charged in any one of the following loan situations. (4.6)

Amount of Loan	No. of Monthly Payments		
	24	36	48
$3000	$160.10	$119.27	$ 99.46
3500	186.79	139.15	116.04
4000	213.47	159.02	132.61
4500	240.15	178.90	149.19
5000	266.84	198.78	165.76

4-43. Construct the investment balance diagram for Problem 4-30. What additional insights into the profitability and liquidity of this new product line do you gain? (4.9)

4-44. A $20,000 ordinary life insurance policy for a 22-year-old female can be obtained for annual premiums of approximately $250. This type of policy (ordinary life) would pay a death benefit of $20,000 in exchange for annual premiums of $250 that are paid during the lifetime of the insured person. If the average life expectancy of a 22-year-old female is 77 years, what interest rate establishes equivalence between cash outflows and inflows for this type of insurance policy? Assume that all premiums are paid on a beginning-of-year basis and that the last premium is paid on the female's 76th birthday. (4.6)

4-45. Evaluate the acceptability of the following project with all methods discussed in Chapter 4. Let MARR = ϵ = 15% per year, maximum acceptable θ = 5 years, and maximum acceptable θ' = 6 years.

Project: R137-A
Title: Syn-Tree Fabrication
Description: Establish a production facility to manufacture synthetic palm trees for sale to resort areas in Alaska.

Cash Flow Estimates:

Year	Amount (thousands)
0	−$1,500
1	200
2	400
3	450
4	450
5	600
6	900
7	1,100

4-46. A homeowner purchased a split-level dwelling for $50,000 twenty years ago. The monthly payments on her $30,000 loan (30-year mortgage) have been $200.00 at a 7% nominal interest rate compounded monthly. The homeowner has just sold the dwelling for $100,000. Her before-tax rate of return, as an annual effective percentage, is closest to which of the following answers? (4.6)

a. 3.6%
b. 8.5%
c. 5.3%
d. 1.5%

4-47. With reference to the following cash flow diagram:

a. What is the breakeven life [θ'] of this project? (4.8)
b. What is the breakeven interest rate (i')? (4.6)
c. Draw the investment balance diagram. (4.9)

4-48. The Going Aircraft Corporation is manually producing a certain subassembly at a direct labor cost of $100,000 per year. This manual work can be totally automated so that $80,000 in direct labor and $20,000 in indirect labor and overhead costs will be saved each year. Annual maintenance for the automated system will be $10,000, and its market (salvage) value will be $7,000 at any time in the future. The system's useful life is 5 to 10 years, inclusive.

a. If the firm's MARR is 15% per year, develop a graph that shows how much money can be spent on the automated equipment. (*Hint:* Plot the PW of positive cash flows versus the useful life). (4.3)
b. When $N = 6$ years and $P = \$344,000$, what is the simple payback period? (4.8)

4-49. Consider this cash flow diagram:

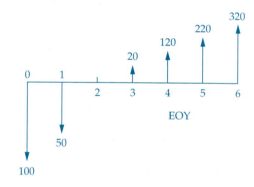

a. If the MARR is 15% per year, is this project financially profitable? (4.3)
b. Calculate the simple payback period, θ. (4.8)
c. Calculate the discounted payback period, θ'. (4.8)

4-50. Advanced manufacturing technology (AMT) typically exhibits net annual revenues that increase over a fairly long period of time. In the long run an AMT project may be profitable as measured by IRR, but its simple payback period may be unacceptable. Evaluate this AMT project when the company MARR is 15% per year and its maximum allowable payback period is three years: (4.6, 4.8)

Capital investment at time 0	−$100,000
Net revenues in year k	$20,000+
	$10,000 \cdot (k - 1)$
Market (salvage) value	$10,000
Life	5 years

a. The IRR equals _____. Use linear interpolation to determine the IRR.

b. The simple payback period equals _____.

4-51. A company has the opportunity to take over a redevelopment project in an industrial area of a city. No immediate investment is required, but it must raze the existing buildings over a 4-year period and at the end of the fourth year invest $2,400,000 for new construction. It will collect all revenues and pay all costs for a period of 10 years, at which time the entire project, and properties thereon, will revert to the city. The net cash flow is estimated to be as follows:

Year End	Net Cash Flow
1	$ 500,000
2	300,000
3	100,000
4	− 2,400,000
5	150,000
6	200,000
7	250,000
8	300,000
9	350,000
10	400,000

Tabulate the PW versus the interest rate and determine whether multiple IRRs exist. If so, use the ERR method when ϵ = 8% per year to determine a rate of return. (4.7)

4-52. A certain project has net receipts equaling $1,000 now, has costs of $5,000 at the end of the first year, and earns $6,000 at the end of the second year.

a. Show that multiple rates of return exist for this problem when using the IRR method (i' = 100%, 200%). (Appendix 4-A)

b. If an external reinvestment rate of 10% is available, what is the rate of return for this project using the ERR method? (4.7)

4-53. The prospective exploration for oil in the outer continental shelf by a small, independent drilling company has produced a rather curious pattern of cash flows, as follows:

End of Year	Net Cash Flow
0	−$ 520,000
1–10	+ 200,000
10	− 1,500,000

The $1,500,000 expense at the end of year 10 will be incurred by the company in dismantling the drilling rig.

a. Over the 10-year period, plot PW versus the interest rate (i) in an attempt to discover whether multiple rates of return exist. (4.6)

b. Based on the projected net cash flows and results in part (a), what would you recommend regarding the pursuit of this project? Customarily, the company expects to earn at least 20% per year on invested capital before taxes. Use the ERR method. (4.7)

4-54. *Brain Teaser*
In 1996 interest rates in the United States were declining, while interest rates in Germany were increasing. Suppose that interest rates in Germany turn out to be twice those in the United States. Develop a graphical model to explain how the ratio of CR costs in Germany to CR costs in the United States *varies* with (a) the magnitude of the interest rates in the United States, (b) the length of the study period, and (c) the balance between initial investment and terminal salvage value. What general observations can you make regarding how reduced interest rates and longer study periods affect the opportunity for productivity growth and increased competitiveness in the United States? (4.5)

APPENDIX 4-A

The Multiple Rate of Return Problem with the IRR Method

Whenever the IRR method is used and the cash flows reverse sign (from net cash outflow to net cash inflow or the opposite) more than once over the study period, one should be alert to the rather remote possibility that either no interest rate or multiple interest rates may exist. Actually, the *maximum* number of possible internal rates of return in the $(-1, \infty)$ interval for any given project is equal to the number of cash flow reversals during the study period. *The simplest way to check for multiple IRRs is to plot equivalent worth (e.g., PW) against the interest rate.* If the resulting plot crosses the interest rate axis more than once, multiple IRRs are present and another equivalence method is recommended for determining project acceptability.

As an example, consider the following project for which the IRR is desired.

EXAMPLE 4-A-1

Plot the present worth versus interest rate for the following cash flows. Are there multiple IRRs? If so, what do they mean?

Year, k	Net Cash Flow	$i\%$	PW($i\%$)
0	$500	0	$250
1	−1,000	10	150
2	0	20	32
3	250	30	~0
4	250	40	−11
5	250	62	~0
		80	24

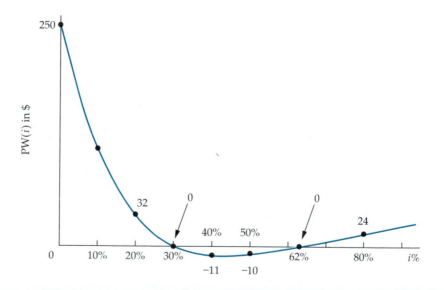

Thus, the PW of the net cash flows equals zero at interest rates of about 30% and 62%, so multiple IRRs do exist. Whenever there are multiple IRRs, which is seldom, it is likely that none is correct.

In this situation, the ERR method (see Section 4.7) could be used to decide whether the project is worthwhile. Or, we usually have the option of using an equivalent worth method. In Example 4-A-1, if the external reinvestment rate (ϵ) is 10% per year, we see that the ERR is 12.4%:

$$\$1,000(P/F, 10\%, 1)(F/P, i'\%, 5) = \$500(F/P, 10\%, 5) + \$250(F/A, 10\%, 3)$$

$$(P/F, 10\%, 1)(F/P, i', 5) = 1.632$$

$$i' = 0.124 \ (12.4\%)$$

In addition PW(10%)= $105, so both the ERR and PW methods indicate that this project is acceptable.

EXAMPLE 4-A-2

Use the ERR method to analyze the cash flow pattern shown in the following table. The IRR is indeterminant (none exists), so the IRR is not a workable procedure. The external reinvestment rate (ϵ) is 12% per year, and the MARR equals 15%.

Year	Cash Flows
0	$5,000
1	− 7,000
2	2,000
3	2,000

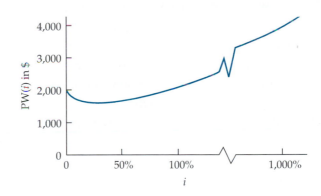

SOLUTION

The ERR method provides these results:

$$\$7,000(P/F, 12\%, 1)(F/P, i'\%, 3) = \$5,000(F/P, 12\%, 3)$$
$$+ \$2,000(F/P, 12\%, 1) + \$2,000$$

$$(F/P, i', 3) = 1.802$$

$$i' = 21.7\%$$

Thus, the ERR is greater than the MARR. Hence, the project having this cash flow pattern would be acceptable. The PW at 15% is equal to $1,740.36, which confirms the acceptability of this project.

Comparing Alternatives

*T*he primary objective of Chapter 5 is to develop and demonstrate methodology for the economic analysis and comparison of mutually exclusive alternatives.

The following topics are discussed in this chapter:

Basic concepts for comparing alternatives

The study (analysis) period

Useful lives equal to the study period

Useful lives are different among the alternatives

The capitalized worth method

Mutually exclusive combinations of projects

5.1 Introduction

Most engineering projects can be accomplished by more than one feasible design alternative. When the selection of one of these alternatives excludes the choice of any of the others, the alternatives are called *mutually exclusive*. Typically, the alternatives being considered require the investment of different amounts of capital, and their annual revenues and costs may vary. Sometimes the alternatives may have different useful lives. Because different levels of investment normally produce varying economic outcomes, we must perform an engineering economy study to determine which one of the mutually exclusive alternatives is preferred and, consequently, how much capital should be invested.

A seven-step procedure for accomplishing engineering economy studies was discussed in Chapter 1. In this chapter, we address Step 5 (analysis and comparison of the feasible alternatives) and Step 6 (selection of the preferred alternative) of the analysis procedure, and we compare mutually exclusive alternatives on the basis of economic considerations alone.

Five of the basic methods discussed in Chapter 4 for analyzing cash flows are used in the analyses in this chapter (PW, AW, FW, IRR, and ERR). These methods provide a *basis for economic comparison* of alternatives for an engineering project or related undertaking. When correctly applied, these methods result in the correct selection of a preferred alternative from a set of mutually exclusive alternatives. The comparison of mutually exclusive alternatives using the benefit/cost ratio method is discussed in Chapter 6.

5.2 Basic Concepts for Comparing Alternatives

Principle 1 (Chapter 1) emphasized that a choice (decision) is among alternatives. Such choices must incorporate the fundamental purpose of capital investment; namely, to obtain at least the MARR for each dollar invested. In practice, there are usually a limited number of feasible alternatives to consider. The problem of deciding which mutually exclusive alternatives should be selected is made easier if we adopt this rule based on Principle 2 in Chapter 1: *The alternative that requires the minimum investment of capital and produces satisfactory functional results will be chosen unless the incremental capital associated with an alternative having a larger investment can be justified with respect to its incremental savings (or benefits).*

Under this rule, we consider the alternative that requires the least investment of capital to be the *base alternative*. The investment of additional capital over that required by the base alternative usually results in increased capacity, increased quality, increased revenues, decreased operating expenses, or increased life. Therefore, before additional money is invested, it must be shown that each avoidable increment of capital can pay its own way relative to other available investment opportunities.

> In summary, *if* the extra benefits obtained by investing additional capital are better than those that could be obtained from investment of the same capital elsewhere at the MARR, the investment should be made. If this is not the case, we obviously would not invest more than the minimum amount of capital required, including the possibility of nothing at all. Stated simply, our rule will keep as much capital as possible invested at a rate of return equal to or greater than the MARR.

This basic policy for the comparison of mutually exclusive alternatives can be demonstrated with two examples. The *first* example involves an investment situation. Alternatives A and B are two mutually exclusive *investment alternatives* with estimated net cash flows,* as shown. *Investment alternatives are those with*

---

*In this book, the terms *net cash flow* and *cash flow* will be used interchangeably when referring to periodic cash inflows and cash outflows for an alternative.

initial (or front-end) capital investment(s) that produce positive cash flows from increased revenue, savings through reduced costs, or both. The useful life of each alternative is four years.

	Alternative	
	A	**B**
Capital investment	−$60,000	−$73,000
Annual revenues less expenses	22,000	26,225

The cash flow diagrams for Alternatives A and B, and for the year-by-year differences between them (i.e., B minus A), are shown in Figure 5-1. These diagrams typify those for investment alternatives. In this first example, at MARR = 10% per year, the PW values are:

$$PW(10\%)_A = -\$60{,}000 + \$22{,}000(P/A, 10\%, 4) = \$9{,}738$$

$$PW(10\%)_B = -\$73{,}000 + \$26{,}225(P/A, 10\%, 4) = \$10{,}131$$

Since the PW_A is greater than 0 at $i = $ MARR, it would be selected *unless* the additional (incremental) capital associated with Alternative B is justified. In this case, Alternative B is preferred to A because it has a greater PW value. Hence, *the extra benefits obtained by investing the additional $13,000 of capital in B* (diagram 3, Figure 5-1), have a net present equivalent value of $10,131 − $9,738 = $393.

FIGURE 5-1
Cash Flow Diagrams for Alternatives
A and B and Their Difference

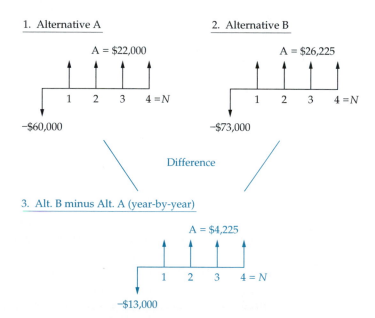

That is,

$$PW(10\%)_{Diff} = -\$13,000 + \$4,225(P/A, 10\%, 4) = \$393$$

and the additional capital invested in B is justified.

The *second* example involves a cost situation. Alternatives C and D are two mutually exclusive *cost alternatives* with estimated net cash flows as shown over a three-year life. *Cost alternatives are those with all negative cash flows except for a possible positive cash flow element from disposal of assets at the end of the project's useful life.* This situation occurs when the organization must take some action and the decision involves the most economical way of doing it (e.g., the addition of environmental control capability to meet new regulatory requirements).

End of Year	Alternative	
	C	D
0	−$380,000	−$415,000
1	− 38,100	− 27,400
2	− 39,100	− 27,400
3	− 40,100	− 27,400
3[a]	0	26,000

[a]Market value.

The cash flow diagrams for Alternatives C and D, and for the year-by-year differences between them (i.e., D minus C), are shown in Figure 5-2. These diagrams typify those for cost alternatives. In this "must take action" situation, Alternative C, which has the lesser capital investment, would be selected *unless* the additional (incremental) capital associated with Alternative D is justified. With the greater capital investment, Alternative D in this illustration has smaller annual expenses. Otherwise, it would not be a feasible alternative (it would not be logical to invest more capital in an alternative without obtaining additional revenues or savings). Note in diagram 3, Figure 5-2, that the difference between two feasible cost alternatives is an investment alternative.

In this second example, at MARR = 10% per year, the $PW(10\%)_C = -\$477,077$ and the $PW(10\%)_D = -\$463,607$. Alternative D is preferred to C because it has the lesser PW of costs. Hence, *the lower annual expenses obtained by investing the additional $35,000 of capital in Alternative D* have a net present worth of $-\$463,607 - (-\$477,077) = \$13,470$. That is, the $PW(10\%)_{Diff} = \$13,470$ and the additional capital invested in Alternative D is justified.

5.3 The Study (Analysis) Period

The study (analysis) period, sometimes called the *planning horizon*, is the selected time period over which mutually exclusive alternatives are compared. The determination of the study period for a decision situation may be influenced by several

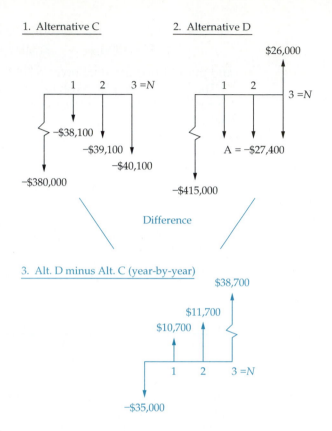

FIGURE 5-2

Cash Flow Diagrams for Alternatives C and D
and Their Difference

factors–for example, the service period required, the useful life* of the shorter-lived alternative, the useful life of the longer-lived alternative, company policy, and so on. *The key point is that the selected study period must be appropriate for the decision situation under investigation.*

The useful lives of alternatives being compared, relative to the selected study period, can involve two situations:

1. Case 1: Useful lives are the same for all alternatives and equal to the study period.
2. Case 2: Useful lives are different among the alternatives and at least one does not match the study period.

> Unequal lives among alternatives somewhat complicate their analysis and comparison. To make engineering economy studies in such cases, we adopt the rule of comparing mutually exclusive alternatives over the same period of time. The repeatability assumption and the coterminated assumption are the two types of assumptions used for these comparisons.

---

*The useful life of an asset is the time period during which it is kept in productive use in a trade or business.

The *repeatability assumption* involves two main conditions.

1. The study period over which the alternatives are being compared is either indefinitely long or equal to a common multiple of the lives of the alternatives.
2. The economic consequences that are estimated to happen in an alternative's initial life span will also happen in all succeeding life spans (replacements).

Actual situations in engineering practice seldom meet both conditions. This has tended to limit the use of the repeatability assumption, except in those situations where the difference between the annual worth of the first life cycle and the annual worth over more than one life cycle of the assets involved is quite small.[*]

The *coterminated assumption* uses a finite and identical study period for all alternatives. This planning horizon, combined with appropriate adjustments to the estimated cash flows, puts the alternatives on a common and comparable basis. For example, if the situation involves providing a service, the same time period requirement applies to each alternative in the comparison. To force a match of cash flow durations to the cotermination time, adjustments are made to cash flow estimates of alternatives having useful lives different from the study period. For example, if an alternative has a useful life shorter than the study period, the estimated annual cost of contracting for the activities involved might be used during the remaining years. Similarly, if the useful life of an alternative is longer than the study period, a re-estimated market value is normally used as a terminal cash flow at the end of a project's coterminated life.

5.4 Case 1: Useful Lives Are Equal to the Study Period

When the useful life of an alternative is equal to the selected study period, adjustments to the cash flows are not required. In this section, we discuss the comparison of mutually exclusive alternatives using equivalent worth methods and rate of return methods when the useful lives of all alternatives are equal to the study period.

5.4.1 Equivalent Worth Methods

In Chapter 4, we learned that the equivalent worth methods convert all relevant cash flows into equivalent present, annual, or future amounts. When these methods are used, consistency of alternative selection results from this equivalency relationship. Also, the economic ranking of mutually exclusive alternatives will be the same using the three methods. Consider the general case of two alternatives, A and B. If

$$PW_A(i) < PW_B(i)$$

---

[*]T. G. Eschenbach and A. E. Smith, "Violating the Identical Repetition Assumption of EAC," *Proceedings, International Industrial Engineering Conference* (May 1990), The Institute of Industrial Engineers, Norcross, Georgia, pp. 99–104.

then

$$PW_A(i)(A/P,i,N) < PW_B(i)(A/P,i,N)$$

and

$$AW_A(i) < AW_B(i)$$

Similarly,

$$PW_A(i)(F/P,i,N) < PW_B(i)(F/P,i,N)$$

and

$$FW_A(i) < FW_B(i)$$

> The most straightforward technique in comparing mutually exclusive alternatives when using equivalent worth methods, and when all useful lives are equal to the study period, is to determine the equivalent worth of each alternative based on total investment at i = MARR. Then, for investment alternatives, the one with the greatest positive equivalent worth is selected. And, in the case of cost alternatives, the one with the least negative equivalent worth is selected.

EXAMPLE 5-1

Three mutually exclusive investment alternatives for implementing an office automation plan in an engineering design firm are being considered. The study period is 10 years, and the useful lives of all three alternatives are also 10 years. Market values of all alternatives are assumed to be zero at the end of their useful lives. If the firm's MARR is 10% per year, which alternative should be selected in view of the following estimates?

	Alternative		
	A	B	C
Capital investment	−$390,000	−$920,000	−$660,000
Net annual revenues less expenses	69,000	167,000	133,500

SOLUTION OF EXAMPLE 5-1 BY THE PW METHOD

$$PW(10\%)_A = -\$390,000 + \$69,000(P/A, 10\%, 10) = \$33,977$$

$$PW(10\%)_B = -\$920,000 + \$167,000(P/A, 10\%, 10) = \$106,148$$

$$PW(10\%)_C = -\$660,000 + \$133,500(P/A, 10\%, 10) = \$160,304$$

Based on the PW method, Alternative C would be selected because it has the largest PW value ($160,304). The order of preference is C > B > A, where C > B means C is preferred to B.

SOLUTION OF EXAMPLE 5-1 BY THE AW METHOD

$$AW(10\%)_A = -\$390,000(A/P, 10\%, 10) + \$69,000 = \$5,547$$
$$AW(10\%)_B = -\$920,000(A/P, 10\%, 10) + \$167,000 = \$17,316$$
$$AW(10\%)_C = -\$660,000(A/P, 10\%, 10) + \$133,500 = \$26,118$$

Alternative C is again chosen because it has the largest AW value ($26,118).

SOLUTION OF EXAMPLE 5-1 BY THE FW METHOD

$$FW(10\%)_A = -\$390,000(F/P, 10\%, 10) + \$69,000(F/A, 10\%, 10) = \$88,138$$
$$FW(10\%)_B = -\$920,000(F/P, 10\%, 10) + \$167,000(F/A, 10\%, 10) = \$275,342$$
$$FW(10\%)_C = -\$660,000(F/P, 10\%, 10) + \$133,500(F/A, 10\%, 10) = \$415,801$$

Based on the FW method, the choice is again Alternative C because it has the largest FW value ($415,801). For all three methods (PW, AW, and FW) in this example, notice that C > B > A because of the equivalency relationship among the methods.

EXAMPLE 5-2

A company is planning to install a new automated plastic-molding press. Four different presses are available. The initial capital investments and annual expenses for these four mutually exclusive alternatives are as follows:

	Press			
	P1	**P2**	**P3**	**P4**
Capital investment	−$24,000	−$30,400	−$49,600	−$52,000
Useful life (years)	5	5	5	5
Annual expenses				
Power	− 2,720	− 2,720	− 4,800	− 5,040
Labor	− 26,400	− 24,000	− 16,800	− 14,800
Maintenance	− 1,600	− 1,800	− 2,600	− 2,000
Property taxes and insurance	− 480	− 608	− 992	− 1,040
Total annual expenses	−$31,200	−$29,128	−$25,192	−$22,880

Each press will produce the same number of units. Therefore, revenue is independent of the press selected and we can solve the problem as a cost-type situation (i.e., revenue can be ignored). However, because of different degrees of automation, the presses require different amounts and grades of labor and also have different annual operation and maintenance expenses. None is expected to have a market value at the end of its useful life, and the selected study period is five years. Any additional capital invested is expected to earn at least 10% per year before taxes. Which press should be chosen?

SOLUTION OF EXAMPLE 5-2 BY THE PW METHOD

When cost alternatives are compared using the PW method, the alternative that has the least negative PW value is economically the most desirable. Table 5-1 shows the analysis by the PW method. The alternative with the least negative PW of costs is P4 (−$138,734). The preference among the feasible alternatives in *decreasing* order is P4, P2, P1, and P3. As previously discussed, this rank ordering is identical for all equivalent worth methods when they are correctly applied.

TABLE 5-1   Comparison of Four Molding Presses Using the PW Method (Example 5-2)

	Press			
	P1	P2	P3	P4
PW of				
Capital investment	−$ 24,000	−$ 30,400	−$ 49,600	−$ 52,000
Annual expenses:				
(Total annual expenses)				
× (P/A, 10%, 5)	− 118,273	− 110,418	− 95,498	− 86,734
Total PW	−$142,273	−$140,818	−$145,098	−$138,734

TABLE 5-2   Comparison of Four Molding Presses Using the AW Method (Example 5-2)

	Press			
	P1	P2	P3	P4
AW of				
Annual expenses	−$31,200	−$29,128	−$25,192	−$22,880
Capital recovery amount:				
(Investment) × (A/P, 10%, 5)	− 6,331	− 8,020	− 13,084	− 13,718
Total AW	−$37,531	−$37,148	−$38,276	−$36,598

TABLE 5-3   Comparison of Four Molding Presses Using the FW Method (Example 5-2)

	Press			
	P1	P2	P3	P4
FW of				
Capital investment:				
(Investment) × (F/P, 10%, 5)	−$ 38,652	−$ 48,959	−$ 79,881	−$ 83,746
Annual expenses:				
(Total annual expenses)				
× (F/A, 10%, 5)	− 190,479	− 177,829	− 153,800	− 139,685
Total FW	−$229,131	−$226,788	−$233,689	−$223,431

SOLUTION OF EXAMPLE 5-2 BY THE AW METHOD
Table 5-2 shows the analysis by the AW method. The economic criterion is to choose that alternative with the least negative AW of costs, which is P4 (−$36,598).

SOLUTION OF EXAMPLE 5-2 USING THE FW METHOD
Table 5-3 shows the analysis by the FW method. Once again, alternative P4 is chosen because it has the least negative FW of costs (−$223,431).

5.4.2 Rate of Return Methods

The basic concepts discussed in Section 5.2 are directly applied when using rate of return methods to evaluate mutually exclusive alternatives. The best alternative produces satisfactory functional results and requires the minimum investment of capital, unless a larger investment can be justified with respect to the incremental benefits and costs it produces. This can be reduced to three rules when applying rate of return methods:

1. Each increment of capital must justify itself by producing a sufficient rate of return on that increment.
2. Compare a higher investment alternative against a lower investment alternative only when the latter is acceptable.
3. Select the alternative that requires the largest investment of capital as long as the incremental investment is justified by benefits that earn at least the MARR. This maximizes equivalent worth on total investment at $i = $ MARR.

These rules can be implemented using the *incremental investment analysis technique* with rate of return methods.* First, however, we will discuss the *inconsistent ranking problem* that can occur with incorrect use of rate of return methods in the comparison of alternatives.

5.4.2.1. The Inconsistent Ranking Problem In Section 5.2, we discussed a small investment project involving two alternatives, A and B. The cash flow for each alternative is restated here, as well as the cash flow (incremental) difference.

	Alternative		Difference
	A	**B**	**Δ (B − A)**
Capital investment	−$60,000	−$73,000	−$13,000
Annual revenues less expenses	22,000	26,225	4,225

The useful life of each alternative and the study period is four years. Also, assume that the MARR = 10% per year. The IRR and PW(10%) of each alternative are

---

*The benefit-cost (B/C) ratio method is also applied using the incremental investment analysis technique when comparing mutually exclusive alternatives. This is discussed in Chapter 6.

as follows:

Alternative	IRR	PW (10%)
A	17.3%	$ 9,738
B	16.3	10,131

If, at this point, a choice were made based on maximizing the IRR of the total cash flow, Alternative A would be selected. But, based on maximizing the PW of the total investment at $i =$ MARR, Alternative B is preferred. Obviously, here we have an inconsistent ranking of the two mutually exclusive investment alternatives.

Now that we know Alternative A is acceptable (IRR > MARR; PW at MARR > 0), we will analyze the incremental cash flow between the two alternatives, which we shall refer to as Δ(B − A). The IRR of this increment, IRR$_\Delta$, is 11.4%. This is greater than the MARR of 10%, and the incremental investment of $13,000 is justified. This outcome is confirmed by the PW of the increment, PW$_\Delta$(10%), which is equal to $393. Thus when the IRR of the incremental cash flow is used, versus the IRR of the total cash flow for each alternative, the ranking of A and B is consistent with that based on the PW on total investment.

The fundamental role that the incremental net cash flow, Δ(B − A), plays in the comparison of two alternatives (where B has the greater capital investment) is based on the relationship:

$$\text{Cash flow of B} = \text{cash flow of A} + \text{cash flow of the difference}$$

Clearly, the cash flow of B is made up of two parts. The first part is equal to the cash flow of Alternative A, and the second part is the incremental cash flow between A and B. Obviously, if the equivalent worth of the difference is greater than zero at $i =$ MARR, then Alternative B is preferred. Otherwise, given that Alternative A is justified (an acceptable *base alternative*), Alternative A is preferred. It is always true that if PW$_\Delta \geq 0$, then IRR$_\Delta \geq$ MARR. Therefore, in this example Alternative B is preferred to A.

Figure 5-3 illustrates how ranking errors can occur when a selection among mutually exclusive alternatives is based wrongly on maximization of IRR on the total cash flow. When the MARR lies to the left of IRR$_\Delta$ (11.4% in this case), an incorrect choice will be made by selecting an alternative that maximizes internal rate of return. This is because the IRR method assumes reinvestment of cash flows at the calculated rate of return (17.4% and 16.3%, respectively, for Alternatives A and B in this case) whereas the PW method assumes reinvestment at the MARR (10%). In this example, reinvestment can occur at a rate as high as 11.4% without causing a reversal in preference (ranking).

Figure 5-3 shows our previous results with PW$_B$ > PW$_A$ at MARR $=$ 10%, even though IRR$_A$ > IRR$_B$. Also, the figure shows how to *avoid this ranking inconsistency by examining the IRR of the increment, IRR$_\Delta$*, which correctly leads to the selection of Alternative B, the same as the PW method.

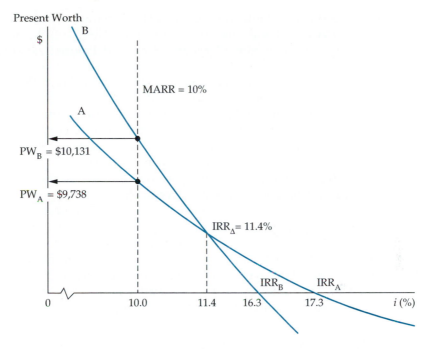

FIGURE 5-3 Illustration of the Ranking Error in Studies Using the Internal Rate of Return Method

5.4.2.2 The Incremental Investment Analysis Procedure

> We recommend the incremental investment analysis procedure to avoid incorrect ranking of mutually exclusive alternatives when using rate of return methods. We will use this procedure in the remainder of the book.

The incremental analysis procedure for the comparison of mutually exclusive alternatives is summarized in three basic steps:

1. Arrange (order) the feasible alternatives based on increasing capital investment.*
2. Establish a base alternative.
   a. Cost alternatives—The first alternative (least capital investment) is the base.
   b. Investment alternatives—If the first alternative is acceptable (IRR > MARR; PW, FW, or AW at MARR > 0), select it as the base. If the first alternative

---

*This ranking rule assumes a logical set of mutually exclusive alternatives. That is to say, for investment or cost alternatives, increased initial investment results in additional economic benefits, whether from added revenues, reduced costs, or a combination of both. Also this rule assumes that for any nonconventional investment cash flow, the PW, AW, or FW analysis method would be used instead of IRR. Simply stated, a nonconventional investment cash flow involves multiple sign changes or positive cash flow at time 0, or both. For a more detailed discussion of ranking rules, see C. S. Park and G. P. Sharp-Bette, *Advanced Engineering Economy* (New York: John Wiley and Sons, 1990).

is not acceptable, choose the next alternative in order of increasing capital investment and check the profitability criterion (PW, etc.) values. Continue until an acceptable alternative is obtained. If none is obtained, the do-nothing alternative is selected.

3. Use iteration to evaluate differences (incremental cash flows) between alternatives until no more alternatives exist.

   a. If the incremental cash flow between the next (higher capital investment) alternative and the current selected alternative is acceptable, choose the next alternative as the current best alternative. Otherwise, retain the last acceptable alternative as the current best.

   b. Repeat, and select as the preferred alternative the last one for which the incremental cash flow was acceptable.

EXAMPLE 5-3

Suppose that we are analyzing the following six mutually exclusive alternatives for a project (arranged in ascending order of initial investment) using the IRR method. The useful life of each alternative is 10 years, and the MARR is 10% per year. Also, net annual revenues less expenses vary among all alternatives. If the study period is 10 years, and the salvage (market) values are 0, which alternative should be chosen?

| | **Alternative** | | | | | |
	A	B	C	D	E	F
Capital investment	−$900	−$1,500	−$2,500	−$4,000	−$5,000	−$7,000
Annual revenues less expenses	150	276	400	925	1,125	1,425

SOLUTION (IRR METHOD)

For each of the feasible alternatives, the IRR on the total cash flow can be computed by determining the interest rate at which the PW equals zero. We can also solve for the IRR by determining the interest rate at which the FW or AW is equal to zero (use of AW is illustrated for Alternative A):

$$0 = -\$900(A/P, i'\%, 10) + \$150; \quad i'\% = ?$$

By trial and error, we determine that $i'\% = 10.6\%$. In the same manner, the IRRs of all the alternatives are computed and summarized:

	A	B	C	D	E	F
IRR on total cash flow	10.6%	13.0%	9.6%	19.1%	18.3%	15.6%

At this point, *only Alternative C is unacceptable* and can be eliminated from the comparison because its IRR is less than the MARR of 10% per year. Also, A is the base alternative from which to begin the incremental investment analysis

TABLE 5-4   Comparison of Five Acceptable Investment Alternatives Using the IRR Method (Example 5-3)

Increment Considered	A	Δ (B − A)	Δ (D − B)	Δ (E − D)	Δ (F − E)
Δ Capital investment	−$900	−$600	−$2,500	−$1,000	−$2,000
Δ Annual revenues less expenses	$150	$126	$649	$200	$300
IRR$_\Delta$	10.6%	16.4%	22.6%	15.1%	8.1%
Is increment justified?	Yes	Yes	Yes	Yes	No

procedure because it is the feasible alternative with the lowest investment cost whose IRR (10.6%) is equal to or greater than the MARR (10%). This preanalysis of the feasibility of each alternative is not required by the incremental analysis procedure. However, it is useful when analyzing a larger set of mutually exclusive alternatives. You can immediately eliminate nonfeasible alternatives, as well as easily identify the base alternative.

As discussed in Section 5.4.2.1, it is not necessarily correct to select the alternative that maximizes the IRR on total cash flow. That is to say, Alternative D may not be the best choice, since maximization of IRR does not guarantee maximization of equivalent worth on total investment at the MARR, and resultant maximization of an organization's future wealth to its owners. Therefore, to make the correct choice, we must examine each increment of investment capital to see if it will pay its own way. Table 5-4 provides the analysis of the five remaining alternatives, and the IRRs on incremental cash flows are again computed using the AW method.

From Table 5-4, it is apparent that Alternative E will be chosen because it requires the largest investment for which the last increment of investment capital is justified. That is, we desire to invest additional increments of the $7,000 presumably available for this project as long as each avoidable increment of investment can earn 10% per year or better.

It was assumed in Example 5-3 (and in all other examples involving mutually exclusive alternatives, unless noted to the contrary) that available capital for a project *not* committed to one of the feasible alternatives is invested in some other project where it will earn a return equal to the MARR. Therefore, in this case the $2,000 left over by selecting Alternative E instead of F is assumed to earn 10% per year elsewhere, which is more than we could obtain by investing it in F.

In summary, three errors commonly made in this type of analysis are to choose the feasible alternative (1) with the highest overall IRR on total cash flow, or (2) with the highest IRR on an incremental capital investment, or (3) with the largest capital investment that has an IRR greater than or equal to the MARR. None of these criteria are generally correct. For instance, in Example 5-3 one might erroneously choose Alternative D rather than E because the IRR for the increment from B to D is 22.6% and that from D to E is only 15.1% (error 2). A more obvious error, as previously discussed, is the temptation to maximize the IRR on total cash flow and select Alternative D (error 1). The third error would be committed by selecting Alternative F for the reason that it has the largest total investment with an IRR greater than the MARR (15.6% > 10%).

The equivalent worth methods may also be applied using the incremental analysis procedure to compare mutually exclusive alternatives. The ranking of the alternatives will be consistent with that of the equivalent worth values based on total investment for each alternative. The ranking will also be consistent with that of the rate of return methods when using incremental analysis. When the equivalent worth of an investment cash flow is greater than zero at i = MARR, its IRR is greater than the MARR. Therefore, the equivalent worth methods, using incremental investment analysis, can be used as a *screening method* for the IRR method. That is, the same decisions are made concerning additional increments of capital investment. These points are included in Example 5-4.

EXAMPLE 5-4

The estimated capital investment and the annual expenses (based on 1,500 hours of operation per year) for four alternative designs of a diesel-powered air compressor are shown, as well as the estimated market value for each design at the end of the common five-year useful life. The perspective (Principle 3, Chapter 1) of these cost estimates is that of the typical user (construction company, plant facilities department, government highway department, and so on). The study period is five years, and the MARR is 20% per year. One of the designs must be selected for the compressor, and each design provides the same level of service. Based on this information, (1) determine the preferred alternative design using the IRR method, and (2) show that the PW method (i = MARR), using the incremental analysis procedure, results in the same decision. Observe that this example is a *cost type situation with four mutually exclusive cost alternatives.* The following solution demonstrates the use of the incremental analysis procedure to compare cost alternatives.

| | Design Alternative | | | |
	D1	D2	D3	D4
Capital investment	−$100,000	−$140,600	−$148,200	−$122,000
Annual expenses	− 29,000	− 16,900	− 14,800	− 22,100
Useful life (years)	5	5	5	5
Market value	10,000	14,000	25,600	14,000

SOLUTION

The first step is to arrange (order) the four mutually exclusive cost alternatives based on their increasing investment costs. Therefore, the order of the alternatives for incremental analysis is D1, D4, D2, and D3.

Since these are cost alternatives, the one with the least capital investment, D1, is the base alternative. Therefore, the base alternative will be preferred unless additional increments of capital investment can produce cost savings (benefits) that lead to a return equal to or greater than the MARR.

TABLE 5-5 Comparison of Four Cost (Design) Alternatives Using the IRR and PW Methods with Incremental Analysis (Example 5-4)

Increment Considered	Δ (D4 − D1)	Δ (D2 − D4)	Δ (D3 − D4)
Δ Capital investment	−$22,000	−$18,600	−$26,200
Δ Annual expense savings	6,900	5,200	7,300
Δ Market value	4,000	0	11,600
Useful life (years)	5	5	5
IRR_Δ	20.5%	12.3%	20.4%
Is increment justified?	Yes	No	Yes
$PW_\Delta(20\%)$	$243	−$3,049	$293
Is increment justified?	Yes	No	Yes

Incremental Investment Analysis   Selection

Increment of Investment	Capital Investment	IRR_Δ	Design	Capital Investment
Δ (D3 − D4)	$26,200	20.4% (Accept)		
Δ (D2 − D4)	$18,600	12.3% (Reject)		
Δ (D4 − D1)	$22,000	20.5% (Accept)	D3*	$148,200
D1	$100,000	Base Alternative*		

*Since these are cost alternatives, the IRR can not be determined.

FIGURE 5-4
Representation of Capital Investment Increments and IRR on Increments Considered in Selecting Design 3 (D3) in Example 5-4

The first incremental cash flow to be analyzed is that between designs D1 and D4, Δ(D4 − D1). The results of this analysis, and of subsequent differences between the cost alternatives, are summarized in Table 5-5, and the incremental investment analysis for the IRR method is illustrated in Figure 5-4. These results show the following:

1. The incremental cash flows between the cost alternatives are, in fact, investment alternatives.
2. The first increment, Δ(D4 − D1), is justified (IRR_Δ = 20.5% is greater than MARR = 20%, and $PW_\Delta(20\%)$ = $243 > 0); the increment Δ(D2 − D4) is not justified; and the last increment, Δ(D3 − D4)—not Δ(D3 − D2) because Design D2 has already been shown to be unacceptable—is justified, resulting in the selection of Design D3 for the air compressor. It is the highest investment for which each increment of investment capital is justified from the user's perspective.

3. The same capital investment decision results from the IRR method and the PW method using the incremental analysis procedure, because when the equivalent worth of an investment at i = MARR is greater than zero, its IRR is greater than the MARR (from the definition of the IRR; Chapter 4).

The external rate of return (ERR) method was explained in Chapter 4. Also, in Appendix 4A, the ERR method was illustrated as a substitute for the IRR method when analyzing a nonconventional investment type of cash flow. In Example 5-5, the ERR method is applied using the incremental investment analysis procedure to compare the mutually exclusive alternatives for an engineering improvement project.

EXAMPLE 5-5

In an automotive parts plant, an engineering team is analyzing an improvement project to increase the productivity of a flexible manufacturing center. The estimated net cash flows for the three feasible alternatives being compared are shown in Table 5-6. The analysis period is six years, and the MARR for capital investments at the plant is 20% per year before taxes. Using the ERR method, which alternative should be selected? (ϵ = MARR.)

SOLUTION
The procedure for using the ERR method to compare mutually exclusive alternatives is the same as for the IRR method. The only difference is in the calculation methodology.

Table 5-6 provides a tabulation of the calculation and acceptability of each increment of capital investment considered. Since these three feasible alternatives

TABLE 5-6   Comparison of Three Mutually Exclusive Alternatives Using the ERR Method (Example 5-5)

	Alternative Cash Flows			Incremental Analysis of Alternatives		
End of Period	A	B	C	A[a]	Δ(B − A)	Δ(C − A)
0	−$640,000	−$680,000	−$755,000	−$640,000	−$ 40,000	−$115,000
1	262,000	− 40,000	205,000	262,000	− 302,000	− 57,000
2	290,000	392,000	406,000	290,000	102,000	116,000
3	302,000	380,000	400,000	302,000	78,000	98,000
4	310,000	380,000	390,000	310,000	70,000	80,000
5	310,000	380,000	390,000	310,000	70,000	80,000
6	260,000	380,000	324,000	260,000	120,000	64,000

Incremental analysis:		
Δ PW of negative cash flow amounts	−640,000	−291,657	−162,498
Δ FW of positive cash flow amounts	2,853,535	651,091	685,082
ERR	28.3%	14.3%	27.1%
Is increment justified?	Yes	No	Yes

[a]The net cash flow for Alternative A, which is the incremental cash flow between making no change ($0) and implementing Alternative A.

are a mutually exclusive set of investment alternatives, the first cash flow analyzed is the one with the least investment cost, Alternative A. For this increment of investment, the PW of the negative cash flow amounts (at $i = \epsilon\%$) is just the $-\$640,000$ investment cost. Therefore, the ERR for Alternative A is the following:

$$|-\$640,000|(F/P, i'\%, 6) = \$262,000(F/P, 20\%, 5) + \cdots + \$260,000$$
$$= \$2,853,535$$
$$(F/P, i'\%, 6) = (1 + i')^6 = \$2,853,535/\$640,000 = 4.4586$$
$$(1 + i') = (4.4586)^{1/6} = 1.2829$$
$$i' = 0.2829, \text{ or ERR} = 28.3\%$$

Using an MARR = 20% per year, this increment of investment is justified and Alternative A is an acceptable base alternative. By using similar calculations, the increment $\Delta(B - A)$, earning 14.3%, is not justified and the increment $\Delta(C - A)$, earning 27.1%, is justified. Therefore, Alternative C is the preferred alternative for the improvement project.

By this point in the chapter, three key observations are clear concerning the comparison of mutually exclusive alternatives: (1) equivalent worth methods are computationally less cumbersome to use, (2) both the equivalent worth and rate of return methods, if used properly, will consistently recommend the best alternative, but (3) rate of return methods may not produce correct choices if the analyst or the manager insists on maximizing the rate of return on the total cash flow. That is, incremental analysis must be used with rate of return methods to ensure that the best alternative is selected.

To reinforce these points further, consider the assignment given to Cynthia Jones in Example 5-6.

EXAMPLE 5-6

The owner of a downtown parking lot has retained an architectural engineering firm to determine whether it would be financially attractive to construct an office building on the site now being used for parking. If retained for parking, improvements are required for its continued use. Cynthia Jones, a newly hired civil engineer and member of the project team, has been requested to perform the analysis and offer a recommendation. The data she has assembled on the four feasible, mutually exclusive alternatives developed by the project team are summarized here:

Alternative	Capital Investment (Including Land)	Net Annual Income
P. Keep existing parking lot, but improve	$-\$ 200,000$	$\$ 22,000$
B1. Construct one-story building	$- 4,000,000$	$600,000$
B2. Construct two-story building	$- 5,550,000$	$720,000$
B3. Construct three-story building	$- 7,500,000$	$960,000$

(a) The study period selected is 15 years. For each alternative, the property has an estimated residual value at the end of 15 years that is *equal* to the capital investment shown. The owner of the parking lot prefers the information from the IRR method, but the firm's manager has always insisted on a PW analysis. Therefore, she decides to make the analysis by using both methods. If the MARR equals 10% per year, which alternative should Cynthia recommend?

(b) Repeat (a) but assume that all alternatives have an estimated residual value of 50% of the investment cost shown.

SOLUTION

(a) The PWs of each alternative are compared as follows:

$$PW(10\%)_P = -\$200,000 + \$22,000(P/A, 10\%, 15) + \$200,000(P/F, 10\%, 15)$$
$$= \$15,214$$

$$PW(10\%)_{B1} = -\$4,000,000 + \$600,000(P/A, 10\%, 15) + \$4,000,000(P/F, 10\%, 15)$$
$$= \$1,521,260$$

$$PW(10\%)_{B2} = -\$5,550,000 + \$720,000(P/A, 10\%, 15) + \$5,550,000(P/F, 10\%, 15)$$
$$= \$1,255,062$$

$$PW(10\%)_{B3} = -\$7,500,000 + \$960,000(P/A, 10\%, 15) + \$7,500,000(P/F, 10\%, 15)$$
$$= \$1,597,356$$

To maximize PW, Cynthia would recommend the construction of a three-story office building (Alternative B3).

The IRR method is now applied to the analysis of this project:

	Mutually Exclusive Alternatives			
	P	B1	B2	B3
Capital investment	−$200,000	−$4,000,000	−$5,550,000	−$7,500,000
Net annual income	22,000	600,000	720,000	960,000
Residual value	200,000	4,000,000	5,550,000	7,500,000
IRR[a]	11%	15%	13%	12.8%

[a] When initial investment equals residual (market) value, the IRR = net annual income ÷ initial investment.

Cynthia observes that IRR on total cash flow is maximized for Alternative B1, *but she knows this is not necessarily the correct choice.* She notes that all alternatives have IRRs that exceed MARR = 10%; therefore, Alternative P (which is the feasible alternative with the lowest investment cost whose IRR ≥ MARR) is an acceptable base alternative from which to begin the analysis of the incremental cash flows. The next step (Table 5-7) is to examine the IRRs on these incremental cash flows after ranking the alternatives from the lowest capital investment to the highest.

Based on this analysis, Cynthia would recommend that the owner of the parking lot seriously consider constructing a three-story office building (Alternative B3).

TABLE 5-7   Example 5-6 with Residual Value Equal to the Initial Capital Investment (IRR Method)

| | | Incremental Analysis of Alternatives | | |
	P^a	$\Delta(B1 - P)$	$\Delta(B2 - B1)$	$\Delta(B3 - B1)$
Δ Capital investment	−$200,000	−$3,800,000	−$1,550,000	−$3,500,000
Δ Annual income	22,000	578,000	120,000	360,000
Δ Residual value	200,000	3,800,000	1,550,000	3,500,000
IRR$_\Delta$	11%	15.2%	7.7%	10.3%
Decision	Keep parking lot	Accept one-story bldg., reject parking lot	Keep one-story bldg., reject two-story bldg.	Accept three-story bldg., reject one-story bldg.

^{a}The net cash flow for Alternative P, which is the incremental cash flow between making no change ($0) and implementing Alternative P.

(b) When the residual value of the property is estimated to be one-half of the capital investment, each alternative's PW can be quickly recomputed:

$$PW_P = -\$200,000 + \$22,000(P/A, 10\%, 15) + \$100,000 \ (P/F, 10\%, 15)$$
$$= -\$8,726$$

Similarly,

$$PW(10\%)_{B1} = \$1,042,460$$
$$PW(10\%)_{B2} = \$590,727$$
$$PW(10\%)_{B3} = \$699,606$$

Under this assumption for the residual value, the one-story building (Alternative B1) would be recommended.

Again, the IRR method requires more time and computation effort:

| | Mutually Exclusive Alternatives | | | |
	P	B1	B2	B3
Capital investment	−$200,000	−$4,000,000	−$5,550,000	−$7,500,000
Net annual income	22,000	600,000	720,000	960,000
Residual value	100,000	2,000,000	2,775,000	3,750,000
IRR^a	9.3%	13.8%	11.6%	11.4%

^{a}For example, the IRR of Alternative P is computed as follows: $0 = -\$200,000 + \$22,000 \ (P/A, i'\%, 15) + \$100,000 \ (P/F, i'\%, 15)$; $i'\% = ?$ By trial and error, $i' = 9.3\%$.

Alternative P is now unacceptable ($9.3\% < 10\%$) and cannot serve as the base alternative from which to proceed with the incremental analysis procedure. However, Alternative B1 is acceptable and has the lowest capital investment of the

TABLE 5-8   Example 5-6 with Residual Value Equal to One-Half of the Initial Capital Investment (IRR Method)

	Incremental Analysis of Alternatives		
	B1[b]	**Δ(B2 − B1)**	**Δ(B3 − B1)**
Δ Capital investment	−$4,000,000	−$1,550,000	−$3,500,000
Δ Annual income	600,000	120,000	360,000
Δ Residual value	2,000,000	775,000	1,750,000
IRR$_Δ^a$	13.8%	5.5%	8.5%
Decision	Accept one-story bldg.	Keep one-story bldg., reject two-story bldg.	Keep one-story bldg., reject three-story bldg.

[a]For instance, the IRR of Δ(B2 − B1) is determined as follows: $0 = -\$1{,}550{,}000 + \$120{,}000(P/A, i'\%, 15) + \$775{,}000(P/F, i'\%, 15); i' = 5.5\%$.

[b]The net cash flow for Alternative B1, which is the incremental cash flow between making no change ($0) and implementing Alternative B1.

remaining three feasible alternatives, so the incremental analysis would proceed as shown in Table 5-8.

Finally, Cynthia concludes that the one-story building is also best by using the IRR method when the estimated residual value is one-half of the capital investment. At this point, she tells her manager: "If I ever have to repeat this sort of analysis involving mutually exclusive alternatives, I'm going to insist on using an equivalent worth method such as PW or having a better computer program." The manager agrees and comments later to an associate that Cynthia is on her way up in the organization.

5.5 Case 2: Useful Lives Are Different Among the Alternatives

When the useful lives of mutually exclusive alternatives are different, the *repeatability assumption* may be used in their comparison if the study period can be infinite in length or a common multiple of the useful lives. This assumes that the economic estimates for an alternative's initial useful life cycle will be repeated in all subsequent replacement cycles. As we discussed in Section 5.3, this condition is more robust for practical application than it may appear. Another viewpoint is to consider the repeatability assumption as a modeling convenience for the purpose of making a current decision. *When this assumption is applicable to a decision situation, it simplifies comparison of the mutually exclusive alternatives.* One solution method often used is to compute the AW of each alternative over its useful life, and select the one with the best value (i.e., the alternative with the largest positive AW value for investment alternatives, and the one with the least negative AW value for cost alternatives).

If the repeatability assumption is not applicable to a decision situation, then an appropriate study period needs to be selected (*coterminated assumption*). This is the approach most frequently used in engineering practice. Often, one or more of the useful lives will be shorter or longer than the selected study period. When this is the case, cash flow adjustments based on additional assumptions need to be used *so all the alternatives are compared over the same study period*. The following guidelines apply to this situation:

1. (Useful life) < (Study period)
   a. Cost alternatives: Since each cost alternative has to provide the same level of service over the study period, contracting for the service or leasing the needed equipment for the remaining years may be appropriate. Another potential course of action is to repeat part of the useful life of the original alternative, and then use an estimated market value to truncate it at the end of the study period.
   b. Investment alternatives: The assumption used is that all cash flows will be reinvested in other opportunities available to the firm at the MARR to the end of the study period. A convenient solution method is to calculate the FW of each mutually exclusive alternative at the end of the study period. The PW can also be used for investment alternatives since the FW at the end of the study period, say N, of each alternative is its PW times a common constant $(F/P, i\%, N)$.
2. (Useful life) > (Study period): The most common technique is to truncate the alternative at the end of the study period using an estimated market value. This assumes that the disposable assets will be sold at the end of the study period at that value.

> The underlying principle, as discussed in Section 5.3, is to compare the mutually exclusive alternatives being considered in a decision situation over the same study (analysis) period.

EXAMPLE 5-7

The following data have been estimated for two mutually exclusive investment alternatives, A and B, associated with a small engineering project for which revenues as well as expenses are involved. They have useful lives of four and six years, respectively. If the MARR = 10% per year, show which feasible alternative is more desirable by using equivalent worth methods. Use the repeatability assumption.

	A	B
Capital investment	−$3,500	−$5,000
Annual revenue	1,900	2,500
Annual expenses	−645	−1,020
Useful life (years)	4	6
Market value at end of useful life	0	0

Repeatability Assumption, Example 5-7, Least Common
Multiple of Useful Lives Is 12 years

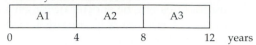

Three cycles of alternative A:

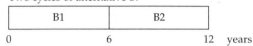

Two cycles of alternative B:

Coterminated Assumption, Example 5-8, 6-Year Analysis Period.

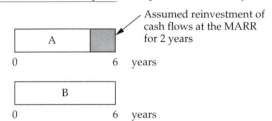

FIGURE 5-5

Illustration of Repeatability Assumption
(Example 5-7), and Coterminated
Assumption (Example 5-8)

SOLUTION

The least common multiple of the useful lives of Alternatives A and B is 12
years. Using the repeatability assumption and a 12-year study period, the first
like (identical) replacement of Alternative A would occur at the end of year four,
and the second would be at the end of year eight. For Alternative B, one like
replacement would occur at the end of year six. This is illustrated in Part 1 of
Figure 5-5.

SOLUTION OF EXAMPLE 5-7 BY THE AW METHOD

The like replacement of assets assumes that the economic estimates for the
initial useful life cycle will be repeated in each subsequent replacement cycle.
Consequently, the AW will have the same value for each cycle (and the study
period) and is calculated over the useful life of each alternative.

$$AW(10\%)_A = -\$3,500(A/P, 10\%, 4) + (\$1,900 - \$645) = \$151$$
$$AW(10\%)_B = -\$5,000(A/P, 10\%, 6) + (\$2,500 - \$1,020) = \$332$$

Based on the AW method we would select Alternative B because it has the larger
value ($332).

SOLUTION OF EXAMPLE 5-7 BY THE PW METHOD

The PW (or FW) solution must be based on the total study period (12 years) and
not determined for only the useful life of each alternative. The PW of the initial
useful life cycle will be different than the PW of subsequent replacement cycles,
even though the economic estimates for each cycle are the same, because of timing
differences in the cash flow amounts.

$$PW(10\%)_A = -\$3{,}500 - \$3{,}500[(P/F, 10\%, 4) + (P/F, 10\%, 8)]$$
$$+ (\$1{,}900 - \$645)(P/A, 10\%, 12)$$
$$= \$1{,}028$$

$$PW(10\%)_B = -\$5{,}000 - \$5{,}000(P/F, 10\%, 6)$$
$$+ (\$2{,}500 - \$1{,}020)(P/A, 10\%, 12)$$
$$= \$2{,}262$$

Based on the PW method, we would again select Alternative B because it has the larger value ($2,262).

EXAMPLE 5-8

Suppose that Example 5-7 is modified such that an analysis period of 6 years is used (coterminated assumption) instead of 12 years, which was based on repeatability and the least common multiple of the useful lives. Perhaps the responsible manager did not agree with the repeatability assumption and wanted a six-year analysis period because it is the planning horizon used in the company for small investment projects.

SOLUTION

An assumption used for an investment alternative (when useful life is less than the study period) is that all cash flows will be reinvested by the firm at the MARR until the end of the study period. This assumption applies to Alternative A, which has a four-year useful life (two years less than the study period), and it is illustrated in Part 2 of Figure 5-5. We use the FW method to analyze this situation.

$$FW(10\%)_A = [-\$3{,}500(F/P, 10\%, 4) + (\$1{,}900 - \$645)(F/A, 10\%, 4)](F/P, 10\%, 2)$$
$$= \$847$$

$$FW(10\%)_B = -\$5{,}000(F/P, 10\%, 6) + (\$2{,}500 - \$1{,}020)(F/A, 10\%, 6)$$
$$= \$2{,}561$$

Based on the FW of each alternative at the end of the six-year study period, we would select Alternative B because it has the larger value ($2,561).

EXAMPLE 5-9

You are a member of an engineering project team that is designing a new processing facility. Your present design task involves the portion of the catalytic system which requires pumping a hydrocarbon slurry that is corrosive and contains abrasive particles. For final analysis and comparison, you have selected two fully lined slurry pump units, of equal output capacity, from different manufacturers. Each unit has the larger diameter impeller required and an integrated electric motor with solid state controls. Both units will provide the same level of service (support) to the catalytic system but have different useful lives and costs.

	Pump Model	
	SP240	**HEPS9**
Capital investment	−$33,200	−$47,600
Annual expenses:		
Electrical energy	−2,165	−1,720
Maintenance	−$1,100 in year 1, and increasing −$500/yr thereafter	−$500 in year 4, and increasing −$100/yr thereafter
Useful life (years)	5	9
Salvage value	0	5,000

The new processing facility is needed by your firm at least as far into the future as the strategic plan forecasts operating requirements. The MARR, before taxes, is 20% per year. Based on this information, which model slurry pump should you select?

SOLUTION

The repeatability assumption is a logical choice for this analysis, and a study period, either infinite or 45 years (least common multiple of the useful lives) in length can be used. With repeatability, the AW over the initial *useful life* of each alternative is the same as its AW over the length of either study period.

$$AW(20\%)_{SP240} = -\$33,200(A/P, 20\%, 5) - \$2,165 - [\$1,100 + \$500(A/G, 20\%, 5)]$$
$$= -\$15,187$$

$$AW(20\%)_{HEPS9} = -\$47,600(A/P, 20\%, 9) + \$5,000(A/F, 20\%, 9)$$
$$- \$1,720 - [\$500(P/A, 20\%, 6)$$
$$+ \$100(P/G, 20\%, 6)] \times (P/F, 20\%, 3) \times (A/P, 20\%, 9)$$
$$= -\$13,622$$

You should select pump model HEPS9, since the AW over its useful life (nine years) has the lesser negative value (−$13,622).

As additional information, the following two points support choosing the repeatability assumption in Example 5-9:

1. The repeatability assumption is commensurate with the long planning horizon for the new processing facility, and with the design and operating requirements of the catalytic system.
2. If the initial estimated costs change for future pump replacement cycles, a logical assumption is that the ratio of the AW values for the two alternatives will remain approximately the same. Competition between the two manufacturers should cause this to happen. Hence, the pump selected (model HEPS9) should continue to be the preferred alternative.

If redesigned or new models of slurry pumps become available, however, another study to analyze and compare all feasible alternatives is required before a replacement of the selected pump occurs.

The guideline given previously in this section, when an alternative's useful life is greater than the study period, is to truncate the alternative at the end of the study period and use an estimated market value for the disposable assets. The next example illustrates application of this guideline.

EXAMPLE 5-10

Suppose that Example 5-9 is modified such that the pump selection task is for an existing processing facility, identical replacement of the present slurry pump in the catalytic system is not a feasible alternative, and the firm's strategic plan includes closing the facility in five years. Also assume the operating requirements for the pump in the existing facility are the same as needed for the new facility in Example 5-9, and the two models (SP240 and HEPS9) are the best alternatives available for replacing the present unit. The estimated market value of pump model HEPS9 in five years is $15,000, and the firm's MARR remains 20% per year. Which pump model should be selected for this replacement action?

SOLUTION

The study period selected for the analysis is five years (coterminated assumption), which is the same as the remaining operating period planned by the firm for the existing facility. The five-year useful life of pump model SP240 is equal to the study period. Its AW over this period, as calculated in Example 5-9, is −$15,187. The AW of pump model HEPS9, using the estimated market value in five years, is

$$AW(20\%)_{HEPS9} = -\$47{,}600(A/P, 20\%, 5) + \$15{,}000(A/F, 20\%, 5) - \$1{,}720$$
$$- [\$500(P/F, 20\%, 4) + \$600(P/F, 20\%, 5)] \times (A/P, 20\%, 5)$$
$$= -\$15{,}783$$

Based on the AW of each alternative over the five-year study period, pump model SP240 should be selected because it has the lesser negative value (−$15,187).

EXAMPLE 5-11

A company requires four additional forklift trucks to support a regional warehouse. The shutdown of this warehouse is anticipated in eight years. Two mutually exclusive alternatives selected for detailed analysis, Stackhigh and S-2000, are models made by different manufacturers. Each model will provide a comparable level of service. The following estimated information for each alternative is based on the requirement for *four* forklifts (excluding labor expenses, which are the same for each model):

	Alternative	
	Stackhigh	S-2000
Capital investment	−$184,000	−$242,000
Annual expenses	− 30,000	− 26,700
Useful life (years)	5	7
Market value at end of useful life	17,000	21,000

The manager of the warehouse has directed that an eight-year study period be used. The MARR for the company is 15% per year. It is estimated that future leasing of four forklifts of comparable capability from an equipment service company would have a total cost (excluding labor, but including maintenance and other annual expense items) of $104,000 per year ($26,000 per forklift) based on a three-year lease and a total cost of $134,000 per year ($33,500 per forklift) based on a one-year lease. Under the assumption that leasing would be used at the end of the useful life to provide a full eight years of comparable service, which model should be selected based on the PW of the incremental cash flow? Confirm the selection based on the ERR method of analysis (assume ϵ = MARR = 15% per year).

SOLUTION

The cash flow table for this situation is shown in Table 5-9. Since these are mutually exclusive cost alternatives, and one must be selected, the first one (Stackhigh) is the base alternative. The incremental cash flow between the two alternatives is shown in column 3, and its PW at MARR = 15% per year is as follows:

$$\text{PW}(15\%)_\Delta = -\$58,000 + \$3,300(P/A, 15\%, 4) - \$13,700(P/F, 15\%, 5)$$
$$+ \$77,300(P/F, 15\%, 6) + \$98,300(P/F, 15\%, 7)$$
$$- \$30,000(P/F, 15\%, 8)$$
$$= \$5,171$$

TABLE 5-9   Cash Flow Table for Example 5-11

	(1)	(2)	(3)
	Alternative		**Increment**
End of Year	**Stackhigh**	**S-2000**	Δ[(S-2000) − (Stackhigh)]
0	−$184,000	−$242,000	−$58,000
1	− 30,000	− 26,700	3,300
2	$\downarrow^a$	$\downarrow^a$	$\downarrow^a$
3			
4	− 30,000		3,300
5	− 13,000[b]		− 13,700
6	− 104,000	− 26,700	77,300
7	− 104,000	− 5,700[b]	98,300
8	− 104,000	− 134,000	− 30,000

[a]Continuing uniform series.
[b]Includes salvage value.

Since the PW of the incremental cash flow is greater than zero at $i = $ MARR, the S-2000 model should be selected. The increment of additional capital investment required versus the Stackhigh model is economically justified.

The ERR of the incremental cash flow, with $\epsilon = $ MARR, is calculated using the three-step procedure discussed in Chapter 4.

Find the PW of the negative cash flow amounts using $\epsilon = 15\%$ per year:

$$PW_{Neg} = -\$58{,}000 - \$13{,}700(P/F, 15\%, 5) - \$30{,}000(P/F, 15\%, 8)$$
$$= -\$74{,}619$$

Find the FW of the positive cash flow amounts using $\epsilon = 15\%$ per year:

$$FW_{Pos} = \$3{,}300(F/A, 15\%, 4)(F/P, 15\%, 4) + \$77{,}300(F/P, 15\%, 2)$$
$$+ \$98{,}300(F/P, 15\%, 1)$$
$$= \$244{,}095$$

Find the interest rate, i'_Δ, such that the absolute value of the PW_{Neg} is equivalent to the FW_{Pos}:

$$|-\$74{,}619|(F/P, i'_\Delta, 8) = \$244{,}095$$
$$(1 + i'_\Delta)^8 = 3.27122$$
$$ERR_\Delta = i'_\Delta = 0.1597, \text{ or } \sim 16\% \text{ per year}$$

Because the ERR of the incremental cash flow is greater than the MARR, the S-2000 model should be selected. This confirms our previous choice based on the PW_Δ at $i = $ MARR.

5.5.1 The Imputed Market Value Technique

In Example 5-10, we used an estimated residual, or net market, value at the end of five years for the model HEPS9 slurry pump to truncate its useful life (nine years) at the end of the study period (five years). *Obtaining a current estimate from the marketplace* for a piece of equipment or another type of asset is the preferred procedure in engineering practice when a market value at time T < (useful life) is required. This approach, however, may not be feasible in some cases. For example, a type of asset may have low turnover in the marketplace and information for recent transactions is not available. Hence, it is sometimes necessary to estimate the market value for an asset without current and representative historical data.

The *imputed market value* technique, which is sometimes called the *implied* market value, can be used for this purpose as well as for comparison with marketplace values when current data are available. The estimating procedure used in the technique is based on logical assumptions about the value of the remaining life for an asset. If an imputed market value is needed for a piece of equipment, say at the end of year T < (useful life), the estimate is calculated based on the sum of two parts as follows:

$$MV_T = [\text{EW at end of year T of remaining capital recovery amounts}]$$
$$+ [\text{EW at end of year T of original market value at end of useful life}]$$

where EW means equivalent worth at i = MARR.

The next example uses information from Examples 5-9 and 5-10 to illustrate the technique, and then to compare results.

EXAMPLE 5-12

Suppose that Example 5-10 is modified such that another market value for pump model HEPS9, at the end of year five, is developed using the imputed market value technique. The same question is again asked, which pump model (SP240 or HEPS9) should be selected for replacement of the current pump in the catalytic system? The MARR remains 20% per year and the study period remains five years.

SOLUTION

An imputed market value for pump model HEPS9, at the end of year five, is developed first. The original information from Example 5-9 will be used for this purpose: capital investment = $47,600, useful life = nine years, and market value = $5,000 at the end of useful life.

Compute the EW at end of year five of the remaining CR amounts (Equation 4-5):

$$EW_{CR} = [\$47,600(A/P, 20\%, 9) - \$5,000(A/F, 20\%, 9)] \times (P/A, 20\%, 4)$$
$$= \$29,949$$

Compute the EW at end of year five, based on the original MV at end of useful life:

$$EW_{MV} = \$5,000(P/F, 20\%, 4) = \$2,412$$

Then, the new market value estimate at the end of year five is as follows:

$$MV_5 = EW_{CR} + EW_{MV}$$
$$= \$29,949 + \$2,412 = \$32,361$$

By using this new market value estimate, the AW of pump model HEPS9 *over the five-year study period* is recalculated as follows:

$$AW(20\%)_{HEPS9} = -\$47,600(A/P, 20\%, 5) + \$32,361 (A/F, 20\%, 5) - \$1,720$$
$$- [\$500(P/F, 20\%, 4) + 600(P/F, 20\%, 5)] \times (A/P, 20\%, 5)$$
$$= -\$13,449$$

Because the AW over five years for pump model SP240 (as calculated in Example 5-9 is −$15,187, pump model HEPS9 should be selected because it has the lesser negative value (−$13,449).

The imputed market value for pump model HEPS9 at the end of year five ($32,361) used in Example 5-12 is greater than the estimated market value of $15,000 used for the pump in Example 5-10. The increase ($17,361) was enough to reverse the selection for the replacement action, based on a five-year study period, from pump model SP240 (Example 5-10) to model HEPS9.

Also notice that $AW_{HEPS9} = -\$13,449$ in Example 5-12 when an imputed market value ($32,361) and a five-year study period are used. This result is very close to $AW = -\$13,622$ for the same pump over its useful life (nine years) in Example 5-9 when the repeatability assumption is utilized. The difference ($-\$172$) is due to the maintenance expenses being a deferred uniform *gradient* sequence. If the maintenance expenses had been equal annual amounts over both study periods, the two AW values for the pump would have been exactly the same.

In general, when the annual expenses (e.g., energy, maintenance, and so on) over the useful life of an asset are the same as for a truncated study period that is less than the useful life, the AW values over both periods will be equal. As shown in Examples 5-9 and 5-12, the repeatability assumption, or the use of an imputed market value to truncate useful life at the end of a shorter study period, provide essentially the same AW results.

> In summary, utilizing the repeatability assumption for Case 2 reduces to the simple rule of "comparing alternatives over their useful lives using the AW method, at $i = $ MARR." However, this simplification does not usually apply when a study period, selected to be shorter or longer than the common multiple of lives (coterminated assumption), is more appropriate for the decision situation. When utilizing the coterminated assumption, cash flows of alternatives need to be adjusted to terminate at the end of the study period. Adjusting these cash flows usually requires estimating the market value of assets at the end of the study period or extending service to the end of the study period through leasing or some other assumption.

5.6 Comparison of Alternatives Using the Capitalized Worth Method

One special variation of the PW method discussed in Chapter 4 involves determining the present worth of all revenues and/or expenses over an infinite length of time. This is known as the *capitalized worth* (CW) method. If expenses only are considered, results obtained by this method are sometimes referred to as *capitalized cost*. This is a convenient basis for comparing mutually exclusive alternatives when the period of needed service is indefinitely long and the repeatability assumption is applicable.

The CW of a perpetual series of end-of-period uniform payments A, with interest at $i\%$ per period, is $A(P/A, i\%, \infty)$. From the interest formulas, it can be seen that $(P/A, i\%, N) \rightarrow 1/i$ as N becomes very large. Thus, $CW = A/i$ for such a series, as can also be seen from the relation

$$CW = PW_{N \rightarrow \infty} = A(P/A, i\%, \infty) = A\left[\lim_{N \rightarrow \infty} \frac{(1+i)^N - 1}{i(1+i)^N}\right] = A\left(\frac{1}{i}\right)$$

Hence, the CW of a project with interest at $i\%$ per year is the annual equivalent of the project over its useful life divided by i.

The AW of a series of payments of amount $X at the end of each kth period with interest at $i\%$ per period is $X(A/F, i\%, k)$. The CW of such a series can thus be calculated as $X(A/F, i\%, k)/i$.

EXAMPLE 5-13

Suppose that a firm wishes to endow an advanced manufacturing processes laboratory at a university. The endowment principal will earn interest that averages 8% per year, which will be sufficient to cover all expenditures incurred in the establishment and maintenance of the laboratory for an indefinitely long period of time (forever). Cash requirements of the laboratory are estimated to be $100,000 now (to establish it), $30,000 per year indefinitely, and $20,000 at the end of every fourth year (forever) for equipment replacement.

(a) For this type of problem, what study period (N) is, practically speaking, defined to be "forever"?

(b) What amount of endowment principal is required to establish the laboratory and then earn enough interest to support the remaining cash requirements of this laboratory forever?

SOLUTION

(a) A practical approximation of "forever" (infinity) is dependent on the interest rate. By examining the $(P/A, i\%, N)$ factor as N increases, we observe that this factor approaches a value of $1/i$. For $i = 8\%$ ($1/i = 12.5000$), note that the $(P/A, 8\%, N)$ factor equals 12.4943 when $N = 100$. Therefore, $N = 100$ is essentially forever (∞) when $i = 8\%$. As the interest rate gets larger, the approximation of forever drops dramatically. For instance, when $i = 20\%$ ($1/i = 5.0000$), forever can be approximated using about 40 years; the $(P/A, 20\%, N)$ factor equals 4.9966 when $N = 40$.

(b) The CW of the cash requirements is synonymous with the endowment principal initially needed to establish and then support the laboratory forever. By using the relationship that CW = (equivalent annual cost)/i, we can compute the amount of the endowment:

$$CW = \frac{-\$100{,}000(A/P, 8\%, \infty) - \$30{,}000 - \$20{,}000(A/F, 8\%, 4)}{0.08}$$

$$= \frac{-\$8{,}000 - \$30{,}000 - \$4{,}438}{0.08}$$

$$= -\$530{,}475$$

where the factor value for $(A/P, 8\%, \infty)$ is given in Table C-11 (Appendix C) as equal to 0.08000.

Another way of considering the amount of endowment principal needed in this example is to have enough to establish the facility ($100,000) and then have enough principal left in the fund to earn a return that will meet the annual maintenance costs ($30,000) and the periodic replacement of equipment needs ($20,000 at the end of each fourth year). Using this logic, we have

$$CW = -\$100,000 - \left[\frac{\$30,000 + \$20,000(A/F, 8\%, 4)}{0.08} \right]$$

$$= -\$100,000 - \left[\frac{(\$30,000 + \$4,438)}{0.08} \right]$$

$$= -\$530,475$$

which, of course, is the same CW amount as our previous calculation.

EXAMPLE 5-14

A selection is to be made between two structural designs. Because revenues do not exist (or can be assumed to be equal), only negative cash flow amounts (costs) and the market value at the end of useful life are estimated, as follows:

	Structure M	Structure N
Capital investment	−$12,000	−$40,000
Market value	0	10,000
Annual expenses	−2,200	−1,000
Useful life (years)	10	25

Using the repeatability assumption and the CW method of analysis, determine which structure is better if the MARR is 15% per year.

SOLUTION
The annual equivalent value (AW) over the useful life of each alternative structure, at MARR = 15% per year, is calculated as follows:

$$AW(15\%)_M = -\$12,000(A/P, 15\%, 10) - \$2,200 = -\$4,592$$

$$AW(15\%)_N = -\$40,000(A/P, 15\%, 25) + \$10,000(A/F, 15\%, 25) - \$1,000$$
$$= -\$7,141$$

Then, the CWs of structures M and N are as follows:

$$CW(15\%)_M = \frac{AW_M}{i} = \frac{-\$4,592}{0.15} = -\$30,613$$

$$CW(15\%)_N = \frac{AW_N}{i} = \frac{-\$7,141}{0.15} = -\$47,607$$

Based on the CW of each structural design, alternative M should be selected because it has the lesser negative value (−$30,613).

5.7 Defining Mutually Exclusive Investment Alternatives in Terms of Combinations of Projects

It is helpful to categorize investment opportunities (projects) into three major groups as follows:

1. *Mutually exclusive:* at most one project out of the group can be chosen.
2. *Independent:* the choice of a project is independent of the choice of any other project in the group, so that all or none of the projects may be selected or some number in between.
3. *Contingent:* the choice of a project is conditional on the choice of one or more other projects.

It is common for decision makers to be faced with sets of mutually exclusive, independent, and/or contingent investment projects. For example, a construction contractor might be considering investing in a dump truck, and/or a backhoe, and/or expansion of the headquarters office building. For each of these investment projects, there may be two or more mutually exclusive alternatives (i.e., brands of dump trucks, types of backhoes, and designs for expansion of the office building). While the choice of a design for the office building is probably independent of that of either dump trucks or backhoes, the choice of any type of backhoe may be contingent (conditional) on the decision to purchase a dump truck.

A general approach, then, requires that all projects be listed and that all the feasible combinations of projects be enumerated. *Such combinations of projects will then be mutually exclusive.* Each combination of projects is mutually exclusive because each is unique and the acceptance of one combination of investment projects precludes the acceptance of any of the other combinations. The total net cash flow of each combination is determined simply by adding, period by period, the cash flows of each project included in the mutually exclusive combination being considered.

For example, suppose that we have three projects: A, B, and C. Each project can be selected once or not at all (i.e., multiple project A's are not possible). If the projects themselves are all mutually exclusive, then the four possible mutually exclusive combinations are shown in binary form in Table 5-10. If, by chance, the firm feels that one of the projects must be chosen (i.e., it is not permissible to turn down all projects), then mutually exclusive combination one would be eliminated from consideration.

TABLE 5-10   Combinations of Three Mutually Exclusive Projects[a]

Mutually Exclusive Combination	Project			Explanation
	X_A	X_B	X_C	
1	0	0	0	Accept none
2	1	0	0	Accept A
3	0	1	0	Accept B
4	0	0	1	Accept C

[a]For each investment project there is a binary variable X_j that will have the value 0 or 1 indicating that project j is rejected (0), or accepted (1). Each row of binary numbers represents an investment alternative in terms of a combination of projects (mutually exclusive combination). This convention is used throughout this book.

TABLE 5-11　Mutually Exclusive Combinations of Three Independent Projects

Mutually Exclusive Combination	Project			Explanation
	X_A	X_B	X_C	
1	0	0	0	Accept none
2	1	0	0	Accept A
3	0	1	0	Accept B
4	0	0	1	Accept C
5	1	1	0	Accept A and B
6	1	0	1	Accept A and C
7	0	1	1	Accept B and C
8	1	1	1	Accept A, B, and C

If the three projects are independent, there are eight mutually exclusive combinations, as shown in Table 5-11.

To illustrate one of the many possible instances of contingent projects, suppose that A is contingent on the acceptance of both B and C and that C is contingent on the acceptance of B. Now there are four mutually exclusive combinations: (1) do nothing, (2) B only, (3) B and C, and (4) A, B, and C.

Suppose that a company is considering two independent sets of mutually exclusive projects. That is, projects A1 and A2 are mutually exclusive, as are B1 and B2. However, the selection of any project from the set of projects A1 and A2 is independent of the selection of any project from the set of projects B1 and B2. *Independent* means that the choice of a project from the A set does not affect the choice from the B set. For example, the decision problem may be to select at most one dump truck out of two models being considered and to select at most one design for expansion of the office building out of two designs being considered. Table 5-12 shows all mutually exclusive combinations for this situation.

TABLE 5-12　Mutually Exclusive Combinations for Two Independent Sets of Mutually Exclusive Projects

Mutually Exclusive Combination	Project				Explanation
	X_{A1}	X_{A2}	X_{B1}	X_{B2}	
1	0	0	0	0	Accept none
2	1	0	0	0	Accept A1
3	0	1	0	0	Accept A2
4	0	0	1	0	Accept B1
5	0	0	0	1	Accept B2
6	1	0	1	0	Accept A1 and B1
7	1	0	0	1	Accept A1 and B2
8	0	1	1	0	Accept A2 and B1
9	0	1	0	1	Accept A2 and B2

TABLE 5-13   Example 5-15 (AW Method)

Project	(1) Net Annual Cash Flow	(2) Capital Recovery Amount (Cost)	(3) = (1) + (2) AW
E1	$2,300	−$1,000	$1,300
E2	2,800	− 3,166	− 366
E3	4,067	− 3,957	110

EXAMPLE 5-15

Given the following three independent engineering projects for improving energy efficiency, determine which should be chosen using the AW method. The MARR = 10% per year, and there is no budget limitation on total investment funds available for this type of project.

Project	Capital Investment, I	Net Annual Cash Flow	Useful Life (yrs)	Market Value (at end of life)
E1	−$10,000	$2,300	5	$10,000
E2	− 12,000	2,800	5	0
E3	− 15,000	4,067	5	0

SOLUTION

As shown in Table 5-13 projects E1 and E3, having positive AWs, would be satisfactory for investment, but project E2 would not. The same indication of satisfactory projects and the unsatisfactory project would be obtained using other equivalent worth methods or the internal rate of return method. Because there is no budget limitation on total investment funds available, both projects E1 and E3 would be recommended for implementation.

Example 5-16 illustrates how to enumerate the mutually exclusive project combinations (investment alternatives) from sets of projects that have all three basic relationships among them (mutually exclusive, independent, and contingent), and then to select an optimal set (*portfolio*) under a capital investment budget constraint.

EXAMPLE 5-16

The following are five proposed projects being considered by an engineer in an integrated transportation company for upgrading an intermodal shipment transfer facility for less than truckload lots of consumer goods. The interrelationships among the projects, and their respective cash flows for the coming budgeting period, are as shown. Some of the projects are mutually exclusive, as noted, and B1 and B2 are independent of C1 and C2. Also, certain projects are dependent on others that may be included in the final portfolio. Using the PW method and

MARR = 10% per year, determine what combination of projects is best if the capital to be invested is (a) unlimited, and (b) limited to $48,000.

Project B1 ⎱ mutually exclusive and independent of C set
Project B2 ⎰

Project C1 ⎱ mutually exclusive and dependent (contingent) on the acceptance
Project C2 ⎰ of B2

Project D   contingent on the acceptance of C1

SOLUTION

The PW for each project by itself is shown in the right-hand column of Table 5-14. As a sample calculation, the PW for project B1 is

$$-\$50,000 + \$20,000(P/A, 10\%, 4) = \$13,400$$

The mutually exclusive project combinations are shown in Table 5-15. Project C1 has not been eliminated from further consideration because project D is contingent on it.

The combined cash flows and the PW for each mutually exclusive combination are shown in Table 5-16. Examination of the right-hand column reveals that mutually exclusive combination 6 has the highest PW if capital available (in year 0) is unlimited, as specified in part (a). If, however, capital available is limited to $48,000, as specified in part (b), mutually exclusive combinations 2 and 6 are not

TABLE 5-14   Project Cash Flows and PWs (Example 5-16)

	Cash Flow ($000s) for End of Year					PW ($000s) at
Project	0	1	2	3	4	MARR = 10%/yr
B1	−50	20	20	20	20	13.4
B2	−30	12	12	12	12	8.0
C1	−14	4	4	4	4	−1.3
C2	−15	5	5	5	5	0.8
D	−10	6	6	6	6	9.0

TABLE 5-15   Mutually Exclusive Project Combinations (Example 5-16)

Mutually Exclusive Combination	Project				
	B1	B2	C1	C2	D
1	0	0	0	0	0
2	1	0	0	0	0
3	0	1	0	0	0
4	0	1	1	0	0
5	0	1	0	1	0
6	0	1	1	0	1

TABLE 5-16   Combined Project Cash Flows and PWs (Example 5-16)

Mutually Exclusive Combination	Cash Flow ($000s) for End of Year					Invested Capital ($000s)	PW ($000s) at MARR = 10%/yr
	0	1	2	3	4		
1	0	0	0	0	0	0	0
2	−50	20	20	20	20	50	13.4
3	−30	12	12	12	12	30	8.0
4	−44	16	16	16	16	44	6.7
5	−45	17	17	17	17	45	8.9
6	−54	22	22	22	22	54	15.7

feasible. Of the remaining mutually exclusive combinations, 5 is best, which means that a portfolio consisting of projects B2 and C2 would be selected with a PW = $8,888.

For problems that involve a relatively small number of projects, the general technique just presented for arranging various types of projects into mutually exclusive combinations is computationally practical. However, for larger numbers of projects, the number of mutually exclusive combinations becomes quite large, and a computer program should be used to perform the calculations.

In many problems involving selections among independent projects, different revenues (or savings) and useful lives are present. Because these projects are typically nonrepeating, it is assumed that cash flows of shorter-lived projects are reinvested at the MARR over a period of time corresponding to the life of the longest-lived project (Section 5.5). The following example illustrates this assumption.

EXAMPLE 5-17

A large corporation is considering the funding of three independent, nonrepeating projects for enlarging freshwater harbors supporting its operations in three areas of the country. Its available capital investment budget this year for such projects is $200 million, and the firm's MARR is 10% per year. In view of the following data, which project(s), if any, should be funded?

Project	Capital Investment, I	Net Annual Benefits, A	Useful Life, N	PW = I + $A(P/A, 10\%, N)$
H1	−$93,000,000	$13,000,000	15 years	$ 5,879,300
H2	− 55,000,000	9,500,000	10 years	3,373,700
H3	− 71,000,000	10,400,000	30 years	27,039,760

SOLUTION

Based on the PW values, each project is economically justified. Hence, the eight feasible mutually exclusive combinations of the three independent projects (reference the general case that was enumerated in Table 5-11) need to be evaluated. The total PW of each combination can be used for this purpose. The total FW of each combination of projects, at the end of the longest-lived project (30 years), is its PW times a common constant, $(F/P, 10\%, 30)$, and will result in the same selection.

A review of the investment cost and PW values for each of the three projects indicates that only the three mutually exclusive combinations involving two of the projects need to be considered. The capital investment budget constraint will not allow all three projects to be implemented, the do nothing alternative is not preferred because each project adds wealth to the firm. Furthermore, each of the three combinations of two projects are within the budget constraint and will add more wealth to the firm than a single project alone. Since projects H1 and H3 have the greatest positive PW values, this combination should be selected. The total PW of this combination is $32,919,060, and its total FW at the end of 30 years is $32,919,060 $(F/P, 10\%, 30)$ = $574,417,850. It is assumed that the remaining $36,000,000 of the $200,000,000 capital investment budget will be invested by the firm in other projects earning at least the MARR = 10% per year.

5.8 Spreadsheet Applications

Because of the repetitive nature of the previous calculations, spreadsheets can be very useful when comparing mutually exclusive alternatives. Given the net cash flow profile for each alternative being considered, we can use the spreadsheet's financial functions (as described in Section 4.12) to compute the equivalent worth measures of merit for each alternative. We can also use a spreadsheet to analyze alternatives with the incremental IRR and ERR procedures.

The analysis of five alternatives (Alpha, Beta, Gamma, Delta, and Theta) using the equivalent worth methods is shown in Figure 5-6. The equivalent worth measures are calculated based on the net cash flow profile presented. The alternative having the largest equivalent worth (Beta) is identified as the recommended alternative. The formulas for the highlighted cells are shown below.

Cell	Contents
C11	= NPV($B$1, B5:B9) + B4
C12	= PMT($B$1, 5, −(NPV($B$1, B5:B9) + B4))
C13	= FV($B$1, 5, PMT($B$1, 5, (NPV($B$1, B5:B9) + B4)))
C14	= IF(C14 = MAX(B11:F11),"Recommend","")

To analyze alternatives using the rate of return methods, we need to perform an incremental analysis. Although there is no financial function for computing an incremental rate of return, we can modify the cash flows and use the IRR( ) financial function. The approach is straightforward.

1. Order the remaining alternatives in magnitude of increasing capital investment.
2. Determine the IRR for each alternative to decide if it is greater than or equal to the MARR. Eliminate any unacceptable alternatives from further consideration.[*]

---

[*]This step applies only when we are comparing investment alternatives. Recall that in the case of cost alternatives, the rate of return is typically less than zero.

	A	B	C	D	E	F
1	MARR	10%				
2						
3	EOY	Alpha	Beta	Gamma	Delta	Theta
4	0	$ (8,000)	$ (16,000)	$ (10,000)	$ (13,000)	$ (9,500)
5	1	$ 2,500	$ 5,000	$ 2,800	$ 3,800	$ 2,000
6	2	$ 2,500	$ 5,000	$ 3,200	$ 3,800	$ 2,200
7	3	$ 2,500	$ 5,000	$ 3,400	$ 3,800	$ 2,600
8	4	$ 2,500	$ 5,000	$ 3,700	$ 3,800	$ 2,800
9	5	$ 2,500	$ 6,000	$ 3,800	$ 3,800	$ 3,000
10						
11	PW	$ 1,476.97	$ 3,574.86	$ 2,631.20	$ 1,404.99	$ (135.02)
12	AW	$ 389.62	$ 943.04	$ 694.10	$ 370.63	$ (35.62)
13	FW	$ 2,378.67	$ 5,757.34	$ 4,237.58	$ 2,262.75	$ (217.45)
14			Recommend			

FIGURE 5-6   Spreadsheet for Comparing MEAs Using Equivalent Worth Methods

3. Create a column that determines the difference between the alternative having the least capital investment (the base alternative) and the next most expensive alternative. Remember that the difference is computed by subtracting the lower investment column from the higher investment column, so that the difference column will have a negative cash flow at time 0.
4. Compute the IRR of the difference column. This is the IRR_Δ. Accept the more expensive alternative only if the $IRR_\Delta \geq$ MARR.
5. Repeat the procedure, forming a new difference column for each comparison, until all alternatives have been compared.

The incremental IRR analysis of the five alternatives considered earlier is shown in Figure 5-7. The alternatives have been reordered according to increasing capital investment. Alternative Theta is eliminated from further consideration since it has an IRR < MARR. Alpha is the base alternative since it requires the smallest capital investment and has an IRR > MARR. The next least expensive alternative is Gamma. Comparing Alpha with Gamma, we see that the incremental investment is justified because $IRR_\Delta \geq$ MARR.

The next comparison made is between Gamma and Delta. Comparing the IRR_Δ to the MARR, we find that the incremental investment is not justified. The same conclusion can be reached by noting that the sum of the nondiscounted cash flows is less than the required incremental investment. Finally, we compare Gamma to Beta. Since $IRR_\Delta \geq$ MARR and there are no more alternatives to be considered, alternative Beta is recommended. This recommendation is consistent with the recommendation using the equivalent worth methods (Figure 5-6). Note that

	A	B	C	D	E	F
1	MARR	10%				
2	ε	8%				
3						
4	EOY	Alpha	Theta	Gamma	Delta	Beta
5	0	$ (8,000)	$ (9,500)	$ (10,000)	$ (13,000)	$ (16,000)
6	1	$ 2,500	$ 2,000	$ 2,800	$ 3,800	$ 5,000
7	2	$ 2,500	$ 2,200	$ 3,200	$ 3,800	$ 5,000
8	3	$ 2,500	$ 2,600	$ 3,400	$ 3,800	$ 5,000
9	4	$ 2,500	$ 2,800	$ 3,700	$ 3,800	$ 5,000
10	5	$ 2,500	$ 3,000	$ 3,800	$ 3,800	$ 6,000
11						
12	IRR	16.99%	9.48%	19.29%	14.15%	18.20%
13	ERR	12.89%	8.90%	14.41%	11.39%	13.65%
14						
15	Incremental Analysis					
16						
17	EOY	Alpha:Gamma	Gamma:Delta	Gamma: Beta		
18	0	$ (2,000)	$ (3,000)	$ (6,000)		
19	1	$ 300	$ 1,000	$ 2,200		
20	2	$ 700	$ 600	$ 1,800		
21	3	$ 900	$ 400	$ 1,600		
22	4	$ 1,200	$ 100	$ 1,300		
23	5	$ 1,300	$ -	$ 2,200		
24						
25	IRR Δ	26.28%	-17.20%	16.18%		
26	Decision	Accept	Reject	Accept		
27						
28	ERR Δ	19.80%	-2.15%	12.33%		
29	Decision	Accept	Reject	Accept		

FIGURE 5-7 Spreadsheet for Comparing MEAs Using Rate of Return Methods

Gamma, which has the highest overall IRR, is not selected as the recommended alternative.

This same procedure applies to an ERR analysis of alternatives. We simply specify the reinvestment rate and substitute the MIRR( ) financial function for the IRR( ) function. The results of an incremental ERR analysis (when $\epsilon = 8\%$) are shown at the bottom of Figure 5-7. The formulas for the highlighted cells are shown in the following table.

Cell	Contents
B12	= IRR(B5:B10, $B$1)
B13	= MIRR(B5:B10, $B$1, $B$2)
B18	= D5 − B5
C18	= E5 − D5
D18	= F5 − D5
B25	= IRR(B18:B23, $B$1)
B26	= IF(B25>=$B$1,"Accept","Reject")
C28	= MIRR(C18:C23, $B$1, $B$2)
C29	= IF(C28>=$B$2,"Accept","Reject")

5.9 Summary

Chapter 5 has built on the previous chapters, in which the principles and applications of money–time relationships were developed. Specifically, Chapter 5 has (1) introduced several difficulties associated with selecting the best alternative from a mutually exclusive set of feasible candidates when using time value of money concepts, and (2) demonstrated the application of most of the analysis methods discussed in Chapter 4 to select the preferred alternative. Moreover, alternatives with unequal lives, various types of dependencies, cost-only versus different revenues and costs, and funding constraints were considered in deciding how to maximize the productivity of invested capital based on the MARR. In summary, we learned that choosing the alternative with the largest equivalent worth (or least negative in the case of cost alternatives) using the MARR would produce this desired result.

If a rate of return method is being used to analyze mutually exclusive alternatives, each avoidable increment of additional capital must earn at least the MARR to ensure that the best alternative is chosen. Examples were provided to illustrate correct computational procedures for avoiding the ranking inconsistency that sometimes occurs when equivalent worth and rate of return methods are applied to the same set of mutually exclusive alternatives. We also considered projects with perpetual lives, applying the capitalized worth method of economic evaluation. The chapter concluded by demonstrating the evaluation of combinations of mutually exclusive, independent, and/or contingent projects using these same methods.

5.10 References

Bussey, L. E. *The Economic Analysis of Industrial Projects.* Englewood Cliffs, N.J.: Prentice Hall, 1978.

Fleischer, Gerald A. "Two Major Issues Associated with The Rate of Return Method for Capital Allocation: The 'Ranking Error' and 'Preliminary Selection.'" *The Journal of Industrial Engineering,* Vol. 17, No. 4, April 1966, pp. 202–208.

GRANT, E. L., W. G. IRESON, and R. S. LEAVENWORTH. *Principles of Engineering Economy*, 8th ed. New York: John Wiley & Sons, 1989.

PARK, C. S., and G. P. SHARP-BETTE. *Advanced Engineering Economics*, New York: John Wiley & Sons, 1990.

5.11 Problems

The number in parentheses ( ) that follows each problem refers to the section from which the problem is taken.

5-1. Four mutually exclusive alternatives are being evaluated, and their costs and revenues are itemized in the first table below. (5.4)

a. If the MARR is 12% per year, use the PW method to determine which alternatives are economically acceptable and which one should be selected.

b. If the total capital investment budget available is $200,000, which alternative should be selected?

5-2. In the design of a new facility, the mutually exclusive alternatives in the second table below are under consideration. Assume that the interest rate (MARR) is 15% per year. Use the following to choose the best of these three design alternatives: (5.4)

a. AW method.

b. FW method.

5-3. The Consolidated Oil Company must install antipollution equipment in a new refinery to meet federal clean air standards. Four design alternatives are being considered, which will have capital investment and annual operating expenses as shown in the third table below. Assuming a useful life of 10 years for each design, no market value, and a desired before-tax MARR of 15% per year, determine which design should be selected based on the PW method. Confirm your selection by using the IRR method. (5.4)

5-4. The 21st Century Development Corporation has a 30-year lease on a plot of land. Estimates of the annual expenses and revenues of various types of structures on the property are as shown at the top left of page 230.

	Mutually Exclusive Alternative				(Pr. 5-1)
	I	**II**	**III**	**IV**	
Capital investment	−$100,000	−$152,000	−$184,000	−$220,000	
Annual revenues less expenses	15,200	31,900	35,900	41,500	
Market value	10,000	0	15,000	20,000	
Useful life (years)	10	10	10	10	

	Design 1	**Design 2**	**Design 3**	(Pr. 5-2)
Capital investment	−$28,000	−$16,000	−$23,500	
Annual revenues less expenses	5,500	3,300	4,800	
Market value	1,500	0	500	
Useful life (years)	10	10	10	

	Alternative Design				(Pr. 5-3)
	D1	**D2**	**D3**	**D4**	
Capital investment	−$600,000	−$760,000	−$1,240,000	−$1,600,000	
Annual expenses:					
Power	− 68,000	− 68,000	− 120,000	− 126,000	
Labor	− 40,000	− 45,000	− 65,000	− 50,000	
Maintenance	− 660,000	− 600,000	− 420,000	− 370,000	
Taxes and insurance	− 12,000	− 15,000	− 25,000	− 28,000	

	Capital Investment	Annual Revenues Less Expenses
Apartment house	−$300,000	$69,000
Theater	− 200,000	40,000
Department store	− 250,000	55,000
Office building	− 400,000	76,000

Each structure is expected to have a market value equal to 20% of its capital investment. If the investor requires an MARR of at least 12% per year before taxes on all investments, which structure (if any) should be selected? Use the AW method. (5.4)

5-5. The following cash flow estimates have been developed for two small, mutually exclusive investment alternatives. The MARR = 12% per year. For parts (a) through (d), select the closest answer. (5.4)

End of Year	Alternative 1	Alternative 2
0	−$2,500	−$4,000
1	750	1,200
2	750	1,200
3	750	1,200
4	750	1,200
5	2,750	3,200

a. What is the AW of Alternative 1?
  1. $371   2. −$162   3. $135
  4. $1,338   5. $1,590
b. What is the IRR of Alternative 1?
  1. 12%   2. 31%   3. 16%
  4. 28%   5. 25%
c. What is the IRR of the incremental net cash flow?
  1. 18%   2. 21%   3. 12%
  4. 24%   5. 15%
d. Given your answers for parts (a) through (c), which alternative should be selected?
  1. Alternative 1
  2. Alternative 2
  3. Neither
  4. Both Alternatives 1 and 2

5-6. An electronics company is trying to determine to which new product they should commit their limited capital resources (there is not enough investment capital for both products). The information below shows the estimated net cash flow for each of the two proposed products. If the MARR = 10% per year, show that the same project selection would be made with proper application of (a) the PW method, and (b) the IRR method. (5.4)

End of Year, k	Product 1	Product 2
0	−$150,000	−$520,000
1	50,000	30,000
2	50,000	130,000
3	50,000	230,000
4	50,000	330,000
IRR	12.6%	11.0%

5-7. In the Rawhide Company (a leather products manufacturer), decisions regarding approval of proposals for capital investment are based upon a stipulated MARR of 18% per year before income taxes. The five packaging devices listed in the table at the top of p. 231 were compared assuming a 10-year life and zero market value for each. Which one (if any) should be selected? Make any additional calculations you think are needed to make a comparison using the IRR method. (5.4)

5-8. Work problem 5-3 using the ERR method when ϵ equals 15% per year. (5.4)

5-9. Work problem 5-2 using the IRR method. (5.4)

5-10. Three alternative designs are being considered for a potential improvement project related to the operation of your engineering department. The prospective net cash flows for these alternatives are shown below. The MARR is 15% per year.

End of Year, k	Alternative Net Cash Flows		
	A	B	C
0	−$200,000	−$230,000	−$212,500
1	90,000	108,000	− 15,000
2			122,500
3			
4	a	a	a
5			
6	90,000	108,000	122,500

[a]Continuing uniform cash flow.

	Packaging Equipment					(Pr. 5-7)
	A	**B**	**C**	**D**	**E**	
Capital investment	−$38,000	−$50,000	−$55,000	−$60,000	−$70,000	
Annual revenues less expenses	11,000	14,100	16,300	16,800	19,200	
Rate of return (IRR)	26.1%	25.2%	26.9%	25.0%	24.3%	

5-10. (Continued) Show that the same capital investment decision results from the IRR method and the PW method applied using the incremental investment analysis procedure. (5.4)

5-11. The Fisstura Missile Corporation of America has identified the following mutually exclusive cost alternatives that will satisfy a specified set of performance requirements. If they are willing to invest any required amount as long as the additional investment will prospectively realize an annual rate of return of 15% (MARR), show which alternative should be chosen. Work the problem (a) with the FW method, and (b) with the IRR method. Assume the service lives for each alternative and the study period are 10 years. Market value is zero in all cases. (5.4)

Alternative	Capital Investment	Annual Expenses for Labor and Material
R	−$10,000	−$7,000
S	− 28,000	− 4,000
T	− 15,000	− 5,900
U	− 25,000	− 3,000

5-12. The net cash flows are shown below for three preliminary design alternatives for a heavy-duty industrial compressor. The perspective of the cash flows is that of the typical user. The MARR = 12% per year, and the study period is seven years. Which preliminary design is economically preferred based on (a) the AW method,

End of Year	Alternative Net Cash Flows		
	A	**B**	**C**
0	−$85,600	−$63,200	−$71,800
1	− 7,400	− 12,100	− 10,050
.			
.	a	a	a
.			
7	− 7,400	− 12,100	− 10,050

aContinuing uniform cash flow.

and (b) the ERR method (ϵ = MARR = 12% per year)? (5.4)

5-13. A new highway is to be constructed. Design A calls for a *concrete* pavement costing $90 per foot with a 20-year life; two paved ditches costing $3 per foot each; and three box culverts every mile, each costing $9,000 and having a 20-year life. Annual maintenance will cost $1,800 per mile; the culverts must be cleaned every five years at a cost of $450 each per mile.

Design B calls for a *bituminous* pavement costing $45 per foot with a 10-year life; two sodded ditches costing $1.50 per foot each; and three pipe culverts every mile, each costing $2,250 and having a 10-year life. The replacement culverts will cost $2,400 each. Annual maintenance will cost $2,700 per mile; the culverts must be cleaned yearly at a cost of $225 each per mile; and the annual ditch maintenance will cost $1.50 per foot per ditch.

Compare the two designs on the basis of equivalent worth per mile for a 20-year period. Find the most economical design on the basis of equivalent annual worth and present worth if the MARR is 6% per year. (5.3)

5-14. A designer is evaluating two electric motors for an automated paint booth application. Each motor's output must be 10 horsepower (hp). She estimates the typical user will operate the booth an average of six hours per day for 250 days per year. Past experience indicates that (a) the annual expense for taxes and insurance averages 2.5% of the capital investment, (b) the MARR is 10% per year before taxes, and (c) the capital invested in machinery must be recovered within five years. Motor A costs $850 and has a guaranteed efficiency of 85% at the indicated operating load. Motor B costs $700 and has a guaranteed efficiency of 80% at the same operating load. Electric energy costs the typical user 5.1 cents per kilowatt-hour (kWh), and 1 hp = 0.746 kW. Recall that electrical input to a motor equals output ÷ efficiency.

		Net Cash Flows for			(Pr. 5-15)
End of Year		Alt. I	Alt. II	Alt. III	
0		−$100,000	−$100,000	−$100,000	
1		0	110,000	0	
2		155,000	0	0	
3		0	0	120,000	
4		0	0	0	
5		−21,000	10,000	50,000	
Measure of merit:	PW	$18,595	$8,658	$29,289	
	IRR	19.5%	15.6%	16.2%	
	Simple payback	2 years	1 year	3 years	

	A	B	C	(Pr. 5-16)
Capital investment	−$2,000	−$8,000	−$20,000	
Annual revenues less expenses	600	2,200	3,600	
Market value	0	0	0	
Project life (years)	5	5	10	

5-14. (Continued) Use the IRR method to choose the better electric motor for the design application. Confirm your selection using the PW method. (5.4)

5-15. A recent engineering graduate has encountered a puzzling situation in her analysis of three mutually exclusive alternatives related to a project. Her cash flow data and results with three different analysis methods are shown in the first table above when the MARR is 8% per year. Comment on her results and rework any of the analyses as you would have done them. Which economic measure(s) of merit is/are correct in this situation? (5.4)

5-16. Consider the three small mutually exclusive investment alternatives in the second table above. The feasible alternative chosen must provide service for a 10-year period. The MARR is 12% per year. State all assumptions you make in your analysis. (5.4, 5.5)

5-17. A certain service can be performed satisfactorily by either process R or process S. Process R has a first cost of $8,000, an estimated service life of 10 years, no market value, and annual revenues less expenses of $2,400. The corresponding figures for process S are $18,000, 20 years, market value equal to 20% of the first cost, and $4,000. Assuming a MARR of 15% per year before income taxes, find the FW of each process and specify which you would recommend. Use the repeatability assumption. (5.4)

5-18. Your plant must add another boiler to its steam generating system. Bids have been obtained from two boiler manufacturers as follows:

	Boiler A	Boiler B
Capital investment	−$50,000	−$120,000
Useful life (years)	20	40
Market value (at end of life)	10,000	20,000
Annual expenses: Year 1	−9,000	−6,000
Rate of increase after Year 1	4%/yr	2%/yr

If the MARR is 10% per year, which boiler would you recommend? Use the repeatability assumption. (5.5)

5-19. As the supervisor of a facilities engineering department, you consider mobile cranes to be critical equipment. The purchase of a new medium-sized, truck-mounted crane is being evaluated. The economic estimates for the two best alternatives are at the top of page 233. You have selected the longest useful life (nine years) for the study period and would lease a crane for the final three years under Alternative A. Based on previous experience, the estimated annual leasing cost at that time will be $36,000 per year. The MARR is 15% per year. Show that the same selection is made based on (a) the PW method, (b) the IRR method, and (c) the ERR method (ϵ = MARR = 15%). (5.4, 5.5)

	Alternatives	
	A	B
Capital investment	−$272,000	−$316,000
Annual expenses[a]	−28,800	−19,300
Useful life (years)	6	9
Market value (at end of life)	25,000	40,000

[a]Excludes the cost of an operator, which is the same for both alternatives.

5-20. Suppose that Problem 5-2 is modified such that the estimated annual revenues less expenses for the three designs are as follows:

Design	Annual Revenues Less Expenses
1	$5,500 in year one and increasing $300 per year thereafter.
2	$3,300 in years one and two and increasing at the rate of 10% per year thereafter.
3	$4,800 in years one through four and increasing at the rate of 7% per year thereafter.

Rework Problem 5-2 using the PW method to determine the preferred design. (5.4)

5-21. Consider the following two mutually exclusive alternatives related to an improvement project, and recommend which one (if either) should be implemented using (a) the AW method and (b) the PW method. The MARR = 15% per year, and the study period is 10 years. Assume repeatability is applicable. (5.5)

	Machine	
	A	B
Capital investment	−$20,000	−$30,000
Annual revenues	10,000	14,000
Annual expenses	−4,400	−8,600
Market value	4,000	0
Useful life (years)	5	10

5-22. Select the preferred investment alternative from the mutually exclusive pair shown in the following table based on (a) the repeatability assumption, (b) the coterminated assumption with a four-year study period and the market value of

Alternative 2 (at the end of year four) determined using the imputed market value technique, and (c) the coterminated assumption with an eight-year study period (Alternative 1 would not be repeated). The MARR is 10% per year. (5.5)

End of Year	Alternative 1	Alternative 2
0	−$40,000	−$60,000
1	12,000	10,000
2	12,000	10,000
3	12,000	10,000
4	36,000	10,000
5		10,000
6		10,000
7		10,000
8		10,000
8 (MV)		40,000

5-23. Two models of new backhoes are being considered by the Apex Construction Company to replace an old one. The equipment will be needed for only three years. Cost and other data for the proposed backhoes are as follows:

	Backhoe M	Backhoe N
Capital investment	−$50,000	−$100,000
Annual savings	25,000	60,000
Market value (at end of year three)	20,000	10,000
Useful life (years)	3	5

Over what *range* of values of the MARR is Alternative N preferred to Alternative M? (5.5)

5-24. A piece of production equipment is to be replaced immediately because it no longer meets quality requirements for the end product. The two best alternatives are a used piece of equipment (E1) and a new automated model (E2). The economic estimates for each are shown below.

	Alternative	
	E1	E2
Capital investment	−$14,000	−$65,000
Annual expenses	−14,000	−9,000
Useful life (years)	5	20
Market value (at end of useful life)	8,000	13,000

The MARR is 15% per year.

a. Which alternative is preferred, based on the repeatability assumption? (5.5)

b. Show, for the coterminated assumption with a five-year study period and an imputed market value for Alternative B, that the AW of B remains the same as it was in part (a) (and obviously the selection is the same as in part a). Explain why that occurs in this problem but not in Example 5-12 in Section 5.5.1. (5.5)

5-25. Estimates for a proposed small public facility are as follows. Plan A has a first cost of $50,000, a life of 25 years, a $5,000 market value, and annual maintenance expenses of $1,200. Plan B has a first cost of $90,000, a life of 50 years, no market value, and annual maintenance expenses of $6,000 for the first 15 years and $1,000 per year for years 16 through 50. Assuming interest at 10% per year, compare the two plans by using the CW method. (5.6)

5-26. In the design of a special-use structure, two mutually exclusive alternatives are under consideration. These design alternatives are as follows:

	D1	D2
Capital investment	−$50,000	−$120,000
Annual expenses	−9,000	−5,000
Useful life (years)	20	50
Market value (at end of useful life)	10,000	20,000

If *perpetual service* from the structure is assumed, which design alternative do you recommend? The MARR is 10% per year. (5.6)

5-27. Use the CW method to determine which mutually exclusive bridge design (L or H) to recommend based on the data provided in the table below. The MARR is 15% per year. (5.6)

5-28.

a. What is the capitalized worth, when $i = 10\%$ per year, of $1,500 per year, starting in year

one and continuing forever and $10,000 in year five, repeating every four years thereafter, and continuing forever? (5.6)

b. When $i = 10\%$ per year in this type of problem, what value of N, practically speaking, defines "forever"? (5.6)

5-29. Three independent projects (A, B, and C) are under consideration, and no more than $150,000 in capital investment funds can be spent to implement any combination of them. Project D is dependent on the acceptance of Project A. If the MARR is 15% per year, which feasible combination of projects would you recommend given the cash flows below? (5.7)

	Cash Flow ($000) at End of Year			
Project	0	1	2	3
A	−100	40	40	60
B	−120	25	50	85
C	−30	6	19	11
D	−20	10	10	5

5-30. The Steel Assembly Corporation is trying to decide between two industrial-type cranes. Crane A and Crane B are mutually exclusive, and one of them must be selected.

Crane A has an extension boom that is optional. It would cost an extra $25,000 but would save an estimated $5,000 per year in operating expenses. Crane B comes equipped with an extension boom, but an optional auger attachment can be purchased for $10,000. The auger would save $8,000 per year in drilling setup costs. The before-tax MARR is 15% per year. State your assumptions for an analysis of this situation, and make a recommendation regarding which alternative to select. Adequate capital investment funds are available for either alternative. (See the first table at the top of p. 235.) (5.5)

	Bridge Design L	Bridge Design H	(Pr. 5-27)
Capital investment	−$274,000	−$326,000	
Annual expenses	− 10,000	− 8,000	
Periodic upgrade cost	− 50,000	− 42,000	
	(every sixth year)	(every seventh year)	
Market value	0	0	
Useful life (years)	83	92	

Crane		
	A	B
Capital investment	−$250,000	−$370,000
Annual expenses	−22,000	−8,000
Market value	100,000	125,000
Useful life (years)	15	18

5-31. The alternatives for an engineering project to recover most of the energy presently being lost in the primary cooling stage of a chemical processing system have been reduced to three designs. The estimated capital investment amounts and annual expense *savings* are as follows:

	Design		
EOY	ER1	ER2	ER3
0	−$98,600	−$115,000	−$81,200
1	25,800	29,000	19,750
2			
3	$\bar{f}=6\%^a$	$G=\$150^b$	c
4			
5			
6	34,526	29,750	19,750

^{a}After year one, the annual savings are estimated to increase at the rate of 6% per year.
^{b}After year one, the annual savings are estimated to increase $150 per year.
^{c}Uniform sequence of annual savings.

Assume that the MARR is 12% per year, the study period is six years, and the market value is zero for all three designs. Apply an analysis method *using the incremental analysis procedure* to determine the preferred alternative. (5.4)

5-32. A small company has $20,000 in surplus capital that it wishes to invest in new revenue-producing projects. Three independent sets of mutually exclusive projects have been developed. The useful life of each is five years, and all market values are zero. You have been asked to perform an IRR analysis to select the best combination of projects. If the MARR is 12% per year, which combination of projects would you recommend? (See the table at the top of the next column.) (5.7)

		Capital Project Investment	Net Annual Benefits
Mutually exclusive	A1	−$ 5,000	$1,500
	A2	− 7,000	1,800
Mutually exclusive	B1	− 12,000	2,000
	B2	− 18,000	4,000
Mutually exclusive	C1	− 14,000	4,000
	C2	− 18,000	4,500

5-33. A firm is considering the development of several new products. The products under consideration are listed here; the products in each project group are mutually exclusive.

Project Group	Products	Development Cost	Annual Net Cash Inflow
A	A1	−$ 500,000	$ 90,000
	A2	− 650,000	110,000
	A3	− 700,000	115,000
B	B1	− 600,000	105,000
	B2	− 675,000	112,000
C	C1	− 800,000	150,000
	C2	− 1,000,000	175,000

At most one product from each group will be selected. The firm has a MARR of 10% per year and a capital investment budget limitation on development costs of $2,100,000. The life of all products is assumed to be 10 years. Assume no market values at the end of 10 years. (5.7)

a. List all mutually exclusive combinations (investment alternatives).
b. Using the PW method, determine which combination of alternatives should be selected.

5-34. Three independent investment projects are being considered:

	Project		
	X	Y	Z
Capital investment*	−$100	−$150	−$200
Annual savings*	16.28	22.02	40.26
Useful life (years)	10	15	8
IRR over the useful life	10%	12%	12%

*In thousands of dollars.

	A	B$_1$	B$_2$	C	(Pr. 5-35)
Capital investment	−$30,000	−$22,000	−$70,000	−$82,000	
Annual revenues less expenses	8,000	6,000	14,000	18,000	
Market value	3,000	2,000	5,000	7,000	

The before-tax MARR is 10% per year, so all projects appear to be acceptable. Assume a study period of 15 years. Which project(s) should be chosen if investment funds are limited to $250,000? State any assumptions. (5.7)

5-35. Engineering projects A, B$_1$, B$_2$, and C are being considered with cash flows estimated over 10 years as shown in the table above. Projects B$_1$ and B$_2$ are mutually exclusive, Project C depends upon B$_2$, and Project A depends upon B$_1$. The capital investment budget limit is $100,000, and the MARR is 12% per year. (5.7)
a. List all possible alternatives.
b. Develop the net cash flows for all feasible alternatives.
c. Which investment alternative (combination of projects) should be selected? Use the PW method.

5-36. There is a continuing requirement for standby electrical power at a public utility service facility. Equipment alternative S1 involves an initial cost of $72,000, a 9-year useful life, annual expenses of $2,200 the first year and increasing $300 per year thereafter, and a net market value of $8,400 at the end of the useful life. Alternative S2 has an initial cost of $90,000, a 12-year-useful life, annual expenses of $2,100 the first year and increasing at the rate of 5% per year thereafter, and a net market value of $13,000. The current interest rate is 10% per year. Which alternative is preferred using the capitalized worth method of analysis? (5.6)

5-37. A single-stage centrifugal blower is to be selected for an engineering design application. Suppliers have been consulted, and the choice has been narrowed down to two new models, both made by the same company and both having the same rated capacity and pressure. Both are driven at 3,600 rpm by identical 90-hp electric motors (output).
One blower has a guaranteed efficiency of 72% at full load and is offered installed for $42,000. The other is more expensive because of aerodynamic refinement, which gives it a guaranteed efficiency of 81% at full load.

Except for these differences in efficiency and installed price, the units are equally desirable in other operating characteristics such as durability, maintenance, ease of operation, and quietness. In both cases, plots of efficiency versus amount of air handled are flat in the vicinity of full rated load. The application is such that whenever the blower is running, it will be at full load.
Assume that both blowers have negligible market values and the firm's MARR is 20% per year on a before-tax basis. Develop a formula for calculating how much the user could afford to pay for the more efficient unit. (*Hint:* You need to specify important parameters and use them in your formula, and remember 1 hp = 0.746 kW.) (5.4)

5-38. A study has been made of the most economical height of skyscrapers. This study grew out of experience related to the Empire State Building, whose height was uneconomical at the time it was constructed. Data are summarized in the graph at the top of p. 237 for a theoretical office building of different heights and corresponding investments. The heights for the building were considered to be 8, 15, 22, 30, 37, 50, 63, and 75 stories. If the owners of this building expect at least a 15% per year return on their capital investment, how many stories should be constructed?

5-39. The annual performance report for Ned and Larry's Ice Cream Company praised the firm for its progressive policies but noted that environmental issues like packaging disposal were a concern. In an effort to reduce the effects of consumer disposal of product packaging, the report stated that Ned and Larry's should consider the following proposals:

A. Package all ice cream and frozen yogurt in quarts;
B. Package all ice cream and frozen yogurt in half-gallons.

By packaging the product in containers larger than the current pints, the plastic-coated bleached sulfate board containers will hold more ounces of product per square inch of surface

(Pr. 5-38)

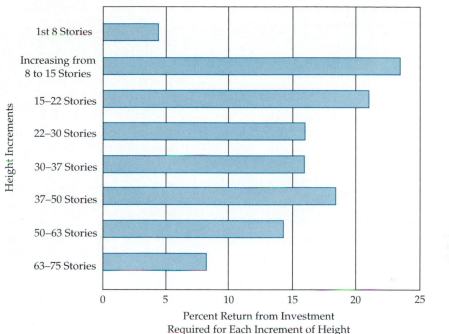

Percent Return from Investment
Required for Each Increment of Height

area. The net result is less discarded packaging per ounce of product consumed. Changing to larger containers requires redesign of the packaging and modification to the filling production line. The existing handling equipment can handle the quarts, but additional equipment will be required to handle half-gallons. Any new equipment purchased for proposals A and B has an expected useful life of six years. The total capital investment for each proposal is shown below. Additional advantages to using larger containers include lower packaging costs per ounce and less handling labor per ounce. The table below summarizes the details of these proposals as well as the current production of pints.

Because Ned and Larry's promotes partnering with suppliers, customers, and the community, they wish to include a portion of the cost to society when evaluating these alternatives. They will consider 50% of the post-consumer landfill cost as part of the costs for each alternative. They have estimated landfill costs to average $20 per cubic yard nationwide.

Assume a MARR of 15% per year, a study period of six years, and that production will remain constant at 10,625,000 gallons per year. Use the IRR method to determine whether Ned and Larry's should package ice cream and frozen yogurt in pints, quarts, or half-gallons.

	(Current) Pints	(A) Quarts	(B) Half-Gallons	
Capital investment	$0	$1,200,000	$1,900,000	(Pr. 5-39)
Packaging cost per gallon	$0.256	$0.225	$0.210	
Handling labor cost per gallon	$0.128	$0.120	$0.119	
Post-consumer landfill contribution from discarded packinging (yd^3/yr)	6,500	5,200	4,050	

5-40. Refer to problem 5-11 and determine the best missile system when the useful lives of the alternatives are:

Alternative	Life
R	10 years
S	15 years
T	5 years
U	5 years

Assume repeatability and zero salvage values for all alternatives.

a. Solve by using an incremental analysis and the IRR method.

b. Check your answer by using the AW method when MARR = 15% per year.

5-41. *Brain Teaser*

Write a computer program that will calculate the AWs of three mutually exclusive electric motor alternatives and select the best alternative based on the assumptions of *repeatability* and *cotermination*. For cotermination, the user must enter estimated market values for the year corresponding to the *shortest life* of the three alternatives. The program will also calculate the incremental rate of return to the nearest 0.1% between any two alternatives under the *cotermination* assumption. The user will select the initial two alternatives for calculating the IRR of the incremental cash flow.

Problem Specifics:

a. Name the program MOTORS and write it in FORTRAN, PASCAL, or C. Provide an executable program MOTORS.EXE on a high-density 3 $1/2$ -inch diskette.

b. The program will first prompt the user for the following information:

- The MARR (10% per year will be entered as 10)

- The output horsepower of the motors (same for all three)
- The number of hours per day the motor will be used (not to exceed 24)
- The number of days per year the motor will be used (not to exceed 365)
- The cost per kilowatt-hour in dollars (remember that 1 hp = 0.746 kW)

c. For each alternative (call them 1, 2, and 3), the user will enter the following:

- Useful life of the motor
- Capital investment
- Market value (if any)
- Efficiency

d. After the user inputs this information, the computer will present a table of economic data that resembles the one shown in problem 5-16. The computer will display the resultant equivalent uniform annual costs for each alternative, along with a brief statement indicating the best choice. Use the repeatability assumption in this part.

e. Next, the computer will determine the shortest of the three lives and prompt the user for estimated market values at that year for all but the shortest alternative. Again, the computer will display a table of economic data, the AWs, and a statement indicating the best choice. This part utilizes the cotermination assumption.

f. In addition to parts a–e, prompt the user to select two alternatives for calculating an incremental rate of return. Display the incremental rate of return and allow the user to calculate another incremental rate of return if desired.

5-42. Refer to column 3 in Table 5-9. Show that multiple IRRs do *not* exist for this incremental net cash flow.

Evaluating Projects with the Benefit/Cost Ratio Method

The objectives of this chapter are (1) to describe many of the unique characteristics of public projects and (2) to learn how to use the benefit/cost (B/C) ratio method as a criterion for project selection. Consideration of both independent projects and mutually exclusive projects is presented.

The following topics are discussed in this chapter:

The perspective and terminology associated with public projects

Self-liquidating and multipurpose projects

Difficulties in evaluating public sector projects

The interest rate that should be used for public projects

The benefit/cost ratio method

Evaluating independent projects by B/C ratios

Comparison of mutually exclusive alternatives

Criticisms and shortcomings of the B/C ratio method

6.1   Introduction

Public projects are those authorized, financed, and operated by federal, state, or local governmental agencies. Such public works are numerous, and although they may be of any size, they are frequently much larger than private ventures. Because they require the expenditure of capital, such projects are subject to the principles of engineering economy with respect to their design, acquisition, and operation. However, because they are public projects, a number of important special factors

TABLE 6-1	Some Basic Differences Between Privately Owned and Publicly Owned Projects	
	Private	**Public**
Purpose	Provide goods and/or services at a profit; Maximize profit or minimize cost	Protect health; Protect lives and property; Provide services (at no profit); Provide jobs
Sources of capital	Private investors and lenders	Taxation; Private lenders
Method of financing	Individual ownership; Partnerships; Corporations	Direct payment of taxes; Loans without interest; Loans at low interest; Self-liquidating bonds; Indirect subsidies; Guarantee of private loans
Multiple purposes	Moderate	Common (e.g., reservoir project for flood control, electrical power generation, irrigation, recreation, education)
Project life	Usually relatively short (5 to 20 years)	Usually relatively long (20 to 60 years)
Relationship of suppliers of capital to project	Direct	Indirect, or none
Nature of "benefits"	Monetary or relatively easy to equate to monetary terms	Often non-monetary, difficult to quantify, difficult to equate to monetary terms
Beneficiaries of project	Primarily, entity undertaking project	General public
Conflict of purposes	Moderate	Quite common (dam for flood control vs. environmental preservation)
Conflict of interests	Moderate	Very common (between agencies)
Effect of politics	Little to moderate	Frequent factors; Short-term tenure for decision makers; Pressure groups; Financial and residential restrictions, etc.
Measurement of efficiency	Rate of return on capital	Very difficult; No direct comparison with private projects

exist that are not ordinarily found in privately financed and operated businesses. The differences between public and private projects are listed in Table 6-1.

As a consequence of these differences, it is often difficult to make engineering economy studies and investment decisions for public works projects in exactly the same manner as for privately owned projects. Different decision criteria are often used, which creates problems for the public (which pays the bill), for those who must make the decisions, and for those who must manage public works projects.

The benefit/cost ratio method, which is normally used for the evaluation of public projects, has its roots in federal legislation. Specifically, the Flood Control Act of 1936 requires that for a federally financed project to be justified, its benefits must be in excess of its costs. In meeting the requirements of this mandate, the B/C method evolved into the calculation of a ratio of project benefits to project costs. Rather than allowing the analyst to apply criteria more commonly used for evaluating private projects (IRR, NPV, etc.), many governmental agencies require the use of the B/C method.

6.2 Perspective and Terminology for Analyzing Public Projects

Before applying the benefit/cost ratio method to evaluate a public project, the appropriate perspective must be established. In conducting an engineering economic analysis of any project, whether it is a public or private undertaking, the proper perspective is to maximize the net benefits to the owners of the enterprise considering the project. This process requires that the question of who owns the project be addressed. Consider, for example, a project involving the expansion of a section of I-80 from four to six lanes. Because the project is paid for primarily with federal funds channeled through the Department of Transportation, we might be inclined to say that the federal government is the "owner." These funds, however, originated from tax dollars—thus the true owners of the project are the taxpayers.

As mentioned previously, the benefit/cost method requires that a ratio of benefits to costs be calculated. Project *benefits* are defined as the favorable consequences of the project to the public, but project *costs* represent the monetary disbursement(s) required of the government. It is entirely possible, however, for a project to have unfavorable consequences to the public. Considering again the widening of I-80, some of the *owners* of the project—farmers along the interstate—would lose a portion of their arable land, along with a portion of their annual revenues. Because this negative financial consequence is borne by (a segment of) the public, it cannot be classified as either a benefit or a cost. The term *disbenefits* is generally used to represent the negative consequences of a project to the public.

EXAMPLE 6-1

A new Convention Center and Sports Complex has been proposed to the Gotham City Council. This public sector project, if approved, will be financed through the issue of municipal bonds. The facility will be located in the City Park near downtown Gotham City, in a wooded area which includes a bike path, a nature trail, and a pond. Because the city already owns the park, no purchase of land is necessary. List separately the project's *benefits*, its *costs*, and any *disbenefits*.

SOLUTION

BENEFITS: Improvement of the image of the downtown area of Gotham City
Potential to attract conferences and conventions to Gotham City
Potential to attract professional sports franchises to Gotham City
Revenues from rental of the facility
Increased revenues for downtown merchants of Gotham City
Use of facility for Civic Events

COSTS:	Architectural design of the facility
	Construction of the facility
	Design and construction of parking garage adjacent to the facility
	Facility operating and maintenance costs
	Facility insurance costs
DISBENEFITS:	Loss of use of a portion of the City Park to Gotham City residents, including the bike path, the nature trail, and the pond
	Loss of wildlife habitat in urban area

6.3   Self-Liquidating Projects

The term *self-liquidating project* is applied to a governmental project that is expected to earn direct revenue sufficient to repay its cost in a specified period of time. Most of these projects provide utility services—for example, the fresh water, electric power, irrigation water, and sewage disposal provided by a hydroelectric dam. Other examples of self-liquidating projects include toll bridges and highways.

As a rule, self-liquidating projects are expected to earn direct revenues that offset their costs, but they are not expected to earn profits or pay income taxes. Although they also do not pay property taxes, in some cases *in lieu* payments are made to state, county, or municipal governments in place of the property and/or franchise taxes that would have been paid had the project been under private ownership. For example, the U.S. government agreed to pay the states of Arizona and Nevada $300,000 each annually for 50 years in lieu of taxes that would have accrued if Hoover Dam had been privately constructed and operated. These in lieu payments are usually considerably less than the actual property and franchise taxes would have been. Furthermore, once such payments are agreed upon, usually at the origination of the project, they are virtually never changed thereafter. These unchanging payments are not the case with property taxes, which are based upon the appraised value of the property.

6.4   Multiple-Purpose Projects

An important characteristic of public sector projects is that many such projects have multiple purposes or objectives. One example of this would be the construction of a dam to create a reservoir on a river (see Figure 6-1). This project would have multiple purposes: (1) assist in flood control, (2) provide water for irrigation, (3) generate electric power, (4) provide recreational facilities, and (5) provide drinking water. Developing such a project to meet more than one objective ensures that greater overall economy can be achieved. Because the construction of a dam involves very large sums of capital and the use of a valuable natural resource—a river—it is likely that the project could not be justified unless it served multiple purposes. This type of situation is generally desirable, but, at the same time, it creates economic and managerial problems due to the overlapping utilization of facilities and the possibility of a conflict of interest between the several purposes and the agencies involved.

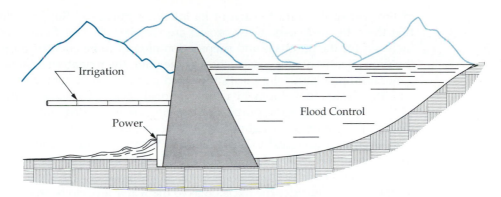

FIGURE 6-1   Schematic Representation of a Multiple-Purpose Project Involving Flood Control, Irrigation, and Power

The basic problems that often arise in evaluating public projects can be illustrated by returning to the dam shown in Figure 6-1. The project under consideration is to be built in the semiarid central portion of California, primarily to provide control against spring flooding resulting from the melting snow in the Sierra Nevadas. If a portion of the water impounded behind the dam could be diverted onto the adjoining land below the dam, the irrigation water would greatly increase the productivity, and thus the value, of that land, which would result in an increase in the nation's resources. So the objectives of the project should be expanded to include both flood control and irrigation.

The existence of a dam with a high water level on one side and a much lower level on the other side also suggests that some of the nation's resources will be wasted unless a portion of the water is diverted to run through turbines, generating electric power. This electricity can be sold to customers in the areas surrounding the reservoir, giving the project the third purpose of generating electric power.

In this semiarid region, the creation of a large reservoir behind the dam would provide valuable facilities for hunting, fishing, boating, swimming, and camping. Thus the project has the fourth purpose of providing recreation facilities. A fifth purpose would be the provision of a steady, reliable supply of drinking water.

Each of the above-mentioned objectives of the project has desirable economic and social value, so what started out a single-purpose project now has five purposes. To fail to meet all five objectives would mean that valuable national resources are being wasted. On the other hand, however, there are certain *disbenefits* to the public that must also be considered. Most apparent of these is the loss of farm land above the dam in the area covered by the reservoir. Other disbenefits might include: (1) the loss of a white-water recreational area enjoyed by canoeing, kayaking, and rafting enthusiasts, (2) the loss of annual deposits of fertile soil in the river basin below the dam due to spring flooding, and (3) the negative ecological impact of obstructing the flow of the river.

If the project is built to serve five purposes, the fact that one dam will serve all of them leads to at least three basic problems. The first of these is the allocation of the cost of the dam to each of its intended purposes. Suppose, for example, that the estimated costs of the project are $35,000,000. This figure includes costs incurred for the purchase and preparation of land to be covered by water above the dam site; the actual construction of the dam, the irrigation system, the power generation plant, and the purification and pumping stations for drinking water; and the design/development of recreational facilities. The allocation of some of these costs to specific purposes is obvious (e.g., the cost of constructing the irrigation system), but what portion of the cost of purchasing and preparing the land should be assigned to flood control? What amounts should be assigned to irrigation, power generation, drinking water, and recreation?

The second basic problem is the conflict of interest among the several purposes of the project. Consider the decision as to the water level to be maintained behind the dam. In meeting the first purpose—flood control—the reservoir should be maintained at a near-empty level to provide the greatest storage capacity during the months of the spring thaw. This lower level would be in direct conflict with the purpose of power generation, which could be maximized by maintaining as high a level as possible behind the dam at all times. Further, maximizing the recreational benefits would suggest that a constant water level be maintained throughout the year. Thus, conflicts of interest arise between the multiple purposes, and compromise decisions must be made. These decisions ultimately affect the magnitude of benefits resulting from the project.

A third problem with multiple-purpose public projects is political sensitivity. Because each of the various purposes, or even the project itself, is likely to be desired or opposed by some segment of the public and by various interest groups that may be affected, inevitably such projects frequently become political issues.* This conflict often has an effect on cost allocations and thus on the overall economy of these projects.

> *The net result of these three factors is that the cost allocations made in multiple-purpose public sector projects tend to be arbitrary.* As a consequence production and selling costs of the services provided also are arbitrary. Because of this fact, they cannot be used as valid yardsticks with which similar private sector projects can be compared to determine the relative efficiencies of public and private ownership.

6.5 Difficulties in Evaluating Public Sector Projects

With all of the difficulties that have been cited in evaluating public sector projects, we may wonder whether or not engineering economy studies of such projects should be attempted. In most cases, economy studies cannot be made in as complete, comprehensive, and satisfactory a manner as in the case of studies of

---

*The construction of the Tellico Dam on the Little Tennessee River was considerably delayed due to two important *disbenefits:* (1) concerns over the impact of the project on the environment of a small fish, the Snail Darter, and (2) the flooding of burial grounds considered sacred by the Cherokee Nation.

privately financed projects. In the private sector, the *costs* are borne by the firm undertaking the project, and the *benefits* are the favorable outcomes of the project accrued by the firm. Any costs and benefits that occur outside of the firm are generally ignored in evaluations unless it is anticipated that those external factors will indirectly impact the firm. But the opposite is true in the case of public-sector projects. In the wording of the Flood Control Act of 1936, "if the benefits to whomsoever they may accrue are in excess of the estimated costs," all of the potential benefits of a public project are relevant and should be considered. Simply enumerating all of the benefits for a large-scale public project is a formidable task! Further, the monetary value of these benefits to all of the affected segments of the public must somehow be estimated. Regardless, decisions regarding the investment of capital in public projects must be made by elected or appointed officials, by managers, or by the general public in the form of referendums. Because of the magnitude of capital and the long-term consequences associated with many of these projects, following a systematic approach for evaluating their worthiness is vital.

There are a number of difficulties inherent in public projects that must be considered in conducting engineering economy studies and making economic decisions regarding those projects. Some of these are listed below:

1. There is no profit standard to be used as a measure of financial effectiveness. Most public projects are intended to be nonprofit.
2. The monetary impact of many of the benefits of public projects is difficult to quantify.
3. There may be little or no connection between the project and the public, which is the owner of the project.
4. There is often strong political influence whenever public funds are used. When decisions regarding public projects are made by elected officials who will soon be seeking re-election, *the immediate benefits and costs are stressed, often with little or no consideration for the more important long-term consequences.*
5. The usual profit motive as a stimulus to promote effective operation is absent, which is not intended to imply that all public projects are ineffective or that managers and employees are not attempting to do their jobs efficiently. But the direct profit stimuli present in privately owned firms are considered to have a favorable impact on project effectiveness in the private sector.
6. Public projects are usually much more subject to legal restrictions than are private projects. For example, the area of operations for a municipally owned power company may be restricted such that power can only be sold within the city limits, regardless of whether a market for any excess capacity exists outside the city.
7. The ability of governmental bodies to obtain capital is much more restricted than with private enterprises.
8. The appropriate interest rate for discounting the benefits and costs of public projects is often controversially and politically sensitive. Clearly, lower interest rates favor long-term projects having major social and/or monetary benefits in the future, whereas higher interest rates promote a short-term outlook whereby decisions are based mostly on initial investments and immediate benefits.

A discussion of several viewpoints and considerations that are often used to establish an appropriate interest rate for public projects is included in the next section.

6.6 What Interest Rate Should Be Used for Public Projects?

When public sector projects are evaluated, interest rates play the same role of accounting for the time value of money as in the evaluation of projects in the private sector. The rationale for the use of interest rates, however, is somewhat different. The choice of an interest rate in the private sector is intended to lead directly to a selection of projects to maximize profit or minimize cost. In the public sector, on the other hand, projects are not usually intended as profit-making ventures. Instead, the goal is the *maximization of social benefits*, assuming that these have been appropriately measured. The choice of an interest rate in the public sector is intended to determine how available funds should best be allocated among competing projects to achieve social goals. The relative differences in magnitude of interest rates between governmental agencies, regulated monopolies, and private enterprises are illustrated in Figure 6-2.

Three main considerations bear on what interest rate to use in engineering economy studies of public sector projects:

1. The interest rate on borrowed capital
2. The opportunity cost of capital to the governmental agency
3. The opportunity cost of capital to the taxpayers

As a general rule, it is appropriate to use the interest rate on borrowed capital as the interest rate for cases in which money is borrowed specifically for the project(s)

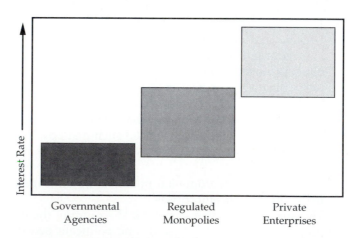

FIGURE 6-2   Relative Differences in Interest Rates for Governmental Agencies, Regulated Monopolies, and Private Enterprise

under consideration. For example, if municipal bonds are issued specifically for the financing of a new school, the effective interest rate on those bonds should be the interest rate.

For public sector projects, the opportunity cost of capital to a governmental agency encompasses the annual rate of *benefit* to either the constituency served by that agency or the composite of taxpayers who will eventually pay for the project. If projects are selected such that the estimated *return* (in terms of benefits) on all accepted projects is higher than that on any of the rejected projects, then the interest rate used in economic analyses is that associated with the best opportunity forgone. If this process is done for all projects and investment capital available within a governmental agency, the result is *an opportunity cost of capital for that governmental agency.* A strong argument against this philosophy, however, is that the different funding levels of the various agencies and the different nature of projects under the direction of each agency would result in different interest rates for each of the agencies, even though they all share a common primary source of funds—taxation of the public.

The third consideration—the opportunity cost of capital to the taxpayers—is based on the philosophy that all government spending takes potential investment capital away from the taxpayers. The taxpayers' opportunity cost is generally greater than either the cost of borrowed capital or the opportunity cost to governmental agencies, and there is a compelling argument for applying the largest of these three rates as the interest rate for evaluating public projects; it is not economically sound to take money away from a taxpayer to invest in a government project yielding benefits at a rate less than what could have been earned by that taxpayer. This argument was supported by a federal government directive issued in 1972— and still in force—by the Office of Management and Budget (OMB).[*] According to this directive, a 10% interest rate should be used in economic evaluations for a wide range of federal projects, with certain exceptions (e.g., a lower rate can be applied in evaluating water resource projects). This 10%, it can be argued, is at least a rough approximation of the average return that taxpayers could earn from the use of that money for private investment.

One additional theory on establishing interest rates for federal projects advocates that the "social discount rate" used in such analyses should be the market-determined *risk-free* rate for private investments.[†] According to this theory, an annual interest rate on the order of 3–4% should be used.

The preceding discussion focuses on the *considerations* that should play a role in establishing an interest rate for public projects. As in the case of the private sector, there is no simple formula for determining the appropriate interest rate for public projects. With the exception of projects falling under the 1972 OMB directive, setting the interest rate is ultimately a policy decision at the discretion of the governmental agency conducting the analysis.

---

[*]Office of Management and Budget, "Discount Rates to be used in Evaluating Time Distributed Costs and Benefits," *Circular No. A-94 (rev.),* March 27, 1972.

[†]Arrow, K. J., and R. C. Lind, "Uncertainty and the Evaluation of Public Investment Decisions," *American Economic Review,* vol. 60, June 1970, pp. 364–378.

6.7 The Benefit/Cost Ratio Method

As the name implies, the benefit/cost ratio method involves the calculation of a ratio of benefits to costs. Whether evaluating a project in the private sector or in the public sector, the time value of money must be considered to account for the timing of cash flows (or benefits) occurring after the inception of the project. Thus the B/C ratio is actually a ratio of *discounted benefits* to *discounted costs*.

Any method for formally evaluating projects in the public sector must consider the worthiness of allocating resources to achieve social goals. For nearly 60 years, the B/C ratio method has been the accepted procedure for making go/no-go decisions on independent projects and for comparing alternative projects in the public sector, even though the other methods discussed in Chapter 4 (PW, AW, IRR, etc.) will lead to identical recommendations, *assuming all these procedures are properly applied.* Accordingly, the purpose of this section is to describe and illustrate the mechanics of the B/C ratio method for evaluating projects.

The B/C ratio is defined as the ratio of the equivalent worth of benefits to the equivalent worth of costs. The equivalent worth measure applied can be present worth, annual worth, or future worth, but customarily, either PW or AW is used. An interest rate for public projects, as discussed in the previous section, is used in the equivalent worth calculations. The benefit/cost ratio is also known as the *savings-investment ratio* (SIR) by some governmental agencies.

Several different formulations of the B/C ratio have been developed. Two of the more commonly used formulations are presented here, illustrating the use of both present worth and annual worth.

Conventional B/C ratio with PW:

$$B/C = \frac{\text{PW(benefits of the proposed project)}}{\text{PW(total costs of the proposed project)}} = \frac{\text{PW(B)}}{\text{I} + \text{PW(O\&M)}} \qquad (6\text{-}1)$$

where PW($\bullet$) = present worth of ($\bullet$)
         B = benefits of the proposed project
         I = initial investment in the proposed project
     O&M = operating and maintenance costs of the proposed project

Modified B/C ratio with PW:

$$B/C = \frac{\text{PW(B)} - \text{PW(O\&M)}}{\text{I}} \qquad (6\text{-}2)$$

The numerator of the modified benefit/cost ratio expresses the equivalent worth of the benefits minus the equivalent worth of the O&M costs, and the denominator includes only the initial investment costs. A project is acceptable when the B/C ratio, as defined in either Equation 6-1 or 6-2, is greater than or equal to 1.0.

Equations 6-1 and 6-2 can be rewritten in terms of equivalent annual worth as follows:

Conventional B/C ratio with AW:

$$B/C = \frac{AW(\text{benefits of the proposed project})}{AW(\text{total costs of the proposed project})} = \frac{AW(B)}{CR + AW(O\&M)} \quad (6\text{-}3)$$

where $AW(\bullet)$ = annual worth of $(\bullet)$
$\quad\quad\quad B$ = benefits of the proposed project
$\quad\quad\quad CR$ = capital recovery amount (i.e., the equivalent annual cost of the initial investment, I, including an allowance for salvage value, if any)
$\quad\quad O\&M$ = operating and maintenance costs of the proposed project

Modified B/C ratio with AW:

$$B/C = \frac{AW(B) - AW(O\&M)}{CR} \quad (6\text{-}4)$$

Note that when using the annual worth approach, the annualized equivalent of any *salvage value* associated with the investment is effectively subtracted from the denominator in the calculation of the capital recovery amount (CR) in Equations 6-3 and 6-4. Similarly, when using the present worth approach to calculate a benefits/cost ratio, it is customary to reduce the investment in the denominator by the discounted equivalent of any salvage value. Equations 6-1 and 6-2 are rewritten to incorporate the salvage value of an investment below:

Conventional B/C ratio with PW, Salvage Value included:

$$B/C = \frac{PW(\text{benefits of the proposed project})}{PW(\text{total costs of the proposed project})} = \frac{PW(B)}{I - PW(S) + PW(O\&M)} \quad (6\text{-}5)$$

where $PW(\bullet)$ = present worth of $(\bullet)$
$\quad\quad\quad B$ = benefits of the proposed project
$\quad\quad\quad\; I$ = initial investment in the proposed project
$\quad\quad\quad S$ = salvage value of investment
$\quad\quad O\&M$ = operating and maintenance costs of the proposed project

Modified B/C ratio with PW, Salvage Value included:

$$B/C = \frac{PW(B) - PW(O\&M)}{I - PW(S)} \quad (6\text{-}6)$$

The resulting B/C ratios for all the above formulations will give consistent results in determining the acceptability of a project (i.e., either $B/C > 1.0$ or $B/C < 1.0$ or $B/C = 1.0$). The conventional B/C ratio will give identical numerical results for both PW and AW formulations; similarly, the modified B/C ratio gives identical numerical results whether PW or AW is used. Although the magnitude of the B/C ratio will differ between conventional and modified B/C, *go/no-go* decisions are not affected by the choice of approach, as shown in Example 6-2.

EXAMPLE 6-2

The city of Bugtussle is considering extending the runways of its Municipal Airport so that commercial jets can use the facility. The land necessary for the runway extension is currently farmland, which can be purchased for $350,000. Construction costs for the runway extension are projected to be $600,000, and the additional annual maintenance costs for the extension are estimated to be $22,500. If the runways are extended, a small terminal will be constructed at a cost of $250,000. The annual operating and maintenance costs for the terminal are estimated at $75,000. Finally, the projected increase in flights will require the addition of two air traffic controllers, at an annual cost of $100,000. Annual *benefits* of the runway extension have been estimated as follows:

$325,000	rental receipts from airlines leasing space at the facility
$65,000	airport tax charged to passengers
$50,000	convenience benefit for residents of Bugtussle
$50,000	additional tourism dollars for Bugtussle

Apply the B/C ratio method with a study period of 20 years and an interest rate of 10% to determine whether the runways at Bugtussle Municipal Airport should be extended.

SOLUTION

Conventional B/C: $B/C = PW(B)/[I + PW(O\&M)]$
Eqn. 6-1 $B/C = \$490,000 \, (P/A, 10\%, 20)/[\$1,200,000 + \$197,500 \, (P/A, 10\%, 20)]$
                $B/C = 1.448$

Modified B/C: $B/C = [PW(B) - PW(O\&M)]/I$
Eqn. 6-2 $B/C = [\$490,000 \, (P/A, 10\%, 20) - \$197,500 \, (P/A, 10\%, 20)]/\$1,200,000$
                $B/C = 2.075$

Conventional B/C: $B/C = AW(B)/[CR + AW(O\&M)]$
Eqn. 6-3 $B/C = \$490,000/[\$1,200,000 \, (A/P, 10\%, 20) + \$197,500]$
                $B/C = 1.448$

Modified B/C: $B/C = [AW(B) - AW(O\&M)]/CR$
Eqn. 6-4 $B/C = [\$490,000 - \$197,500]/[\$1,200,000 \, (A/P, 10\%, 20)]$
                $B/C = 2.075$

As can be seen in the above example, the difference between conventional and modified B/C ratios is essentially due to subtracting the equivalent worth measure of operating and maintenance costs from both the numerator and the denominator of the B/C ratio. In order for the B/C ratio to be greater than 1.0, the numerator must be greater than the denominator. Similarly, the numerator must be less than the denominator for the B/C ratio to be less than 1.0. Subtracting a constant (the equivalent worth of O&M costs) from both numerator and denominator does not alter the *relative* magnitudes of the numerator and denominator. Thus, project acceptability is not affected by the choice of conventional versus modified B/C ratio. This information is stated mathematically below for the case of B/C > 1.0:

Let N = the numerator of the conventional B/C ratio
       D = the denominator of the conventional B/C ratio
     O&M = the equivalent worth of operating and maintenance costs

if B/C = $\dfrac{N}{D}$ > 1.0, *then* N > D.

if N > D, *then* [N − O&M] > [D − O&M], *then* $\dfrac{N - O\&M}{D - O\&M}$ > 1.0.

Note that $\dfrac{N - O\&M}{D - O\&M}$ *is the modified B/C ratio, thus if conventional* B/C > 1.0, *then modified* B/C > 1.0.

Two additional issues of concern are the treatment of *disbenefits* in benefit/cost analyses and the decision as to whether certain cash flow items should be treated as *additional benefits* or as *reduced costs*. The first concern arises whenever disbenefits are formally defined in a B/C evaluation of a public-sector project. An example of the second concern would be a public-sector project proposing to replace an existing asset having high annual operating and maintenance costs with a new asset having lower O&M costs. As will be seen in Sections 6.7.1 and 6.7.2, the final recommendation on a project is not altered by either the approach to incorporating disbenefits or the classification of an item as a reduced cost or an additional benefit.

6.7.1 Disbenefits in the B/C Ratio

In a previous section, disbenefits were defined as negative consequences to the public resulting from the implementation of a public-sector project. The traditional approach for incorporating disbenefits into a benefit/cost analysis is to reduce benefits by the amount of disbenefits (i.e., to subtract disbenefits from benefits in the numerator of the B/C ratio). Alternatively, the disbenefits could be treated as costs (i.e., add disbenefits to costs in the denominator). Equations 6-7 and 6-8 illustrate the two approaches for incorporating disbenefits in the conventional B/C ratio, with benefits, costs, and disbenefits in terms of equivalent AW. (Similar equations could also be developed for the modified B/C ratio or for PW as the measure of equivalent worth.) Again, the *magnitude* of the B/C ratio will be different depending upon which approach is used to incorporate disbenefits, but project acceptability—that is, whether the B/C ratio is >, <, or = 1.0—will not be affected, as shown in Example 6-3.

Conventional B/C ratio with AW, *Benefits* reduced by amount of *Disbenefits*:

$$B/C = \frac{AW(\text{benefits}) - AW(\text{disbenefits})}{AW(\text{costs})} = \frac{AW(B) - AW(D)}{CR + AW(O\&M)} \qquad (6\text{-}7)$$

where AW($\bullet$) = annual worth of ($\bullet$)
    B = benefits of the proposed project
    D = disbenefits of the proposed project
    CR = capital recovery amount (i.e., the equivalent annual cost of the initial investment, I, including an allowance for salvage value, if any)
   O&M = operating and maintenance costs of the proposed project

Conventional B/C ratio with AW, *Costs* increased by amount of *Disbenefits*:

$$\text{B/C} = \frac{\text{AW(benefits)}}{\text{AW(costs)} + \text{AW(disbenefits)}} = \frac{\text{AW(B)}}{\text{CR} + \text{AW(O\&M)} + \underline{\text{AW(D)}}} \quad \text{(6-8)}$$

EXAMPLE 6-3

Refer back to **Example 6-2.** In addition to the benefits and costs, suppose that there are disbenefits associated with the runway extension project. Specifically, the increased noise level from commercial jet traffic will be a serious nuisance to homeowners living along the approach path to the Bugtussle Municipal Airport. The annual disbenefit to citizens of Bugtussle caused by this "noise pollution" is estimated to be $100,000. Given this additional information, reapply the conventional B/C ratio, with equivalent annual worth, to determine whether or not this disbenefit affects your recommendation on the desirability of this project.

SOLUTION

Disbenefits Reduce   B/C = [AW(B) − AW(D)]/[CR + AW(O&M)]
Benefits   B/C = [$490,000 − $100,000]/[$1,200,000 (A/P, 10%, 20) + $197,500]
Eqn. 6-7   **B/C = 1.152**

Disbenefits Treated as   B/C = AW(B)/[CR + AW(O&M) + AW(D)]
Additional Costs   B/C = $490,000/[$1,200,000 (A/P, 10%, 20) + $197,500 + $100,000]
Eqn. 6-8   **B/C = 1.118**

As in the case of conventional and modified B/C ratios, the treatment of disbenefits may affect the magnitude of the B/C ratio, but it has no effect on project desirability in go/no-go decisions. It is left to the reader to develop a mathematical rationale for this, similar to that included in the discussion of conventional versus modified B/C ratios.

6.7.2 Added Benefits Versus Reduced Costs in B/C Analyses

The analyst often needs to classify certain cash flows as either added benefits or reduced costs in calculating a B/C ratio. The questions arise, "How critical is the proper assignment of a particular cash flow as an added benefit or a reduced cost?", and "Is the outcome of the analysis affected by classifying a reduced cost as a benefit?" *An arbitrary decision as to the classification of a benefit or a cost has no impact on project acceptability.* The mathematical rationale for this information is presented below and in Example 6-4.

Let   B = the equivalent annual worth of project benefits
      C = the equivalent annual worth of project costs
      X = the equivalent annual worth of a cash flow (either an added benefit or a reduced cost) not included in either *B* or *C*

if X is classified as an added benefit, then $B/C = \dfrac{B + X}{C}$. Alternatively,

if X is classified as a reduced cost, then $B/C = \dfrac{B}{C - X}$.

Assuming that the project is acceptable, that is, B/C > 1.0,

$$\frac{B + X}{C} > 1.0,$$ which indicates that $B + X > C$, and

$$\frac{B}{C - X} > 1.0,$$ which indicates that $B > C - X$,

which can be restated as: B + X > C.

EXAMPLE 6-4

A project is being considered by the Tennessee Department of Transportation to replace an aging bridge across the Cumberland River on a state highway. The existing two-lane bridge is expensive to maintain and creates a traffic bottleneck because the state highway is four lanes wide on either side of the bridge. The new bridge can be constructed at a cost of $300,000, and estimated annual maintenance costs are $10,000. The existing bridge has annual maintenance costs of $18,500. The annual benefit of the new four-lane bridge to motorists, due to the removal of the traffic bottleneck, has been estimated to be $25,000. Conduct a benefit/cost analysis, using an interest rate of 8% and a study period of 25 years, to determine whether the new bridge should be constructed.

SOLUTION

Treating the reduction in annual maintenance costs as a *Reduced Cost*:

$$B/C = \$25,000/[\$300,000(A/P, 8\%, 25) - (\$18,500 - \$10,000)]$$

$$B/C = 1.275$$

Treating the reduction in annual maintenance costs as an *Increased Benefit*:

$$B/C = [\$25,000 + (\$18,500 - \$10,000)]/[\$300,000(A/P, 8\%, 25)]$$

$$B/C = 1.192$$

Therefore, the decision to classify a cash flow item as an additional benefit or as a reduced cost will affect the magnitude of the calculated B/C ratio, but it will have no effect on project acceptability.

6.8 Evaluating Independent Projects by B/C Ratios

In Chapter 5, independent projects were categorized as groupings of projects for which the choice to select any particular project in the group is *independent* of choices regarding any and all other projects within the group. Thus, it is permissible to select none of the projects, any combination of projects, or all of the projects from an independent group. (Note that this does not hold true under conditions of *capital rationing.* Methods of evaluating otherwise independent projects under capital rationing are discussed in a later section of this chapter.) Because any or all projects from an independent set can be selected, formal comparisons of independent projects are unnecessary. The issue of whether one project is *better* than another is unimportant if those projects are independent; the only criterion for selecting each of those projects is whether or not their respective B/C ratios are equal to or greater than 1.0.

A typical example of an economy study of a federal project—using the conventional B/C ratio method—is a study of a flood control and power project on the White River in Missouri and Arkansas. Considerable flooding and consequent damage had occurred along certain portions of this river, as shown in Table 6-2. In addition, the uncontrolled water flow increased flood conditions on the lower Mississippi River. In this case, there were independent options of building a reservoir *and/or* a channel improvement to alleviate the problem. The cost and benefit summaries for the Table Rock reservoir and the Bull Shoals channel improvement are shown in Table 6-3. The fact that the Bull Shoals channel improvement project has the higher benefit cost ratio is irrelevant; both options are acceptable because their B/C ratios are greater than 1.

Several interesting facts may be noted concerning this study. First, there was no attempt to allocate the cost of the projects between flood control and power production. Second, very large portions of the flood control benefits were shown to be in connection with the Mississippi River and are not indicated in Table 6-2; these

---

TABLE 6-2   Annual Loss as a Result of Floods on Three Stretches of the White River

Item	Annual Value of Loss	Annual Loss per Acre of Improved Land in Floodplain	Annual Loss per Acre for Total Area in Floodplain
Crops	$1,951,714.00	$6.04	$1.55
Farm (other than crops)	215,561.00	0.67	0.17
Railroads and highways	119,800.00	0.37	0.09
Levees[a]	87,234.00	0.27	0.07
Other losses	168,326.00	0.52	0.13
TOTALS:	**$2,542,635.00**	**$7.87**	**$2.01**

[a]Expenditures by the United States for levee repairs and high-water maintenance.

TABLE 6-3   Estimated Costs, Annual Charges, and Annual Benefits for the Table Rock Reservoir and Bull Shoals Channel Improvement Projects

Item	Table Rock Reservoir	Bull Shoals Channel Improvement
Cost of dam and appurtenances, and reservoir:		
Dam, including reservoir-clearing, camp, access railroads and highways, and foundation exploration and treatment	$20,447,000	$25,240,000
Powerhouse and equipment	6,700,000	6,650,000
Power transmission facilities to existing load-distribution centers	3,400,000	4,387,000
Land	1,200,000	1,470,000
Highway relocations	2,700,000	140,000
Cemetery relocations	40,000	18,000
Damage to villages	6,000	94,500
Damage to miscellaneous structures	7,000	500
Total Construction Cost (estimated appropriation of public funds necessary for the execution of the project)	$34,500,000	$38,000,000
Federal investment:		
Total construction cost	$34,500,000	$38,000,000
Interest during construction	$1,811,300	1,995,000
Total	$36,311,300	$39,995,000
Present value of federal properties	1,200	300
Total Federal Investment	**$36,312,500**	**$39,995,300**
Annual charges: interest, amortization, maintenance, and operation	**$1,642,200**	**$1,815,100**
Annual benefits:		
Prevented direct flood losses in White River basin:		
Present conditions	60,100	266,900
Future developments	19,000	84,200
Prevented indirect flood losses owing to floods in White River basin	19,800	87,800
Enhancement in property values in White River valley	7,700	34,000
Prevented flood losses on Mississippi River	220,000	980,000
Annual Flood Benefits	326,000	1,452,900
Power value	1,415,600	1,403,400
Total Annual Benefits	**$1,742,200**	**$2,856,300**
Conventional B/C Ratio = Total Annual Benefits ÷ Annual Charges	1.06	1.57

were not detailed in the main body of the report but were shown in an appendix. Only a moderate decrease in the value of these benefits would have changed the B/C ratio considerably. Third, without the combination of flood control and power generation objectives, neither project would have been economical for either purpose. These facts point to the advantages of multiple purposes for making flood control projects economically feasible and to the necessity for careful enumeration and evaluation of the prospective benefits of a public sector project.

6.9 Comparison of Mutually Exclusive Projects by B/C Ratios

Recall that a group of *mutually exclusive projects* was defined as *a group of projects from which, at most, one project may be selected.* When using an equivalent worth method to select from among a set of mutually exclusive alternatives (MEAs), the "best" alternative can be selected by maximizing the PW (or AW, or FW). Because the benefit/cost method provides a *ratio* of benefits to costs rather than a direct measure of each project's *profit potential*, selecting the project that maximizes the B/C ratio does not guarantee that the best project is selected. In addition to the fact that *maximizing the B/C ratio for mutually exclusive alternatives is **incorrect,*** any attempt to do so would be further confounded by the potential for inconsistent ranking of projects by the conventional B/C ratio versus the modified B/C ratio (i.e., the conventional B/C ratio might favor a different project than would the modified B/C ratio). This phenomenon is illustrated in Example 6-5. (The approach to handling *disbenefits* and/or classifications of cash flow items as *added benefits* versus *reduced costs* could also change the preference for one MEA over another.) As with the rate of return procedures, an evaluation of mutually exclusive alternatives by the B/C ratio requires that an *incremental* benefit/cost analysis be conducted.

EXAMPLE 6-5

The required investments, annual operating and maintenance costs, and annual benefits for two mutually exclusive alternative projects are shown below. Both conventional and modified B/C ratios are included for each project. Note that **Project A** has the greater *conventional B/C,* but **Project B** has the greater *modified B/C.* Given this information, which project should be selected?

	Project A	Project B	
Capital investment	$110,000	$135,000	i = 10%
Annual O&M cost	12,500	45,000	N = 20 yrs
Annual benefit	37,500	80,000	
Conventional B/C:	**1.475**	1.315	
Modified B/C:	1.935	**2.207**	

SOLUTION

The B/C analysis has been conducted improperly. Although each of the B/C ratios shown above is *numerically correct,* a comparison of mutually exclusive alternatives requires that an incremental analysis be conducted.

When comparing mutually exclusive alternatives with the B/C ratio method, *they are first ranked in order of increasing total equivalent worth of costs.* This rank ordering will be identical whether the ranking is based on PW, AW, or FW

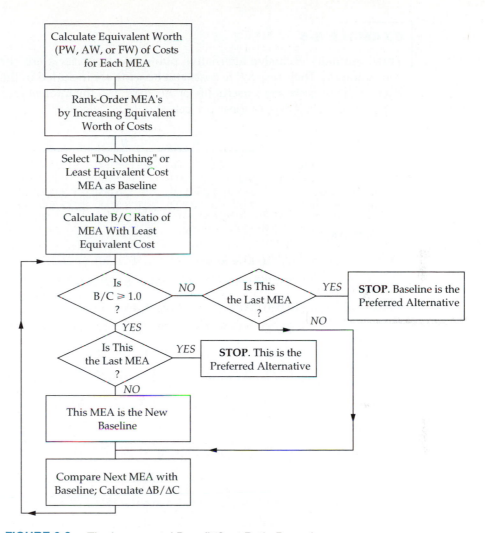

FIGURE 6-3   The Incremental Benefit/Cost Ratio Procedure

of costs. The "do-nothing" alternative is selected as a baseline alternative. The B/C ratio is then calculated for the alternative having the lowest equivalent cost. If the B/C ratio for this alternative is equal to or greater than 1.0, then that alternative becomes the new baseline; otherwise, "do-nothing" remains as the baseline. The next least equivalent cost alternative is then selected, and the difference (Δ) in the respective benefits and costs of this alternative and the baseline is used to calculate an incremental B/C ratio ($\Delta B/\Delta C$). If that ratio is equal to or greater than 1.0, then the higher equivalent cost alternative becomes the new baseline; otherwise the last baseline alternative is maintained. Incremental B/C ratios are determined for each successively higher equivalent cost alternative until the last alternative has been compared. The flowchart of this procedure is included as Figure 6-3, and the procedure is illustrated in Example 6-6.

EXAMPLE 6-6

Three mutually exclusive alternative public works projects are currently under consideration. Their respective costs and benefits are included in the table below. Each of the projects has a useful life of 50 years, and the interest rate is 10 percent per year. Which, if any, of these projects should be selected?

	A	B	C
Capital investment	$8,500,000	$10,000,000	$12,000,000
Annual oper. & maint. costs	750,000	725,000	700,000
Salvage value	1,250,000	1,750,000	2,000,000
Annual benefit	2,150,000	2,265,000	2,500,000

SOLUTION

$PW(Costs, A) = \$8{,}500{,}000 + \$750{,}000(P/A, 10\%, 50)$
$\qquad - \$1{,}250{,}000(P/F, 10\%, 50) = \$15{,}925{,}463$

$PW(Costs, B) = \$10{,}000{,}000 + \$725{,}000(P/A, 10\%, 50)$
$\qquad - \$1{,}750{,}000(P/F, 10\%, 50) = \$17{,}173{,}333$

$PW(Costs, C) = \$12{,}000{,}000 + \$700{,}000(P/A, 10\%, 50)$
$\qquad - \$2{,}000{,}000(P/F, 10\%, 50) = \$18{,}923{,}333$

$PW(Benefit, A) = \$2{,}150{,}000(P/A, 10\%, 50) = \$21{,}316{,}851$

$PW(Benefit, B) = \$2{,}265{,}000(P/A, 10\%, 50) = \$22{,}457{,}055$

$PW(Benefit, C) = \$2{,}750{,}000(P/A, 10\%, 50) = \$24{,}787{,}036$

$B/C(A) = \$21{,}316{,}851/\$15{,}925{,}463 = \mathbf{1.3385} > \mathbf{1.0}$ ∴ **Project A is Acceptable**

$\Delta B/\Delta C(B - A) = (\$22{,}457{,}055 - \$21{,}316{,}851)/(\$17{,}173{,}333 - \$15{,}925{,}463)$
$\qquad = \mathbf{0.9137} < \mathbf{1.0}$ ∴ **Project B not Acceptable**

$\Delta B/\Delta C(C - A) = (\$24{,}787{,}036 - \$21{,}316{,}851)/(\$18{,}923{,}333 - \$15{,}925{,}463)$
$\qquad = \mathbf{1.1576} > \mathbf{1.0}$ ∴ **Project C is Acceptable**

Decision: *Recommend Project C*

It is not uncommon for some of the projects in a set of mutually exclusive public works projects to have different lives. Recall from a previous chapter that the AW criterion can be used to select from among alternatives with different lives as long as the assumption of *repeatability* is valid. Similarly, if a mutually exclusive set of public works projects includes projects with varying useful lives, it may be possible to conduct an incremental B/C analysis using the AW of benefits and costs of the various projects. This analysis is illustrated in Example 6-7.

EXAMPLE 6-7

Two mutually exclusive alternative public works projects are under consideration. Their respective costs and benefits are included in the table below. Project I has an anticipated life of 35 years, and the useful life of Project II has been estimated to be 25 years. If the interest rate is 9%, which, if either, of these projects should be selected?

	Project I	Project II
Capital investment	$750,000	$625,000
Annual oper. & maint. costs	120,000	110,000
Annual benefit	245,000	230,000
Useful life of project (years)	35	25

SOLUTION

$$AW(Costs, I) = \$750,000(A/P, 9\%, 35) + \$120,000 = \$190,977$$

$$AW(Costs, II) = \$625,000(A/P, 9\%, 25) + \$110,000 = \$173,629$$

$$B/C(II) = \$230,000/\$173,629 = \mathbf{1.3247} > \mathbf{1.0} \therefore \textbf{Project II is Acceptable}$$

$$\Delta B/\Delta C(I-II) = (\$245,000 - \$230,000)/(\$190,977 - \$173,629)$$

$$= \mathbf{0.8647} < \mathbf{1.0} \therefore \textbf{Project I not Acceptable}$$

Decision: *Project II should be selected.*

In a previous section, the calculation of B/C ratios for independent projects was presented, and it was stated that the issue of whether one project was superior to another is irrelevant. How, then, can we select from among a set of independent public works projects under conditions of capital rationing? Recall that in Chapter 5 mutually exclusive combinations (MECs) of independent projects were developed when budgetary constraints prevented all of the economically viable projects from being selected. Such an analysis can also be conducted with the B/C ratio, but as with mutually exclusive alternatives, the procedure must be applied incrementally, as illustrated in Example 6-8.

EXAMPLE 6-8

A governmental agency is considering three independent projects, each having 30-year projected useful lives. The current budget for this agency allows not more than $35,000,000 to be spent, in terms of initial investments, and the interest rate is 10% per year. Using the B/C ratio method, which of the projects shown below should be selected?

Project	Initial Investment	Annual Costs	Annual Benefits
A	$12,000,000	$1,250,000	$3,250,000
B	20,000,000	4,500,000	8,000,000
C	10,000,000	750,000	1,250,000
D	14,000,000	1,850,000	4,050,000

SOLUTION

Project C is first removed from further consideration because its B/C ratio is < 1.0. From the remaining three projects, a total of $2^3 = 8$ mutually exclusive combinations can be developed. One of these, the combination of all three remaining projects, is infeasible due to the budget constraint. Comparing the mutually exclusive combinations incrementally, beginning with the combination having the least PW of costs, it is shown that the combination of projects **A** and **B** should be selected.

Project	PW of Costs	PW of Benefits	B/C Ratio	Acceptable?
A	$23,783,643	$30,637,472	1.2882	YES
B	62,421,115	75,415,316	1.2082	YES
C	17,070,186	11,783,643	0.6903	NO
D	31,439,792	38,179,004	1.2144	YES

MEC	Projects	Total Investment	PW of Costs	PW of Benefits	Feasible?
1	do nothing	0	0	0	YES
2	A	$12,000,000	$23,783,643	$30,637,472	YES
3	B	20,000,000	62,421,115	75,415,316	YES
4	D	14,000,000	31,439,792	38,179,004	YES
5	AB	32,000,000	86,204,758	106,052,788	YES
6	AD	26,000,000	55,223,435	68,816,476	YES
7	BD	34,000,000	93,860,907	113,594,319	YES
8	ABD	46,000,000	117,644,550	144,231,791	NO

Incremental Comparison of MEC	Δ PW (Costs)	Δ PW (Benefits)	Δ B/C Ratio	Increment Justified?
1 ⇾ 2	$23,783,643	$30,637,472	1.2882	MEC 2 accepted
2 ⇾ 4	7,656,149	7,541,532	0.9850	MEC 4 not accepted
2 ⇾ 6	31,439,792	38,179,004	1.2144	MEC 6 accepted
6 ⇾ 3	7,197,680	6,598,840	0.9168	MEC 3 not accepted
6 ⇾ 5	30,981,323	37,236,312	1.2019	MEC 5 accepted
5 ⇾ 7	7,656,149	7,541,532	0.9850	MEC 7 not accepted

By applying the B/C ratio incrementally to mutually exclusive combinations of independent projects, the optimal set of projects is shown to be MEC#5. Note that although this approach is acceptable and when properly applied will lead to the selection of the "best" set of projects, the same result can be accomplished in a more straightforward manner by using PW (or AW or FW) and selecting the feasible MEC that maximizes the equivalent worth criterion without the need of an incremental analysis.

6.10 Criticisms and Shortcomings of the Benefit/Cost Ratio Method*

Although the benefit/cost ratio method is firmly entrenched as *the* procedure used by most governmental agencies for the evaluation of public-sector projects, it has been widely criticized over the years. Among the criticisms are that (1) it is often used as a tool for after-the-fact justifications rather than for project evaluation, (2) serious distributional inequities (i.e., one group reaps the benefits while another incurs the costs) may not be accounted for in B/C studies, and (3) qualitative information is often ignored in B/C studies.

Regarding the first criticism, the benefit/cost ratio is considered by many to be a method of using the numbers to support the views and interests of the group paying for the analysis. A congressional subcommittee once supported this critical view with the following conclusion:

> "...the most significant factor in evaluating a benefit-cost study is the name of the sponsor. Benefit-cost studies generally are formulated after basic positions on an issue are taken by the respective parties. The results of competing studies predictably reflect the respective positions of the parties on the issue." (p. 55)

One illustration of such a flawed B/C study is the Bureau of Reclamation's 1967 analysis supporting the proposed Nebraska Mid-State Project. The purpose of this project was to divert water from the Platte River to irrigate crop land, and the computed B/C ratio for this project was 1.24, indicating that the project should be undertaken. This favorable ratio was based, in part, on the following fallacies (p. 53):

1. An unrealistically low interest rate of only 3.125% was used.
2. The analysis used a project life of 100 years, rather than the generally more accepted and supportable assumption of a 50-year life.
3. The analysis counted wildlife and fish benefits among the project's benefits, although in reality, the project would have been devastating to wildlife. During a 30-year history of recorded water flow (1931–1960), the proposed water diversion would have left a substantial stretch of the Platte River dry for over half of that time, destroying fish and waterfowl habitats on 150 miles of the river. This destruction of wildlife habitats would have adversely affected the populations of endangered species including the bald eagle, the whooping crane, and the sandhill crane.

---

*Campen, J. T. *Benefit, Cost, and Beyond: The Political Economy of Benefit-Cost Analysis,* (Cambridge, MA: Ballinger, 1986). All quotes in Section 6.10 are taken from this source.

4. The benefits of increased farm output were based on government support prices that would have required a substantial federal subsidy.

Campen states that "the common core of these criticisms is not so much the fact that benefit-cost analysis is used to justify particular positions but that it is presented as a scientific unbiased method of analysis" (pp. 52–53). In order for an analysis to be fair and reliable, it must be based on an accurate and realistic assessment of all the relevant benefits and costs. Thus, the analysis must be performed by either an unbiased group or by a group that includes representatives from each of the interest groups involved. When the Nebraska Mid-State Project was reevaluated by an impartial third party, for example, a more realistic B/C ratio of only 0.23 was determined. Unfortunately, analyses of public sector projects are sometimes performed by parties who have already formed strong opinions on the worth of these projects.

Another shortcoming of the B/C ratio method is that benefits and costs effectively cancel each other out without regard for who reaps the benefits or who incurs the costs, which does not cause significant difficulties in the private sector where the benefits accrue to and the costs or borne by the owners of the firm. Recall that in the public sector, however, "the benefits to whomsoever they accrue" must be considered, which may result in serious distributional inequities in benefit/cost studies. Two reasons for being critical of this potential lack of distributional equity are that (1) public policy generally "operates to reduce economic inequality by improving the well-being of disadvantaged groups," and (2) there is little concern for the equality or inequality of people who are in the same general economic circumstances (p. 56).

Public policy is generally viewed as one method of reducing the economic inequalities of the poor, residents of underdeveloped regions, and racial minorities. Of course, there are many cases where it is not possible to make distributional judgments, but there are others where it is clear. Conceivably, a project could have adverse consequences for Group A, a disadvantaged group, but the benefits to another group, Group B, might far outweigh the disbenefits to Group A. If the project has a B/C ratio > 1, it will likely be accepted without regard for the consequences to Group A, particularly if Group B includes wealthy, politically influential members.

A lack of consideration of the distributional consequences of a project might also produce inequities for individuals who are in the same approximate economic circumstances. Consider, for example, the inequity shown in the following example:

> Consider a proposal to raise property taxes by 50 percent on all properties with odd-numbered street addresses and simultaneously to lower property taxes by 50 percent on all properties with even-numbered addresses. A conventional benefit-cost analysis of this proposal would conclude that its net benefits were approximately zero, and an analysis of its impact on the overall level of inequality in the size distribution of income would also show no significant effects. Nevertheless, such a proposal would be generally, and rightly, condemned as consisting of an arbitrary and unfair redistribution of income. (p. 56)

Another, more realistic, example of adverse distributional consequences would involve a project to construct a new chemical plant in *Town A*. The chemical plant would employ hundreds of workers in an economically depressed area, but, according to one group of concerned citizens, it would also release hazardous by-products that could pollute the groundwater and a nearby stream that provides most of the drinking water for *Town B*. Thus, the benefits of this project would be the addition of jobs and a boost to the local economy for *Town A*, but *Town B* would have increased water treatment costs and residents may be subject to long-term health hazards and increased medical bills. Unfortunately, a B/C ratio analysis would show only the net monetary effect of the project, without regard to the serious distributional inequities.

In the previously cited example of the Bureau of Reclamation's analysis of the Nebraska Mid-State Project, the problem of using unreliable monetary values for nonmonetary considerations (irreducibles) was highlighted in the discussion of wildlife and fish "benefits." But the outcome of the B/C analysis might also be highly suspect if no attempt had been made to quantify these aspects of the project. When only easily quantifiable information is included in the analysis, the importance of the other factors is totally neglected, which results in projects with mostly monetary benefits being favored, and projects with less tangible benefits being rejected without due consideration. Unfortunately, busy decisionmakers want a single number upon which to base their acceptance or rejection of a project. No matter how heavily the nonmonetary aspects of a project are emphasized during the discussion, most managers will go directly to the bottom line of the report to get a single number and use that to make their decision. A 1980 congressional committee concluded that "whenever some quantification is done—no matter how speculative or limited—the number tends to get into the public domain and the qualifications tend to get forgotten. . . . The number is the thing" (p. 68).

Although the above criticisms have often been leveled at the benefit/cost method itself, problems in the use (and abuse) of the B/C procedure are largely due to the inherent difficulties in evaluating public projects (see section 6.5) and the manner in which the procedure is applied. Note that these same criticisms might be in order for a poorly conducted analysis that relied on an equivalent worth or rate of return method.

▼ 6.11 Spreadsheet Applications

To demonstrate the use of spreadsheets in a benefit cost analysis, we consider the three mutually exclusive public works projects introduced in Example 6-6. Their respective benefits and costs are inputs to the spreadsheet model shown in Figure 6-4. The B/C ratio is computed for each alternative, using both conventional and modified forms of the ratio. Because all of the projects have a B/C > 1, we must perform an incremental analysis to determine the best public works project.

	A	B	C	D
1	MARR	10%		
2	Study Period	50		
3				
4		Project A	Project B	Project C
5	Initial Costs	$ (8,500,000)	$ (10,000,000)	$ (12,000,000)
6	Annual O&M Costs	$ (750,000)	$ (725,000)	$ (700,000)
7	Salvage Value	$ 1,250,000	$ 1,750,000	$ 2,000,000
8	Annual Benefit	$ 2,150,000	$ 2,265,000	$ 2,500,000
9				
10	CR Amount	$ 856,229	$ 1,007,088	$ 1,208,592
11				
12	Conventional B/C Ratio	1.3385	1.3077	1.3099
13	Modified B/C Ratio	1.6351	1.5292	1.4893
14				
15	Incremental Analysis			
16		A:B	A:C	
17	Δ Initial Costs	$ (1,500,000)	$ (3,500,000)	
18	Δ Annual O&M Costs	$ 25,000	$ 50,000	
19	Δ Salvage Value	$ 500,000	$ 750,000	
20	Δ Annual Benefit	$ 115,000	$ 350,000	
21				
22	Δ CR Amount	$ 150,859	$ 352,363	
23				
24	ΔB/ΔC (Conventional)	0.9137	1.1576	
25	ΔB/ΔC (Modified)	0.9280	1.1352	
26	Increment Justified?	No	Yes	

FIGURE 6-4   Spreadsheet for Comparing MEAs Using an Incremental B/C Ratio

The incremental analysis is shown in the lower portion of Figure 6-4. Project A is our base alternative because it has the smallest annual equivalent of costs and a B/C ratio ≥ 1. The first comparison then is between A and B. The incremental benefits and costs are found by subtracting the estimates for Project A from the estimates for Project B. The ratio of the incremental benefits to the incremental costs, $\Delta B/\Delta C$, is less than one, indicating that the increment is not justified.

Project C is then compared to Project A in a similar fashion. The resulting $\Delta B/\Delta C > 1$ tells us that the incremental benefits of Project C outweigh the incremental costs. Because Project C is the last alternative, it is the recommended alternative. Note that (1) the same conclusion is reached regardless if we use the conventional or modified B/C ratio and (2) Project A, which had the highest B/C ratio, is not the recommended project. The formulas for the highlighted cells are given in the following table.

Cell	Contents
B10	= PMT($B$1, $B$2, B5 + B7/(1 + $B$1)^$B$2)
B12	= B8/(B10 − B6)
B13	= (B8 + B6)/B10
B17	= C5 − B5
C17	= D5 − B5
B22	= PMT($B$1, $B$2, B17 + B19/(1 + $B$1)^$B$2)
B24	= B20/(B22 − B18)
B25	= (B20 + B18)/B22
B26	= IF (B24 >= 1, "Yes", "No")

6.12 Summary

From the discussion and examples of public projects presented in this chapter, it is apparent that because of the methods of financing, the absence of tax and profit requirements, and political and social factors, the criteria used in evaluating privately financed projects frequently cannot be applied to public works. Nor should public projects be used as yardsticks with which to compare private projects. Nevertheless, whenever possible, public works should be justified on an economic basis to ensure that the public obtains the maximum return from the tax money that is spent. Whether an engineer is working on such projects, is called upon to serve as a consultant, or assists in the conduct of the benefit/cost analysis, he or she is bound by professional ethics to do his or her utmost to see that the projects and the associated analyses are carried out in the best possible manner and within the limitations of the legislation enacted for their authorization.

The B/C ratio has remained a popular method for evaluating the financial performance of public projects. Both the conventional and modified B/C ratio methods have been explained and illustrated for the case of *independent* and *mutually exclusive* projects. A final note of caution: The best project among a *mutually exclusive* set of projects is not necessarily the one that maximizes the B/C ratio. In this chapter we have seen that an incremental analysis approach to evaluating benefits and costs is necessary to ensure the correct choice.

6.13 References

Au, T., and T. P. Au. *Engineering Economics for Capital Investment Analysis*. Boston: Allyn and Bacon, 1983.

Campen, J. T. *Benefit, Cost, and Beyond*. Cambridge, Mass.: Ballinger, 1986.

Dasgupta, Agit K., and D. W. Pearce. *Cost-Benefit Analysis: Theory & Practice*. New York: Harper & Row, 1972.

Mishan, E. J. *Cost-Benefit Analysis*. New York: Praeger, 1976.

SASSONE, PETER G., and WILLIAM A. SCHAFFER. *Cost-Benefit Analysis: A Handbook.* New York: Academic Press, 1978.

SCHWAB, B., and P. LUSZTIG. "A Comparative Analysis of the Net Present Value and the Benefit-Cost Ratio as Measure of Economic Desirability of Investment," *Journal of Finance,* vol. 24, 1969, pp. 507–516.

▼ 6.14 Problems

6-1.
   a. What is a self-liquidating project? (6.3)
   b. What is a multiple-purpose project? (6.4)
   c. Why is it difficult to make a logical assignment of costs in a multiple-purpose project? (6.4)
   d. Why is it difficult to assess the efficiency of operation in most multiple-purpose projects? (6.4)

6-2. Describe how engineering economy studies that involve public projects differ from those that are conducted by private organizations. (6.2)

6-3. Discuss some of the considerations that go into determining an interest rate to be used in evaluating alternatives in the public sector. (6.6)

6-4. Select a public works project that currently is being proposed or carried out in your area. Obtain the economic analysis for the project. Determine the following: (6.2)
   a. Are any intangibles included in the benefits?
   b. If there are intangibles, would the project be economically justified if the intangible benefits were omitted?
   c. Would most people in the affected area agree on the values assigned to intangible benefits?

6-5. A nonprofit government corporation is considering two alternatives for generating power:

   Alternative A. Build a coal-powered generating facility at a cost of $20,000,000. Annual power sales are expected to be $1,000,000 per year. Annual O&M costs are $200,000 per year. A benefit of this alternative is that it is expected to attract new industry, worth $500,000 per year, to the region.

   Alternative B. Build a hydroelectric generating facility. The capital investment, power sales, and operating costs are $30,000,000, $800,000, and $100,000 per year, respectively. Annual benefits of this alternative are as follows:

Flood control savings	$600,000
Irrigation	$200,000
Recreation	$100,000
Ability to attract new industry	$400,000

The useful life of both alternatives is 50 years. Using an interest rate of 5%, determine which alternative (if either) should be selected according to the conventional B/C ratio method. (6.7, 6.9)

6-6. Five independent projects are available for funding by a certain public agency. The following tabulation shows the equivalent annual benefits and costs for each. (6.8)

Project	Annual Benefits	Annual Costs
A	$1,800,000	$2,000,000
B	5,600,000	4,200,000
C	8,400,000	6,800,000
D	2,600,000	2,800,000
E	6,600,000	5,400,000

   a. Assume that the projects are of the type for which the benefits can be determined with considerable certainty and that the agency is willing to invest money as long as the B/C ratio is at least 1. Which alternatives should be selected for funding?
   b. What is the rank ordering of projects from best to worst?
   c. If the projects involved intangible benefits that required considerable judgment in assigning their values, would your recommendation be affected?

6-7. In the development of a publicly owned, commercial waterfront area, three possible independent plans are being considered. Their costs and estimated benefits are shown at the top of the first column of page 267. (6.8)
   a. Which plan(s) should be adopted, if any, if the controlling board wishes to invest any

(Pr. 6-9)

Alternative	Equivalent Annual Cost of Project	Expected Annual Flood Damage	Annual Benefits
I. No flood control	0	$100,000	0
II. Construct levees	$30,000	80,000	$112,000
III. Build small dam	$100,000	5,000	110,000

	PW ($000s)	
Plan	Costs	Benefits
A	$123,000	$139,000
B	135,000	150,000
C	99,000	114,000

amount required, provided that the B/C ratio on the required investment is at least 1.0?

b. Suppose that 10% of the costs of each plan are reclassified as "disbenefits." What percentage change in the B/C ratio of each plan results from the reclassification?

c. Comment on why the rank orderings in (a) are unaffected by the change in (b).

6-8. Consider the two types of equipment in the following table and determine which choice is better if a firm desires to invest as long as the B/C ratio is greater than or equal to 1. The firm's MARR is 10% per year. Assume repeatability and show all work. (6.9)

	Equipment Type	
	RS-422	RS-511
Capital investment	$500	$1,750
Useful life (years)	6	12
Market (salvage) value	125	375
Annual benefits	238	388
Annual O&M costs	108	113

6-9. Consider the mutually exclusive alternatives at the top of this page. Which alternative

(Pr. 6-7)

would be chosen according to these decision criteria?

a. Maximum benefit
b. Minimum cost
c. Maximum benefits minus costs
d. Largest investment having an incremental B/C ratio larger than 1
e. Largest B/C ratio

Which project should be chosen? (6.9)

6-10.* A river that passes through private lands is formed from four branches of water that flow from a national forest. Some flooding occurs each year, and a major flood generally occurs every few years. If small earthen dams are placed on each of the four branches, the chances of major flooding would be practically eliminated. Construction of one or more dams would reduce the amount of flooding by varying degrees.

Other potential benefits from the dams are the reduction of damages to fire and logging roads in the forest, the value of the dammed water for protection against fires, and the increased recreational use. The table below contains the estimated benefits and costs associated with building one or more dams.

The equation used to calculate B/C ratios is as follows:

$$B/C \ ratio = \frac{\begin{array}{c} \textit{annual flood and fire savings} \\ + \textit{ recreational benefits} \end{array}}{\begin{array}{c} \textit{equivalent annual construction costs} \\ + \textit{maintenance} \end{array}}$$

		Costs			Benefits		(Pr. 6-10)
Option	Dam Sites	Construction	Annual Maintenance	Annual Flood	Annual Fire	Annual Recreation	
A	1	$3,120,000	$ 52,000	$520,000	$ 52,000	$ 78,000	
B	1&2	3,900,000	91,000	630,000	104,000	78,000	
C	1, 2&3	7,020,000	130,000	728,000	156,000	156,000	
D	1, 2, 3&4	9,100,000	156,000	780,000	182,000	182,000	

*Fashioned after a problem from James L. Riggs, Engineering Economics (New York: McGraw-Hill, 1977), pp. 432–434.

Route	Construction Costs	Annual Maintenance Cost	Annual Savings in Fire Damage	Annual Recreational Benefit	Annual Timber Access Benefit	(Pr. 6-11)
A	$185,000	$2,000	$ 5,000	$3,000	$ 500	
B	220,000	3,000	7,000	6,500	1,500	
C	290,000	4,000	12,000	6,000	2,800	

6-10 (Continued) Benefits and costs are to be compared using the AW method with an interest rate of 8%; useful lives of the dams are 100 years. (6.7, 6.9)

a. Which of the four options would you recommend? Show why.

b. If fire benefits are reclassified as reduced costs, would the choice in (a) be affected? Show your work.

6-11. A state-sponsored Forest Management Bureau is evaluating alternative routes for a new road into a formerly inaccessible region. Three mutually exclusive plans for routing the road provide different benefits, as indicated in the table above. The roads are assumed to have an economic life of 50 years, and the interest rate is 8%. Which route should be selected according to the B/C ratio method? (6.7, 6.9)

6-12. An area on the Colorado River is subject to periodic flood damage that occurs, on the average, every two years and results in a $2,000,000 loss. It has been proposed that the river channel should be straightened and deepened, at a cost of $2,500,000, to reduce the probable damage to not over $1,600,000 for each occurrence during a period of 20 years before it would have to be deepened again. This procedure would also involve annual expenditures of $80,000 for minimal maintenance. One legislator in the area has proposed that a better solution would be to construct a flood-control dam at a cost of $8,500,000, which would last indefinitely, with annual maintenance costs of not over $50,000. He estimates that this project would reduce the probable annual flood damage to not over $450,000. In addition, this solution would provide a substantial amount of irrigation water that would produce annual revenue of $175,000 and recreational facilities, which he estimates would be worth at least $45,000 per year to the adjacent populace. A second legislator believes that the dam should be built and that the river channel also should be straightened and deepened, noting that the total cost of $11,000,000 would reduce the probable annual flood loss to not over $350,000 while providing the same irrigation and recreational benefits. If the state's capital is worth 10%, determine the B/C ratios and the incremental B/C ratio. Recommend which alternative should be adopted. (6.9)

6-13. Ten years ago the port of Secoma built a new pier containing a large amount of steel work, at a cost of $300,000, estimating that it would have a life of 50 years. The annual maintenance cost, much of it for painting and repair caused by environmental damage, has turned out to be unexpectedly high, averaging $27,000. The port manager has proposed to the port commission that this pier be replaced immediately with a reinforced concrete pier at a construction cost of $600,000. He assures them that this pier will have a life of at least 50 years, with annual maintenance costs of not over $2,000. He presents the figures below as justification for the replacement, having determined that the net market value of the existing pier is $40,000.

He has stated that because the port earns a net profit of over $3,000,000 per year, the project could be financed out of annual earnings. Thus there would be no interest cost, and an annual saving of $19,000 would be obtained by making the replacement. (6.9)

a. Comment on the port manager's analysis.

b. Make your own analysis and recommendation regarding the proposal.

Annual Cost of Present Pier		Annual Cost of Proposed Pier		(Pr. 6-13)
Depreciation: $300,000/50	$6,000	Depreciation: $600,000/50	$12,000	
Maintenance cost	27,000	Maintenance cost	2,000	
Total	$33,000	Total	$14,000	

6-14. A toll bridge across the Mississippi River is being considered as a replacement for the current I-40 bridge linking Tennessee to Arkansas. Because this bridge, if approved, will become a part of the U.S. Interstate Highway system, the B/C ratio method must be applied in the evaluation. Investment costs of the structure are estimated to be $17,500,000, and $325,000 per year in operating and maintenance costs are anticipated. In addition, the bridge must be resurfaced every fifth year of its 30-year projected life, at a cost of $1,250,000 per occurrence (no resurfacing cost in year 30). Revenues generated from the toll are anticipated to be $2,500,000 in its first year of operation, with a projected annual rate of increase of 2.25% per year due to the anticipated annual increase in traffic across the bridge. Assuming zero market (salvage) value for the bridge at the end of 30 years and an MARR of 10%, should the toll bridge be constructed? (6.7)

6-15. Refer back to **Problem 6-14.** Suppose that the toll bridge can be redesigned such that it will have a (virtually) infinite life. The MARR remains at 10% per year. Revised costs and revenues (benefits) are given below: (6.7, 6.9)

> Capital investment: $22,500,000
> Annual operating and maintenance costs: $250,000
> Resurface cost every 7th year: $1,000,000
> Structural repair cost, every 20th year: $1,750,000
> Revenues (treated as constant — no rate of increase): $3,000,000

a. What is the capitalized worth of the bridge?
b. Determine the B/C ratio of the bridge over an infinite time horizon.
c. Should the initial design (*Pr. 6-14*) or the new design be selected?

6-16. In the aftermath of Hurricane Thelma, the U.S. Army Corps of Engineers is considering two alternative approaches to protect a fresh-water wetland from the encroaching seawater during high tides. The first alternative, the construction of a five-mile long, 20-foot high levee, would have an investment cost of $25,000,000 with annual upkeep costs estimated at $725,000. A new roadway along the top of the levee would provide two major "benefits": (1) improved

recreational access for fishermen, and (2) reduction of the driving distance between the towns at opposite ends of the proposed levee by 11 miles. The annual benefit for the levee has been estimated at $1,500,000. The second alternative, a channel dredging operation, would have an investment cost of $15,000,000. The annual cost of maintaining the channel is estimated at $375,000. There are no documented "benefits" for the channel dredging project. Using an MARR of 8% and assuming a 25-year life for either alternative, apply the incremental benefit cost ratio ($\Delta B/\Delta C$) method to determine which alternative should be chosen. (NOTE: The null alternative, "Do Nothing," is not a viable alternative.) (6.9)

6-17. A tunnel through a mountain is being considered as a replacement for an existing stretch of highway in Southeastern Kentucky. The existing road is a steep, narrow, winding two-lane highway that has been the site of numerous fatal accidents, with an average of 2.05 fatalities and 3.35 "serious injuries" per year. It has been projected that the tunnel will significantly reduce the frequency of accidents, with estimates of not more than 0.15 fatalities and 0.35 "serious injuries" per year. Initial capital investment requirements, including land acquisition, tunnel excavation, lighting, roadbed preparation, etc., have been estimated to be $45,000,000. Annualized upkeep costs for the tunnel will be significantly less than for the existing highway, resulting in an annual savings of $85,000. For purposes of this analysis, a "value per life saved" of $1,000,000 will be applied, along with an estimate of $750,000 for medical costs, disability, etc., per "serious injury." (6.7)

a. Apply the benefit/cost ratio method, with an anticipated life of the tunnel project of 50 years and an interest rate of 8%, to determine whether or not the tunnel should be constructed.
b. Assuming that the cost per "serious injury" is unchanged, determine the "value per life saved" at which the tunnel project is marginally justified (i.e., B/C = 1.0).

6-18. You have been assigned the task of comparing the economic results of three alternative designs for a state government public works project. The estimated values for various

	Alternative Design			(Pr. 6-18)
Factor	1	2	3	
Capital investment	$1,240,000	$1,763,000	$1,475,000	
Salvage value (end of year 15)	90,000	150,000	120,000	
Annual O & M costs	215,000	204,000	201,000	
Annual benefits to user group A	315,000	367,000	355,000	
Annual benefits to other user groups	147,800	155,000	130,500	

economic factors related to the three designs are given above. The MARR being used is 9% and the analysis period is 15 years. (6.9)

a. Use the conventional benefit/cost ratio method, with AW as the equivalent worth measure, to select the preferred design for the project.

b. Use the modified B/C ratio method, with PW as the equivalent worth measure, to select the preferred design for the project.

6-19. The Fox River is bordered on the east by Illinois Route 25 and on the west by Illinois Route 31. Along one stretch of the river, there is a distance of 16 miles between adjacent crossings. An additional crossing in this area has been proposed, and three alternative bridge designs are under consideration. Two of the designs have 25-year useful lives, and the third has a useful life of 35 years. Each bridge must be resurfaced periodically, and the roadbed of each bridge will be replaced at the end of its useful life, at a cost significantly less than initial construction costs. The annual benefits of each design differ on the basis of disruption to normal traffic flow along Routes 25 and 31. Given the information

in the table below, use the B/C ratio method to determine which bridge design should be selected. Assume that the selected design will be used indefinitely, and use an interest rate of 10%. (6.9)

6-20. New legislation is being considered that would effectively stop the manufacture of a certain product. The process used to manufacture this product yields a carcinogenic by-product, dioxin. A plant that employs 5000 workers in *Town A* produces the product, and it has been estimated that the loss of this product will necessitate the layoff of 1200 workers. The wastewater discharge from the plant apparently contains dioxin, as low levels of this carcinogen have been found in a nearby river. This river serves as the primary supply of drinking water in *Town B* 50 miles downstream from the plant. You have been asked to conduct a benefit/cost study of the proposed legislation. As a part of this, you must list all of the associated *benefits, costs,* and *disbenefits.* Include statements addressing the potential distributional inequities for the two possible outcomes (the passage of the legislation or the failure to pass the legislation). (6.10)

	Bridge Design			(Pr. 6-19)
	A	B	C	
Capital investment	$17,000,000	$14,000,000	$12,500,000	
Annual maintenance cost*	12,000	17,500	20,000	
Resurface (every fifth year)*	—	40,000	40,000	
Resurface (every seventh year)*	40,000	—	—	
Bridge replacement cost	3,000,000	3,500,000	3,750,000	
Annual benefit	2,150,000	1,900,000	1,750,000	
Useful life of bridge (yrs)**	35	25	25	

*Cost not incurred in last year of bridge's useful life
**Applies to roadbed only; Structural portion of bridge has indefinite useful life

Depreciation and Income Taxes

The objectives of this chapter are to illustrate some of the concepts and mechanics of depreciation and depletion and to describe their role in after-tax analysis. Additionally, we illustrate the differences between before-tax and after-tax analysis in engineering economy studies.

The following topics are discussed in this chapter:

The nature of depreciation

Classical (historical) depreciation methods

The Modified Accelerated Cost Recovery System (MACRS)

Depletion

Distinctions between different types of taxes

The after-tax MARR

Corporate taxable income

The effective (marginal) income tax rate

Gain or loss on the disposal of an asset

General procedure for making after-tax analyses

Illustrations of after-tax analyses (including cost alternatives and projects with specific financing arrangements)

The after-tax effect of depletion allowances

7.1   Introduction

Taxes have been paid since the dawn of civilization. Interestingly, in the United States a federal income tax did not exist until March 13, 1913, when Congress enacted the Sixteenth Amendment to the Constitution.* Most organizations consider the effect of income taxes on the financial results of a proposed engineering project because income taxes usually represent a significant cash outflow that cannot be ignored in decision making. Based on the previous chapters, we can now describe how income tax liabilities (or credits) and after-tax cash flows are determined in engineering practice. In this chapter, an *after-tax cash flow (ATCF) procedure* will be used in capital investment analysis because it avoids innumerable problems associated with measuring corporate income. This theoretically sound procedure provides a quick and relatively easy way to define a project's financial results.

Because the amount of material included in the Internal Revenue Code (and in state and municipal laws, where they exist) is very extensive, only selected parts of the subject can be discussed within the scope of this textbook. Our focus in this chapter is *federal corporate income taxes* and their general effect on the financial results of proposed engineering projects. The material presented is for educational purposes. In practice, you should seek expert counsel when analyzing a specific project.

Due to the effect of *depreciation* on the after-tax cash flows of a project, this topic is discussed first. The selected material on depreciation will then be used in the remainder of the chapter for accomplishing after-tax analyses of engineering projects.

7.2   Depreciation Concepts and Terminology

> Depreciation is the decrease in value of physical properties with the passage of time and use. More specifically, depreciation is an *accounting concept* that establishes an annual deduction against before-tax income such that the effect of time and use on an asset's value can be reflected in a firm's financial statements. Annual depreciation deductions are intended to "match" the yearly fraction of value used by an asset in the production of income over the asset's actual economic life. The actual amount of depreciation can never be established until the asset is retired from service. Because depreciation is a *noncash cost* that affects income taxes, we must consider it properly when making after-tax engineering economy studies.

Depreciable property is property for which depreciation is allowed under federal, state, or municipal income tax laws and regulations. To determine if depreciation deductions can be taken, the classification of various types of property

---

*During the Civil War, a federal income tax rate of 3% was initially imposed in 1862 to help pay for war expenditures. The federal rate was later raised to 10% but eventually eliminated in 1872.

must be understood. In general, property is depreciable if it meets the following basic requirements:

1. It must be used in business or held to produce income.
2. It must have a determinable useful life (defined in Section 7.2.2), and the life must be longer than one year.
3. It must be something that wears out, decays, gets used up, becomes obsolete, or loses value from natural causes.
4. It is not inventory, stock in trade, or investment property.

Depreciable property is classified as either *tangible* or *intangible*. Tangible property can be seen or touched, and it includes two main types called *personal property* and *real property*. Personal property includes assets such as machinery, vehicles, equipment, furniture, and similar items. In contrast, real property is land and generally anything that is erected on, growing on, or attached to land. Land itself, however, is not depreciable because it does not have a determinable life.

Intangible property is personal property such as a copyright, patent, or franchise. We will not discuss the depreciation of intangible assets in this chapter because engineering projects rarely include this class of property.

A company can begin to depreciate property it owns when the property is *placed in service* for use in the business or for the production of income. Property is considered to be placed in service when it is ready and available for a specific use, even if it is not actually used yet. Depreciation stops either when the cost of placing it in service has been recovered or it is retired from service.

7.2.1 Depreciation Methods and Related Time Periods

The depreciation methods permitted under the Internal Revenue Code have changed with time. In general, the following summary indicates the *primary methods used for property placed in service during three distinct time periods.*

Before 1981 Several methods could be elected for depreciating property placed in service before 1981. The primary methods used were straight-line (SL), declining balance (DB), and sum-of-the-years-digits (SYD). We will refer to these methods, collectively, as the *classical or historical methods* of depreciation.

After 1980 and Before 1987 For federal income taxes, tangible property placed in service during this period must be depreciated using the Accelerated Cost Recovery System (ACRS). This system was implemented by the Economic Recovery Tax Act of 1981 (ERTA).

After 1986 The Tax Reform Act of 1986 (TRA 86) was one of the most extensive income tax reforms in the history of the United States. This act modified the previous ACRS implemented under ERTA and requires the use of the Modified Accelerated Cost Recovery System (MACRS) for the depreciation of tangible property placed in service after 1986.

A description of the classical (historical) methods of depreciation are included in the chapter for several important reasons. They apply directly to property placed in service prior to 1981 as well as for property, such as intangible property (which requires the straight-line method), that does not qualify for ACRS or MACRS. Also, these methods are often specified by the tax laws and regulations of state and municipal governments in the United States, and are used for depreciation purposes in other countries. In addition, as we shall see in Section 7.4, the declining balance and the straight-line methods are used in determining the annual recovery rates under MACRS.

We do not discuss the application of ACRS in the chapter, but readily available Internal Revenue Service (IRS) publications describe its use.* Selected parts of the Modified Accelerated Cost Recovery System, however, are described and illustrated because this system applies to depreciable property in present and future engineering projects.

7.2.2 Additional Definitions

Because this chapter uses many terms that are not generally included in the vocabulary of engineering education and practice, an abbreviated set of definitions is presented here. This list is intended to supplement the previous definitions provided in this section.

Adjusted (cost) basis The original cost basis of the asset, adjusted by allowable increases or decreases, is used to compute depreciation and depletion deductions. For example, the cost of any improvement to a capital asset with a useful life *greater* than one year increases the original cost basis, and a casualty or theft loss decreases it. If the basis is altered, the depreciation deduction may need to be adjusted.

Basis, or cost basis The initial cost of acquiring an asset (purchase price plus any sales taxes), including transportation expenses and other normal costs of making the asset serviceable for its intended use. This amount is also called the *unadjusted cost basis*.

Book value (BV) The worth of a depreciable property as shown on the accounting records of a company. It is the original cost basis of the property, including any adjustments, less all allowable depreciation or depletion deductions. It thus represents the amount of capital that remains invested in the property and must be recovered in the future through the accounting process. The BV of a property may not be a useful measure of its market value. In general, the book value of a property at the end of year k is

$$(\text{Book value})_k = \text{adjusted cost basis} - \sum_{j=1}^{k} (\text{depreciation deduction})_j \quad \textbf{(7-1)}$$

---

*Useful references on material in this chapter, available from the Internal Revenue Service in an annually updated version, are Publication 534 (Depreciation), Publication 334 (Tax Guide for Small Business), Publication 542 (Tax Information on Corporations), and Publication 544 (Sales and Other Dispositions of Assets).

Market value (MV) The amount that will be paid by a willing buyer to a willing seller for a property where each has equal advantage and is under no compulsion to buy or sell. The MV approximates the present value of what will be received through ownership of the property, including the time value of money (or profit).

Recovery period The number of years over which the basis of a property is recovered through the accounting process. For the classical methods of depreciation, this period is normally the useful life. Under MACRS, this period is the *property class* for the General Depreciation System (GDS), and it is the *class life* for the Alternative Depreciation System (ADS)—see Section 7.4.

Recovery rate A percentage (expressed in decimal form) for each year of the MACRS recovery period that is utilized to compute an annual depreciation deduction.

Salvage value (SV) The estimated value of a property at the end of its useful life.* It is the expected selling price of a property when the asset can no longer be used productively by its owner. The term *net salvage value* is used when the owner will incur expenses in disposing of the property, and these cash outflows must be deducted from the cash inflows to obtain a final net SV. When the classical methods of depreciation are applied, an estimated salvage value is initially established and used in the depreciation calculations. Under MACRS, the SV of depreciable property is defined to be zero.

Useful life The expected (estimated) period of time that a property will be used in a trade or business or to produce income. It is not how long the property will last but how long the owner expects to productively use it. Useful life is sometimes referred to as *depreciable life*. Actual useful life of an asset, however, may be different than its depreciable life.

7.3 The Classical (Historical) Depreciation Methods

This section describes and illustrates the straight-line, declining balance, and sum-of-the-years-digits methods of calculating depreciation deductions. As mentioned in Section 7.2, these historical methods continue to apply, directly and indirectly, to the depreciation of property. Also included is a discussion of the units of production method.

7.3.1 Straight-Line (SL) Method

Straight-line depreciation is the simplest depreciation method. It assumes that a constant amount is depreciated each year over the depreciable (useful) life of the asset. The following definitions are used in the equations below. If we define

$$N = \text{depreciable life of the asset in years}$$
$$B = \text{cost basis, including allowable adjustments}$$
$$d_k = \text{annual depreciation deduction in year } k \ (1 \leq k \leq N)$$

---

*We often use the term market value (MV) in place of salvage value (SV).

$$BV_k = \text{book value at end of year } k$$

$$SV_N = \text{estimated salvage value at end of year N}$$

$$d_k^* = \text{cumulative depreciation through year } k$$

then

$$d_k = (B - SV_N)/N \tag{7-2}$$

$$d_k^* = kd_k \text{ for } 1 \le k \le N \tag{7-3}$$

$$BV_k = B - d_k^* \tag{7-4}$$

Note for this method you must have an estimate of the final SV, which will also be the final book value at the end of year N. In some cases, the estimated SV_N may not equal an asset's actual terminal MV.

EXAMPLE 7-1

A new electric saw for cutting small pieces of lumber in a furniture manufacturing plant has a cost basis of $4,000 and a 10-year depreciable life. The estimated SV of the saw is zero at the end of 10 years. Determine the annual depreciation amounts using the straight-line method. Tabulate the annual depreciation amounts and the book value of the saw at the end of each year.

SOLUTION

The depreciation amount, cumulative depreciation, and book value for each year are obtained by applying Equations 7-2, 7-3, and 7-4. Sample calculations for year five are shown below.

$$d_5 = \frac{\$4,000 - 0}{10} = \$400$$

$$d_5^* = \frac{5(\$4,000 - 0)}{10} = \$2,000$$

$$BV_5 = \$4,000 - \frac{5(\$4,000 - 0)}{10} = \$2,000$$

The depreciation and book value amounts for each year are shown below.

EOY, k	d_k	BV_k
0	—	$4,000
1	$400	3,600
2	400	3,200
3	400	2,800
4	400	2,400
5	400	2,000
6	400	1,600
7	400	1,200
8	400	800
9	400	400
10	400	0

7.3.2   Declining Balance (DB) Method

In the declining balance method, sometimes called the *constant percentage method* or the *Matheson formula*, it is assumed that the annual cost of depreciation is a fixed percentage of the BV at the *beginning* of the year. The ratio of the depreciation in any one year to the BV at the beginning of the year is constant throughout the life of the asset and is designated by R ($0 \le R \le 1$). In this method, $R = 2/N$ when a 200% declining balance is being used (i.e., twice the straight line rate of $1/N$), and N equals the depreciable (useful) life of an asset. If the 150% declining balance method is specified, then $R = 1.5/N$. The following relationships hold true for the declining balance method:

$$d_1 = B(R) \tag{7-5}$$

$$d_k = B(1 - R)^{k-1}(R) \tag{7-6}$$

$$d_k^* = B[1 - (1 - R)^k] \tag{7-7}$$

$$BV_k = B(1 - R)^k \tag{7-8}$$

$$BV_N = B(1 - R)^N \tag{7-9}$$

Notice that Equations 7-5 through 7-9 do not contain a term for SV_N.

EXAMPLE 7-2

Rework Example 7-1 with the declining balance method when (a) $R = 2/N$ (200% declining balance method) and (b) $R = 1.5/N$ (150% declining balance method). Again, tabulate the annual depreciation amount and book value for each year.

SOLUTION

Annual depreciation, cumulative depreciation, and book value are determined by using Equations 7-6, 7-7, and 7-8, respectively. Sample calculations for year six are shown below.

(a)

$$R = 2/N = 0.2$$

$$d_6 = \$4{,}000(1 - 0.2)^5(0.2) = \$262.14$$

$$d_6^* = \$4{,}000[1 - (1 - 0.2)^6] = \$2{,}951.42$$

$$BV_6 = \$4{,}000(1 - 0.2)^6 = \$1{,}048.58$$

(b)

$$R = 1.5/N = 0.15$$

$$d_6 = \$4{,}000(1 - 0.15)^5(0.15) = \$266.22$$

$$d_6^* = \$4{,}000[1 - (1 - 0.15)^6] = \$2{,}491.40$$

$$BV_6 = \$4{,}000(1 - 0.15)^6 = \$1{,}508.60$$

The depreciation and book value amounts for each year, when $R = 2/N = 0.2$, are shown at the top of page 278.

200% Declining Balance Method Only

EOY, k	d_k	BV_k
0	—	$4,000
1	$800	3,200
2	640	2,560
3	512	2,048
4	409.60	1,638.40
5	327.68	1,310.72
6	262.14	1,048.58
7	209.72	838.86
8	167.77	671.09
9	134.22	536.87
10	107.37	429.50

7.3.3 Sum-of-the-Years-Digits (SYD) Method

To compute the depreciation deduction by the SYD method, the digits corresponding to the number for each permissible year of life are first listed in reverse order. The sum of these digits is then determined. The depreciation factor for any year is the number from the reverse-ordered listing for that year divided by the sum of the digits. For example, for a property having a depreciable (useful) life of five years, SYD depreciation factors are as follows:

Year	Number of the Year in Reverse Order (digits)	SYD Depreciation Factor
1	5	5/15
2	4	4/15
3	3	3/15
4	2	2/15
5	1	1/15
Sum of the digits	15	

The depreciation for any year is the product of the SYD depreciation factor for that year and the difference between the cost basis (B) and the estimated final SV. The general expression for the annual cost of depreciation for any year k, when N equals the depreciable life of an asset, is

$$d_k = (B - SV_N) \cdot \left[\frac{2(N - k + 1)}{N(N + 1)} \right] \qquad \textbf{(7-10)}$$

The book value at the end of the year k is

$$BV_k = B - \left[\frac{2(B - SV_N)}{N} \right] k + \left[\frac{(B - SV_N)}{N(N + 1)} \right] k(k + 1) \qquad \textbf{(7-11)}$$

and the cumulative depreciation through the kth year is simply

$$d_k^* = B - BV_k \qquad \textbf{(7-12)}$$

EXAMPLE 7-3

Rework Example 7-1 using the sum-of-the-years-digits method. Tabulate the annual depreciation amount and book value for each year.

SOLUTION

With Equations 7-10, 7-11, and 7-12, respectively, the annual depreciation, book value, and cumulative depreciation amounts are obtained. Sample calculations for year four are given below.

$$d_4 = \$4,000 \left[\frac{2(10 - 4 + 1)}{10(11)} \right] = \$509.09$$

$$BV_4 = \$4,000 - \left[\frac{2(\$4,000)}{10} \right] \cdot 4 + \left[\frac{\$4,000}{10(11)} \right] \cdot 4 \cdot 5 = \$1,527.27$$

$$d_4^* = \$4,000 - \$1,527.27 = \$2,472.73$$

Depreciation and book value amounts for each year are shown below.

EOY, k	d_k	BV_k
0	—	$4,000
1	$727.27	3,272.73
2	654.55	2,618.18
3	581.82	2,036.36
4	509.09	1,527.27
5	436.36	1,090.91
6	363.64	727.27
7	290.91	436.36
8	218.18	218.18
9	145.45	72.73
10	72.73	0

7.3.4 Declining Balance with Switchover to Straight-Line

Because the declining balance method never reaches a BV of zero, it is permissible to switch from this method to the straight-line method so that an asset's SV_N will be zero (or some other desired amount). Also, this method is used in calculating the MACRS recovery rates in Table 7-3.

Table 7-1 illustrates a switchover from double declining balance depreciation to straight-line depreciation for Example 7-1. The switchover occurs in the year in which a larger depreciation amount is obtained from the straight-line method. From Table 7-1, it is apparent that $d_6 = \$262.14$. The BV at the end of year six (BV_6) is $1,048.58. Additionally, observe that BV_{10} is $4,000 − $3,570.50 = $429.50 without switchover to the straight-line method in Table 7-1. With switchover, BV_{10} equals 0. It is clear that this asset's d_k, d_k^*, and BV_k in years seven through ten are established from the straight-line method, which permits the full cost basis to be depreciated over the ten-year recovery period.

TABLE 7-1   The 200% Declining Balance Method with Switchover to the Straight-Line Method (Example 7-1)

		Depreciation Method		
Year, k	(1) Beginning-of-Year BV[a]	(2) 200% Declining Balance Method[b]	(3) Straight-Line Method[c]	(4) Depreciation Amount Selected[d]
1	$4,000.00	$ 800.00	> $400.00	$ 800.00
2	3,200.00	640.00	> 355.56	640.00
3	2,560.00	512.00	> 320.00	512.00
4	2,048.00	409.60	> 292.57	409.60
5	1,638.40	327.68	> 273.07	327.68
6	1,310.72	262.14	= 262.14	262.14 (switch)
7	1,048.58	209.72	< 262.14	262.14
8	786.44	167.77	< 262.14	262.14
9	524.30	134.22	< 262.14	262.14
10	262.16	107.37	< 262.14	262.14
		$3,570.50		$4,000.00

[a]Column 1 for year k less column 4 for year k equals the entry in column 1 for year k + 1.
[b]200% (= 2/10) of column 1.
[c]Column 1 divided by the remaining years from the beginning of the year through the tenth year.
[d]Select the larger amount in column 2 or column 3.

7.3.5   Units-of-Production Method

All the depreciation methods discussed to this point are based on elapsed time (years) on the theory that the decrease in value of property is mainly a function of time. When the decrease in value is mostly a function of use, depreciation may be based on a method not expressed in terms of years. The units-of-production method is normally used in this case.

This method results in the cost basis (minus final SV) being allocated equally over the estimated number of units produced during the useful life of the asset. The depreciation rate is calculated as:

$$\text{Depreciation per unit of production} = \frac{B - SV_N}{(\text{Estimated lifetime production in units})}$$

(7-13)

EXAMPLE 7-4

A piece of equipment used in a business has a basis of $50,000 and is expected to have a $10,000 salvage value when replaced after 30,000 hours of use. Find its depreciation rate per hour of use, and find its book value after 10,000 hours of operation.

SOLUTION

$$\text{Depreciation per unit of production} = \frac{\$50,000 - \$10,000}{30,000 \text{ hours}} = \$1.33 \text{ per hour}$$

After 10,000 hours, BV = $50,000 − $\frac{\$1.33}{\text{hour}}$(10,000 hours), or BV = $36,700.

7.4 The Modified Accelerated Cost Recovery System

As we discussed in Section 7.2.1, the Modified Accelerated Cost Recovery System (MACRS) was created by TRA 86 and is now the principal method for computing depreciation deductions for property in engineering projects. MACRS applies to most tangible depreciable property placed in service after December 31, 1986. Examples of assets that cannot be depreciated under MACRS are property you elect to exclude because it is to be depreciated under a method that is not based on a term of years (units-of-production method) and intangible property. Previous depreciation methods have required estimates of useful life (N) and salvage value at the end of useful life (SV_N). Under MACRS, however, SV_N is defined to be 0 and useful life estimates are not used directly in calculating depreciation amounts.

> MACRS consists of two systems for computing depreciation deductions. The main system is called the *General Depreciation System (GDS)* and the second system is called the *Alternative Depreciation System (ADS)*. In general, ADS provides a longer recovery period and uses only the straight-line method of depreciation. Property that is placed in any tax-exempt use and property used predominantly outside the United States are examples of assets that must be depreciated under ADS. Any property that qualifies under GDS, however, can be depreciated under ADS, if elected.

When an asset is depreciated under MACRS, the following information is needed before depreciation deductions can be calculated:

1. The cost basis (B)
2. The date the property was placed in service
3. The property class and recovery period
4. The MACRS depreciation method to be used (GDS or ADS)
5. The time convention that applies (half year)

The first two items were discussed in Section 7.2. Items 3 through 5 are discussed in the following sections.

7.4.1 Property Class and Recovery Period

Under MACRS, tangible depreciable property is categorized (organized) into asset classes. The property in each asset class is then assigned a *class life, GDS recovery period (and property class), and ADS recovery period*. For our use, a partial listing of depreciable assets used in business is provided in Table 7-2. The types of depreciable property grouped together are identified in the second column. Then the class life, GDS recovery period (and property class), and ADS recovery period (all in years) for these assets are listed in the remaining three columns.

Under the General Depreciation System (GDS), the basic information about property classes and recovery periods is as follows:

1. Most tangible personal property is assigned to one of six *personal property classes* (3-, 5-, 7-, 10-, 15-, and 20-year property). The personal property class (in years) is

TABLE 7-2   MACRS Class Lives and Recovery Periods[a]

Asset Class	Description of Assets or Depreciable Assets Used in Business	Class Life	Recovery Period GDS[b]	Recovery Period ADS
00.11	Office furniture and equipment	10	7	10
00.12	Information systems, including computers	6	5	5
00.22	Automobiles, taxis	3	5	5
00.23	Buses	9	5	9
00.241	Light general purpose trucks	4	5	5
00.242	Heavy general purpose trucks	6	5	6
00.26	Tractor units for use over the road	4	3	4
10.0	Mining	10	7	10
13.2	Production of petroleum and natural gas	14	7	14
13.3	Petroleum refining	16	10	16
15.0	Construction	6	5	6
22.3	Manufacture of carpets	9	5	9
24.4	Manufacture of wood products	10	7	10
28.0	Manufacture of chemicals and allied products	9.5	5	9.5
30.1	Manufacture of rubber products	14	7	14
32.2	Manufacture of cement	20	15	20
34.0	Manufacture of fabricated metal products	12	7	12
36.0	Manufacture of electronic components, products, and systems	6	5	6
37.11	Manufacture of motor vehicles	12	7	12
37.2	Manufacture of aerospace products	10	7	10
48.12	Telephone central office equipment	18	10	18
49.13	Electric utility steam production plant	28	20	28
49.21	Gas utility distribution facilities	35	20	35

[a]Partial listing abstracted from *Depreciation*, IRS Publication 534, Table B-1, 1994.
[b] Also the *GDS property class.*

the same as the *GDS recovery period*. Any depreciable personal property that does not "fit" into one of the defined asset classes is depreciated as being in the 7-year property class.

2. Real property is assigned to two *real property classes*—nonresidential real property and residential rental property.
3. The *GDS recovery period* is 39 years for nonresidential real property (31.5 years if placed in service before May 13, 1993) and 27.5 years for residential rental property.

The following is a summary of basic information for the Alternative Depreciation System (ADS):

1. For tangible personal property, the *ADS recovery period* is shown in the last column on the right of Table 7.2 (and is normally the same as the class life of the property; there are exceptions such as those shown in Asset Classes 00.12 and 00.22).
2. Any tangible personal property that does not fit into one of the asset classes is depreciated using a 12-year ADS recovery period.
3. The *ADS recovery period* for nonresidential real property is 40 years.

The use of these rules under the Modified Accelerated Cost Recovery System is discussed further in the next section.

7.4.2 Depreciation Methods, Time Convention, and Recovery Rates

The primary methods used under MACRS for calculating the depreciation deductions over the recovery period of an asset are summarized below.

1. GDS 3-, 5-, 7-, and 10-year personal property classes: The 200% declining balance (DB) method, which switches to the straight-line (SL) method when that method provides a greater deduction. The DB method with switchover to SL was illustrated in Section 7.3.4.
2. GDS 15- and 20-year personal property classes: The 150% DB method, which switches to the SL method when that method provides a greater deduction.
3. GDS nonresidential real and residential rental property classes: The SL method over the fixed GDS recovery periods.
4. ADS: The SL method for both personal and real property over the fixed ADS recovery periods.

TABLE 7-3 GDS Recovery Rates (r_k) for the Six Personal Property Classes

	Recovery Period (and Property Class)					
Year	3-year[a]	5-year[a]	7-year[a]	10-year[a]	15-year[b]	20-year[b]
1	0.3333	0.2000	0.1429	0.1000	0.0500	0.0375
2	0.4445	0.3200	0.2449	0.1800	0.0950	0.0722
3	0.1481	0.1920	0.1749	0.1440	0.0855	0.0668
4	0.0741	0.1152	0.1249	0.1152	0.0770	0.0618
5		0.1152	0.0893	0.0922	0.0693	0.0571
6		0.0576	0.0892	0.0737	0.0623	0.0528
7			0.0893	0.0655	0.0590	0.0489
8			0.0446	0.0655	0.0590	0.0452
9				0.0656	0.0591	0.0447
10				0.0655	0.0590	0.0447
11				0.0328	0.0591	0.0446
12					0.0590	0.0446
13					0.0591	0.0446
14					0.0590	0.0446
15					0.0591	0.0446
16					0.0295	0.0446
17						0.0446
18						0.0446
19						0.0446
20						0.0446
21						0.0223

SOURCE: *Depreciation*. IRS Publication 534. Washington, D.C.: U.S. Government Printing Office, for 1994 tax returns.

[a]These rates are determined by applying the 200% declining-balance method (with switchover to the straight-line method) to the recovery period with the half-year convention applied to the first and last years. Rates for each period must sum to 1.0000.

[b]These rates are determined with the 150% declining-balance method instead of the 200% declining-balance method (with switchover to the straight-line method) and are rounded off to four significant digits.

A *half-year time convention* is used in MACRS depreciation calculations for tangible personal property. This means that all assets placed in service during the year are treated as if use began in the middle of the year, and one-half year of depreciation is allowed. When an asset is disposed of, the half-year convention is also allowed. *If the asset is disposed of before the full recovery period is used, then only half of the normal depreciation deduction can be taken for that year.*

The GDS *recovery rates* (r_k) for the six personal property classes that we will use in our depreciation calculations are listed in Table 7-3 on page 283. The GDS personal property rates in Table 7-3 include the half-year convention as well as switchover from the DB method to the SL method when that method provides a greater deduction. Note that if an asset is disposed of in year $N + 1$ the final BV of the asset will be zero. Furthermore, there are $N + 1$ recovery rates shown for each GDS property class for a recovery period of N years.

The information in Table 7-4 provides a summary of the principal features of the main General Depreciation System under MACRS. Included are some selected special rules about depreciable assets. A flow diagram for computing depreciation deductions under MACRS is shown in Figure 7-1. As indicated in the figure, an

TABLE 7-4   MACRS (GDS) Property Classes and Primary Methods for Calculating Depreciation Deductions

GDS Property Class and Depreciation Method	Class Life	Special Rules
3-year, 200% DB with Switchover to SL	4 years or less	Includes some race horses. Excludes cars and light trucks.
5-year, 200% DB with Switchover to SL	More than 4 years to less than 10	Includes cars and light trucks, semiconductor manufacturing equipment, qualified technological equipment, computer-based central office switching equipment, some renewable and biomass power facilities, and research and development property.
7-year, 200% DB with Switchover to SL	10 years to less than 16	Includes single-purpose agricultural and horticultural structures and railroad track. Includes property not assigned to a property class.
10-year, 200% DB with Switchover to SL	16 years to less than 20	None.
15-year, 150% DB with Switchover to SL	20 years to less than 25	Includes sewage treatment plants, telephone distribution plants, and equipment for two-way voice and data communication.
20-year, 150% DB with Switchover to SL	25 years or more	Excludes real property of 27.5 years or more. Includes municipal sewers.
27.5 year, SL	N/A	Residential rental property.
39-year, SL	N/A	Nonresidential real property.

SOURCE: Arthur Andersen & Company, *Tax Reform 1986: Analysis and Planning*, Chicago, 1986, p. 112. Reproduced with permission of Arthur Andersen & Co.

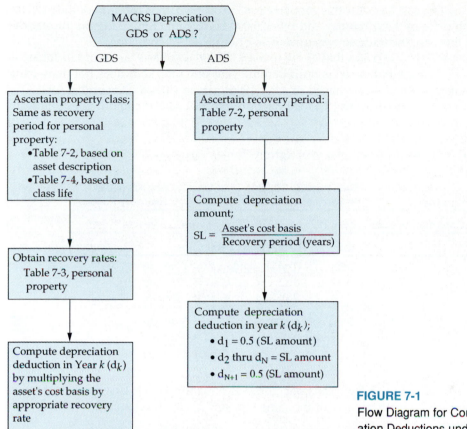

MACRS Depreciation
GDS or ADS ?

GDS   ADS

GDS branch:

Ascertain property class;
Same as recovery
period for personal
property:
• Table 7-2, based on
  asset description
• Table 7-4, based on
  class life

Obtain recovery rates:
   Table 7-3, personal
   property

Compute depreciation
deduction in Year k (d_k)
by multiplying the
asset's cost basis by
appropriate recovery
rate

ADS branch:

Ascertain recovery period:
Table 7-2, personal
property

Compute depreciation
amount;

$$SL = \frac{\text{Asset's cost basis}}{\text{Recovery period (years)}}$$

Compute depreciation
deduction in year k (d_k);
   • $d_1 = 0.5$ (SL amount)
   • d_2 thru $d_N =$ SL amount
   • $d_{N+1} = 0.5$ (SL amount)

FIGURE 7-1

Flow Diagram for Computing Depreciation Deductions under MACRS

important choice is whether the main GDS is to be used for an asset or whether ADS is elected instead (or required). Normally the choice would be to use the GDS for calculating the depreciation deductions.

EXAMPLE 7-5

In May 1996 your company traded in a computer and peripheral equipment, used in its business, that had a book value at that time of $25,000. A new, faster computer system having a fair MV of $400,000 was acquired. Because the vendor accepted the older computer as a trade-in, a deal was agreed to whereby your company would pay $325,000 cash for the new computer system.

(a) What is the GDS property class of the new computer system?

(b) How much depreciation can be deducted each year based on this class life? (Refer to Figure 7-1.)

SOLUTION

(a) The new computer is in Asset Class 00.12 and has a class life of six years (see Table 7-2). Hence, its GDS property class and recovery period is five years.

(b) The cost basis for this property is $350,000, which is the sum of the $325,000 cash price of the computer and the $25,000 book value remaining on the trade-in (in this case the trade-in was treated as a nontaxable transaction).

MACRS (GDS) rates that apply to the $350,000 cost basis are found in Table 7-3. An allowance (half-year) is built into the year one rate, so it does not matter that the computer was purchased in May 1996 instead of, say, November 1996. The depreciation deductions (d_k) for 1996 through 2001 can be computed as follows:

$$d_k = r_k \cdot B; \quad 1 \leq k \leq N + 1 \tag{7-14}$$

and, r_k = recovery rate for year k (from Table 7-3).

Property	Data Placed in Service	Cost Basis	Class Life	MACRS (GDS) Recovery Period
Computer System	May 1996	$350,000	6 years	5 years

Year	Depreciation Deductions
1996	0.20 × $350,000 = $ 70,000
1997	0.32 × 350,000 = 112,000
1998	0.192 × 350,000 = 67,200
1999	0.1152 × 350,000 = 40,320
2000	0.1152 × 350,000 = 40,320
2001	0.0576 × 350,000 = 20,160
	Total $350,000

From Example 7-5 we can conclude that Equation 7-15 is true from the buyer's viewpoint when property of the same type and class is exchanged.

$$\text{Basis} = \text{Actual Cash Cost} + \text{Book Value of the Trade-in} \tag{7-15}$$

To illustrate Equation 7-15, suppose your company has operated an optical character recognition (OCR) scanner for two years. Its book value now is $35,000 and its fair market value is $45,000. The company is considering a new OCR scanner that costs $105,000. Ordinarily, you would trade the old scanner for the new one and pay the dealer $60,000. The basis (B) for depreciation is then $60,000 + $35,000 = $95,000.

EXAMPLE 7-6

A firm purchased and placed in service a new piece of semiconductor manufacturing equipment. The cost basis for the equipment is $100,000. Determine (a) the depreciation charge permissible in the fourth year, (b) the BV at the end of the fourth year, (c) the cumulative depreciation through the third year, and (d) the BV at the end of the fifth year if the equipment is disposed of at that time.

SOLUTION
From Table 7-2, it may be seen that the semiconductor (electronic) manufacturing equipment has a class life of six years and a GDS recovery period of five years. The recovery rates that apply are given in Table 7-3.

(a) The depreciation deduction, or cost recovery allowance, that is allowable in year four (d_4) is 0.1152 ($100,000) = $11,520.

(b) The BV at the end of year four (BV_4) is the cost basis less depreciation charges in years one through four:

$$BV_4 = \$100,000 - \$100,000(0.20 + 0.32 + 0.192 + 0.1152)$$
$$= \$17,280$$

(c) Accumulated depreciation through year three, d_3^*, is the sum of depreciation amounts in years one through three:

$$d_3^* = d_1 + d_2 + d_3$$
$$= \$100,000(0.20 + 0.32 + 0.192)$$
$$= \$71,200$$

(d) The depreciation deduction in year five can only be $(0.5)(0.1152) \cdot$ ($100,000) = $5,760 when the equipment is disposed of prior to year six. Thus the BV at the end of year five is $BV_4 - \$5,760 = \$11,520$.

EXAMPLE 7-7

A large manufacturer of sheet metal products in the Midwest purchased and placed in service a new, modern, computer-controlled flexible manufacturing system for $3.0 million. Because this company would not be profitable until the new technology had been in place for several years, it elected to utilize the Alternative Depreciation System (ADS) under MACRS in computing its depreciation deductions. Thus, the company could slow down its depreciation allowances in hopes of postponing its income tax advantages until it became a profitable concern. What depreciation deductions can be claimed for the new system?

SOLUTION
From Table 7-2, the ADS recovery period for a manufacturer of fabricated metal products is 12 years. Under ADS, the straight-line method with no SV is applied to the 12-year recovery period using half-year convention. Consequently, the depreciation in year one would be

$$\frac{1}{2}\left(\frac{\$3,000,000}{12}\right) = \$125,000$$

Depreciation deductions in years 2 through 12 would be $250,000 each year, and depreciation in year 13 would be $125,000. Notice how the half-year convention extends depreciation deductions over 13 years ($N + 1$).

7.5 A Comprehensive Depreciation Example

We now consider an asset for which depreciation is computed by the historical and MACRS (GDS) methods previously discussed. Be careful to observe the differences in the mechanics of each method, as well as the differences in the annual depreciation amounts themselves. Also, we compare the present worths at $k = 0$ of selected depreciation methods when MARR $= 10\%$ per year. As we shall see later in this chapter, depreciation methods that result in larger present worths (of the depreciation amounts) are preferred by a firm that wants to reduce the present worth of its *income taxes paid* to the government.

EXAMPLE 7-8

The La Salle Bus Company has decided to purchase a new bus for $85,000 with a trade-in of their old bus. The old bus has a book value of $10,000 at the time of the trade-in. The new bus will be kept for ten years before being sold. Its estimated salvage value at that time is expected to be $5,000.

First, we must calculate the cost basis. The basis is the original purchase price of the bus plus the book value of the old bus that was traded in (Equation 7-15). Thus the basis is $85,000 + $10,000, or $95,000. We need to look at Table 7-2 and find buses, which are asset class 00.23. Hence, we find that buses have a nine-year class life, over which we depreciate the bus with historical methods discussed in Section 7.3, and a five-year GDS recovery period.

SOLUTION: STRAIGHT-LINE METHOD

For the straight-line method, we use the class life of nine years even though the bus will be kept for ten years. By using equations 7-2 and 7-4, the following information is obtained:

$$d_k = \frac{\$95,000 - \$5,000}{9 \text{ years}} = \$10,000 \qquad \text{for } k = 1 \text{ to } 9$$

Straight-Line Method

EOY k	d_k	BV_k
0	—	$95,000
1	$10,000	85,000
2	10,000	75,000
3	10,000	65,000
4	10,000	55,000
5	10,000	45,000
6	10,000	35,000
7	10,000	25,000
8	10,000	15,000
9	10,000	5,000

Notice that no depreciation was taken after year nine because the class life was only nine years. Also, the final BV was the estimated SV, and the BV will remain at $5,000 from the last year until the bus is sold.

SOLUTION: DECLINING-BALANCE METHOD

To demonstrate this method, we will use the 200% declining-balance equations. With Equations 7-6 and 7-8, we calculate:

$$R = 2/9 = 0.2222$$

$$d_1 = \$95{,}000(0.2222) = \$21{,}111$$

$$d_5 = \$95{,}000(1 - 0.2222)^{5-1}(0.2222) = \$7{,}726$$

$$BV_5 = \$95{,}000(1 - 0.2222)^5 = \$27{,}040$$

200% Declining-Balance Method

EOY k	d_k	BV_k
0	—	$95,000
1	$21,111	73,889
2	16,420	57,469
3	12,771	44,698
4	9,932	34,765
5	7,726	27,040
6	6,009	21,031
7	4,674	16,357
8	3,635	12,722
9	2,827	9,895

SOLUTION: SUM-OF-THE-YEARS-DIGITS METHOD

Once again, we will use the class life of nine years. The SYD depreciation amounts are as follows:

EOY k	Number of the Year in Reverse Order	SYD Depreciation Factor	$d_k =$ $(B - SV_N) * \text{Factor}$	BV_k
0	—	—	—	$95,000
1	9	9/45	$18,000.00	77,000
2	8	8/45	16,000.00	61,000
3	7	7/45	14,000.00	47,000
4	6	6/45	12,000.00	35,000
5	5	5/45	10,000.00	25,000
6	4	4/45	8,000.00	17,000
7	3	3/45	6,000.00	11,000
8	2	2/45	4,000.00	7,000
9	1	1/45	2,000.00	5,000
	Sum = 45			

Using Equations 7-10 and 7-11 we compute the following:

$$d_5 = (\$95{,}000 - \$5{,}000)\left[\frac{2(9 - 5 + 1)}{9(9 + 1)}\right] = \$10{,}000$$

$$BV_5 = \$95{,}000 - \frac{2(\$95{,}000 - \$5{,}000)}{9}(5) + \frac{(\$95{,}000 - \$5{,}000)5(5 + 1)}{9(9 + 1)} = \$25{,}000$$

SOLUTION: DECLINING BALANCE WITH SWITCHOVER TO STRAIGHT-LINE DEPRECIATION

To illustrate the mechanics of Table 7-1 for this example, we first specify that the bus will be depreciated by the 200% declining-balance method ($R = 2/N$).

Because declining-balance methods never reach a zero BV, suppose that we further specify that a switchover to straight-line depreciation will be made to ensure a BV of $5,000 at the end of the vehicle's nine-year class life.

EOY k	Beginning-of-Year BV	200% Declining-Balance Method	Straight-Line Method ($BV_9 = \$5,000$)	Depreciation Amount Selected
1	$95,000	$21,111	$10,000	$21,111
2	73,889	16,420	8,611	16,420
3	57,469	12,771	7,496	12,771
4	44,698	9,933	6,616	9,933
5	34,765	7,726	5,953	7,726
6	27,040	6,009	5,510	6,009
7	21,031	4,674	5,344	5,344[a]
8	15,687	3,635	5,344	5,344
9	10,344	2,827	5,344	5,344

[a]Switchover occurs in year seven.

SOLUTION: MACRS (GDS) WITH HALF-YEAR CONVENTION

To demonstrate the GDS with the half-year convention, we will change the La Salle bus problem so that the bus is now sold in year five in part (a) and in year six for part (b).

(a) Selling Bus in Year Five:

EOY k	Factor	d_k	BV_k
0	—	—	$95,000
1	0.2000	$19,000	76,000
2	0.3200	30,400	45,600
3	0.1920	18,240	27,360
4	0.1152	10,944	16,416
5	0.0576	5,472	10,944

(b) Selling Bus in Year Six:

EOY k	Factor	d_k	BV_k
0	—	—	$95,000
1	0.2000	$19,000	76,000
2	0.3200	30,400	45,600
3	0.1920	18,240	27,360
4	0.1152	10,944	16,416
5	0.1152	10,944	5,472
6	0.0576	5,472	0.00

Notice that when we sold the bus in year five before the recovery period had ended, we took only half of the normal depreciation. The other years (years one through four) were not changed. When the bus was sold in year six, at the end of the recovery period, we did not divide the last year amount by 2.

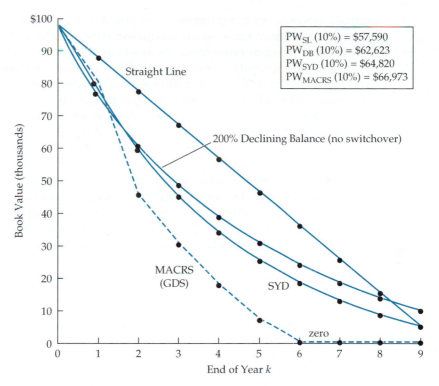

FIGURE 7-2   BV Comparisons for Selected Methods of Depreciation in Example 7-8
(Note: The bus is assumed to be sold in year 6 for the MACRS-GDS
method.)

Selected methods of depreciation, illustrated in Example 7-8, are compared in
Figure 7-2. In addition, the PW (10%) of each method is shown in Figure 7-2.
Because large present worths of depreciation deductions are generally viewed
as desirable, it is clear that the MACRS method is very attractive to profitable
companies.

7.6 Depletion

When natural resources are being consumed in producing products or services,
the term *depletion* is used to indicate the decrease in value of the resource base that
has occurred. The term is commonly used in connection with mining properties,
oil and gas wells, timberlands, and so on. In any given parcel of mineral property,
for example, there is a definite quantity of ore, oil, or gas available. As some of the
resource is extracted and sold, the reserve decreases and the value of the property
normally diminishes.

However, there is a difference in the manner in which the amounts recovered
through depletion and depreciation must be handled. In the case of depreciation, the property involved usually may be replaced with similar property
when it has become fully depreciated. In the case of depletion of mineral or other

natural resources, such replacement usually is not possible. Once the gold has been removed from a mine, or the oil from an oil well, it cannot be replaced. Thus, in a manufacturing or other business in which depreciation occurs, the principle of maintenance of capital is practiced, and the amounts charged for depreciation expense are reinvested in new equipment so that the business may continue in operation indefinitely. On the other hand, in the case of a mining or other mineral industry, the amounts charged as depletion cannot be used to replace the sold natural resource, and the company, in effect, may sell itself out of business, bit by bit, as it carries out its normal operations. Such companies frequently pay out to the owners each year the amounts recovered as depletion. Thus, the annual payment to the owners is made up of two parts: (1) the profit that has been earned and (2) a portion of the owner's capital that is being returned, marked as depletion. In such cases, if the natural resource were eventually completely consumed, the company would be out of business, and the stockholder would hold stock that was theoretically worthless but would have received back all of his or her invested capital.

In the operation of many natural resource businesses, the depletion funds may be used to acquire new properties, such as new mines and oil-producing properties, and thus give continuity to the enterprise.

There are two ways to compute depletion allowances: (1) the cost method and (2) the percentage method. The cost method applies to all types of property subject to depletion and is the more widely used method. Under the cost method a *depletion unit* is determined by dividing the adjusted cost basis of a property by the number of units remaining to be mined or harvested. (Units may be feet of timber, tons of ore, etc.) The deduction (depletion allowance) for a given tax year is then calculated as the product of number of units *sold* during the year and the depletion unit, in dollars.

In practice, depletion may also be based on a percentage of the year's income in accordance with IRS regulations. Depletion allowances on mines and other natural deposits, including geothermal deposits, may be computed as a percentage of gross income, provided that the amount charged for depletion does not exceed 50% of the *net income* (100% for oil and gas property) before deduction of the depletion allowance. The percentage method can be used for most types of metal mines, geothermal deposits, and coal mines, but not for timber. Generally, the use of percentage depletion for oil and gas is not allowed except for certain domestic oil and gas production. Some examples of depletion allowances[*] are as follows:

Sulfur and uranium; domestically mined lead, zinc, nickel, and asbestos	22%
Gold, silver, copper, iron ore, and oil shale from U.S. deposits; geothermal wells in the United States	15%
Coal, lignite, and sodium chloride	10%
Clay, gravel, sand, and stone	5%

---

[*]Depletion allowances are established by the IRS and may be revised with new federal income tax legislation.

It is possible that the total amount charged for depletion over the life of a property under this procedure may be far more than the original cost. When the percentage method applies to a property, depletion allowances must be calculated by both the cost method and the percentage method. The larger allowance may be taken and used to reduce the basis of the property for purposes of refiguring the depletion unit as necessary. Figure 7-3 provides the logic for determining whether percentage or cost depletion is allowable in a given tax year.

Example 7-9 illustrates the cost method of determining a depletion allowance.

EXAMPLE 7-9

The WGS Zinc Company recently bought an ore-bearing parcel of land for $2,000,000. The recoverable reserves in the mine were estimated to be 500,000 tons.

(a) If 75,000 tons of ore were mined during the first year and 50,000 tons were sold, what was the depletion allowance for year one?

(b) Suppose at the end of year one reserves were reevaluated and found to be only 400,000 tons. If 50,000 additional tons are sold in the second year, what is the depletion allowance for year two?

SOLUTION

(a) The depletion unit is $2,000,000/500,000$ tons = $4.00 per ton. A depletion allowance, based on units sold, for year one is 50,000 tons ($4.00/ton) = $200,000.

(b) The adjusted cost basis at the beginning of the second year would be $2,000,000 − $200,000 = $1,800,000. The depletion unit would be $1,800,000/400,000$ tons = $4.50/ton, and a depletion allowance of 50,000 tons ($4.50/ton) = $225,000 is permissible in year two.

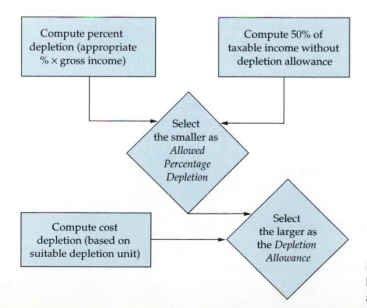

FIGURE 7-3
Logic for Determining Whether Percentage or Cost Depletion Is Allowed

7.7 Introduction to Income Taxes

Up to this point, there has been no consideration of income taxes in our discussion of engineering economy, except for the influence of depreciation and other types of deductions. By not complicating our studies with income tax effects, we have placed primary emphasis on basic engineering economy principles and methodology. However, there is a wide variety of capital investment problems in which income taxes do affect the choice among alternatives, and after-tax studies are essential.

> In the remainder of this chapter we shall be concerned with how income taxes affect a project's estimated cash flows. Income taxes resulting from the profitable operation of a firm are usually taken into account in evaluating engineering projects. The reason is quite simple: income taxes associated with a proposed project may represent a major cash outflow that should be considered together with other cash inflows and outflows in assessing the overall economic profitability of that project.

There are other taxes discussed in Section 7.7.1 that are not directly associated with the income-producing capability of a new project, but they are usually negligible when compared with federal and state income taxes. *When other types of taxes are included in engineering economy studies, they are normally deducted from gross revenues, as any other operating expense would be, in determining the before-tax cash flows that we considered in Chapters 4 and 5.*

The mystery behind the sometimes complex computation of income taxes is reduced when we recognize that income taxes paid are just another type of expense, but income taxes saved (through depreciation, expenses, and direct tax credits) are identical to other kinds of reduced expenses (e.g., savings in maintenance expenses).

The basic concepts underlying federal and state income tax laws and regulations that apply to most economic analyses of capital investments generally can be understood and applied without difficulty. This discussion of income taxes is not intended to be a comprehensive treatment of the subject. Rather, we utilize some of the more important provisions of the federal Tax Reform Act of 1986 (TRA 86), followed by illustrations of a general procedure for computing the net after-tax cash flow (ATCF) for an engineering project and conducting after-tax economic analyses. Where appropriate, important changes to TRA 86 enacted by the Omnibus Budget Reconciliation Act of 1993 (OBRA 93) are also included in this chapter.

7.7.1 Distinctions Between Different Types of Taxes

Before discussing the consequences of income taxes in engineering economy studies, we need to distinguish between income taxes and several other types of taxes.

1. *Income taxes* are assessed as a function of gross revenues minus allowable deductions. They are levied by the federal, most state, and occasionally municipal governments.
2. *Property taxes* are assessed as a function of the value of property owned, such as land, buildings, equipment, and so on, and the applicable tax rates. Hence, they are independent of the income or profit of a firm. They are levied by municipal, county, and/or state governments.
3. *Sales taxes* are assessed on the basis of purchases of goods and/or services, and are thus independent of gross income or profits. They are normally levied by state, municipal, or county governments. Sales taxes are relevant in engineering economy studies only to the extent that they add to the cost of items purchased.
4. *Excise taxes* are federal taxes assessed as a function of the sale of certain goods or services often considered nonnecessities, and are hence independent of the income or profit of a business. Although they are usually charged to the manufacturer or original provider of the goods or services, the cost is passed on to the purchaser.

Income taxes are usually the most significant type of tax encountered in engineering economic analysis.

7.7.2 The Before-Tax and After-Tax Minimum Attractive Rates of Return

In the preceding chapters we have treated income taxes as if they are not applicable, or we have taken them into account, in general, by using a before-tax MARR, which is larger than the after-tax MARR. An approximation of the before-tax MARR requirement, which includes the effect of income taxes, for studies involving only before-tax cash flows can be obtained from the following relationship:

$$(\text{Before-tax MARR})[(1 - \text{effective income tax rate})] \cong \text{after-tax MARR}$$

Thus,

$$\text{Before-tax MARR} \approx \frac{\text{After-tax MARR}}{(1 - \text{effective income tax rate})} \qquad \textbf{(7-16)}$$

Determining the effective income tax rate for a firm is discussed in Section 7.8.

This approximation is exact if the asset is nondepreciable and there are no gains or losses on disposal, tax credits, or other types of deductions involved. Otherwise, these factors affect the amount and timing of income tax payments, and some degree of error is introduced into the relationship in Equation 7-16.

7.7.3 Taxable Income of Corporations (Business Firms)

At the end of each tax year, a corporation must calculate its net (i.e., taxable) before-tax income or loss. Several steps are involved in this process, beginning with the calculation of *gross income*. Gross income represents the gross profits from operations (revenues from sales minus the cost of goods sold) plus income from dividends, interest, rent, royalties, and gains (or losses) on the sale or exchange of capital assets. The corporation may deduct from gross income all ordinary and necessary operating expenses, including interest, to conduct the business *except capital investments*. Deductions for depreciation are permitted each tax period as a means of consistently and systematically recovering capital investment. Consequently, allowable expenses and deductions may be used to determine taxable income, as shown in Equation 7-17.

$$\text{Taxable income} = \text{gross income} - \text{all expenses except capital investments}$$
$$- \text{depreciation (depletion) deductions} \qquad \textbf{(7-17)}$$

This taxable income is also referred to as *net income before taxes (NIBT)*. When income taxes are subtracted, the remainder is called the *net income after taxes (NIAT)*. In summary,

$$\text{Net income after taxes} = \left\{ \begin{array}{c} \text{taxable income} \\ \text{(i.e., NIBT)} \end{array} \right\} - \text{income taxes} \qquad \textbf{(7-18)}$$

EXAMPLE 7-10

A company generates $1,500,000 of gross income during its tax year and incurs operating expenses of $800,000. Interest payments on borrowed capital amount to $48,000. The total depreciation deductions for the tax year equal $114,000. What is the taxable income (NIBT) of this firm?

SOLUTION
Based on Equation 7-17, this company's taxable income for the tax year would be

$$\$1,500,000 - \$800,000 - \$48,000 - \$114,000 = \$538,000$$

7.8 The Effective (Marginal) Corporate Income Tax Rate

The federal corporate income tax rate structure is shown in Table 7-5. Depending on the taxable income bracket that a firm is in for a tax year, the marginal federal rate can vary from 15% to a maximum of 39%. However, note that the weighted average tax rate at taxable income = $335,000 is 34%, and the weighted average tax rate at taxable income = $18,333,333 is 35%. Therefore, if a corporation has a taxable income for a tax year greater than $18,333,333, federal taxes are computed using a flat rate of 35%.

EXAMPLE 7-11

Suppose that a firm for a tax year has a gross income of $5,270,000, expenses (excluding capital) of $2,927,500, and depreciation deductions of $1,874,300. What would be its taxable income and federal income tax for the tax year based on Equation 7-17 and Table 7-5?

TABLE 7-5 Corporate Federal Income Tax Rates

If taxable income is:		The tax is:	
Over	but not over		of the amount over
0	$ 50,000	15%	0
$ 50,000	75,000	$ 7,500 + 25%	$ 50,000
75,000	100,000	13,750 + 34%	75,000
100,000	335,000	22,250 + 39%	100,000
335,000	10,000,000	113,900 + 34%	335,000
10,000,000	15,000,000	3,400,000 + 35%	10,000,000
15,000,000	18,333,333	5,150,000 + 38%	15,000,000
18,333,333	....	6,416,667 + 35%	18,333,333

SOURCE: *Tax Information on Corporations*, IRS Publication 542, 1994.

SOLUTION

Taxable income = gross income − expenses − depreciation deductions

= $5,270,000 − $2,927,500 − $1,874,300

= $468,200

Income tax = 15% of first $50,000	$7,500
+ 25% of next $25,000	6,250
+ 34% of next $25,000	8,500
+ 39% of next $235,000	91,650
+ 34% of remaining $133,200	45,288
Total	$159,188

The total tax liability in this case is $159,188. As an added note, we could have used a flat rate of 34% in this example because the federal weighted average tax rate at taxable income = $335,000 is 34%. The remaining $133,200 of taxable income above this amount is in a 34% tax bracket (Table 7-5). So we have 0.34($468,200) = $159,188.

Although the tax laws and regulations of most of the states (and some municipalities) with income taxes have the same basic features as the federal laws and regulations, there is significant variation in income tax rates. State income taxes are in most cases much less than federal taxes and often can be closely approximated as ranging from 6% to 12% of taxable income. No attempt will be made to discuss the details of state income taxes. However, to illustrate the calculation of an effective income tax rate (t) for a large corporation based on the consideration of both federal and state income taxes, assume that the applicable federal income tax rate is 35% and the state income tax rate is 8%. Further assume the common case in which taxable income is computed the same way for both types of taxes, except that state income taxes are deductible from taxable income for federal tax purposes but federal income taxes are not deductible from taxable income for state tax purposes. Based on these assumptions, the general expression for the effective income tax rate is

$$t = \text{state rate} + \text{federal rate}(1 - \text{state rate}) \qquad \textbf{(7-19)}$$

and, in this example, the effective income tax rate for the corporation would be

$$t = 0.08 + 0.35(1 - 0.08) = 0.402, \text{or approximately } 40\%$$

In this chapter, we will often use an effective corporate income tax rate of *approximately* 40% as a representative value that includes state income taxes.

The effective income tax rate on *increments* of taxable income is of importance in engineering economy studies. This concept is illustrated in Figure 7-4, which plots the federal income tax rates and brackets listed in Table 7-5 and shows the added (incremental) taxable income and federal income taxes that would result from a proposed engineering project. In this case, the corporation is assumed to have a taxable income for their tax year greater than $18,333,333. However, the same concept applies to a smaller firm with less taxable income for its tax year, which is illustrated in Example 7-12.

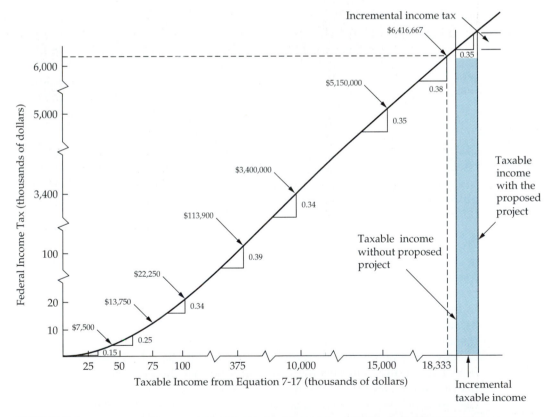

FIGURE 7-4   The Federal Income Tax Rates for Corporations (Table 7-5) with Incremental Income Tax for a Proposed Project (Assumes, in this case, corporate taxable income without project > $18,333,333)

EXAMPLE 7–12

A small corporation is expecting an annual taxable income of $45,000 for its tax year. It is considering an additional capital investment of $100,000 in an engineering project, which is expected to create an added annual net cash flow (revenues minus expenses) of $35,000 and an added annual depreciation deduction of $20,000. What is the corporation's federal income tax liability (a) without the added capital investment and (b) with the added capital investment?

SOLUTION

		Rate	Amount
(a)	*Income Taxes*		
	On first $45,000	15%	$6,750
		Total	$6,750
(b)	*Taxable Income*		
	Before added investment		$45,000
	+ added net cash flow		+35,000
	− depreciation deduction		−20,000
		Net total	$60,000
	Income Taxes on $60,000	Rate	Amount
	On first $50,000	15%	$7,500
	On next $10,000	25%	2,500
		Total	$10,000

The increased income tax liability from the investment is $3,250.

As an added note, the change in tax liability can usually be determined more readily by an incremental approach. For instance, this example involved changing the taxable income from $45,000 to $60,000 as a result of the new investment. Thus, the change in income taxes for the tax year could be calculated as follows:

$$\text{First } \$50,000 - \$45,000 = \$5,000 \text{ at } 15\% = \$\ \ 750$$
$$\text{Next } \$60,000 - \$50,000 = \$10,000 \text{ at } 25\% = \ \ 2,500$$
$$\text{Total} \quad \$3,250$$

The average federal income tax rate on the additional $35,000 − $20,000 = $15,000 of taxable income is calculated as ($3,250/$15,000) = 0.2167, or 21.67%.

In addition to lowering the maximum rate on corporate taxable income from 46% to 35%, TRA 86 (as changed by OBRA 93) created a new *alternative minimum tax (AMT)* system that is intended to ensure that any corporation with economic income pays a minimum amount of federal income tax. Now corporations must compute their income tax liability as illustrated in this section, and many must also compute their AMT according to a rather complex set of rules that is beyond the scope of our discussion. Corporations now pay the *maximum* income tax resulting from the system using rates in Table 7-5 or from the AMT system. It is generally acknowledged that the AMT is the most far-reaching and complex business tax provision in TRA 86.

7.9 Gain (Loss) on the Disposal of an Asset

When a *depreciable asset* (tangible personal or real property, Section 7.2) is sold, the market value is seldom equal to its book value (Equation 7-1). In general, the gain (loss) on sale of depreciable property is the fair market value minus its book value at that time. That is,

$$[\text{gain (loss) on disposal}]_N = MV_N - BV_N \qquad \textbf{(7-20)}$$

When the sale results in a gain, it is often referred to as *depreciation recapture*. The tax rate for the gain (loss) on disposal of depreciable personal property is usually the same as for ordinary income or loss, which is the effective income tax rate, t.

When a *capital asset* is sold or exchanged, the gain (loss) is referred to as a *capital gain (loss)*. Examples of capital assets are stocks, bonds, gold, silver, and other metals, as well as real property such as a home. Because engineering economic analysis seldom involves an actual capital gain (loss), the more complex details of this situation are not discussed further.

EXAMPLE 7-13

A corporation sold a piece of equipment during the current tax year for $78,600. The accounting records show that its cost basis, B, is $190,000 and the accumulated depreciation is $139,200. Assume that the effective income tax rate is 40%. Based on this information, what is (a) the gain (loss) on disposal, (b) the tax liability (or credit) resulting from this sale, and (c) the tax liability (or credit) if the accumulated depreciation was $92,400 instead of $139,200?

SOLUTION

(a) The book value at time of sale is $190,000 − $139,200 = $50,800. Therefore, the gain (loss) on disposal is $78,600 − $50,800 = $27,800.

(b) The tax owed on this gain is −0.40($27,800) = −$11,120.

(c) With $d_k^* = \$92,400$, the book value at the time of sale is $190,000 − $92,400 = $97,600. The gain is $78,600 − $97,600 = −$19,000. The tax credit resulting from this loss on disposal is −0.40(−$19,000) = $7,600.

7.10 General Procedure for Making After-Tax Economic Analyses

After-tax economic analyses can be performed by exactly the same methods as before-tax analyses. The only difference is that ATCFs are used in place of before-tax cash flows (BTCFs) by including expenses (or savings) due to income taxes and then making equivalent worth calculations *using an after-tax MARR*. The tax rates and governing regulations may be complex and subject to changes, but once those rates and regulations have been translated into their effect on ATCFs, the remainder of the after-tax analysis is relatively straightforward. To formalize the procedure, let

R_k = revenues (and savings) from the project; this is the cash inflow from the project during period k

E_k = cash outflows during year k for deductible expenses and interest

d_k = sum of all noncash, or book, costs during year k, such as depreciation and depletion

t = effective income tax on *ordinary* income (federal, state, and other); t is assumed to remain constant during the study period

T_k = income taxes paid during year k

$ATCF_k$ = ATCF from the project during year k

Because the NIBT (i.e., taxable income) is $(R_k - E_k - d_k)$, the *ordinary income tax liability* when $R_k > (E_k + d_k)$ is computed with Equation 7-21:

$$T_k = -t(R_k - E_k - d_k) \tag{7-21}$$

The NIAT (Equation 7-18) is then simply taxable income (i.e., net income before taxes) minus the tax liability amount determined by Equation 7-21:

$$NIAT_k = \underbrace{(R_k - E_k - d_k)}_{\text{taxable income}} - \underbrace{t(R_k - E_k - d_k)}_{\text{income taxes}}$$

or

$$NIAT_k = (R_k - E_k - d_k)(1 - t) \tag{7-22}$$

The ATCF associated with a project equals the NIAT plus noncash items such as depreciation:

$$ATCF_k = NIAT_k + d_k \tag{7-23}$$
$$= (R_k - E_k - d_k)(1 - t) + d_k \tag{7-24}$$

or

$$ATCF_k = (1 - t)(R_k - E_k) + td_k \tag{7-25}$$

In many economic analyses of engineering projects, ATCFs in year k are computed in terms of $BTCF_k$ (i.e., year k before-tax cash flows):

$$BTCF_k = R_k - E_k \tag{7-26}$$

Thus,

$$ATCF_k = BTCF_k + T_k^* \tag{7-27}$$
$$= (R_k - E_k) - t(R_k - E_k - d_k)$$
$$= (1 - t)(R_k - E_k) + td_k \tag{7-28}$$

Equations 7-25 and 7-28 are obviously identical.

---

*In Figure 7-5 we use $-t$ in column D, so subtraction of income taxes in Equation 7-27 is accomplished.

Tabular headings to facilitate the computation of after-tax cash flows with Equations 7-21 and 7-28 are as follows:

Year	(A) BTCF	(B) Depreciation	(C) = (A) − (B) Taxable Income	(D) = −t(C) Cash Flow for Income Taxes	(E) = (A) + (D) ATCF
k	$R_k − E_k$	d_k	$R_k − E_k − d_k$	$−t(R_k − E_k − d_k)$	$(1 − t)(R_k − E_k) + td_k$

Column A consists of the same information used in before-tax analyses, namely, the cash revenues (or savings) less the deductible expenses. Column B contains depreciation that can be claimed for tax purposes. Column C is the taxable income, or amount subject to income taxes. Column D contains the income taxes paid (or saved). Finally, column E shows the ATCFs to be used directly in after-tax economic analyses.

A summary of the process of determining NIAT and ATCF during each year of an N-year study period is provided in Figure 7-5. NIAT is well understood in most companies, and it can be easily obtained from Figure 7-5 for making presentations to upper-level management. The format of Figure 7-5 is used extensively throughout the remainder of this chapter, and it provides a convenient way to organize data in after-tax studies.

The column headings of Figure 7-5 indicate the arithmetic operations for computing columns C, D, and E when $k = 1, 2, \ldots, N$. When $k = 0$ and $k = N$, capital investments are usually involved, and their tax treatment (if any) is illustrated in the examples that follow. The table should be used with the conventions of + for cash inflow or savings and − for cash outflow or opportunity forgone.

EXAMPLE 7-14

If the revenue from a project is $10,000 during a tax year, out-of-pocket expenses are $4,000, and depreciation deductions for income tax purposes are $2,000, what is the ATCF when $t = 0.40$? What is the NIAT?

SOLUTION
From Equation 7-24, we have

$$\text{ATCF} = (1 − 0.4)(\$10{,}000 − \$4{,}000 − \$2{,}000) + \$2{,}000 = \$4{,}400$$

The same result can be obtained with Equation 7-25 or 7-28:

$$\text{ATCF} = (1 − 0.4)(\$10{,}000 − \$4{,}000) + 0.4(\$2{,}000) = \$4{,}400$$

Equation 7-25 shows clearly that depreciation contributes a credit of $t \cdot d_k$ to the after-tax cash flow in operating year k. The NIAT, from Equation 7-23, is $4,400 − $2,000 = $2,400.

The ATCF attributable to depreciation (a tax savings) is td_k in year k. After income taxes, an expense becomes $(1 − t)E_k$.

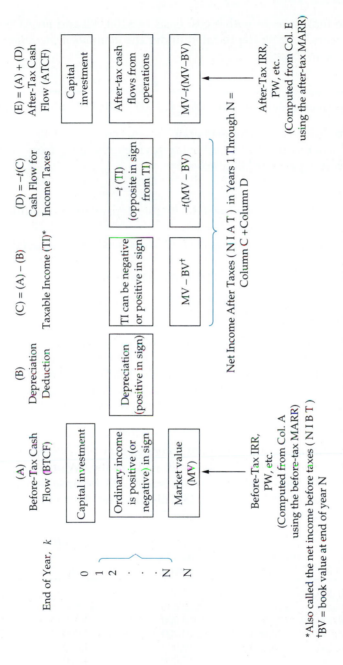

FIGURE 7-5 General Format (Worksheet) for After-Tax Analysis; Determining the ATCF (and NIAT)

EXAMPLE 7-15

Suppose that an asset with a cost basis of $100,000 and an ADS recovery period of five years is being depreciated under the *alternate depreciation system (ADS) of MACRS* as follows:

Year	1	2	3	4	5	6
Depreciation deduction	$10,000	$20,000	$20,000	$20,000	$20,000	$10,000

If the firm's effective income tax rate remains constant at 40% during this six-year period, what is the PW of after-tax savings resulting from depreciation when MARR = 10% per year (after taxes)?

SOLUTION

The PW of tax credits (savings) due to this depreciation schedule is

$$\text{PW}(10\%) = \sum_{k=1}^{6} 0.4d_k(1.10)^{-k} = \$4,000(0.9091) + \$8,000(0.8264)$$

$$+ \cdots + \$4,000(0.5645) = \$28,948$$

EXAMPLE 7-16

The asset in Example 7-15 is expected to produce net cash inflows (net revenues) of $30,000 per year during the six-year period, and its terminal market value is negligible. If the effective income tax rate is 40%, how much can a firm afford to spend for this asset and still earn the MARR? What is the meaning of any excess in affordable amount over the $100,000 cost basis given in Example 7-15?

SOLUTION

After income taxes, the PW of net revenues is $(1 - 0.4)(\$30,000) \cdot (P/A, 10\%, 6) = \$18,000(4.3553) = \$78,395$. After adding to this the PW of tax savings computed in Example 7-15, the affordable amount is $107,343. Because the capital investment is $100,000, the net PW equals $7,343. This same result can be obtained by using the general format (worksheet) of Figure 7-5:

EOY	(A) BTCF	(B) Depreciation Deduction	(C) = (A) − (B) Taxable Income	(D) = −0.4(C) Income Taxes	(E) = (A) + (D) ATCF
0	−$100,000				−$100,000
1	30,000	$10,000	$20,000	−$8,000	22,000
2	30,000	20,000	10,000	−4,000	26,000
3	30,000	20,000	10,000	−4,000	26,000
4	30,000	20,000	10,000	−4,000	26,000
5	30,000	20,000	10,000	−4,000	26,000
6	30,000	10,000	20,000	−8,000	22,000
Total	$80,000		Total $80,000		PW(10%) of ATCF = $7,343

7.11 Illustration of Computations of ATCFs

The following problems (Examples 7-17, 7-18, 7-19, and 7-20) illustrate the computation of ATCFs, as well as many common situations that affect income taxes. All problems include the assumption that income tax expenses (or savings) occur at the same time (year) as the revenue or expense that gives rise to the taxes. For purposes of comparing the effects of various situations, the after-tax IRR and/or PW are computed for each example. We can observe from the results of Examples 7-17 and 7-19 that the faster (i.e., earlier) the depreciation deduction is, the more favorable the after-tax IRR and PW will become.

EXAMPLE 7-17

Certain new machinery when placed in service is estimated to cost $180,000. It is expected to *reduce* net annual operating expenses by $36,000 per year for 10 years and to have a $30,000 MV at the end of the tenth year. (a) Develop the before-tax and after-tax cash flows and (b) calculate the before-tax and after-tax IRR. Assume the firm is in the federal taxable income bracket of $335,000 to $10,000,000, and the state income tax rate is 6%. State income taxes are deductible from federal taxable income. This machinery is in the MACRS (GDS) five-year property class. (c) Calculate the after-tax PW when the *after-tax* MARR = 10% per year. In this example the study period is ten years but the class life of the machinery is six years (which includes the carryover effect of the half-year convention).

SOLUTION

(a) Table 7-6 applies the format illustrated in Figure 7-5 to calculate the BTCF and ATCF for this example. In column D the effective income tax rate is very close to 0.38 (from Equation 7-19) based on the information just provided.

(b) The before-tax IRR is computed from column A:

$$0 = -\$180,000 + \$36,000(P/A, i'\%, 10) + \$30,000(P/F, i'\%, 10)$$

By trial and error, $i' = 16.1\%$.

The entry in the last year is shown to be $30,000 because the machinery will have this estimated MV. However, the asset was depreciated to zero with the GDS method. Therefore, when the machine is sold at the end of year ten, there will be $30,000 of *recaptured depreciation,* or gain on disposal (Equation 7-20), which is taxed at the effective income tax rate of 38%. This tax entry is shown in column D (EOY 10).

By trial and error, the after-tax IRR for Example 7-17 is found to be 12.4%.

(c) When MARR = 10% per year is inserted into the PW equation at the bottom of Table 7-6 it can be determined that the after-tax PW of this investment is $17,209.

TABLE 7-6 ATCF Analysis of Example 7-17

End of Year, k	(A) BTCF	(B) Cost Basis	GDS Recovery Rate	Deduction	(C) = (A) − (B) Taxable Income	(D) = −0.38(C) Cash Flow for Income Taxes	(E) = (A) + (D) ATCF
			Depreciation Deduction				
0	−$180,000	—	—	—			−$180,000
1	36,000	$180,000 ×	0.2000 =	$36,000	0	0	36,000
2	36,000	180,000 ×	0.3200 =	57,600	−21,600	+ 8,280	44,208
3	36,000	180,000 ×	0.1920 =	34,560	1,440	− 547	35,453
4	36,000	180,000 ×	0.1152 =	20,736	15,264	− 5,800	30,200
5	36,000	180,000 ×	0.1152 =	20,736	15,264	− 5,800	30,200
6	36,000	180,000 ×	0.0576 =	10,368	25,632	− 9,740	26,260
7–10	36,000	0		0	36,000	−13,680	22,320
10	30,000				30,000[a]	−11,400[b]	18,600
	Total $210,000			Total $180,000			Total $130,201

[a] Depreciation recapture = $MV_{10} - BV_{10} = \$30,000 - 0 = \$30,000$ (gain on disposal).

[b] Tax on depreciation recapture = $30,000(0.38) = $11,400.

After-tax IRR: Set PW of column E = 0 and solve for i' in the following equation.

$$0 = -\$180,000 + \$36,000(P/F, i', 1) + \$44,208(P/F, i', 2) + \$35,453(P/F, i', 3) + \$30,200(P/F, i', 4) + \$30,200(P/F, i', 5) + \$26,260(P/F, i', 6)$$
$$+ \$22,320(P/A, i', 4)(P/F, i', 6) + \$18,600(P/F, i', 10); \text{ IRR} = 12.4\%$$

PW (10%) = $ 17,209

If the machinery in Example 7-17 had been classified in the ten-year MACRS (GDS) property class instead of five-year property class, depreciation deductions would be slowed down in the early years of the study period and shifted into later years, as shown in Table 7-7. Compared to Table 7-6, entries in columns C, D, and E of Table 7-7 are less favorable in the sense that a fair amount of ATCF is deferred until later years, producing a lower after-tax IRR and PW. For instance, the PW is reduced from $17,209 in Table 7-6 to $9,136 in Table 7-7. The basic difference between Table 7-6 and Table 7-7 is the *timing of the ATCF*, which is a function of the timing and magnitude of the depreciation deductions. In fact, the curious reader can confirm that the sums of entries in columns A through E of Tables 7-6 and 7-7 are nearly the same (except for the half-year of depreciation only in year ten of Table 7-7). The timing of cash flows does, of course, make a difference!

> Depreciation does not affect BTCF. Fast (accelerated) depreciation produces a larger PW of tax savings than does the same amount of depreciation claimed later in an asset's life.

A minor complication is introduced in ATCF analyses when the study period is shorter than an asset's MACRS recovery period (e.g., for a five-year recovery period, the study period is five years or less). In such cases, we shall assume throughout this book that the asset is sold for its MV in the last year of the study period. Due to the half-year convention, only one-half of the normal MACRS depreciation can be claimed in the year of disposal or end of the study period, so there will usually be a difference between an asset's BV and its MV. Resulting income tax adjustments will be made at the time of the sale (see last row in Table 7-7) unless the asset in question is not sold but instead kept for standby service. In such a case, depreciation deductions usually continue through the end of the asset's MACRS recovery period. Our assumption of project termination at the end of the study period makes good economic sense, as illustrated in Example 7-18.

EXAMPLE 7-18

A highly specialized piece of optical character recognition equipment has a first cost of $50,000. If this equipment is purchased, it will be used to produce income (through rental) of $20,000 per year for only four years. At the end of year four, the equipment will be sold for a negligible amount. Estimated annual expenses for upkeep are $3,000 during each of the four years. The MACRS (GDS) recovery period for the equipment is seven years, and the firm's effective income tax rate is 40%.

(a) If the after-tax MARR is 7% per year, should the equipment be purchased?

(b) Rework the problem, assuming that the equipment is placed on standby status such that depreciation is taken over the full MACRS recovery period.

TABLE 7-7 Reworked Example 7-17 with Machinery in the Ten-Year MACRS (GDS) Property Class

End of Year, k	(A) BTCF	(B) Depreciation Deduction — Cost Basis	×	GDS Recovery Rate	=	Deduction	(C) = (A) − (B) Taxable Income	(D) = −0.38(C) Cash Flow for Income Taxes	(E) = (A) + (D) ATCF
0	−$180,000	—		—		—			−$180,000
1	36,000	$180,000	×	0.1000	=	$18,000	$18,000	−$ 6,840	29,160
2	36,000	180,000	×	0.1800	=	32,400	3,600	− 1,368	34,632
3	36,000	180,000	×	0.1440	=	25,920	10,080	− 3,830	32,170
4	36,000	180,000	×	0.1152	=	20,736	15,264	− 5,800	30,200
5	36,000	180,000	×	0.0922	=	16,596	19,404	− 7,374	28,626
6	36,000	180,000	×	0.0737	=	13,266	22,734	− 8,639	27,361
7	36,000	180,000	×	0.0655	=	11,790	24,210	− 9,200	26,800
8	36,000	180,000	×	0.0655	=	11,790	24,210	− 9,200	26,800
9	36,000	180,000	×	0.0656	=	11,808	24,192	− 9,193	26,807
10	36,000	180,000	×	0.0655/2	=	5,895	30,105	− 11,440	24,560
10	30,000						18,201[a]	− 6,916	23,084

Total $130,196

PW (10%) ≈ $9,136

IRR = 11.2%

[a]Gain on disposal $= MV_{10} - BV_{10} = \$30,000 - \left(\dfrac{0.0655}{2} + 0.0328\right)(\$180,000) = \$18,201.$

SOLUTION
(a)

End of Year, k	(A) BTCF	(B) Depreciation Deduction	(C) = (A) − (B) Taxable Income	(D) = −0.4(C) Cash Flow for Income Taxes	(E) = (A) + (D) ATCF
0	−$50,000				−$50,000
1	17,000	$ 7,145	$ 9,855	−$3,942	13,058
2	17,000	12,245	4,755	− 1,902	15,098
3	17,000	8,745	8,255	− 3,302	13,698
4	17,000	3,123[a]	13,877	− 5,551	11,449
4	0		−18,742[b]	7,497	7,497

[a]Half-year convention applies with disposal in year four.
[b]Remaining BV.

PW(7%) = $1,026. Because the PW > 0, the equipment should be purchased.

(b)

End of Year, k	(A) BTCF	(B) Depreciation Deduction	(C) Taxable Income	(D) Cash Flow for Income Taxes	(E) ATCF
0	−$50,000				−$50,000
1	17,000	$ 7,145	$ 9,855	−$3,942	13,058
2	17,000	12,245	4,755	−1,902	15,098
3	17,000	8,745	8,255	−3,302	13,698
4	17,000	6,245	10,755	−4,302	12,698
5	0	4,465	−4,465	1,786	1,786
6	0	4,460	−4,460	1,784	1,784
7	0	4,465	−4,465	1,786	1,786
8	0	2,230	−2,230	892	892
8	0				0

PW(7%) = $353, so the equipment should be purchased.

The present worth is $673 higher in part (a), which equals the PW of deferred depreciation deductions in part (b). A firm would select the situation in part (a) if it had a choice.

An illustration of determining ATCFs for a somewhat more complex, though realistic, capital investment opportunity is provided in Example 7-19.

EXAMPLE 7-19

The Ajax Semiconductor Company is attempting to evaluate the profitability of adding another integrated circuit production line to its present operations. The company would need to purchase two or more acres of land for $275,000 (total).

The facility would cost $60,000,000 and have no net MV at the end of five years. The facility could be depreciated using a GDS recovery period of five years. An increment of working capital would be required, and its estimated amount is $10,000,000. Gross income is expected to increase by $30,000,000 per year for five years, and operating expenses are estimated to be $8,000,000 per year for five years. The firm's effective income tax rate is 40%. (a) Set up a table and determine the ATCF for this project. (b) What is the NIAT in year three? (c) Is the investment worthwhile when the after-tax MARR is 12% per year?

SOLUTION

(a) The format recommended in Figure 7-5 is followed in Table 7-8 to obtain ATCFs in years zero through five. Acquisition of land as well as additional working capital are treated as nondepreciable capital investments whose MVs at the end of year five are estimated to equal their first costs. (In economic evaluations, it is customary to assume that land and working capital do not inflate in value during the study period because they are "nonwasting" assets.) By using Equation 7-24, we are able to compute ATCF in year three (for example) to be

$$\text{ATCF}_3 = (\$30,000,000 - \$8,000,000 - \$11,520,000)(1 - 0.40) + \$11,520,000$$
$$= \$17,808,000$$

(b) The NIAT in year three can be determined with Equation 7-22:

$$\text{NIAT}_3 = (\$30,000,000 - \$8,000,000 - \$11,520,000)(1 - 0.40) = \$6,288,000$$

This can also be obtained directly from Table 7-8 by adding the year three entries from columns C and D: $10,480,000 - $4,192,000 = $6,288,000.

(c) The depreciable property in Example 7-19 ($60,000,000) will be disposed of for $0 at the end of year five, and a loss on disposal of $6,912,000 will be claimed at the end of year five. Only a half-year of depreciation ($3,456,000) can be claimed

---

TABLE 7-8   After-Tax Analysis of Example 7-19

End of Year, k	(A) BTCF	(B) Depreciation Deduction	(C) = (A) − (B) Taxable Income	(D) = −0.4(C) Cash Flow for Income Taxes	(E) = (A) + (D) ATCF
0	−$60,000,000 − 10,000,000 − 275,000				−$70,275,000
1	22,000,000	$12,000,000	$10,000,000	−$4,000,000	18,000,000
2	22,000,000	19,200,000	2,800,000	− 1,120,000	20,880,000
3	22,000,000	11,520,000	10,480,000	− 4,192,000	17,808,000
4	22,000,000	6,912,000	15,088,000	− 6,035,200	15,964,800
5	22,000,000	3,456,000	18,544,000	− 7,417,600	14,582,400
5	10,275,000[a]		−6,912,000[b]	2,764,800[b]	13,039,800

[a]MV of working capital and land.
[b]Because BV_5 of the production facility is $6,912,000 and net $MV_5 = 0$, a loss on disposal would be taken at EOY 5.

as a deduction in year five, and the BV is $6,912,900 at the end of year five. Because the selling price (MV) is zero, the loss on disposal equals our BV of $6,912,000. As seen from Figure 7-5, a tax credit of 0.40($6,912,000) = $2,764,800 is created at the end of year five. The after-tax IRR is obtained from entries in column E of Table 7-8 and is found to be 12.5%. The after-tax PW equals $936,715 at MARR = 12% per year. Based on economic considerations, this integrated circuit production line should be recommended because it appears to be quite attractive.

In the following example the after-tax comparison of mutually exclusive alternatives involving costs only is illustrated.

EXAMPLE 7-20

An engineering consulting firm can purchase a fully configured computer-aided design (CAD) workstation for $20,000. It is estimated that the useful life of the workstation is seven years, and its MV in seven years should be $2,000. Operating expenses are estimated to be $40 per eight-hour workday, and maintenance will be performed under contract for $8,000 per year. The MACRS (GDS) property class is five years, and the effective income tax rate is 40%.

As an alternative, sufficient computer time can be leased from a service company at an annual cost of $20,000. If the after-tax MARR is 10% per year, how many workdays per year must the workstation be needed in order to justify *leasing* it?

SOLUTION

This example involves an after-tax evaluation of purchasing depreciable property versus leasing it. We are to determine how much the workstation must be utilized so that the lease option is a good economic choice. A *key* assumption is that the cost of engineering design time (i.e., operator time) is unaffected by whether the workstation is purchased or leased. Variable operations expenses associated with ownership result from the purchase of supplies, utilities, and so on. Hardware and software maintenance cost is contractually fixed at $8,000 per year. It is further assumed that the maximum number of working days per year is 250.

Lease fees are treated as an annual expense, and the consulting firm (the lessee) may *not* claim depreciation of the equipment to be an additional expense. (The leasing company presumably has included the cost of depreciation in its fee.) Determination of ATCF for the lease option is relatively straightforward and is not affected by how much the workstation is utilized:

$$(\text{After-tax expense of the lease})_k = -\$20,000(1 - 0.40) = -\$12,000; \quad k = 1, \ldots, 7$$

ATCFs for the purchase option involve expenses that are fixed (not a function of equipment utilization) in addition to expenses that vary with equipment usage. If we let X equal the number of working days per year that the equipment is utilized, the variable cost per year of operating the workstation is $40X$. The after-tax analysis of the purchase alternative is shown in Table 7-9.

TABLE 7-9 After-Tax Analysis of Purchase Alternative (Example 7-20)

End of Year, k	(A) BTCF	(B) Depreciation Deduction[a]	(C) = (A) − (B) Taxable Income	(D) = −t(C) Cash Flow for Income Taxes	(E) = (A) + (D) ATCF
0	−$20,000				−$20,000
1	−40X − 8,000	$4,000	−$40X − $12,000	$16X + $4,800	−24X − 3,200
2	−40X − 8,000	6,400	− 40X − 14,400	16X + 5,760	−24X − 2,240
3	−40X − 8,000	3,840	− 40X − 11,840	16X + 4,736	−24X − 3,264
4	−40X − 8,000	2,304	− 40X − 10,304	16X + 4,122	−24X − 3,878
5	−40X − 8,000	2,304	− 40X − 10,304	16X + 4,122	−24X − 3,878
6	−40X − 8,000	1,152	− 40X − 9,152	16X + 3,661	−24X − 4,339
7	−40X − 8,000	0	− 40X − 8,000	16X + 3,200	−24X − 4,800
7	2,000		2,000	−800	1,200

[a]Depreciation deduction$_k$ = $20,000 × (GDS recovery rate).

The after-tax annual worth of purchasing the workstation is

$$AW(10\%) = -\$20,000(A/P, 10\%, 7) - \$24X - [\$3,200(P/F, 10\%, 1) + \cdots$$
$$+ \$4,800(P/F, 10\%, 7)](A/P, 10\%, 7) + \$1,200(A/F, 10\%, 7)$$
$$= -\$24X + -\$7,511$$

To solve for X, we equate the after-tax annual worth of both alternatives:

$$-\$12,000 = -\$24X - \$7,511$$

Thus $X = 187$ days per year. Therefore, if the firm expects to utilize the CAD workstation in its business *more than* 187 days per year, the equipment should be leased. The graphic summary of Example 7-20 shown in Figure 7-6 provides the rationale for this recommendation. The importance of the workstation's estimated utilization, in workdays per year, is now quite apparent.

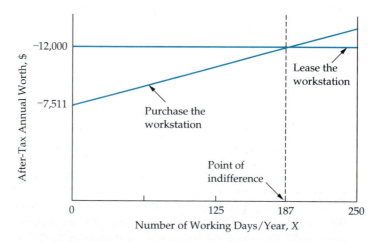

FIGURE 7-6 Summary of Example 7-20

7.12 After-Tax Analyses That Include Specific Financing Arrangements

Engineering economy problems encountered thus far have implicitly assumed that the firm's pool of equity and debt capital has been used for investment in engineering and business projects. Engineers and other technical personnel generally do not concern themselves with how funds in a firm's treasury have been sourced because their mission is to ascertain the best (most profitable) uses of the capital, given that it is available. Consequently, this book has dealt with engineering projects from the standpoint of a firm's *overall* pool of investment capital rather than any specific financing arrangement for the project. As a result, we have focused on *project* cash flows and have excluded financial cash flows specific to the project. Hence, the interest rate (i.e., the MARR) we use for time-value-of-money calculations typically includes the cost of borrowed money and equity in a particular firm's overall pool of capital.

By using project cash flows (after taxes) to evaluate investment alternatives, we avoid some serious errors that can arise when explicit treatment of interest expenses is attempted. For instance, if interest is deducted from gross revenues to arrive at operating income, double counting of interest expenses will occur when the MARR includes a component for "return on borrowed funds." *In general, it is advisable to exclude interest expenses from cash flows and to discount at the firm's after-tax MARR.* Nevertheless, there are special situations where the analysts must consider an investment problem that includes financing specific to the project.

In this section we shall demonstrate how to analyze a capital investment from the viewpoint of the *equity investor* (i.e., the stockholder). This viewpoint may be desired when a particular undertaking is quite risky and equity investors insist on an explicit assessment of the project's profitability in terms of *their* investment. To accomplish this, a project's ATCFs must be evaluated from the viewpoint of a firm's stockholders by using an interest rate equal to the *opportunity cost of equity capital* (Chapter 12). This can be done simply by adding a column for "Loan and Interest Cash Flow" to our general worktable format, as is illustrated in Example 7-21.

EXAMPLE 7-21

This example is fashioned after Example 7-17. Additional information is given concerning capital that is borrowed to purchase the machinery, and it is desired to determine the after-tax PW of equity capital. The problem is restated with additional information as follows.

Certain new machinery is estimated to cost $180,000 installed. It is expected to reduce net annual operating expenses by $36,000 per year (not including interest) for 10 years and to have a $30,000 MV at the end of the tenth year. Fifty thousand dollars of the investment is to be *borrowed* at 10% interest payable annually, with *all* the principal to be repaid at the end of the tenth year. To simplify the analysis, the machinery is depreciated with the straight-line method over 10 years using SV = $30,000 in the depreciation calculations.

Determine the after-tax cash flow for this machinery from the stockholders' (equity) viewpoint and calculate the after-tax profitability of equity capital ($130,000). The opportunity cost of equity capital has been set at 18% per year.

SOLUTION

Table 7-10 shows the recommended tabular format, with the column headings indicating how the calculations are made. Notice that interest on the money borrowed for the project is deductible as a business expense and hence reduces before-tax savings to $31,000. From the results of the right-hand column, the after-tax PW of the equity investment can be calculated:

$$PW(18\%) = -\$130,000 + \$24,920(P/A, 18\%, 10)$$
$$- \$20,000(P/F, 18\%, 10)$$
$$= -\$21,832$$

Because the PW is negative, this investment appears to be a poor one based on ATCFs to owners (stockholders) of the firm.

EXAMPLE 7-22

A firm is considering purchasing an asset for $10,000, with half of this amount coming from retained earnings (equity) and the other half being borrowed for three years at an effective interest rate of 12% per year. Uniform annual payments, consisting of interest and loan principal, will be utilized to repay the $5,000 loan. To simplify the example, straight-line depreciation over a three-year period will be used, and the asset's MV at this time is expected to be $4,000. Before-tax net revenues minus expenses from the asset are estimated to be $5,000 per year, which *does not* include the cost of borrowed money. Finally, the firm's effective income tax rate is 50%.

If the opportunity cost of capital to equity investors is 20% per year (after taxes), is this a profitable investment from the stockholder's viewpoint?

SOLUTION

The annual amount of the uniform loan repayment is $5,000(A/P, 12\%, 3) = $2,081.75. As described in Chapters 3 and 4, this capital recovery amount consists of repayment of borrowed money (principal) and interest on the unpaid principal at the beginning of each year. Because interest is a deductible business expense but repayment of loan principal is not, a schedule of annual loan principal and interest must be developed as follows (to the nearest dollar):

End of Year, k	Interest	Principal
1	$5,000(0.12) = $600	$2,082 − $600 = $1,482
2	(5,000 − 1,482)(0.12) = 422	2,082 − 422 = 1,660
3	(5,000 − 1,482 − 1,660)(0.12) = 223	2,082 − 223 = 1,859

TABLE 7-10 Computations to Determine the After-Tax Equity Cash Flow for Example 7-21

End of Year, k	(A) BTCF	(B) Depreciation Deduction	(C) Loan Cash Flow (C1) Interest	(C2) Principal	(D) = (A) − (B) + (C1) Taxable Income	(E) = −0.38(D) Cash Flow for Income Taxes	(F) = (A) + (C1) + (C2) + (E) Equity ATCF
0	−$180,000			+$50,000			−$130,000
1–10	+ 36,000[a]	$15,000	−$5,000	0	+$16,000	−$6,080	+ 24,920
10	+ 30,000[b]			− 50,000		0[b]	− 20,000

[a]Interest expense has not yet been deducted.
[b]The MV of the machinery at the end of year ten equals its BV. Thus, there is no gain or loss on disposal.

The after-tax equity cash flow for the asset can now be determined, as indicated in Table 7-11.

The PW of column F at 20% per year is $763. Because the PW is greater than zero, this investment is marginally attractive.

After-tax evaluations based on equity investment frequently utilize an annual schedule of loan interest and principal such as the one illustrated in Example 7-22.

EXAMPLE 7-23

In Example 7-22, what is the PW of interest payments in the loan repayment schedule after income taxes ($t = 0.50$) are considered? The after-tax MARR is 20% per year.

SOLUTION

Because interest is deductible from taxable income for legitimate business purposes, its after-tax cost in year k is $(1 - t)(-\text{Int}_k)$, and the PW in Example 7-22 is

$$\text{PW}(20\%) = \sum_{k=1}^{3}(1 - t)(-\text{Int}_k)(1.20)^{-k}$$

$$= 0.5(-\$600)(0.8333) + 0.5(-\$422)(0.6944) + 0.5(-\$223)(0.5787)$$

$$= -\$461.03$$

7.13 The After-Tax Effect of Depletion Allowances

Income from investment in certain natural resources is subject to depletion allowances before income taxes are computed (Section 7.6). Under certain conditions, notably where the taxpayer is in a relatively high income tax bracket, depletion provisions in the tax law can provide considerable economic advantages.

As an example, consider the case of a profitable corporation that has a net taxable income of $600,000. During the tax year, the firm spends $400,000 to drill and develop a geothermal well that has an estimated reservoir of 10,000,000 gallons of water. Hot water is produced and sold at $0.20 per gallon in accordance with the schedule shown in column 2 of Table 7-12 to produce the gross income shown in column 3. Column 4 shows the net cash flow after production costs have been deducted.

The depletion allowance that can be deducted in a given year may be based on a fixed *percentage* of the gross income (15% for geothermal wells), provided that the deduction *does not exceed 50% (100% for oil and gas property) of the net income before such a deduction* (column 5). Depletion, computed in this manner, is shown in column 7 of Table 7-12. Another method is to base depletion on the estimated investment cost of the product. In this case the estimated 10,000,000 gallons of water in the well cost $400,000. Such *cost depletion* may, if desired, be charged at the rate of $0.04 per gallon, shown in column 6 of the same table.

TABLE 7-11 After-Tax Equity Cash Flow Analysis for Example 7-22.

End of Year, k	(A) BTCF	(B) Depreciation Deduction	(C) Loan Cash Flow (C1) Interest[a]	(C) Loan Cash Flow (C2) Principal[b]	(D) = (A) − (B) + (C1) Taxable Income	(E) = −0.5(D) Cash Flow for Income Taxes	(F) = (A) + (C1) + (C2) + (E) Equity ATCF
0	−$10,000			+$5,000			−$5,000
1	5,000	$2,000	−$600	− 1,482	$2,400	−$1,200	1,718[c]
2	5,000	2,000	− 422	− 1,660	2,578	− 1,289	1,629
3	5,000	2,000	− 223	− 1,859	2,777	− 1,389	1,529
3	4,000						4,000

[a]Interest is deductible from column A when computing taxable income (column D).
[b]Principal is an after-tax cost deducted from column F.
[c]$5,000 − $600 − $1,482 − $1,200 = $1,718.

318

TABLE 7-12 Capital Recovery Provided by a Geothermal Well, with Cost and Percentage Depletion Allowances Used in Computation of Income Taxes

(1) End of Year, k	(2) Gallons of Water Sold	(3) Gross Income (Cash Flow)	(4) Net Income	(5) 50% of Net Income	(6) Cost Depletion at $0.04 per Gallon[a]	(7) Depletion Allowance at 15% of Gross Income	(8) [= (4) − either (6) or (7)] Taxable Income[b]	(9) [= −0.40(8)] Income Taxes	(10) [= (4) + (9)] ATCF
1	700,000	$140,000	$80,000	$40,000	$28,000	$21,000	$52,000	−$20,800	$59,200
2	600,000	120,000	70,000	35,000	24,000	18,000	46,000	− 18,400	51,600
3	450,000	90,000	48,000	24,000	18,000	13,500	30,000	− 12,000	36,000
4	200,000	40,000	24,000	12,000	8,000	6,000	16,000	− 6,400	17,600
5	50,000	10,000	2,500	1,250	2,000	1,500	500	− 200	2,300

[a]Cost depletion summary: Yr. 1 $\dfrac{700,000}{10,000,000}(\$400,000) = \$28,000$

Yr. 2 $\dfrac{600,000(\$400,000 - \$28,000)}{10,000,000 - 700,000} = \$24,000$

Yr. 3 $\dfrac{450,000(\$372,000 - \$24,000)}{9,300,000 - 600,000} = \$18,000$

Yr. 4 $\dfrac{200,000(\$348,000 - \$18,000)}{8,700,000 - 450,000} = \$8,000$

Yr. 5 $\dfrac{50,000(\$330,000 - \$8,000)}{8,250,000 - 200,000} = \$2,000$

[b]In computing taxable income, the larger allowance in column 6 or column 7 is chosen as long as percentage depletion does not exceed 50% of column 4. If cost depletion exceeds percentage depletion, then cost depletion must be used during that particular year.
Note: Percentage depletion is generally not allowed for oil and gas properties.

The taxable income resulting from the most favorable application of depletion allowances (by the cost method or the percentage method) is shown in column 8. The advantage of percentage depletion allowances stems from the fact that the total depletion that can be claimed often exceeds the depreciable capital investment. However, this advantage is not shown in this particular situation because of the relatively small fraction of reservoir capacity that is sold in years one through five. In fact, the cost depletion allowance in column 6 is consistently better (higher) than the fixed percentage allowance in column 7. When cost depletion exceeds percentage depletion, the cost depletion allowance must be used in computing taxable income, given that the basis of the property has not been exhausted. It is also noteworthy that cost depletion is not limited to 50% of net income shown in column 5, which is determined before depletion deductions are considered. (Recall that Figure 7-3 summarized the procedure for determining allowable depletion.)

As the firm has a net taxable income of $600,000 even before returns from the geothermal well are considered, we shall assume that the total incremental tax rate (t) is 40%, thus giving the income tax shown in column 9 of Table 7-12. Column 10 shows the net ATCF provided to the investor for years one through five of the well's operation. The remaining 8,000,000 gallons of hot water would presumably be sold over the subsequent 10 to 15 years of the well's operation.

7.14 Summary

In this chapter, we have presented important aspects of TRA 86 and OBRA 93 relating to depreciation, depletion, and income taxes. It is essential to understand these topics so that correct after-tax engineering economy evaluations of proposed projects may be conducted. Depreciation and income taxes are also integral parts of subsequent chapters in this book. For instance, Chapter 9 examines the interrelationship between inflation and income taxes, and Chapter 10 describes how before- and after-tax sensitivity studies are performed. Various models for evaluating before- and after-tax replacement decisions are presented in Chapter 11.

In this chapter, many concepts regarding current federal income tax laws were described. For example, topics such as taxable income, effective income tax rates, taxation of ordinary income, and gains and losses on disposal of assets were explained. A general format for pulling together and organizing all these apparently diverse subjects was presented in Figure 7-5. This format offers the student or practicing engineer a means of collecting on one worksheet information that is required for determining ATCFs and properly evaluating the after-tax financial results of a proposed capital investment. Figure 7-5 was then employed in numerous examples. The student's challenge is now to use this worksheet in organizing data presented in problem exercises at the end of this and subsequent chapters and to answer questions regarding the after-tax desirability of the proposed undertaking(s).

7.15 References

AMERICAN TELEPHONE AND TELEGRAPH COMPANY, Engineering Department. *Engineering Economy,* 3rd ed. New York: American Telephone and Telegraph Company, 1977.

ARTHUR ANDERSEN & CO. *Tax Reform 1986: Analysis and Planning*, Subject File AA3010, Item 27. St. Louis, Mo., 1986.

COMMERCE CLEARING HOUSE, INC. *Explanation of Tax Reform Act of 1986*. Chicago, 1987.

Engineering Economist, The. A quarterly journal jointly published by the Engineering Economy Division of the American Society for Engineering Education and the Institute of Industrial Engineers. Published by IIE, 25 Technology Park, Atlanta, Ga. 30092.

LASSER, J. K. *Your Income Tax*. New York: Simon & Schuster (see the latest edition).

SMITH, G. W. *Engineering Economy: The Analysis of Capital Expenditures*, 4th ed. Ames: Iowa State University Press, 1987.

U.S. DEPARTMENT OF THE TREASURY. *Tax Guide for Small Business*, IRS Publication 334, Washington, D.C.: U.S. Government Printing Office (revised annually).

———. *Depreciation*, IRS Publication 534, Washington, D.C.: U.S. Government Printing Office (revised annually).

———. *Sales and Other Dispositions of Assets*, IRS Publication 544, Washington, D.C.: U.S. Government Printing Office (revised annually).

———. *Investment Income and Expenses*, IRS Publication 550, Washington, D.C.: U.S. Government Printing Office (revised annually).

———. *Basis of Assets*, IRS Publication 551, Washington, D.C.: U.S. Government Printing Office (revised annually).

———. *Tax Information on Corporations*, IRS Publication 542, Washington, D.C.: U.S. Government Printing Office (revised annually).

7.16   Problems

The number in parentheses ( ) that follows each problem refers to the section from which the problem is taken.

7-1. Define depreciation. (7.2)

7-2. How are depreciation deductions different from other production or service expenses such as labor, material, and electricity? (7.2)

7-3. What conditions must a property satisfy to be considered depreciable? (7.2)

7-4. Explain the difference between real and personal property. (7.2)

7-5. Explain the difference between tangible and intangible property. (7.2)

7-6. Explain how the cost basis of depreciable property is determined. (7.2)

7-7. What is the depreciation deduction, using each of the following methods, for the second year for an asset that costs $35,000 and has an estimated MV of $7,000 at the end of its seven-year useful life? Assume its MACRS class life is also seven years. (a) SYD, (b) 200% declining-balance, (c) GDS (MACRS), and (d) ADS (MACRS). (7.3, 7.4)

7-8. Your company has purchased a large new truck-tractor for over-the-road use (asset class 00.26). It has a basic cost of $180,000. With additional options costing $15,000, the cost basis for depreciation purposes is $195,000. Its market value at the end of five years is estimated as $40,000. Assume it will be depreciated under the GDS: (7.4)

a. The cumulative depreciation through the end of year three is closest to
  1. $195,000   **2.** $187,775   **3.** $180,000
  4. $151,671   **5.** $180,551

b. The MACRS depreciation in year four is most nearly
  1. 0   **2.** $13,350   **3.** $14,450
  4. $31,150   **5.** $45,400

c. The book value at the end of year two is most nearly
  1. $33,000   **2.** $36,000   **3.** $42,000
  4. $43,000   **5.** $157,000

7-9. Why would a business elect, under MACRS, to use the ADS rather than the GDS? (7.4)

7-10. Develop a logic flow chart for determining the optimal year to switch from 200% declining-balance to straight-line depreciation. (7.3)

7-11. A company purchased a machine for $15,000. It paid sales taxes and shipping costs of $1,000 and nonrecurring installation costs amounting to $1,200. At the end of three years, the company had no further use for the machine, so it spent $500 to have the machine dismantled and was able to sell the machine for $1,500. (7.3)

a. What is the cost basis for this machine?

b. The company had depreciated the machine on a straight-line basis, using an estimated useful life of five years and $1,000 SV. By what amount did the depreciation deductions fail to cover the actual depreciation?

7-12. An asset for drilling was purchased and placed in service by a petroleum production company. Its cost basis is $60,000 and it has an estimated MV of $12,000 at the end of an estimated useful life of 14 years. Compute the depreciation amount in the third year and the BV at the end of the fifth year of life by each of these methods: (7.3, 7.4)

a. The straight-line method.

b. The SYD method.

c. The 200% declining-balance method with switchover to straight line.

d. The GDS.

e. The ADS.

7-13. An optical scanning machine was purchased for $150,000 in the current tax year (year one). It is to be used for reproducing blueprints of engineering drawings, and its MACRS class life is nine years. The estimated MV of this machine at the end of ten years is $30,000. (7.4)

a. What is the GDS recovery period of the machine?

b. Based on your answer to (a), what is the depreciation deduction in year four?

c. What is the BV at the beginning of year five?

7-14. A piece of construction equipment (asset class 15.0) was purchased by the Jones Construction Company. The cost basis was $300,000.

a. Determine the GDS and ADS depreciation deductions for this property. (7.4)

b. Compute the difference in present worth of the two sets of depreciation deductions in (a) if $i = 12\%$ per year. (7.5)

7-15. A bowling alley costs $500,000 and has an estimated life of ten years. Its estimated salvage value at the end of year ten is $20,000.

a. Determine the depreciation for years one through ten using: (i) the straight-line method, (ii) the 200% declining-balance method, and (iii) the MACRS method (GDS class life = ten years). A table containing some of the depreciation values is provided below. Complete the table below. (7.5)

b. Compute the present worth of the depreciation deductions (at EOY 0) for each of the three methods. The MARR is 10% per year.

c. If a large present worth in part (b) is desirable, what do you conclude regarding which method is preferred?

7-16. During its current tax year (year one) a pharmaceutical company purchased a mixing tank that had a fair market price of $120,000. It replaced an older, smaller mixing tank that had a BV of $15,000. Because a special promotion was underway, the old tank was used as a trade-in for the new one, and the cash price (including delivery and installation) was set at $99,500. The MACRS class life for the new mixing tank was 9.5 years. (7.4, 7.3)

a. Under the GDS, what is the depreciation deduction in year three?

EOY	Straight-Line Method	Declining-Balance Method	MACRS Method	(Pr. 7-15)
1		$100,000	$71,450	
2		$80,000		
3				
4				
5				
6		$32,768	$44,600	
7		$26,214	$44,650	
8				
9				
10	$48,000			

b. Under the GDS, what is the BV at the end of year four?

c. If 200% declining-balance depreciation had been applied to this problem, what would be the cumulative depreciation through the end of year four?

7-17. A special-purpose machine is to be depreciated as a linear function of use (units-of-production method). It costs $25,000 and is expected to produce 100,000 units and then be sold for $5,000. Up to the end of the third year it had produced 60,000 units, and during the fourth year it produced 10,000 units. What is the depreciation deduction for the fourth year and the BV at the end of the fourth year? (7.3)

7-18. A gold mine that is expected to produce 30,000 ounces of gold is purchased for $2,400,000. The gold can be sold for $450 per ounce; however, it costs $265 per ounce for mining and processing costs. If 3,500 ounces are produced this year, what will be the depletion allowance for (a) unit depletion, and (b) percentage depletion? (7.6)

7-19. A marble quarry is estimated to contain 900,000 tons of stone, and the ZARD Mining Company just purchased this quarry for $1,800,000. If 100,000 tons of marble can be sold each year and the average selling price per ton is $8.60, calculate the first year's depletion allowance for (a) the cost depletion method and (b) percentage depletion at 5% per year. ZARD's net income before deduction of a depletion allowance is $350,000. (7.6)

7-20. A gas well in Oklahoma has reserves of 2,000,000 mcf in the ground. The initial cost basis was $800,000, and during the first year of operation a depletion allowance of $280,000 was taken. At the beginning of the second year of operation, the reserves were reestimated to be 1,400,000 mcf. What is the new value of the depletion unit with the cost method? (7.6)

7-21. Consider a firm that had a taxable income of $90,000 in the current tax year, and total gross revenues of $220,000. Based on this information, answer these questions: (7.7, 7.11)

a. How much federal income tax was paid for the tax year?

b. What was the net income after taxes (NIAT)?

c. What was the total amount of deductible expenses (e.g., materials, labor, fuel, interest) and depreciation deductions claimed in the tax year?

7-22. For the Surefire Automatic Casting Company, gross revenues for the tax year amounted to $7,800,000. Operating expenses were $4,900,000 and depreciation deductions were $1,200,000. There was no interest on borrowed money. (7.7, 7.11)

a. How much federal income tax was paid for the tax year?

b. What was the NIAT?

c. What was the firm's ATCF?

7-23. Your company is contemplating the purchase of a large stamping machine. The machine will cost $180,000. With additional transportation and installation costs of $5,000 and $10,000, the cost basis for depreciation purposes is $195,000. Its MV at the end of five years is estimated as $40,000. For simplicity, assume that this machine is in the three-year MACRS (GDS) property class. The justifications for this machine include $40,000 savings per year in labor and $30,000 per year in reduced materials. The before-tax MARR is 20% per year, and the effective income tax rate is 40%. (7.4, 7.7, 7.8, 7.9, 7.11)

a. The NIAT at the end of year one is most nearly

   a. −$13,000   **b.** $3,000   **c.** $23,000
   d. $68,000   **e.** $130,000

b. The GDS depreciation in year four is most nearly

   a. 0   **b.** $13,350   **c.** $14,450
   d. $31,150   **e.** $45,400

c. The BV at the end of year two is most nearly

   a. $33,000   **b.** $36,000   **c.** $42,000
   d. $43,000   **e.** $157,000

d. The *total* BTCF in year five is most nearly (assuming that you sell the machine at the end of year five)

   a. $9,000   **b.** $40,000   **c.** $70,000
   d. $80,000   **e.** $110,000

e. The taxable income for year three is most nearly

   a. $5,010   **b.** $16,450   **c.** $28,880
   d. $41,120   **e.** $70,000

f. The PW of the *after-tax savings* from the machine, *in labor and materials only* (neglecting the first cost, depreciation, and the MV), is most nearly (using the after-tax MARR)

   a. $12,000   **b.** $95,000   **c.** $151,000
   d. $184,000   **e.** $193,000

g. Assume that the stamping machine will now be used for only three years due to the loss of several government contracts. The MV at the end of year three is $50,000. What is the income tax owed at the end of year three due to depreciation recapture (gain on disposal)?

 a. $8,444   **b.** $14,220   **c.** $21,111
 d. $35,550   **e.** $20,000

7-24. If the incremental federal income tax rate is 34% and the incremental state income tax rate is 6%, what is the effective combined income tax rate (t)? If state income taxes are 12% of taxable income, what is the value of t? (7.8)

7-25. Select the most correct response for each of these statements (T for true; F for false).

 a. **T F** Income taxes are most commonly levied on gross income.

 b. **T F** The IRR before income taxes is independent of the depreciation method used.

 c. **T F** The BV of a depreciable asset is the amount that has been charged off as depreciation deductions.

 d. **T F** A higher MARR should be used in after-tax engineering economy studies than in before-tax studies.

7-26. A corporation has estimated that its taxable income will be $57,000 in the current tax year. It has the opportunity to invest in a project that is expected to add $8,000 to this taxable income. How much federal tax will be owed *with* and *without* the proposed venture? (7.8)

7-27. Storage tanks to hold a highly corrosive chemical are currently made of material Z26. The capital investment in a tank is $30,000, and its useful life is eight years. Your company manufactures electronic components, and uses the alternative depreciation system (ADS) under MACRS to calculate depreciation deductions for these tanks. The net MV of the tanks at the end of their useful life is zero. When a tank is four years old, it must be relined at a cost of $10,000. This cost is not depreciated and can be claimed as an expense during year four.

 Instead of purchasing the tanks, they can be leased. A contract for up to 20 years of storage tank service can be written with the Rent-All Company. If your firm's after-tax MARR is 12% per year, what is the greatest annual amount that you can afford to pay for tank leasing without causing purchasing to be the more economical alternative? Your firm's effective income tax rate is 40%. State any assumptions you make. (7.4, 7.10)

7-28. A centerless grinder can be purchased new for $18,000. It will have an eight-year useful life and no terminal MV. Reductions in operating expenses (savings) from the machine will be $8,000 for the first four years and $3,000 for the last four years. Depreciation will be by the MACRS method (GDS recovery period of five years). A used grinder can be bought for $8,000 and will have no scrap value (MV) in eight years. It will save a constant $3,000 per year over the eight-year period and will be depreciated $1,000 per year for eight years. The effective income tax rate is 40%. Determine the PW of the *incremental* after-tax cash flow. Based on this economic measure, which alternative is preferred? Let the MARR be 12% per year after taxes. (7.4., 7.10)

7-29. A firm expects for the next several years to have annual taxable income in the $100,000 to $335,000 tax rate bracket. A new project is proposed that will raise revenues by $30,000 and increase the cost of sales by $10,000. If this new project necessitates a total capital investment of $50,000, and has zero MV at the end of its six-year life, what is the IRR after federal income taxes are paid? Assume no state taxes and that MACRS depreciation is used (GDS with a recovery period of five years). (7.4, 7.8, 7.10)

7-30. You have a piece of equipment with a present BV of $192,000. Next year's depreciation will be $96,000. You can sell the equipment now for $80,000 or you can sell it one year from now for the same amount. If you do not sell it now, you will definitely sell it next year. How much BTCF must the equipment produce over the next year (assume that it all comes at the end of the year) to justify keeping the equipment for one more year? Assume an after-tax MARR of 15% per year. Depreciation recapture, if any, is taxed at an effective income tax rate of 40%. (7.10)

7-31. A firm must decide between two system designs, S1 and S2, shown on the next page. Their effective income tax rate is 40%, and MACRS (GDS) depreciation is used. If the after-tax desired return on investment is 10% per year, which design should be chosen? State your assumptions. (7.10)

	Design	
	S1	**S2**
Capital investment	$100,000	$200,000
GDS recovery period (years)	5	5
Useful life (years)	7	6
Market value at end of useful life	$ 30,000	$ 50,000
Annual revenues less expenses over useful life	$ 20,000	$ 40,000

7-32. The owners of a small TV repair shop are planning to invest in some new circuit-testing equipment. The details of the proposed investment are as follows (study period = five years): (7.4, 7.10)

Purchase price = $5,000
Market value = $0 (end of five years)
Extra revenue = $2,000 per year
Extra expenses = $800 per year
Useful life = 5 years (also equal to the GDS and ADS recovery periods under MACRS)
Effective income tax rate = 15%

a. If the GDS is used, calculate the PW of ATCF when the MARR (after taxes) is 12% per year. Should the equipment be purchased?
b. Use the ADS (depreciation method) and calculate the PW of the ATCF, and recommend whether the new equipment should be purchased. (After-tax MARR is 12% per year.)

7-33. Your firm can purchase a machine for $12,000 to replace a rented machine. The rented machine costs $4,000 per year. The machine that you are considering would have a useful life of eight years and a $5,000 MV at the end of its useful life. By how much could annual operating expenses increase and still provide a return of 10% per year after taxes? The firm is in the 40% income tax bracket, and revenues produced with either machine are identical. Assume that alternate MACRS (ADS) depreciation is utilized to recover the investment in the machine, and that the ADS recovery period is five years. (7.4, 7.10)

7-34. An injection molding machine can be purchased and installed for $90,000. It is in the seven-year GDS property class, and is expected to be kept in service for eight years. It is believed that $10,000 can be obtained when the machine

is disposed of at the end of year eight. The net annual *value added* (i.e., revenues less expenses) that can be attributed to this machine is constant over eight years and amounts to $15,000. An effective income tax rate of 40% is used by the company, and the after-tax MARR equals 15% per year. (7.4, 7.10)

a. What is the approximate value of the company's before-tax MARR?
b. Determine the GDS depreciation amounts in years one through eight.
c. What is the taxable income at the end of year eight that is related to capital investment?
d. Set up a table and calculate the ATCF for this machine.
e. Should a recommendation be made to purchase the machine?

7-35. Your company has purchased equipment (for $50,000) that will reduce materials and labor costs by $14,000 each year for N years. After N years there will be no further need for the machine, and because the machine is specially designed, it will have no MV at any time. However, the IRS has ruled that you must depreciate the equipment on a straight-line basis with a tax life of five years. If the effective income tax rate is 40%, what is the minimum number of years your firm could operate the equipment to earn 10% per year after taxes on its investment? (7.10)

7-36. Entropy Enterprise, Ltd., is considering a $100,000 heat recovery incinerator that is expected to cause a net reduction in out-of-pocket costs of $30,000 per year for six years (the study period). No market value is expected. The incinerator is in the MACRS (GDS) five-year property class. The effective income tax rate is 34%. Determine the PW of the ATCF when the after-tax MARR is 8% per year. (7.4, 7.11)

7-37. The following information is for a proposed project that will provide the capability to produce a specialized product estimated to have a short market (sales) life:

- Capital investment is $1,000,000 (this includes land and working capital).
- The cost of depreciable property, *which is part* of the $1,000,000 total estimated project cost, is $420,000.
- Assume, for simplicity, that the depreciable property is in the MACRS (GDS) three-year property class.

- The analysis period is three years.
- Annual operating and maintenance expenses are $636,000 in the first year and they increase at the rate of 6% per year (i.e., $\bar{f} = 6\%$) thereafter (see geometric gradient, Chapter 3).
- Estimated MV of depreciable property from the project at the end of three years is $280,000.
- Federal income tax rate = 34%; state income tax rate = 4%.
- MARR (after taxes) is 10% per year.

Based on an after-tax analysis using the PW method, what minimum amount of equivalent uniform annual revenue is required to justify the project economically? (7.10, 7.11)

7-38. Your company has to obtain some new production equipment for the next six years, and leasing is being considered. You have been directed to perform an after-tax study of the leasing approach. The pertinent information for the study is as follows:

Lease costs: First year, $80,000; second year, $60,000; third through sixth years, $50,000 per year. Assume that a six-year contract has been offered by the lessor that fixes these costs over the six-year period. Other costs (not covered in the contract) are $4,000 per year, and the effective income tax rate is 40%.

a. Develop the annual ATCFs for the leasing alternative.

b. If the MARR after taxes is 8% per year, what is the AW for the leasing alternative? (7.10, 7.11)

7-39.

a. Suppose that you have just completed the mechanical design of a high-speed automated palletizer that requires a capital investment of $3,000,000. The existing palletizer is quite old and has no MV. The new palletizer can be depreciated under MACRS (GDS five-year property class), and the equipment's MV at the end of seven years is estimated to be $300,000. The effective income tax rate is 40%. One million pallets will be handled by the palletizer each year during the seven-year expected project life.

What net annual savings per pallet (i.e., total savings less expenses) will have to be generated by the palletizer to justify this purchase in view of an after-tax MARR of 15% per year? (7.4, 7.10)

b. Referring to the situation described in part (a), suppose that you have estimated that the following incremental savings and expenses will be realized after the automated palletizer is installed:

1. Twenty-five operators will no longer be needed. Each operator earns an average of $15,000 per year in direct wages, and company fringe benefits are 30% of direct wages.

2. Property taxes and insurance amounting to 5% of the palletizer's installed cost will have to be paid over and above those paid for the present system.

3. Annual maintenance expenses relative to the present system will increase by $15,000.

4. Orders will be filled more efficiently because of the automated palletizer (less waste, quicker response). These savings are estimated at $100,000 per year.

The effective income tax rate is 40%, and the equipment's life, depreciation method, after-tax MARR, and so on, are the same as those in part (a). If the existing system has no MV, what is the maximum amount of money that can be invested in the automated palletizer (in view of the savings and costs just itemized) so that a 15% per year after-tax return on this investment is realized? (7.4, 7.10, 7.11)

7-40. Faultless Faucets, Inc., has been advised by a federal regulatory agency that it should purchase a particular asset for $10,000. The useful life of the asset is expected to be five years. The GDS (MACRS) recovery period is three years, and management of this company believes that the asset will have a net MV of $2,000 at the end of its five-year useful life. The annual operating and maintenance expenses are $2,000 the first year, and they increase by $200 per year thereafter. This asset will be purchased entirely with borrowed money costing 10% annually. The principal of the loan and interest will be repaid in equal end-of-year payments (as an annuity) over the five-year period. With equity MARR = 12% per year and $t = 0.40$, calculate the following: (7.4, 7.10, 7.12)

a. The GDS depreciation deduction in year four.

b. The income tax on depreciation recapture in year five.

c. The after-tax cash flow at end of year two.

7-41. A firm is considering the introduction of a new product. The marketing department has estimated that the product can be sold over a period of five years at a price of $7.00 per unit. Sales are estimated to be 10,000 units the first year and will increase by 2,000 units each year thereafter. Manufacturing equipment necessary to produce the item will cost $200,000. It is estimated that this equipment can be sold for $50,000 at the end of year five. MACRS depreciation (GDS five-year recovery period) will be used. The equipment will be financed by borrowing $160,000, and this loan *principal* is to be repaid in five equal end-of-year payments of $32,000 each. Interest at 10% per year will be paid on the loan principal outstanding at the beginning of each year. The $40,000 balance will be financed from equity funds. Operating and maintenance costs (not including taxes) will be $50,000 the first year and decrease by $3,000 each year thereafter. The firm's effective tax rate is 40%, and the equity MARR is 15% per year after taxes. Calculate the PW of the after-tax equity cash flow, assuming that the corporation is profitable in its other activities. (7.4, 7.10, 7.12)

7-42. Refer to Figure 7-3 and the depletion example in Table 7-12. Suppose that in years six through ten of this well's operation, hot water can be sold for $0.22 per gallon, a constant 1,000,000 gallons per year can be sold, and the depletion allowance is 22%. The firm's expected net income, before any depletion allowance has been deducted, is $80,000 per year (in column 4 of Table 7-12). If the effective income tax rate remains at 40%, what is the net cash flow after taxes in years six through ten? (7.13)

7-43. A large mineral deposit in Wyoming is estimated to contain 1,000,000 tons of a mineral whose percentage depletion allowance is 22%. A mining company has made an initial investment of $40,000,000 to recover this ore, and the market price for the ore is $175 per ton. The company's after-tax MARR is 12% per year, and its effective income tax rate is 40%. It is anticipated that the ore will be sold at the rate of 100,000 tons per year and that operating expenses, exclusive of depletion deductions, will be approximately $9,000,000 per year. (7.13)

a. Determine the ATCF for this mining venture when percentage depletion (or cost depletion, if appropriate) is used.

b. Determine the PW of after-tax cash flow in part (a).

7-44. Complete the table on p. 327, which lists capital recovery provided by a mining operation, with cost percentage depletion allowances used in computation of income taxes (refer to Figure 7-3). Initial reserves are 200,000 units. (7.13)

(Pr. 7-44)

(1) End of Year, k	(2) Units Sold	(3) Gross Income (Cash Flow)	(4) Net Income	(5) 50% of Net Income	(6) Cost Depletion at $4.00 per Unit	(7) Depletion Allowance at 22% of Gross Income	(8) [= (4) − either (6) or (7)] Taxable Income	(9) [= −0.40(8)] Income Tax	(10) [= (4) + (9)] Net ATCF
1	70,000	$1,400,000	$800,000	$400,000	?	?	?	?	?
2	60,000	1,200,000	700,000	350,000	?	?	?	?	?
3	45,000	900,000	480,000	240,000	?	?	?	?	?
4	20,000	400,000	240,000	120,000	?	?	?	?	?
5	5,000	85,000	25,000	12,500	?	?	?	?	?

327

8

Estimating Cash Flows

*T*he objectives of this chapter are (1) to discuss an integrated approach used to develop cash flows for the alternatives being analyzed in a study, and (2) to delineate and illustrate selected techniques that will be useful in making such estimates.

The following topics are discussed in this chapter:

An integrated approach for developing cash flows

Definition of a work breakdown structure

The cost and revenue structure

Estimating techniques (models)

Description of the learning curve effect

Estimating total product costs and selling price

Estimating cash flows for a typical small project

Developing cash flows (a case study)

8.1 Introduction

In Chapter 1, we discussed the engineering economic analysis procedure in terms of seven steps, which are listed here:

1. Recognition and formulation of the problem.
2. Development of the feasible alternatives.
3. Development of the net cash flow (and other prospective outcomes) for each alternative.
4. Selection of a criterion (or criteria) for determining the preferred alternative.
5. Analysis and comparison of the alternatives.

6. Selection of the preferred alternative.

7. Performance monitoring and postevaluation of results.

In Chapters 3 through 7, the methodology needed to accomplish Steps 4, 5, and 6 was developed and demonstrated. In this chapter, we return to Step 3.

> Because engineering economy studies deal with outcomes that extend into the future, estimating the future cash flows for feasible alternatives is a critical step in the analysis procedure. A decision based on the analysis is economically sound only to the extent that these cost and revenue estimates are representative of what subsequently will occur.

In Step 1 of the procedure, the need for doing an analysis was identified; the specific situation (improvement opportunity, design project, new venture, etc.) was explicitly defined; the desired outcomes in terms of goals, objectives, and other results were developed; and any special conditions and constraints that needed to be met were delineated. Then, in Step 2, the feasible alternatives to be analyzed in the engineering economy study were selected and described using the systems approach.

Thus, in Step 3 the *alternatives* to be analyzed have *already been selected* and the differences between them *already highlighted*. Other important information (results to be achieved and requirements to be met) that is needed in the analysis is available from the first two steps.

Applying the concepts and methodology discussed in this chapter is an important part of engineering practice. A commercial building project is used as the basis for some of the examples in Chapter 8. Any other engineering project, such as the expansion of a chemical processing plant or the design of an electrical distribution system switching center, could have been chosen.

8.2 An Integrated Approach

An integrated approach to developing the net cash flows for the feasible project alternatives (Step 3) is shown in Figure 8-1. We will use the term *project* to refer to the undertaking that is the subject of the analysis. This integrated approach includes three basic components:

1. *Work breakdown structure (WBS).* This is a technique for explicitly defining, at successive levels of detail, the work elements of a project and their interrelationships (sometimes called a *work element structure*).

2. *Cost and revenue structure (classification).* The delineation of the cost and revenue categories and elements that will be estimated in developing the cash flows.

3. *Estimating techniques (models).* Selected mathematical models used to estimate the future costs and revenues during the analysis period.

These three basic components, together with integrating procedural steps, provide an organized approach for developing the cash flows for the alternatives.

As shown in Figure 8-1, the integrated approach begins with a description of the project in terms of a WBS. This project WBS is used to describe the project and each alternative's unique characteristics in terms of design, labor, material requirements, and so on. Then these variations in design, resource requirements, and other characteristics are reflected in the estimated future costs and revenues (net cash flow) for that alternative.

To estimate future costs and revenues for an alternative, the perspective (viewpoint) of the cash flow must be established and an estimating baseline and analysis period defined. We discussed the perspective of a cash flow in Chapter 1. Normally cash flows are developed from the owner's viewpoint.

The net cash flow for an alternative represents what is estimated to happen to future revenues and costs from the perspective being used. Therefore, the estimated changes in revenues and costs associated with an alternative have to be relative to a baseline that is consistently used for all the alternatives being compared. This baseline is defined and applied in either of two ways.

The first method is the *total revenue and cost approach*. That is, the no-change (do nothing) alternative is explicitly included in the set of alternatives, and the total revenues and costs for it are estimated. Thus, when the total cost and revenue baseline approach is used, the net cash flow for the no-change alternative represents the projected revenues and costs of the current operation or situation. Similarly, the net cash flow for each of the other feasible alternatives is estimated.

The second method often used is the *differential* approach. Using this approach, the cash flow for the no-change alternative is defined as zero whether or not it is one of the feasible alternatives. The cash flow for each of the other feasible alternatives then represents the estimated differences (changes) in revenues and costs relative to the current situation (no-change alternative). Whichever estimating baseline approach is used in a study, it must be consistently applied for all feasible alternatives. *A common error is to inadvertently use both baseline definitions when developing the individual cash flows. For example, the total revenue and cost approach might be used in estimating maintenance costs for the no-change alternative, but in the other alternatives these costs might be estimated by using differences from current operations.*

Before developing the cash flows, other procedural steps need to be accomplished. First, decide what level(s) of the WBS to use for developing the cost and revenue estimates. The purpose of the study will be a primary factor in this decision. If the study is a project feasibility analysis, cost and revenue estimating will be less accurate than in the detailed economic analysis that will be used to make the final decision about a project (this is discussed further in Section 8.2.3).

Next, organize cost and revenue information from sources internal and external to the organization and assemble the relevant data for the study. Use these data, together with selected estimating techniques (models), to develop the future estimates.

8.2.1 The Work Breakdown Structure (WBS)

We briefly defined a work breakdown structure (work element structure) in Section 8.2, and identified it as the first basic component in an integrated approach to developing cash flows.

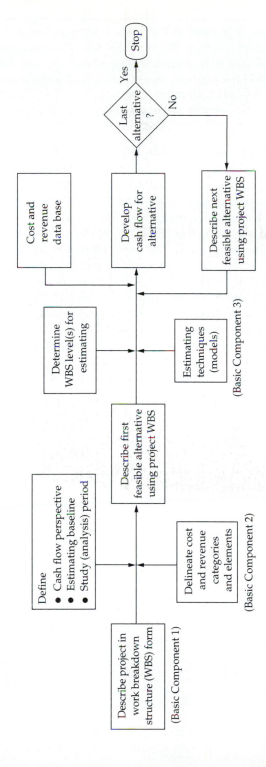

FIGURE 8-1 Integrated Approach for Developing the Cash Flows for Alternatives

This technique is a basic tool in project management and is a vital aid in an engineering economy study. The WBS serves as a framework for defining all project work elements and their interrelationships, collecting and organizing information, developing relevant cost and revenue data, and integrating project management activities. If a WBS does not exist and the project is of reasonable size, the first step in preparing cash flows for the alternatives should be to develop one.

The WBS is essential in ensuring the inclusion of all work elements, eliminating duplications and overlaps between work elements, avoiding nonrelated activities, and preventing other errors that could be introduced into the study. A WBS description dictionary is often prepared for large projects to ensure that each work element in the hierarchy is uniquely defined.

Figure 8-2 shows a diagram of a typical four-level work breakdown structure. It is developed from the top (project level) down in successive levels of detail. The project is divided into its major work elements (Level 2). These major elements are then divided to develop Level 3, and so on. For example, an automobile (first level of the WBS) can be divided into second-level components (or work elements) such as the chassis, drive train, and electrical system. Then each second-level component of the WBS can be subdivided further into third-level elements. The drive train, for example, can be subdivided into third-level components such as the engine, differential, and transmission. This process is continued until the desired detail in the definition and description of the project or system is achieved.

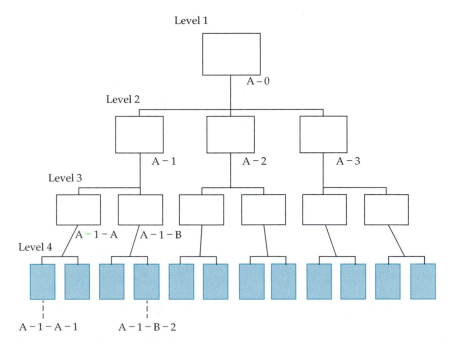

FIGURE 8-2   The WBS Diagram

Different numbering schemes may be used. The objectives of numbering are to indicate the interrelationships of the work elements in the hierarchy and to facilitate the manipulation and integration of data. The scheme illustrated in Figure 8-2 is an alphanumeric format. Another scheme often used is all numeric—Level 1: 1-0; Level 2: 1-1, 1-2, 1-3; Level 3: 1-1-1, 1-1-2, 1-2-1, 1-2-2, 1-3-1, 1-3-2; and so on (i.e., similar to the organization of this book). Usually, the level is equal (except for Level 1) to the number of characters indicating the work element.

Other characteristics of a project WBS are as follows:

1. Both functional (e.g., planning) and physical (e.g., foundation) work elements are included in it.
   a. Typical functional work elements are logistical support, project management, marketing, engineering, and systems integration.
   b. Physical work elements are the parts that make up a structure, product, piece of equipment, weapon system, or similar item; they require labor, materials, and other resources to produce or construct.
2. The content and resource requirements for a work element are the sum of the activities and resources of related subelements below it.
3. A project WBS usually includes recurring (e.g., maintenance) and nonrecurring (e.g., initial construction) work elements.

EXAMPLE 8-1

You have been appointed by your company to manage a project involving construction of a small commercial building with two floors of 15,000 gross square feet each. The ground floor is planned for small retail shops, and the second floor is planned for offices. Develop the first three levels of a representative WBS adequate for all project efforts from the time the decision was made to proceed with the design and construction of the building until initial occupancy is completed.

SOLUTION

There would be variations in the WBSs developed by different individuals for a commercial building. However, a representative three-level WBS is shown in Figure 8-3. Level 1 is the total project. At Level 2, the project is divided into seven major physical work elements and three major functional work elements. Then each of these major elements is divided into subelements as required (Level 3). The numbering scheme used in this example is all numeric.

8.2.2 The Cost and Revenue Structure

The second basic component of the integrated approach for developing cash flows (Figure 8-1) is the cost and revenue structure. This structure is used to identify and categorize the costs and revenues that need to be included in the analysis. Detailed data are developed and organized within this structure for use with the estimating techniques of Section 8.3 to prepare the cash flow estimates.

8.2.2.1 Using the Life-Cycle Concept and the WBS The life-cycle concept was discussed and illustrated in Chapter 2. The life cycle is divided into two

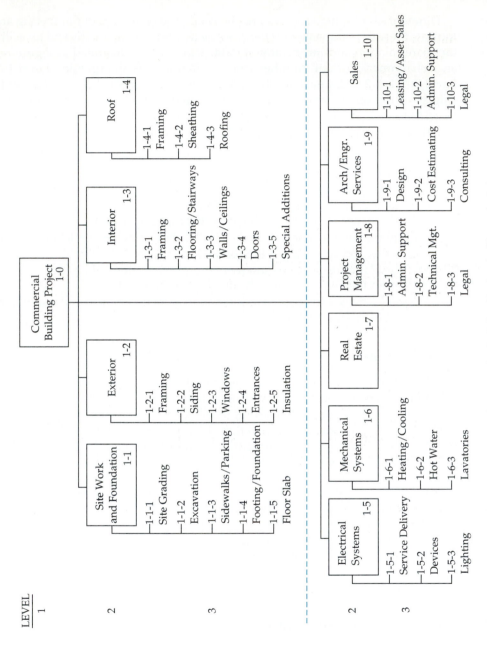

FIGURE 8-3 WBS (Three Levels) for Commercial Building Project in Example 8-1

general time periods: the acquisition phase and the operation phase. It begins with initial identification of the economic need or want (the requirement) and ends with retirement or disposal. Thus, it is intended to encompass all present and future costs and revenues.

The life-cycle concept and the WBS are important aids in developing the cost and revenue structure for a project. The life cycle defines a maximum time period and establishes a range of cost and revenue elements that need to be considered in developing cash flows. The WBS focuses the analyst's effort on the specific functional and physical work elements of a project, and on its related costs and revenues.

Ideally, the study period for a project is the life cycle of the product, structure, system, or service involved. This permits all relevant costs and revenues, both present and future, to be fully considered in decision making. Also, it makes possible the *explicit trade-off between initial costs during the acquisition phase and all subsequent costs and revenues during the operation phase* in analyzing the alternatives.

However, the accuracy of cost and revenue estimates decreases with increases in the length of the study period. Also, the effort required to develop cash flows increases with the length of the study period. Thus, a time horizon for the study period is selected to balance these factors and provide a sound basis for decision making.

8.2.2.2 Estimates Needed for a Typical Engineering Economy Study
As discussed in the previous section, judgment is required, based on the decision situation, to determine the study period, and thus how far into the future to estimate costs and revenues in an engineering economy study. This judgment should also weigh which cost and revenue elements are the most important and deserve more detailed study and which elements, even if drastically misjudged, will not produce significant changes in the estimated cash flows.

> Perhaps the most serious source of errors in developing cash flows is overlooking important categories of costs and revenues. The cost and revenue structure, prepared in tabular or checklist form, is a good means of preventing such oversights. Technical familiarity with the project is essential in ensuring the completeness of the structure, as are using the life-cycle concept and the WBS in its preparation.

Following is a brief listing of some categories of costs and revenues that are typically needed in an engineering economy study, together with a discussion of how estimates might be obtained. Some of these terms were discussed in Chapter 2.

1. *Capital investment* consists of two principal categories:
   a. *Fixed capital investment*, such as for feasibility studies, design and engineering, land purchase and improvement, buildings, equipment, installation, promotional and legal fees, and startup costs.
   b. *Working capital*, such as for inventories, accounts receivable, cash for wages, materials, and other accounts payable. Working capital is a revolving fund needed to get a project started and to meet subsequent obligations.

Normally, it is assumed that some or all working capital can be recovered by the end of the life of a project.

2. *Labor costs* are a function of skill level, labor supply, and time required. Standards for the normal amount of output per labor hour have been developed for many classes of work. Standard times combined with expected wage rates provide a reasonable estimate of labor costs for repetitive jobs. Labor costs for specialized work can be predicted from bid estimates or quotes by agencies offering the service. Remember that labor costs should include fringe benefits as well as direct wages.

3. *Material costs* are dependent on the project or operating situation; for example, an *output* of Operation A may be an *input* to Operation B. Material costs are those associated with the physical substance(s) being worked on or transformed.

4. *Maintenance costs* are the ordinary costs required for the upkeep of property and minor changes that are required for its more efficient use. Maintenance costs tend to increase with the age of an asset.

5. *Property taxes and insurance* are usually expressed as an annual percentage of the capital investment in economic comparisons.

6. *Quality (and scrap) costs* depend upon the types of products and associated quality standards, as well as upon the abilities of the work force, learning time, and rework possibilities.

7. *Overhead costs* are those costs that cannot be conveniently and practicably charged to particular products or services, and thus are normally prorated among the products or cost centers on some arbitrary basis. *Guard against using these arbitrary allocations in economic analyses,* because the differences in overhead costs brought about by various alternatives being considered are rarely described by these rates. In general, consider each cost element included in the overhead and estimate how much it is affected by each alternative.

8. *Disposal costs* are those nonrecurring costs associated with shutting down an operation and the retirement, disposal, or sale of assets to serve the best interests of the owners. These costs are estimated based on the projected labor, material, and other costs needed to accomplish the disposal activities.

9. *Revenues* are cash inflows (receipts) from all potential sources. Revenue differences between the alternatives need to be carefully considered. They are projected based on current market conditions, expected future changes in the market for the products and services involved, the company's expected market share, and pricing based on competition. The number of cost categories in a typical study usually far exceeds the number of revenue categories, and the oversight of a revenue source could cause a serious error in the results. Thus, the revenue impact of each alternative needs to be addressed and properly reflected in the cash flows. In addition, the study assumptions regarding differences in revenue between the alternatives should be explicitly stated before beginning an analysis.

10. *Salvage or market values* are typically a function of the useful life of assets and, in the case of long lives, are relatively unimportant to the study result. They are usually projected from current information about the salvage and market value of similar assets.

8.2.3 Estimating Techniques (Models)

The third basic component of the integrated approach (Figure 8-1) involves estimating techniques (models). These techniques, together with the detailed cost and revenue data, are used to develop individual cash flow estimates and the net cash flow for each alternative.

> The purpose of estimating is to develop cash flow projections—*not to produce exact data* about the future, which is virtually impossible. Neither a preliminary estimate nor a final estimate is expected to be exact; rather, it should adequately suit the need at a reasonable cost.

Cost and revenue estimates can be classified according to detail, accuracy, and their intended use as follows:

1. *Order of magnitude estimates:* used in the planning and initial evaluation stage of a project.
2. *Semidetailed or budget estimates:* used in the preliminary or conceptual design stage of a project.
3. *Definitive (detailed) estimates:* used in the detailed engineering/construction stage of a project.

Order of magnitude estimates are used in selecting the feasible alternatives for the study. They typically provide accuracy in the range of ±30 to 50% and are developed through semiformal means such as conferences, questionnaires, and generalized equations applied at Level 1 or 2 of the WBS.

Budget (semidetailed) estimates are compiled to support the preliminary design effort and decision making during this project period. Their accuracy usually lies in the range of ±15%. These estimates differ in the fineness of cost and revenue breakdowns and the amount of effort spent on the estimate. Estimating equations applied at Levels 2 and 3 of the WBS are normally used.

Detailed estimates are used as the basis for bids and to make detailed design decisions. Their accuracy is about ±5%. They are made from specifications, drawings, site surveys, vendor quotations, and in-house historical records, and are usually done at Level 3 and successive levels in the WBS.

Thus, it is apparent that a cost or revenue estimate can vary from a "back of the envelope" calculation by an expert to a very detailed and accurate prognostication of the future prepared by a project team. The level of detail and accuracy of estimates should depend on

1. Time and effort available and justified by the importance of the study.
2. Difficulty of estimating the items in question.
3. Methods or techniques employed.
4. Qualifications of the estimator(s).
5. Sensitivity of study results to particular factor estimates.

As estimates become more detailed, accuracy typically improves but the cost of estimating increases dramatically. This general relationship is shown in Figure 8-4

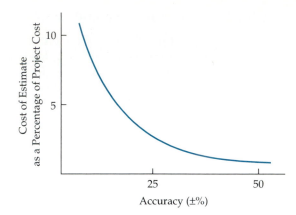

FIGURE 8-4

Accuracy of Cost and Revenue Estimates Versus the Cost of Making Them

and illustrates the idea that cost and revenue estimates should be prepared with full recognition of how accurate a particular study requires them to be.

Regardless of how estimates are made, individuals who use them should recognize that the estimates will be in error to some extent, even if sophisticated estimation techniques are used. However, estimation errors can be minimized.

8.2.3.1 Sources of Estimating Data The number of information sources useful in cost and revenue estimating is too great to list completely. The following four major sources of information are listed roughly in order of importance:

1. Accounting records.
2. Other sources within the firm.
3. Sources outside the firm.
4. Research and development.

1. *Accounting records.* A prime source of information for economic analyses; however, they are often not suitable for direct, unadjusted use.

A brief discussion of the accounting process and information is given in Appendix A. In its most basic sense, accounting consists of a series of procedures for keeping a detailed record of monetary transactions between established categories of assets, each of which has an accepted interpretation useful for its own purposes. The data generated by the accounting function are often inherently misleading for engineering economic analyses, not only because they are based on past results but also because of the following limitations:

a. The accounting system is rigidly categorized. Categories of various types of assets, liabilities, net worth, income, and expenses for a given firm may be perfectly appropriate for operating decisions and financial summaries, but rarely are they fully appropriate to the needs of economic analyses and decision making involving engineering design and project alternatives.

b. Standard accounting conventions cause misstatements of some types of financial information to be built into the system. These misstatements tend to be based on the philosophy that management should avoid overstating the value of its assets or understating the value of its liabilities and should therefore assess them very conservatively.

c. Accounting data often have illusory precision and implied authoritativeness. Although it is customary to present data to the nearest dollar or the nearest cent, the records are not nearly that accurate in general.

In summary, accounting records are a good source of historical data, but have some limitations when used in making prospective estimates for engineering economic analyses. Moreover, accounting records rarely contain direct statements of incrementing costs or opportunity costs, both of which are essential in most engineering economic analyses.

2. *Other sources within the firm.* The typical firm has a number of people and records that may be excellent sources of estimating information. Examples of functions within firms that keep records useful to economic analyses are engineering, sales, production, quality, purchasing, and personnel.

3. *Sources outside the firm.* There are numerous sources outside the firm that can provide helpful information. The main problem is in determining those that are most beneficial for particular needs. The following is a listing of some commonly used outside sources:

a. *Published information.* Technical directories, buyer indexes, U.S. government publications, reference books, and trade journals offer a wealth of information. For instance, *Standard and Poor's Industry Surveys* gives monthly information regarding key industries. *The Statistical Abstract of the United States* is a remarkably comprehensive source of cost indexes and data. The Bureau of Labor Statistics publishes many periodicals that are good sources of labor costs, such as the *Monthly Labor Review, Employment and Earnings, Current Wage Developments, Handbook of Labor Statistics,* and the *Chartbook on Wages, Prices and Productivity.* A buyer's index of manufacturers is *Thomas' Register of American Manufacturers,* which can be used to obtain addresses of vendors in order to inquire for prices. An annual construction cost handbook, *Building Construction Cost Data,* is published by the R. S. Means Company in Kingston, Massachusetts. It includes standard crew sizes, unit prices, and prevailing wage rates for various regions of the country. The R. S. Means Company also publishes other volumes of cost data on mechanical and electrical work, repair and remodeling, and selected cost indexes.

b. *Personal contacts* are excellent potential sources. Vendors, salespeople, professional acquaintances, customers, banks, government agencies, chambers of commerce, and even competitors are often willing to furnish needed information on the basis of a serious and tactful request.

4. *Research and development (R&D).* If the information is not published and cannot be obtained by consulting someone who knows, the only alternative may be to undertake R&D to generate it. Classic examples are developing a pilot plant and undertaking a test market program. These activities are usually expensive and may not always be successful; thus, this final step is taken only in connection with very important decisions and when the sources mentioned earlier are known to be inadequate.

The evaluation of the market and the business environment for large new capital projects is, along with the related estimating of project sales, product prices, and so on, a major area of analysis. R. F. de la Mare provides a good summary discussion of economic forecasting and market analysis related to large investment projects, and of the incorporation of revenue estimates into cash flows.*

8.2.3.2 How Estimates Are Accomplished Estimates can be prepared in a number of ways, for example, by:

1. A *conference* of various people who are thought to have good information or bases for estimating the quantity in question. A special version of this is the *Delphi method* (previously discussed in Chapter 1), which involves cycles of questioning and feedback in which the opinions of individual participants are kept anonymous.
2. *Comparison* with similar situations or designs about which there is more information and from which estimates for the alternatives under consideration can be extrapolated. This is sometimes called *estimating by analogy*. The comparison method may be used to approximate the cost of a design or product that is new. This is done by taking the cost of a more complex design for a similar item as an upper bound and the cost of a less complex item of similar design as a lower bound. The resulting approximation may not be very accurate, but the comparison method does have the virtue of setting bounds that might be useful for decision making.
3. *Using quantitative techniques*, which do not always have standardized names. Some selected techniques, with the names used being generally suggestive of the approaches, are discussed in the next section.

8.3 Selected Estimating Techniques (Models)

The estimating models discussed in this section are applicable for order-of-magnitude estimates and for many semidetailed or budget estimates. They are useful in the initial selection of feasible alternatives for further analysis and in the conceptual or preliminary design phase of a project. Sometimes these models can be used in the detailed design phase of a project to reduce the number of engineering estimates based on bills of material, standard costs, and other detailed information.

8.3.1 Indexes

Costs and prices† vary with time for a number of reasons, including 1) technological advances; 2) availability of labor and materials; and 3) inflation. An *index*

---

*R. F. de la Mare, *Manufacturing Systems Economics: The Life-Cycle Cost and Benefits of Industrial Assets* (London: Holt, Rinehart and Winston, 1982), pp. 123–149.

†The terms *cost* and *price* are often used together. The cost of a product or service is the total of the resources, direct and indirect, required to produce it. The price is the value of the good or service in the marketplace. In general, price is equal to cost plus a profit.

is a dimensionless number that indicates how a cost or a price has changed with time (typically escalated) with respect to a base year. Indexes provide a convenient means for developing present and future cost and price estimates from historical data. An estimate of the cost or selling price of an item in year n can be obtained by multiplying the cost or price of the item at an earlier point in time (year k) by the ratio of the index value in year n ($\bar{I}_n$) to the index value in year k ($\bar{I}_k$):

$$C_n = C_k \left(\frac{\bar{I}_n}{\bar{I}_k} \right) \tag{8-1}$$

where k = reference year (e.g., 1990) for which cost or price of item is known
n = year for which cost or price is to be estimated ($n > k$)
C_n = estimated cost or price of item in year n
C_k = cost or price of item in reference year k

Equation 8-1 is sometimes referred to as the *ratio technique* of updating costs and prices. Use of this technique allows the cost or potential selling price of an item to be taken from historical data with a specified base year and updated with an index. This concept can be applied at the lower levels of a WBS to estimate the cost of equipment, materials, and labor, as well as at the top level of a WBS to estimate the total project cost of a new facility, bridge, and so on.

EXAMPLE 8-2

A certain index for the cost of purchasing and installing utility boilers is keyed to 1974, where its baseline value was arbitrarily set at 100. Company XYZ installed a 50,000 lb/hr boiler in 1989 for $350,000 when the index had a value of 312. This same company must install another boiler of the same size in 1996. The index in 1996 is 468. What is the approximate cost of the new boiler?

SOLUTION
In this example, n is 1996 and k is 1989. From Equation 8-1, an approximate cost of the boiler in 1996 is

$$C_{1996} = \$350,000(468/312) = \$525,000.$$

Many indexes are periodically published, including the *Engineering News-Record* Construction Index, which incorporates labor and material costs, and the Marshall and Stevens cost index. The *Statistical Abstract of the United States* publishes government indexes on yearly materials, labor, and construction costs. The Bureau of Labor Statistics publishes the *Producer Prices and Price Indexes* and the *Consumer Price Index Detailed Report*. Indexes of cost and prices changes are frequently used in engineering economy studies.

8.3.2 Unit Technique

The *unit technique* involves using a "per unit factor" that can be estimated effectively. Examples are as follows:

Capital cost of plant per kilowatt of capacity.
Revenue per mile.
Fuel cost per kilowatt-hour generated.
Annual savings per 500 operating hours.
Capital cost per installed telephone.
Revenue per customer served.
Temperature loss per 1,000 feet of steam pipe.
Operating cost per mile.
Revenue per case.
Maintenance cost per hour.
Construction cost per square foot.
Revenue per thousand pounds.

Such factors, when multiplied by the appropriate unit, give a total estimate of cost, savings, or revenue.

As a simple example, suppose that we need a preliminary estimate of the cost of a particular house. Using a unit factor of, say, $55 per square foot and knowing that the house is approximately 2,000 square feet, we estimate its cost to be $55 × 2,000 = $110,000.

While the unit technique is very useful for preliminary estimating purposes, such average values can be misleading. In general, more detailed methods will result in greater estimation accuracy.

8.3.3 Factor Technique

The *factor technique* is an extension of the unit method, within a basic segmenting strategy, in which one sums the product of several quantities or components and adds these to any components estimated directly. That is,

$$C = \sum_{d} C_d + \sum_{m} f_m U_m \qquad \text{(8-2)}$$

where C = cost being estimated
C_d = cost of the selected component d that is estimated directly
f_m = cost per unit of component m
U_m = number of units of component m

As a simple example, suppose that we need a slightly refined estimate of the cost of a house consisting of 2,000 square feet, two porches, and a garage. Using a unit factor of $50 per square foot, and $5,000 per porch and $8,000 per garage for the two directly estimated components, we can calculate the total estimate as

$$(\$5,000 \times 2) + \$8,000 + (\$50 \times 2,000) = \$118,000$$

The factor technique is particularly useful when the complexity of the estimating situation does not require a WBS but several different parts are involved. Example 8-3 and the product cost-estimating example in Section 8.4.1 further illustrate this technique.

EXAMPLE 8-3

The detailed design of the commercial building described in Example 8-1 affects the utilization of the gross square feet (and, thus, the net rentable space) available on each floor. Also, the size and location of the parking lot and the prime road frontage available along the property may offer some additional revenue sources. As project manager, analyze the potential revenue impacts of these considerations.

SOLUTION

The first floor of the building has 15,000 gross square feet of retail space, and the second floor has the same amount planned for office use. Based on discussions with the sales staff, you develop the following additional information:

1. The retail space should be designed for two different uses—60% for a restaurant operation and 40% for a retail clothing store.
2. There is a high probability that all the office space on the second floor will be leased to one client.
3. An estimated 20 parking spaces can be rented on a long-term basis to two existing businesses that adjoin the property. Also, one spot along the road frontage can be leased to a sign company for erection of a billboard without impairing the primary use of the property.

Based on this information, you estimate annual project revenue ($\hat{R}$) as follows:

$$\hat{R} = W(r_1)(12) + Y(r_2)(12) + \sum_{j=1}^{3} S_j(u_j)(d_j)$$

where W = number of parking spaces
Y = number of billboards
r_1 = rate per month per parking space = \$22
r_2 = rate per month per billboard = \$65
j = index of type of building space use
S_j = space (gross square feet) being used for purpose j
u_j = space j utilization factor (% net rentable)
d_j = rate per (rentable) square foot per year of building space used for purpose j

Then

$$\hat{R} = [20(\$22)(12) + 1(\$65)(12)] + [9{,}000(0.79)(\$23)$$

$$+ 6{,}000(0.83)(\$18) + 15{,}000(0.89)(\$14)]$$

$$= \$6{,}060 + 440{,}070 = \$446{,}130$$

A breakdown of the annual estimated project revenue in Example 8-3 shows that

1.4% is from miscellaneous revenue sources
98.6% is from leased building space

From a detailed design perspective, changes in annual project revenue due to changes in building space utilization factors can be easily calculated. For example, an average 1% improvement in the ratio of rentable square feet to gross square feet would change the annual revenue as follows:

$$\Delta\hat{R} = \sum_{j=1}^{3} S_j(u_j + 0.01)(d_j) - (\$446{,}130 - \$6{,}060)$$

$$= \$445{,}320 - \$440{,}070$$

$$= \$5{,}250 \text{ per year}$$

8.3.4   Estimating Relationships

Cost and price estimating relationships are mathematical models that describe the cost or price of an item (e.g., a product, good, service, or activity) as a function of one or more independent variables. Various statistical and other mathematical techniques are used to develop estimating relationships. For example, simple linear regression and multiple linear regression models, which are standard statistical methods for estimating the value of a dependent variable (the unknown quantity) as a function of one or more independent variables, are often used to develop estimating relationships.

8.3.4.1   Power-Sizing Technique

The *power-sizing technique*, which is sometimes referred to as an *exponential model*, is frequently used for costing industrial plants and equipment. This method recognizes that cost varies as some power of the change in capacity or size. That is,

$$(C_A \div C_B) = (S_A \div S_B)^X \tag{8-3}$$

$$C_A = C_B(S_A/S_B)^X \tag{8-4}$$

where C_A = cost for plant A $\left.\right\}$ (both in \$ as of the point in time for
C_B = cost for plant B $\left.\right.$ which the estimate is desired)
S_A = size of plant A $\left.\right\}$
S_B = size of plant B $\left.\right.$ (both in same physical units)
X = *cost-capacity factor* to reflect economies of scale*

EXAMPLE 8-4

Suppose that it is desired to make a preliminary estimate of the cost of building a 600-MW fossil fuel power plant. It is known that a 200-MW plant cost \$100 million 20 years ago when the appropriate cost index was 400, and that cost index is now 1,200.

---

*May be calculated/estimated from experience. See W. R. Park, *Cost Engineering Analysis* (New York: John Wiley & Sons, 1973), p. 137, for typical factors. For example, $X = 0.68$ for nuclear generating plants and 0.79 for fossil-fuel generating plants.

SOLUTION

The power-sizing model estimate, with $X = 0.79$, is

Cost now of 200-MW plant: $100 million $\times$ (1,200 $\div$ 400) = $300 million (call it C_B)

Cost now of 600-MW plant: $C_A \div$ $300 million = $(600 \div 200)^{0.79}$ (call it C_A)

$$C_A = \$300 \text{ million} \times 2.38 = \$714 \text{ million}$$

8.3.4.2　Learning and Improvement

A *learning curve* is a mathematical model that explains the phenomenon of increased worker efficiency and improved organizational performance with repetitive production of a good or service. The learning curve is sometimes called an *experience curve* or a *manufacturing progress function*; fundamentally, it is an estimating relationship. The learning (improvement) curve effect was first observed in the aircraft and aerospace industries with respect to labor hours per unit.* However, it applies in many different situations. For example, the learning curve effect can be used in estimating the professional hours expended by an engineering staff to accomplish successive detailed designs within a family of products, as well as in estimating the labor hours required to assemble automobiles.

The basic concept of learning curves is that some input resources (e.g., energy costs, labor hours, material costs, engineering hours) decrease, on a per output unit basis, as the number of units produced increases. Most learning curves are based on the assumption that a constant percentage reduction occurs in, say, labor hours as the number of units produced is *doubled*. For example, if 100 labor hours are required to produce the first output unit, and a 90% learning curve is assumed, then 100(0.9) = 90 labor hours would be required to produce the second unit. Similarly, $100(0.9)^2 = 81$ labor hours would be needed to produce the fourth unit, $100(0.9)^3 = 72.9$ hours to produce the eighth unit, and so on. Therefore, a 90% learning curve results in a 10% reduction in labor hours each time the production quantity is doubled.

The assumption of a constant percentage reduction in the amount of an input resource used (per output unit) each time the number of output units is doubled can be used to develop a mathematical model for the learning (improvement) function. Let

u = the output unit number
Z_u = the number of input resource units needed to produce output unit number u
K = the number of input resource units needed to produce the first output unit
s = the learning curve slope parameter expressed as a decimal fraction (for a 90% learning curve, $s = 0.9$)

Then,

$$Z_u = Ks^a \quad \text{where } a = 0, 1, 2, 3, \ldots$$

---

*T. P. Wright, "Factors Affecting the Cost of Airplanes," *Journal of Aeronautical Sciences*, Vol. 3, No.4 (February 1936).

and,

$$u = 2^a$$

Thus,

$$\log Z_u - \log K = a(\log s)$$

and,

$$\log u = a(\log 2)$$

Or,

$$a = \frac{\log Z_u - \log K}{\log s} = \frac{\log u}{\log 2}$$

and,

$$\log Z_u - \log K = n(\log u) \quad \text{where } n = \frac{\log s}{\log 2}$$

Now, by taking the antilog of both sides, we have

$$\frac{Z_u}{K} = u^n$$

Or,

$$Z_u = Ku^n \tag{8-5}$$

EXAMPLE 8-5

The Mechanical Engineering department has a student team that is designing a formula car for national competition. The time required for the team to assemble the first car is 100 hours. Their improvement (or learning rate) is 0.8, which means that as output is doubled, their time to assemble a car is reduced by 20%. Use this information to determine (a) the time it will take the team to assemble the 10th car, (b) the *total time* required to assemble the first 10 cars, and (c) the estimated *cumulative average* assembly time for the first 10 cars.

SOLUTION

(a) From Equation 8-5 and assuming a proportional decrease in assembly time for output units between doubled quantities, we have,

$$Z_{10} = 100(10)^{\log 0.8 / \log 2}$$

$$= 100(10)^{-0.322}$$

$$= \frac{100}{2.099} = 47.6 \text{ hours}$$

(b) The total time to produce x units, T_x, is given by

$$T_x = \sum_{u=1}^{x} Z_u = \sum_{u=1}^{x} (Ku^n) = K \sum_{u=1}^{x} u^n \tag{8-6}$$

Using Equation 8-6,

$$T_{10} = 100 \sum_{u=1}^{10} u^{-0.322} = 100[1^{-0.322} + 2^{-0.322} + \cdots + 10^{-0.322}] = 631 \text{ hours}$$

(c) The cumulative average time for x units, C_x, is given by

$$C_x = T_x/x \qquad (8\text{-}7)$$

Using Equation 8-7,

$$C_{10} = T_{10}/10 = 631/10 = 63.1 \text{ hours}$$

EXAMPLE 8-6

The Betterbilt Construction Company designs and builds residential family homes. The purchasing manager for the company has developed a strategy whereby all the construction materials for a home are purchased from the same large supplier, but competitive bidding among a few firms is used to select the supplier for each home.

The company is ready to construct, in sequence, 16 new homes of 2,400 square feet each. The same basic design, with minor changes, will be used for each home. The successful bid for the construction materials in the first home is $64,800, or $27 per square foot. The purchasing manager believes, based on past experience, that several actions can be taken to reduce material costs by 8% each time the number of homes constructed doubles. Based on this information, (a) what is the estimated cumulative average material cost per square foot for the first five homes, and (b) what is the estimated material cost per square foot for the last (16th) home?

SOLUTION

(a) Based on the constant reduction rate of 8% each time the number of homes constructed doubles, a 92% learning curve applies to the situation. The cumulative average material cost for the first five homes is developed in the following table (assuming a proportional decrease in material costs for homes between doubled quantities).

(A) Home	(B) Material Cost[a] per Ft^2	(C) Cumulative Sum	(D) = (C) ÷ (A) Cumulative Average Cost per Ft^2
1	$27.00	$ 27.00	$27.00
2	24.84	51.84	25.92
3	23.66	75.50	25.17
4	22.85	98.35	24.59
5	22.25	120.60	24.12

[a]From Equation 8-5; for example $Z_3 = \$27(3)^{\log 0.92/\log 2} = \$23.66$.

(b) From Equation 8-5:

$$Z_{16} = \$27(16)^{\log 0.92/\log 2}$$

$$= \$27(16)^{-0.1203}$$

$$= \frac{\$27}{1.3959} = \$19.34 \text{ per square foot}$$

8.4 Estimating Total Product Costs and Selling Price

Manufacturers are faced with the problem of making a product that can be sold at a competitive price so that they can make a reasonable profit. The price of their product is based on the overall cost of making the item plus a built-in profit. Some companies that make a variety of products do not have a precise idea of exactly what each product costs—to find out might be prohibitively expensive—but they still need estimates to help them make decisions about what to produce and how to price their products.

As discussed in Chapter 2, product costs are classified as direct or indirect. Direct costs are easily assignable to a specific product, while indirect costs are not easily allocated to a certain product. For instance, direct labor would be the wages of a machine operator; indirect labor would be supervision.

Manufacturing costs have a distinct relationship to production volume in that they may be fixed, variable, or step-variable. Generally, administrative costs are fixed regardless of volume, material costs vary directly with volume, and equipment cost is a step function of production level.

The primary costs within the manufacturing expense category include engineering and design, development costs, tooling, manufacturing labor, materials, supervision, quality control, reliability and testing, packaging, plant overhead, general and administrative, distribution and marketing, financing, taxes, and insurance. Where do we start?

A detailed estimate is required. Therefore, we need drawings, specifications, production schedules, historical records of the company's labor cost, a bill of materials, and the process plan. The process plan describes all operations that must be done to a product and the labor hours involved.

Engineering and design costs consist of design, analysis, and drafting, together with miscellaneous charges such as reproductions. The engineering cost may be allocated to a product on the basis of how many engineering labor hours are involved. Other major types of costs that must be estimated are as follows:

Tooling costs, which consist of repair and maintenance plus the cost of any new equipment.
Manufacturing labor costs, which are determined from standard data, historical records, or the accounting department.

Materials costs, which can be obtained from historical records, vendor quotations, and the bill of materials. Scrap allowances must be included.

Supervision, which is a fixed cost based on the salaries of supervisory personnel.

Plant overhead, which includes utilities, maintenance, and repairs. As discussed in Chapter 2, there are various methods used to allocate overhead, such as in proportion to direct labor dollars, or direct labor hours, or machine hours.

Administrative costs, which are often included with the factory overhead (or burden).

8.4.1 A Manufacturing Cost Estimating Example

The following simple example shows the general procedure for making a per unit product cost estimate and illustrates the use of a typical spreadsheet form of the cost structure for preparing the estimate.

The spreadsheet in Figure 8-5 shows the determination of the cost of a throttle assembly. Column A shows typical cost elements that contribute to total product cost. The list of cost elements can easily be modified to meet a company's needs. This spreadsheet allows for per unit estimates (column B), factor estimates (column C), and direct estimates (column D). The shaded rows are selected subtotals.

Typically, direct labor costs are estimated via the unit technique. The manufacturing process plan is used to estimate the total number of direct labor hours required per output unit. This quantity is then multiplied by the composite labor rate to obtain the total direct labor cost. In this example, 36.48 direct labor hours are required to produce 50 throttle assemblies and the composite labor rate is $10.54 per hour, which yields a total direct labor cost of $384.50.

Indirect costs, such as quality control and planning labor, are often allocated to individual products using factor estimates. Estimates are obtained by expressing the cost as a percentage of another cost. In this example, planning labor and quality control are expressed as 12% and 11% of direct labor cost (row A), respectively. This gives a total labor cost of $472.93. Factory overhead and general and administrative expenses are estimated as percentages of the total labor cost (row D).

Entries for cost elements for which direct estimates are available are placed in column D. The total production materials cost for the 50 throttle assemblies is $167.17. A direct estimate of $28.00 applies to the outside manufacture of required components. The subtotal of cost elements at this point is $1,235.62.

Packing costs are estimated as 5% of all previous costs (row I), giving a total direct charge of $1,297.41. The cost of other miscellaneous direct charges are figured in as 1% of the current subtotal (row K). This results in a total manufacturing cost of $1,310.38 for the entire lot of 50 throttle assemblies. The manufacturing cost per assembly is $1,310.38/50 = $26.21.

As mentioned earlier in this section, the price of a product is based on the overall cost of making the item plus a built-in profit. The bottom of the spreadsheet in Figure 8-5 shows the computation of unit selling price based on this strategy. In this example, the desired profit (often called the profit margin) is 10% of the unit manufacturing cost, which corresponds to a profit of $2.62 per throttle assembly. The total selling price of a throttle assembly is then $26.21 + $2.62 = $28.83.

	Column A	Column B		Column C		Column D	Column E
		Unit Estimate		Factor Estimate		Direct Estimate	Row Total
		Unit	Cost/Unit	Factor	Estimate of Row		
A:	Factory Labor	36.48	$ 10.54				$ 384.50
B:	Planning Labor			12%	A		46.14
C:	Quality Control			11%	A		42.29
D:	TOTAL LABOR						472.93
E:	Factory Overhead			105%	D		496.58
F:	General & Admin. Expense			15%	D		70.94
G:	Production Material					$ 167.17	167.17
H:	Outside Manufacture					28.00	28.00
I:	SUBTOTAL						1,235.62
J:	Packing Costs			5%	I		61.78
K:	TOTAL DIRECT CHARGE						1,297.41
L:	Other Direct Charge			1%	K		12.97
M:	Facility Rental						--
N:	TOTAL MANUFACTURING COST						1,310.38
O:	Quantity (lot size)						50
P:	MANUFACTURING COST / UNIT						26.21
Q:	Profit/Fee			10%	P		2.62
R:	UNIT SELLING PRICE						$ 28.83

FIGURE 8-5   Manufacturing Cost Estimating Spreadsheet

350

8.4.2 Design and Target Costing

Typically, American firms determine an initial estimate of a new product's selling price using the bottom-up approach described in the previous section. That is, the estimated selling price is obtained by accumulating relevant fixed and variable costs and then adding a profit margin, which is a percentage of total costs. This process is often termed *design to price*. The estimated selling price is then used by the marketing department to determine whether or not the new product can be sold.

In contrast, Japanese firms apply the concept of *target costing*, which is a top-down approach. The focus of target costing is "what *should* the product cost" instead of "what *does* the product cost." As shown in Figure 8-6, target costing is initiated by conducting market surveys to determine the selling price of the best competitor's product. A target cost is obtained by deducting the desired return on sales (ROS) from the best competitor's selling price. ROS is typically expressed as a percentage of price.

$$\text{Target Cost} = \text{Competitor's Price}(1 - \text{ROS}) \tag{8-8}$$

This target cost is obtained prior to the design of the product and is used as a goal for engineering design, procurement, and production.

The preliminary engineering design process is initiated concurrently with target cost determination and utilizes conventional tools such as work breakdown structure and cost estimating to prepare the bottom-up total manufacturing cost projection discussed in the previous subsection. The total manufacturing cost represents an *initial* appraisal of what it would cost the firm to design and manufacture the product being considered. The total manufacturing cost is then compared to the top-down target cost. If the total manufacturing cost is *more* than the target cost, then the design must be fed back into *engineering* to challenge the value and the functionality of the design and attempt to reduce the cost of the design (i.e., *design to cost*). This iterative process is a key feature of the design-to-cost procedure. If the total manufacturing cost can be made less than the target cost, the design process continues into detailed design, culminating in the final design to be produced. If the total manufacturing cost cannot be reduced to the target cost, the firm should seriously consider abandoning the product.

The spreadsheet in Figure 8-7 illustrates the use of the manufacturing cost estimating spreadsheet to compute both a target cost and the cost reductions necessary to achieve the target cost. In this example, the target cost per throttle assembly is based on a competitor's selling price of $27.50 and a desired return on sales of 10%. Thus, the target cost is $(1 - 0.1)(\$27.50) = \$24.75$. Since the total manufacturing cost (determined to be $26.21 in Figure 8-5) is greater than the target cost ($24.75), we must work backwards from the total manufacturing cost, changing values of a chosen (single) cost element to the level required to reduce the cost to the desired target. This can be accomplished by trial and error (manually manipulating the spreadsheet values) or by making use of the "solver" feature of the software package (if one is available). Figure 8-7 shows one possible result of this process. If the process of assembling the throttles could be made more efficient such that total direct labor requirements are reduced to 34.07 hours (instead

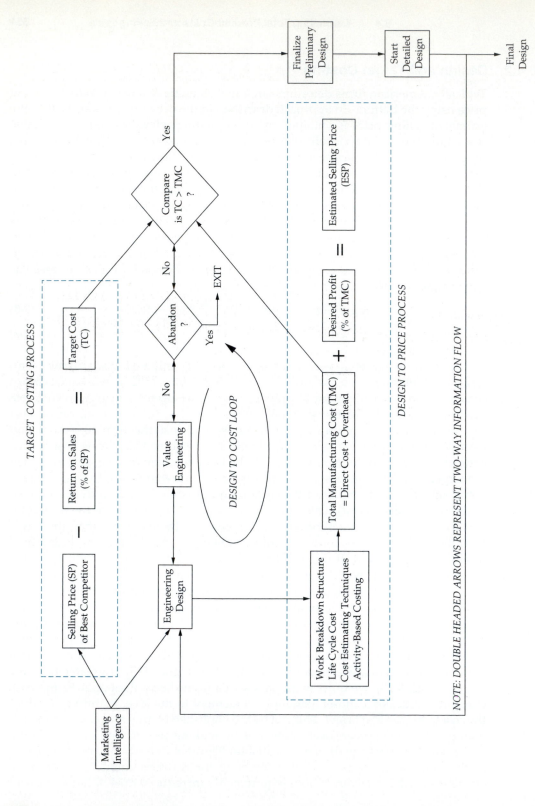

FIGURE 8-6 The Concept of Target Costing and Its Relationship to Design

A	B	C	D	E	F	G	H
1							
2	Column A	Column B		Column C		Column D	Column D
3		Unit Estimate		Factor Estimate		Direct	Row
4	MANUFACTURING COST ELEMENTS	Units	Cost/Unit	Factor	of Row	Estimate	Total
5							
6	A: Factory Labor	34.07	$ 10.54				$ 359.10
7	B: Planning Labor			12%	A		43.09
8	C: Quality Control			11%	A		39.50
9	D: TOTAL LABOR						441.70
10	E: Factory Overhead			105%	D		463.78
11	F: General & Admin. Expense			15%	D		66.25
12	G: Production Material					$ 167.17	167.17
13	H: Outside Manufacture					28.00	28.00
14	I: SUBTOTAL						1,166.90
15	J: Packing Costs			5%	I		58.35
16	K: TOTAL DIRECT CHARGE						1,225.25
17	L: Other Direct Charge			1%	K		12.25
18	M: Facility Rental						-
19	N: TOTAL MANUFACTURING COST						1,237.50
20	O: Quantity (lot size)						50
21	P: MANUFACTURING COST / UNIT						24.75
22							
23	Competitor's Selling Price	$ 27.50					
24	Desired Return on Sales	10%					
25	Target Cost	$ 24.75					

FIGURE 8-7   Manufacturing Cost Estimating and Target Costing

of 36.48), the target cost could be realized. Through this iterative process of comparing the effects of various cost inputs and evaluating the feasibility of changing them to achieve the target cost, one is able to develop a cost-effective design.

8.5 Estimating Cash Flows for a Typical Small Project

We will consider a small project typical of those often encountered in practice. To what extent does the integrated approach (Figure 8-1) apply when the project is not large and complex? The answer is that it applies regardless of the size and complexity of the project. However, several adjustments can be made to reduce the level of detail to fit the specific situation.

1. *WBS*. The number of levels and scope of the WBS can normally be significantly reduced for a small project. Sometimes the WBS can be combined with the cost and revenue structure into a worksheet for developing the estimates (see Example 8-7). The important point is that this initial component of the integrated

approach needs to be explicitly evaluated for the specific project. A WBS in the proper form and scope will facilitate the economic analysis of any project.

2. *Cost and revenue structure.* The number of cost and revenue categories and elements required can be reduced for most small projects. This second component, however, still needs to be considered in detail. For example, the number of operating and maintenance cost elements that may need to be included, even in a small project, can be quite extensive.

3. *Estimating techniques (models).* Estimating future costs and revenues is usually less complex with small projects. The techniques discussed in Sections 8.2.3 through 8.3, however, will still need to be used.

These three basic components of the integrated approach apply regardless of the size of the project. Their application to small projects, however, is reduced in scope along with the information database required. In any engineering economy study, it is necessary to (1) define the cash flow perspective, (2) determine the estimating baseline, and (3) establish the length of the analysis (study) period. These parts of the approach do not vary with project size.

EXAMPLE 8-7

Your company is involved in the manufacture of transmission components and axles for heavy-duty trucks and is a major supplier to three truck manufacturing plants. Just-in-time inventory concepts are used at each of the three manufacturing plants. Therefore, price competitiveness, reliable delivery to meet plant production schedules, and quality of the delivered product are essential to maintain the company's position as a supplier to the three plants. Meeting such customer expectations is critical to increasing the company's market share. Consequently, a project is being considered to replace some existing equipment with new automated equipment for the production of axles.

One of the feasible alternatives involves new equipment manufactured by Company A. Describe the development of its before-tax cash flow using the integrated approach in Figure 8-1. Discuss potential sources and compilation of necessary data as appropriate (all details do not need to be given). Some basic data related to the project are the following:

1. The equipment acquisition cost is $2,650,000 (including computer software and primary installation costs) if purchased from Company A. Other miscellaneous installation costs of $83,000 would be expensed in the first year of operation (i.e., not included in the cost basis of the equipment).

2. The analysis (study) period established by the company for this type of investment is six years.

SOLUTION

The new automated equipment, if purchased from Company A, is a complete system; that is, the hardware and software do not need to be broken down further to define the system explicitly for cost and revenue estimating. Therefore, the WBS level that can be used for estimating is the total project (i.e., Level 1 of the WBS). As a result, the WBS and the cost and revenue structure can be combined in a single worksheet. Thus, in this situation a separate, detailed WBS is not required.

The cash flow perspective that should be used in this project is that of the company (owners). Since this is a project that will upgrade an existing operation, the best estimating baseline is the current operation, and the differential approach (Section 8.2) should be used. Thus, cost data from the present operation plus those obtained from the manufacturer (Company A) are the primary sources of data for estimating purposes. The estimating techniques to be used are determined by the database that is available.

A representative worksheet is shown in Figure 8-8 for summarizing the costs and revenues needed to develop the net cash flow over six years when equipment is purchased from Company A. The estimate of capital investment is based primarily on data from the manufacturer (cost of the equipment and computer software).

A. Nonrecurring Costs and Revenues	Costs	Revenues
1. Capital Investment		
a. Hardware (including computer equipment)	$2,195,000	
b. Computer software	185,700	
c. Primary installation	269,300	
d. Other installation costs	83,000	
e. Working capital	28,400	
f. Project engineering and management	172,500	
	Total: $2,933,900	
2. Revenue		
a. Sale of present equipment (year 0)		$185,000
b. Sale of new equipment (year 6)		310,000

B. Recurring Annual Costs and Revenues	Costs	Revenue or Reduced Costs
1. Operational and Maintenance (O&M) Costs		
a. Direct Costs		
Labor		$201,000
Material		58,000
Other direct costs		44,600
b. Indirect Costs		
Labor/overtime		14,300
Materials and supplies		
Cost of quality (during production)		32,000
Tooling/fixtures		11,500
Maintenance	$18,600	
Utilities	4,200	
Property taxes and insurance	28,900	
Other indirect costs		5,900
2. Revenue		
Increased sales		525,000
	Total: $51,700	$892,300

FIGURE 8-8 Project Cost and Revenue Estimating Worksheet for Example 8-7

Internal estimates developed by the project engineering group are used for the other cost elements (installation costs, working capital, etc.).

The increased revenue estimate, based on additional market share (sales volume) as a result of the project, would be developed by the sales staff. The estimated MV for the equipment being replaced and the new equipment at the end of six years could be developed using data obtained from firms involved in the resale of this type of equipment. The operational and maintenance costs would be estimated from present operating experience and expected new equipment performance data supplied by Company A.

Based on the cost and revenue estimates shown on the worksheet, the estimated six-year BTCF for the alternative involving purchase of the new equipment from Company A is:

End of Year	BTCF (Company A)	
0	−$2,748,900	(= −$2,933,900 + $185,000)
1	840,600	
2	840,600	
3	840,600	
4	840,600	
5	840,600	
6	1,179,000	(= $892,300 + $310,000 + $28,400 − $51,700)

The cash flow amounts in year zero and year six include revenue from the disposal of assets of $185,000 and $310,000, respectively, as indicated on the worksheet. Also, the amount at the end of year six includes recovery of working capital.

8.6 Developing Cash Flows (A Case Study)

In the previous sections of this chapter, we discussed the work breakdown structure, cost and revenue structure, and estimating techniques related to an integrated approach for developing cash flows. In this section, we will demonstrate developing the cash flows for a large project (the commercial building project which was the subject of Examples 8-1 and 8-3).

As project manager, you want an updated, semidetailed estimate (Section 8.2.3) of the project net cash flow for 15 years of operation after construction is completed and full tenant occupancy of the building occurs. Assume the present time is the beginning of 1996. The semidetailed estimate will be based on the project as defined by the WBS in Example 8-1 and the floor space utilization assumptions and annual revenue estimates in Example 8-3. The planned schedule and assumptions related to construction, occupancy, and disposal of the building are:

Year	Schedule/Assumptions
1(1996)	Real estate purchased; design and engineering services—75% completed; construction—30% completed.
2	Design and engineering and construction—100% completed; four months of full tenant occupancy.
3–17	Average occupancy: 90% first floor and 95% second floor.
17	Building is sold at the end of the year.

8.6.1 Capital Investment

The capital investment consists of two major components: (1) the construction costs of the office building and related facilities, and (2) other project costs incurred until initial occupancy is completed. The construction costs for this case study project are shown in Table 8-1 and described below. Then in Table 8-2, the construction cost data are combined with the other estimated cost elements to develop the total capital investment for the project. This cost is then divided between 1996 (year one of the project) and 1997 (year two) in accordance with the project schedule and assumptions.

- The unit cost data shown in Table 8-1 are the result of the combined experience in the geographical location of the company and the architectural-engineering firm which has provided engineering and design services in this type of construction project for 24 years.* The square footage amounts are based on the project WBS definition and available design data. The general and administrative (G and A) overhead costs of the construction contractor plus profit are shown under the cost element "G and A Overhead and Profit." This cost element is estimated at 24% of total labor and material costs.

The total estimated project construction cost, including sidewalks, parking, contractor overhead, and profit, is $1,175,710. On the basis of 30,000 gross square feet of building space, the average cost per square foot is $39.19.

In Table 8-2, the development of the total capital investment using the factor technique for the project is shown. The other cost elements involved are estimated and added to the construction cost. Then, the total for each cost element is distributed between 1996 and 1997 based on the project schedule. The following information applies to the data in Table 8-2:

1. In the construction and operation of other office buildings, the company has been averaging an initial working capital investment of 3% of initial construction material costs (Table 8-1).

$$\text{working capital} = 0.03(\$596,190) = \$17,885 \text{ (rounded)}$$

2. The real estate cost ($262,000) at this point is an actual cost. That is, the company acquired the property immediately prior to this updating of the estimated project cash flow.

3. The project management costs shown in the table are based on a detailed estimate of the personnel time, office space, travel, and so on during the 20 months from the beginning of the project (January 1, 1996) until full occupancy and routine operations are scheduled to be achieved (August 31, 1997). The total estimated project management costs are $81,600, with 60% distributed to year one (12 months) and 40% to year two (8 months).

4. The cost of A–E (engineering and design) services is based on a contract rate of 8% of total construction costs:

$$\text{A–E services} = 0.08(\$1,175,710) = \$94,060 \text{ (rounded)}$$

---

*Sources of this type of data outside an organization were discussed in Section 8.2.3.1.

TABLE 8-1 Semidetailed Estimate of Construction Costs for the Commercial Building Project

Construction Cost Elements

WBS Elements	Labor/Installation[a]			Material[a]			Labor and Materials	G and A Overhead and Profit	Total
	$/Ft²	Ft²	Total	$/Ft²	Ft²	Total			
1.1 Site work and foundation	$2.67	18,000	$ 48,060	$2.48	18,000	$ 44,640	$ 92,700	$ 22,250	$ 114,950
1.1.3 Sidewalks/Parking	0.88	22,000	19,360	1.05	22,000	23,100	42,460	10,190	52,650
1.2 Exterior	3.64	30,000	109,200	6.38	30,000	191,400	300,600	72,140	372,740
1.3 Interior	2.72	30,000	81,600	5.50	30,000	165,000	246,600	59,180	305,780
1.4 Roof	1.19	15,000	17,850	1.61	15,000	24,150	42,000	10,080	52,080
1.5 Electrical Systems	1.73	30,000	51,900	1.71	30,000	51,300	103,200	24,770	127,970
1.6 Mechanical Systems	0.80	30,000	24,000	3.22	30,000	96,600	120,600	28,940	149,540
Subtotals:			351,970			596,190	948,160	227,550	
						Total Construction Cost:			$1,175,710

[a]Construction overhead costs related to labor and materials are included in the dollar per gross square foot unit rates used for these cost elements (this construction overhead is similar to factory overhead in Section 8.4).

TABLE 8-2	Semidetailed Estimate of Total Capital Investment for the Commercial Building Project

WBS Element(s)	Cost Element	Total	Distribution of Cost	
			Year 1 (1996)	Year 2 (1997)
1.1–1.6	working capital	$ 17,885		$ 17,885
1.1–1.6	construction (plus overhead)	1,175,710	$352,710	823,000
1.7	real estate	262,000	262,000	
1.8	project management	81,600	48,960	32,640
1.9	A-E services	94,060	70,545	23,515
1.10	sales (leasing)	44,615		44,615
	TOTALS	$1,675,870	$734,215	$941,655

5. The real estate management subsidiary of the company handles leasing of the available space. The estimated sales cost to achieve initial full occupancy is 10% of the first 12 months of rental income. The annual building rental income was estimated in Example 8-3 to be $446,130.

$$\text{sales (leasing) costs} = 0.10(\$446,130) = \$44,615 \text{ (rounded)}$$

Thus, the updated estimate of the total capital investment for the project is $1,675,870. Of this total, $734,215 is estimated to occur in 1996 and $941,655 in 1997. On the basis of 30,000 gross square feet of building space, the average cost per square foot based on the total project capital investment is $55.86.

8.6.2 Annual Operation and Maintenance Cost

A semidetailed estimate of annual operation and maintenance (O and M) costs for an office building is normally based on unit rates (dollars per gross square foot) with general and administrative (G and A) overhead costs added separately. The unit rates usually include indirect (overhead) costs associated with labor and materials. Since the company has significant experience in operating office buildings in the geographical location involved, historical data are the basis of the updated estimates.

The estimated annual O and M cost, except for the G and A overhead cost, is as follows:

Area	Ft^2	Annual Unit Cost ($/Ft^2)	Total
Office Building	30,000	$2,450	$73,500
Sidewalks/Parking	20,000	0.228	4,560
			$78,060

The company G and A overhead costs associated with operating the building are based on the annual operating and maintenance costs. The G and A rate applied on this basis by the company is 19%:

$$\text{annual G and A overhead costs} = 0.19(\$78,060) = \$14,830 \text{ (rounded)}$$

and

$$\text{total annual O and M cost} = \$78,060 + 14,830 = \$92,890$$

The operating and maintenance cost for the initial four months of occupancy in 1997 would be $\frac{1}{3}(\$92,890) = \$30,963$.

8.6.3 Annual Revenue and Leasing Fees

From the solution to Example 8-3, the estimated annual project revenue ($\hat{R}$) on the basis of 100% building occupancy is $446,070. This total annual revenue is made up of $253,170 from leasing the first floor, $186,900 from leasing the second floor, and $6,060 from leasing 20 parking spaces and one billboard location. Thus, the estimated revenue in 1997 for the initial four months of operation is $446,130(0.33) = $147,223.

For project years 3 through 17 (1998-2012), the estimated annual project revenue, based on an average first floor occupancy rate of 90% and a second floor rate of 95%, is

$$\hat{R} = 0.90(\$253,170) + 0.95(\$186,900) + \$6,060$$

$$= \$411,468$$

The real estate management subsidiary of the company charges an 8% fee based on the annual revenue for handling all annual leasing arrangements for the project. Therefore, the annual leasing fee is

$$\text{annual fee} = 0.08(\$411,468) = \$32,917$$

8.6.4 Asset Sale Revenue and Disposal Costs

The project plan is that the office building will be sold by the company at the end of project year 17 (2012). Estimated revenue from the sale of the asset is 80% of the original construction cost of the building and related facilities, plus the original cost of land and working capital. The estimated cost of selling the property is 7% of the total sales price. Thus, the estimated revenue and disposal costs associated with the sale of the asset are

$$\text{asset sale revenue} = 0.80(\$1,175,710) + 262,000 + 17,885$$

$$= \$1,220,453$$

and

$$\text{sales (disposal) cost} = 0.07(\$1,220,453) = \$85,432$$

8.6.5 Compilation of the Updated Project Net Cash Flow

The project before-tax net cash flow (BTCF) based on the updated semidetailed analysis is shown in Table 8-3. In columns 1–6, the annual cost and revenue cash

TABLE 8-3 Updated Before-Tax Net Cash Flow (BTCF) for the Commercial Building Project

	(1)	(2)	(3)	(4)	(5)	(6)	(7)	(8)
End of Year	**Cost Estimates**			**Revenue**		**Net BTCF**	**Revenue Growth Adjustment[a] $(1.045)^{k-2}$**	**Net BTCF (Adjusted)**
k	**First Cost**	**O & M**	**Sales/Fees**	**Annual**	**Asset Sale**			
1 (1996)	−$734,215					−$ 734,215	1.0	−$ 734,215
2 (1997)	− 941,655	−$30,963		$147,223		− 825,395	1.0	− 825,395
3		− 92,890	−$32,917	411,468		285,661	1.0450	298,516
4							1.09203	311,949
5							1.14117	325,988
6							1.19252	340,658
7							1.24619	355,988
8							1.30227	372,007
9		*b*	*b*	*b*		*b*	1.36087	388,747
10							1.42211	406,241
11							1.48610	424,522
12							1.55298	443,625
13							1.62286	463,589
14							1.69589	484,450
15							1.77221	506,250
16							1.85195	529,311
17 (2012)		− 92,890	− 32,917	411,468		285,661	1.93528	552,834
17 (2012)			− 85,349		$1,220,453	1,135,104	1.0[c]	1,135,104

[a]For $3 \leq k \leq 17$
[b]The arrow indicates a uniform cash flow amount for the years indicated.
[c]Selling price, land, and working capital not affected by revenue growth.

flows previously estimated in Sections 8.6.1-8.6.4 are compiled into the project net cash flow. However, the managers of the company believe that prudent cost control in operating the office building and renegotiation of leases on an annual basis will result in the project BTCF, starting in 1999, further increasing at the rate of 4.5% per year. Thus, the net BTCF (column 6), adjusted for this additional revenue growth (column 7), is shown in column 8.

Using a before-tax MARR of 20%, we find that the PW of the adjusted semi-detailed estimate of the project BTCF in column 8 is $35,566. Therefore, the project meets the company's economic criterion.

8.7 Summary

Developing the cash flow for each alternative in a study is a pivotal step in the engineering economic analysis procedure. An integrated approach for developing cash flows includes three major components: (1) a WBS definition of the project, (2) a cost and revenue structure that identifies all the cost and revenue elements involved in the study, and (3) estimating techniques (models). Other considerations such as the length of the analysis period, the perspective and estimating baseline for the cash flows, and a cost and revenue database are illustrated in Figure 8-1 and discussed in the chapter.

The WBS is a powerful technique for defining all the work elements and their interrelationships for a project. It is a basic tool in project management and is a vital aid in an engineering economy study. Understanding this technique and its applications are important in engineering practice.

The development of a cost and revenue structure will help to ensure that a cost element or a source of revenue is not overlooked in the analysis. The life-cycle concept and the WBS are used in developing this structure for a project.

Estimating techniques (models) are used to develop the cash flows for the alternatives as they are defined by the WBS. Thus, the estimating techniques form a bridge between the WBS and detailed cost and revenue data and the estimated cash flows for the alternatives.

8.8 References

Engineering News-Record. Published monthly by McGraw-Hill Book Co., New York.

JELEN, F. C., and J. H. BLACK. *Cost and Optimization Engineering,* 2nd ed. New York: McGraw-Hill Book Co., 1983.

MATTHEWS, L. M. *Estimating Manufacturing Costs: A Practical Guide for Managers and Estimators.* New York: McGraw-Hill Book Co., 1983.

OSTWALD, P. F. *Engineering Cost Estimating,* 3rd ed. Englewood Cliffs, NJ: Prentice-Hall, 1992.

PARK, W. R., and D. E. JACKSON. *Cost Engineering Analysis: A Guide to Economic Evaluation of Engineering Projects,* 2nd ed. New York: John Wiley & Sons, 1984.

PETERS, M. S., and K. D. TIMMERHAUS. *Plant Design and Economics for Chemical Engineers,* 2nd ed. New York: McGraw-Hill Book Co., 1968.

STEWART, R. D. *Cost Estimating.* New York: John Wiley & Sons, 1982.

STEWART, R. D., and R. M. WYSKIDA, eds. *Cost Estimators' Reference Manual.* New York: John Wiley & Sons, 1987.

8.9 Problems

The number in parentheses ( ) that follows each problem refers to the section from which the problem is taken.

8-1. Two of the basic components of an integrated approach for developing cash flows for the feasible alternatives of a project are (1) the WBS and (2) the cost and revenue structure. (8.2)
a. Describe the concept of each component.
b. Explain the applications of each component in an engineering economy study.

8-2. Visually examine a lawnmower for home use that is (a) nonriding, (b) approximately 21 inches in cutting width, and (c) powered with a 3.5- to 5.0-hp, air-cooled engine. Develop a WBS for this product through Level 3. (8.2)

8-3. You are planning to build a new home with approximately 2,000 to 2,500 gross square feet of living space on one floor. In addition, you are planning an attached two-car garage (with storage space) of approximately 600 gross square feet. Develop a cost and revenue structure for designing and constructing, operating (occupying) for 10 years, and then selling the home at the end of the tenth year. (8.2)

8-4. Why might a company's purchasing department be a good source of estimates for equipment required 2 years from now for a modernized production line? Which department(s) could provide good estimates of labor costs and maintenance associated with operating the production line? What information might the accounting department provide that would bear upon the incremental costs of the modernized line? (8.2)

8-5. Suppose that your brother-in-law has decided to start a company that produces synthetic lawns for lazy homeowners. He anticipates starting production in 18 months. In estimating future cash flows of the company, which of the following would be relatively easy versus relatively difficult to obtain? Also, suggest how each might be estimated with reasonable accuracy. (8.2)
a. Cost of land for a 10,000-square-foot building.
b. Cost of the building (cinder block construction).
c. Initial working capital.

d. Total investment (first) cost.
e. First year's labor and material costs.
f. First year's sales revenues.

8-6. Manufacturing equipment that was purchased in 1991 for $200,000 must be replaced in 1996. What is the estimated cost of the replacement based on the following equipment cost index? (8.3)

Year	Index	Year	Index
1991	223	1994	257
1992	238	1995	279
1993	247	1996	293

8-7. The purchase price of a natural gas–fired commercial boiler (capacity X) was $181,000 eight years ago. Another boiler of the same basic design, except with capacity $1.42X$, is currently being considered for purchase. If it is purchased, some optional features presently costing $28,000 would be added for your application. If the cost index was 162 for this type of equipment when the capacity X boiler was purchased, and is 221 now, and the applicable cost capacity factor is 0.8, what is your estimate of the purchase price for the new boiler? (8.3)

8-8. In a new industrial park, telephone poles and lines must be installed. It has been estimated that 10 miles of telephone lines will be needed, and each mile of line costs $14,000 (includes labor). In addition, a pole must be placed every 40 yards on the average to support the lines, and the cost of the pole and its installation is $210. What is the estimated cost of the entire job? (8.3)

8-9. If an ammonia plant that produces 500,000 pounds per year cost $2,500,000 to construct eight years ago, what would a 1,500,000-pound-per-year plant cost now? Suppose that the construction cost index has increased an average rate of 12% per year for the past eight years and that the cost-capacity factor (X) to reflect economy of scale is 0.65. (8.3)

8-10. The structural engineering design section within the engineering department of a regional electrical utility corporation has developed several standard designs for a group of similar transmission line towers. The detailed design for each tower is based on one of the standard designs. A transmission line project involving 50 towers has been approved. The estimated number of engineering hours needed to accomplish the first detailed tower design is 126. Assuming a 95% learning curve, (a) what is your estimate of the number of engineering hours needed to design the eighth tower and to design the last tower in the project, and (b) what is your estimate of the cumulative average hours required for the five designs? (8.3)

8-11. The standard labor hours (8.74) per output unit for Product A were based on a time study of the 32nd unit. If the learning curve, based on previous experience with similar products, is 85%, (a) what was the number of hours required for the first unit and (b) what is the estimated number of hours needed for the 1,000th output unit? (8.3)

8-12. The overhead costs for a company are presently $X per month. The management team of the company, in cooperation with the employees, is ready to implement a comprehensive improvement program to reduce these costs. If you (a) consider an observation of actual overhead costs for one month analogous to an output unit, (b) estimate the overhead costs for the first month of program implementation to be 1.15X due to extra front-end effort, and (c) consider a 90% improvement curve applicable to the situation, what is your estimate of the percentage reduction in present overhead costs per month after 30 months of program implementation? (8.3)

8-13. Refer to problem 8-3. You have decided to build a one-floor home with 2,450 gross square feet of living space. Also, the attached two-car garage (with storage space) will have 615 gross square feet of area.
a. Develop a WBS (through Level 3) defining the work elements involved in the design and construction of the home. (8.2)
b. Develop a semidetailed estimate of your investment (first) cost associated with the project until the time of your initial occupancy of the home. (*Note:* Your instructor will provide you with additional information to assist with this part of the problem.)(8.3)

8-14. Using the costing worksheet provided in this chapter (Figure 8-5), estimate the unit cost and selling price of manufacturing metal wire cutters in lots of 100 when these data have been obtained: (8.4)

Factory labor	4.2 hours at $11.15/hour
Factory overhead	150% of labor
Outside manufacture	$74.87
Production material	$26.20
Packing costs	7% of factory labor
Desired profit	12%

8-15. You have been asked to *estimate the per unit selling price* of a new line of widgets. Pertinent data are as follows:

Direct labor rate	$15.00 per hour
Production material	$375 per 100 widgets
Factory overhead	125% of factory labor
Packing costs	75% of factory labor
Desired profit	20% of total manufacturing cost

Past experience has shown that an 80% learning curve applies to the labor required for producing widgets. The time to complete the first widget has been estimated to be 1.76 hours. Use the estimated time to complete the 50th widget as your standard time for the purpose of estimating the unit selling price. (8.3, 8.4)

8-16. A personal computer company is trying to bring a new model of PC ("Pentium Plus") to the market. According to the marketing department, the best selling price for a similar model from a world-class competitor is $2,500 per computer. The company wants to sell at the same price as its best competitor. The cost breakdown of the new model is shown below:

Assembling time for the first unit:	1.00 hr
Handling time:	10% of Assembling Time
Direct labor rate:	$15/hr
Planning & liaison:	10% of Direct Labor
Quality control:	50% of Direct Labor
Factory overhead:	200% of Total Labor
General & Admin. expense:	300% of Total Labor

Direct material cost:	$200/computer
Outside manufacture:	$2,000/computer
Packing cost:	10% of Total Labor
Facility rental:	10% of Total Labor
Profit:	20% of Total Manufacturing Cost
Number of units:	20,000

Since the company mainly produces sub-assemblies purchased from other manufacturers and repackages the product, the direct material cost is estimated at only $200 per computer. Direct labor time consists of handling time and assembling time. The company estimates the learning curve for assembling the new model is 95%.

How can the company reduce its cost to meet the best competitors price? (8.3, 8.4)

8-17. A conventional chemical plant is being planned with the capacity to produce 3 million lbs. of product annually. Raw materials costs for the product are $0.45/lb, labor costs are $0.40/lb, and utility costs are $0.25/lb. Overhead costs are 75% of the labor costs. When pricing products, the company applies a 30% profit margin. A similar plant with a capacity of 2 million lbs. of product per year was constructed five years ago at a cost of $6 million when the appropriate cost index was 124. The current cost index value is 179 and the cost-capacity factor for this type of plan is 0.7. Annual sales estimates for the new plant are 2.8 million lb/yr for the first 10 years and 2.5 million lb/yr for the following five years. (8.3, 8.4)

a. Develop the net cash flows for the initial construction of the plant and the first 15 years of operation.

b. If the company's before-tax MARR = 10%, should the new facility be profitable?

8-18. Use Figure 8-1 and the information and results of problem 8-13 to develop an estimated before-tax net cash flow for 10 years of ownership of the new home. Assume sale (disposal) of the home at the end of 10 years. Obtain (locally) representative operating, repair, resale, and other data related to home ownership as needed to support development of the 10-year cash flow. Indicate the cost-estimating techniques used in estimating cash flows. State any assumptions you make. (8.6)

8-19. You have been assigned as technical manager of the design team for a plant expansion project (the order of magnitude estimate of the project cost is $10,000,000). Describe how the engineering economic analysis part of the design effort should be performed. What would be some of your major concerns? (8.2)

8-20. Why can Step 3 of the engineering economic analysis procedure be called the pivotal step in the procedure? (8.1)

8-21. Your company is planning to produce and sell high-density double-sided computer disks with 2 MB storage capacity. The disks are produced by installing a magnetic film into a plastic cartridge. A total of three operations need to be performed:

(1) Cut out disks from magnetic film. The magnetic film is bought in rolls which cost $90 each. From each roll, 2,000 circular disks can be cut out. One person is needed to operate and supervise the cut-out machine. Installing a new roll takes 8 minutes and cutting out 2,000 circular disks takes 25 minutes.

(2) Apply disk control center-pieces. The disk control center-pieces cost $0.12 per unit. One person is required to apply the center-pieces to the magnetic disks. Applying the first center-piece takes 30 seconds, and for the remaining center-pieces an 80% learning curve is applicable.

(3) Insert into plastic cartridges. The plastic cartridges cost $0.15 per unit. One person is needed to supervise the disk-insertion operation. This operation is done automatically by a machine which can insert 1,500 disks per hour. The film, center-pieces, and cartridges are purchased from an outside manufacturer. A total of 10,000 disks are to be produced. Other relevant cost data are:

The direct labor rate is $15.00 per hour.
Planning labor is 15% of factory labor.
Quality control is 30% of factory labor.
Factory overhead is 800% of total labor.
General and administrative expenses are 50% of total labor.
Packing costs are 100% of total labor.
The profit margin is 15% of total manufacturing cost.

a. Based on this information, estimate the unit selling price for a single disk. (8.4)

b. Compute the target cost when the best competitor's selling price is $0.60 per disk and a 15% return on sales is desired. (8.4)

c. Investigate and report on any cost reduction alternatives that can be implemented to reach the target cost. (8.4)

8.22. Brain Teaser

You have been asked to prepare a quick estimate of the construction cost for a coal-fired electricity generating plant and facilities. A work breakdown structure (levels one through three) is shown on page 367. You have the following information available.

A coal-fired generating plant twice the size of the one you are estimating was built in 1977. The 1977 boiler (1.2) and boiler support system (1.3) cost $110 million. The cost index for boilers was 110 in 1977; it is 492 in 1997. The cost capacity factor for similar boilers and support systems is 0.9.

The 600 acre site is on property you already own, but improvements (1.1.1) and roads (1.1.2), will cost $2,000 per acre and railroads (1.1.3) will cost $3,000,000. Project integration (1.9) is projected to cost 3% of all other construction costs.

The security systems (1.5.4) are expected to cost $1,500 per acre, based on recent (1997) construction of similar plants. All other support facilities and equipment (1.5) elements are to be built by Viscount Engineering. Viscount Engineering has built the support facilities and equipment elements for two similar generating plants. Their experience is expected to reduce labor requirements substantially; a 90% learning curve can be assumed. Viscount built the support facilities and equipment on their first job in 95,000 hours. For this project, Viscount's labor will be billed to you at $60/hour. Viscount estimates materials for the construction of the support facilities and equipment elements (except 1.5.4) will cost you $15,000,000.

The coal storage facility (1.4) for the coal-fired generating plant built in 1977 cost $5 million. Although your plant is smaller, you require the same size coal storage facility as the 1977 plant. You assume you can apply the cost index for similar boilers to the coal storage facility.

What is your estimated 1997 cost for building the coal-fired generating facility? Summarize your calculations in a cost estimating worksheet, and state the assumptions you make.

PROJECT: Coal-Fired Electricity Generating Plant and Facilities   (Pr. 8-22)

Line No.	Title	WBS Element Code
001	Coal-Fired Power Plant	1.
002	Site	1.1
003	Land Improvements	1.1.1
004	Roads, Parking, and Paved Areas	1.1.2
005	Railroads	1.1.3
006	Boiler	1.2
007	Furnace	1.2.1
008	Pressure Vessel	1.2.2
009	Heat Exchange System	1.2.3
010	Generators	1.2.4
011	Boiler Support System	1.3
012	Coal Transport System	1.3.1
013	Coal Pulverizing System	1.3.2
014	Instrumentation & Control	1.3.3
015	Ash Disposal System	1.3.4
016	Transformers & Distribution	1.3.5
017	Coal Storage Facility	1.4
018	Stockpile Reclaim System	1.4.1
019	Rail Car Dump	1.4.2
020	Coal Handling Equipment	1.4.3
021	Support Facilities & Equipment	1.5
022	Hazardous Waste Systems	1.5.1
023	Support Equipment	1.5.2
024	Utilities & Communications System	1.5.3
025	Security Systems	1.5.4
026	Project Integration	1.9
027	Project Management	1.9.1
028	Environmental Management	1.9.2
029	Project Safety	1.9.3
030	Quality Assurance	1.9.4
031	Test, Start-Up, & Transition Management	1.9.5

Inflation and Price Changes

*W*hen the monetary unit does not have a constant value in exchange for goods and services in the marketplace, and when future price changes are expected to be significant, an undesirable choice among competing alternatives can be made if price change effects are not included in an engineering economic analysis (before taxes and after taxes). The objectives of this chapter are to (1) introduce a methodology for dealing with inflation and price changes, (2) develop and illustrate proper techniques to account for these effects in engineering economic analysis, and (3) discuss the relationship of these concepts to foreign exchange rates.

The following topics are discussed in this chapter:

Terminology and basic concepts

The relationship between actual (current) dollars and real (constant) dollars

Use of combined (nominal) versus real interest rates

Differential price inflation

Modeling price changes with geometric sequences of cash flows

Application strategy for use of actual and real dollar analysis

A comprehensive example

Foreign exchange rates

9.1 General Price Inflation

In earlier chapters, we assumed that prices for goods and services in the marketplace remain relatively unchanged over extended periods of time. Unfortunately, this is not generally a realistic assumption.

General price inflation, which is defined here as an increase in the prices paid for goods and services bringing about a reduction in the purchasing power of the monetary unit, is a business reality that can affect the economic comparison of alternatives. The history of price changes shows that price inflation is much more common than general price *deflation,* which involves a decrease in prices with an increase in the purchasing power of the monetary unit. The concepts and methodology discussed in this chapter, however, apply to any price changes.

One measure of price changes in our economy (and an estimate of general price inflation) is the Consumer Price Index (CPI). Tabulated by the U.S. government, the CPI is a composite price index that measures price changes in food, shelter, medical care, transportation, apparel, and other selected goods and services used by individuals and families. As shown in Figure 9-1, the CPI increased from an annual average of 38.8 in 1970 to 148.2 in 1994. For the *base period* 1982–84, the CPI was 100, the average for that period; that is, $(96.5 + 99.6 + 103.9)/3 = 100$. The CPI and its related annual inflation rates for the period 1970–94 are listed in Table 9-1. The annual inflation rates were calculated as follows:

$$(\text{CPI annual inflation rate})_k = \frac{(\text{CPI})_k - (\text{CPI})_{k-1}}{(\text{CPI})_{k-1}}(100)$$

For example, the CPI annual inflation rate for 1994 is

$$\frac{(\text{CPI})_{1994} - (\text{CPI})_{1993}}{(\text{CPI})_{1993}} = \frac{148.2 - 144.5}{144.5}(100) = 2.56\%$$

In addition to the CPI, the federal government develops other price indexes from the data it gathers and analyzes. The primary agencies involved in the development of price indexes are the Department of Labor (Bureau of Labor

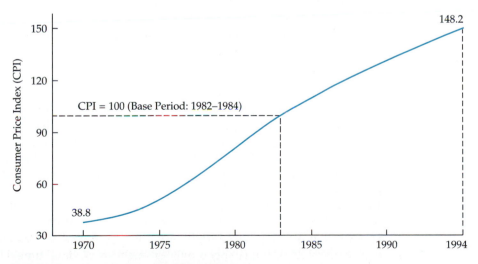

FIGURE 9-1 Annual Average CPI, 1970–94

TABLE 9-1   CPI and Related Annual Inflation Rate (%), 1970–1994

Year	Annual Average CPI	CPI Annual Inflation Rate (%)	Year	Annual Average CPI	CPI Annual Inflation Rate (%)
1970	38.8	5.72	1983	99.6	3.21
1971	40.5	4.38	1984	103.9	4.32
1972	41.8	3.21	1985	107.6	3.56
1973	44.4	6.22	1986	109.6	1.86
1974	49.3	11.04	1987	113.6	3.65
1975	53.8	9.13	1988	118.3	4.14
1976	56.9	5.76	1989	124.0	4.82
1977	60.6	6.50	1990	130.7	5.40
1978	65.2	7.59	1991	136.2	4.21
1979	72.6	11.35	1992	140.3	3.01
1980	82.4	13.50	1993	144.5	2.99
1981	90.9	10.32	1994	148.2	2.56
1982	96.5	6.16			

SOURCE: CPI Detailed Report (January, 1995), U.S. Department of Labor, Bureau of Labor Statistics (Washington, D.C., U.S. Government Printing Office)

Statistics) and the Department of Commerce (Bureau of Economic Analysis). The Producer Price Index (PPI) and the Implicit Price Index for the Gross National Product (IPI-GNP) are two indexes, in addition to the CPI, that are often used to estimate general price inflation. These indexes are based on historical data but can be used for short-term forecasting. Longer-term forecasts of price changes, including information on specific goods and services, may be purchased from private firms engaged in the business of providing economic forecasting services.

9.2 Terminology and Basic Concepts

To facilitate the development and discussion of the methodology for including price changes of goods and services in engineering economy studies, we need to define and discuss some terminology and basic concepts. The dollar is used as the monetary unit in this book except when discussing foreign exchange rates.

1. *Actual dollars (A$):* The number of dollars associated with a cash flow (or a noncash flow amount such as depreciation) as of the time it occurs. For example, people typically anticipate their salaries two years in advance in terms of actual dollars. Sometimes A$ are referred to as *current* dollars, *then-current* dollars, and *inflated* dollars, and their relative purchasing power is affected by general price inflation.
2. *Real dollars (R$):* Dollars expressed in terms of the same purchasing power relative to a particular time. For instance, the future unit prices of goods or services that are changing rapidly are often estimated in real dollars (relative to some base year) to provide a consistent means of comparison. Sometimes R$ are termed *constant* dollars.

3. *General price inflation rate (f):* A measure of the change in the purchasing power of a dollar during a specified period of time. The general price inflation rate (applies also to a deflation situation) is defined by a selected, broadly based index of market price changes. In engineering economic analysis, the rate is projected for a future time interval and usually is expressed as an effective annual rate. Many large organizations have their own selected index that reflects the particular business environment in which they operate.

4. *Combined (nominal) interest rate (ic):* The money paid for the use of capital, normally expressed as a nominal rate (%) that includes a market adjustment for the anticipated general price inflation rate in the economy. Thus, it is a *market interest rate* and represents the time value change in future cash flows that takes into account both the potential real earning power of money and the estimated general price inflation in the economy.

5. *Real interest rate (ir):* The money paid for the use of capital, normally expressed as a nominal rate (%) that does *not* include a market adjustment for the anticipated general price inflation rate in the economy. It represents the time value change in future cash flows based only on the potential real earning power of money. It is sometimes called the *inflation-free* interest rate.

6. *Base time period (b):* The reference or base time period used to define the purchasing power of real (constant) dollars. Often, in practice, the base time period is designated as the time of the engineering economic analysis, or reference time 0 (i.e., $b = 0$). However, b can be any designated point in time.

With an understanding of these definitions, we can delineate and illustrate some useful relationships that are important in engineering economic analysis.

9.2.1 The Relationship Between Actual Dollars and Real Dollars

> The relationship between actual dollars (A$) and real dollars (R$) is defined in terms of the general price inflation rate; that is, it is a function of f.

Actual dollars as of any point in time, k, can be converted into real dollars of constant market purchasing power as of any base time period, b, by the following relationship

$$(R\$)_k = (A\$)_k \left(\frac{1}{1+f}\right)^{k-b} = (A\$)_k(P/F, f\%, k - b) \tag{9-1}$$

for a given b value. This relationship between actual dollars and real dollars applies to the unit prices, or costs of fixed amounts of individual goods or services, used to develop the individual cash flows. The designation for a specific type of cash flow, j, would be included as follows:

$$(R\$)_{k,j} = (A\$)_{k,j} \left(\frac{1}{1+f}\right)^{k-b} = (A\$)_{k,j}(P/F, f\%, k - b) \tag{9-2}$$

for a given b value, where the terms R$\$_{k,j}$ and A$\$_{k,j}$ are the unit price, or cost for a fixed amount, of good or service j in time period k in real dollars and actual dollars, respectively.

EXAMPLE 9-1

Suppose that your salary is $35,000 in year one, will increase at 6% per year through year four, and is expressed in actual dollars as follows:

End of Year, k	Salary (A$)
1	$35,000
2	37,100
3	39,326
4	41,685

If the general price inflation rate (f) is expected to average 8% per year, what is the real dollar equivalent of these actual dollar salary amounts? Assume that the base time period is year one $(k = 1)$.

SOLUTION

By using Equation 9-2, the real dollar salary equivalents are readily calculated for the base point in time, $b = 1$:

Year	Salary (R$ in Year 1)		
1	$35,000(P/F, 8\%, 0)$	=	$35,000
2	$37,100(P/F, 8\%, 1)$	=	34,351
3	$39,326(P/F, 8\%, 2)$	=	33,714
4	$41,685(P/F, 8\%, 3)$	=	33,090

In year one (the designated base time period for the analysis), the annual salary in actual dollars remained unchanged when converted to real dollars. *This illustrates an important point: In the base time period (b), the purchasing power of an actual dollar and a real dollar is the same; that is,* $R\$_{b,j} = A\$_{b,j}$. This example also illustrates the results when the actual annual rate of increase in salary (6% in this example) is less than the general price inflation rate (f). As you can see, the actual dollar salary cash flow shows a reasonable increase, but a decrease in the real dollar salary cash flow occurs (and thus a decrease in market purchasing power).

EXAMPLE 9-2

An engineering project team is analyzing the potential expansion of an existing production facility. Different design alternatives are being considered. The estimated after-tax cash flow (ATCF) in actual dollars for one alternative is shown in column 2 of Table 9-2. If the general price inflation rate (f) is estimated to be 5.2% per year during the eight-year analysis period, what is the real dollar ATCF equivalent to the actual dollar ATCF? The base time period is year zero $(b = 0)$.

TABLE 9-2	ATCFs for Example 9-2		
(1) End-of-Year, k	(2) ATCF (A$)	(3) $(P/F, f\%, k - b)$ $= [1/(1.052)^{k-0}]$	(4) ATCF (R$), $b = 0$
0	−$172,400	1.0	−$172,400
1	− 21,000	0.9506	− 19,963
2	51,600	0.9036	46,626
3	53,000	0.8589	45,522
4	58,200	0.8165	47,520
5	58,200	0.7761	45,169
6	58,200	0.7377	42,934
7	58,200	0.7013	40,816
8	58,200	0.6666	38,796

SOLUTION

The application of Equation 9-1 is shown in column 3 of Table 9-2. The ATCF in real dollars shown in column 4 has purchasing power in each year equivalent to the original ATCF in actual dollars (column 2).

9.2.2 The Correct Interest Rate to Use in Engineering Economy Studies

In general, the interest rate that is appropriate for equivalence calculations in engineering economy studies depends on the type of cash flow estimates:

Method	If Cash Flows Are in Terms of	Then the Interest Rate to Use Is
A	Actual dollars (A$)	Combined interest rate, i_c
B	Real dollars (R$)	Real interest rate, i_r

This table should make intuitive sense as follows. If one is estimating cash flows in terms of actual (inflated) dollars, the combined interest rate (market interest rate with inflation component) is used. Similarly, if one is estimating cash flows in terms of real dollars, the real (inflation-free) interest rate is used. Thus, one can make economic analyses in either the actual or real dollar domains with equal validity, provided that the appropriate interest rate is used for equivalence calculations.

It is important to be consistent in using the correct interest rate for the type of analysis (actual or real dollars) being done. The two mistakes commonly made are as follows:

Interest Rate (MARR)	Type of Analysis	
	A$	R$
i_c	(Correct)	Mistake 1 Bias is against capital investment
i_r	Mistake 2 Bias is toward capital investment	(Correct)

In Mistake 1 the combined interest rate (i_c), which includes an adjustment for the general price inflation rate (f), is used in equivalent worth calculations for cash flows estimated in real dollars. Because real dollars have constant purchasing power expressed in terms of the base time period (b) and do not include the effect of general price inflation, we have an inconsistency. There is a tendency to develop future cash flow estimates in terms of dollars with purchasing power at the time of the study (that is, real dollars with $b = 0$), and then to use the combined interest rate in the analysis. The result of Mistake 1 is a bias against capital investment. The cash flow estimates in real dollars for a project are numerically lower in value than actual dollar estimates with equivalent purchasing power (assuming that $f > 0$). Additionally, the i_c value (which is greater than the i_r value that should be used) further reduces (understates) the equivalent worth of the results of a proposed capital investment.

In Mistake 2 the cash flow estimates are in actual dollars, which include the effect of general price inflation (f), but the real interest rate (i_r) is used for equivalent worth calculations. Since the real interest rate does not include an adjustment for general price inflation, we again have an inconsistency. The effects of this mistake, in contrast to those in Mistake 1, result in a bias toward capital investment by overstating the equivalent worth of the future results.

9.2.3 The Relationship Among the Combined and Real Interest Rates and the General Inflation Rate

Equation 9-1 showed that the relationship between an actual dollar amount and a real dollar amount of equal purchasing power at time k is a function of the general inflation rate (f). It is desirable to do engineering economy studies in terms of either actual dollars or real dollars. Thus, the relationship between the two dollar domains is important, as well as the relationship among i_c, i_r, and f, so that the equivalent worth of a cash flow is equal in the base time period when either an actual or a real dollar analysis is used.

Consider a future cash flow in actual dollars at time k, $(A\$)_k$. The equivalent worth (EW) of this amount in the base time period (b), using the combined interest rate, is

$$(EW)_b = (A\$)_k \frac{1}{(1 + i_c)^{k-b}}$$

The future cash flow amount in real dollars at time k, $(R\$)_k$, with the same purchasing power amount is

$$(R\$)_k = (A\$)_k \frac{1}{(1 + f)^{k-b}}$$

The equivalent worth of this real dollar amount in the base time period, using the real interest rate (i_r), is

$$(EW)_b = (R\$)_k \frac{1}{(1 + i_r)^{k-b}} = (A\$)_k \frac{1}{(1 + f)^{k-b}} \left[\frac{1}{(1 + i_r)^{k-b}} \right]$$

Because these two equivalent worths in the base time period have to equal each other, the following relationship among i_c, i_r, and f results:

$$(A\$)_k \frac{1}{(1 + i_c)^{k-b}} = (A\$)_k \frac{1}{(1 + f)^{k-b}} \left[\frac{1}{(1 + i_r)^{k-b}} \right]$$

And,

$$(1 + i_c)^{k-b} = [(1 + f)(1 + i_r)]^{k-b}$$

$$1 + i_c = (1 + f)(1 + i_r) \qquad (9\text{-}3)$$

$$i_c = i_r + f + i_r(f) \qquad (9\text{-}4)$$

$$i_r = \frac{i_c - f}{1 + f} \qquad (9\text{-}5)$$

Thus, the combined interest rate (Equation 9-4) is the sum of the real interest rate (i_r) and the general price inflation rate (f), plus the product of those two terms. Also, as shown in Equation 9-5, the real interest rate (i_r) can be calculated from the combined interest rate and the general price inflation rate. Similarly, based on Equation 9-5, the IRR of a real dollar cash flow is related to the IRR of an actual dollar cash flow (with the same purchasing power each period) as follows: $\text{IRR}_r = (\text{IRR}_c - f)/(1 + f)$.

EXAMPLE 9-3

If a company borrowed $100,000 today to be repaid at the end of three years at a combined (market) interest rate of 11%, what is the actual dollar amount owed at the end of three years, the real rate of return to the lender, and the real dollar amount equivalent in purchasing power to the actual dollar amount at the end of the third year? Assume that the base or reference year is now ($b = 0$), and the general price inflation rate (f) is 5% per year.

SOLUTION

In three years, the company will owe the original $100,000 plus interest that has accumulated, in actual dollars:

$$(A\$)_3 = (A\$)_0 (F/P, i_c\%, 3) = \$100,000(F/P, 11\%, 3) = \$136,763$$

With this payment, the actual internal rate of return, IRR_c, to the lender is 11%. Thus, the real rate of return to the lender can be calculated based on Equation 9-5:

$$\text{IRR}_r = \frac{0.11 - 0.05}{1.05} = 0.05714, \text{ or } 5.714\%$$

The real interest rate in this example is the same as IRR_r. Using this value for i_r, the real dollar amount equivalent in purchasing power to the actual dollar amount owed is

$$(R\$)_3 = (R\$)_0 (F/P, i_r\%, 3) = \$100,000(F/P, 5.714\%, 3) = \$118,140$$

This amount can be verified by the following calculation based on Equation 9-1:

$$(R\$)_3 = (A\$)_3(P/F, f\%, 3) = \$136,763(P/F, 5\%, 3) = \$118,140$$

EXAMPLE 9-4

In Example 9-1 your salary was projected to increase at the rate of 6% per year, and the general price inflation rate was expected to be 8% per year. Your resulting estimated salary for the four years in actual and real dollars was as follows:

End of Year, k	Salary (A$)	Salary (R$), $b = 1$
1	$35,000	$35,000
2	37,100	34,351
3	39,326	33,714
4	41,685	33,090

What is the equivalent worth (EW) of the four-year actual and real dollar salary cash flows at the end of year one (base year) if your personal MARR is 10% per year (i_c)?

SOLUTION
(a) Actual dollar salary cash flow:

$$EW(10\%)_1 = \$35,000 + 37,100(P/F, 10\%, 1) + 39,326(P/F, 10\%, 2) + 41,685(P/F, 10\%, 3)$$
$$= \$132,545$$

(b) Real dollar salary cash flow:

$$i_r = \frac{i_c - f}{1 + f} = \frac{0.10 - 0.08}{1.08} = 0.01852, \text{ or } 1.852\%$$

$$EW(1.852\%)_1 = \$35,000 + 34,351\left(\frac{1}{1.01852}\right)^1 + 33,714\left(\frac{1}{1.01852}\right)^2 + 33,090\left(\frac{1}{1.01852}\right)^3$$
$$= \$132,545$$

Thus, we obtain the same equivalent worth at the end of year one (the base time period) for both the actual dollar and real dollar four-year salary cash flows when the appropriate interest rate is used for the equivalence calculations.

9.2.4 Fixed and Responsive Annuities

Whenever future cash flows are predetermined by contract, as in the case of a bond or a fixed annuity, these amounts do not respond to general price inflation. In cases where the future amounts are not predetermined, however, they may respond to general price inflation. The degree of response varies from case to case.

TABLE 9-3   Illustration of Fixed and Responsive Annuities with General Price Inflation Rate of 6% per Year

| End of Year k | Fixed Annuity | | Responsive Annuity | |
	In Actual Dollars	In Equivalent Real Dollars[a]	In Actual Dollars	In Equivalent Real Dollars[a]
1	$2,000	$1,887	$2,120	$2,000
2	2,000	1,780	2,247	2,000
3	2,000	1,679	2,382	2,000
4	2,000	1,584	2,525	2,000
5	2,000	1,495	2,676	2,000
6	2,000	1,410	2,837	2,000
7	2,000	1,330	3,007	2,000
8	2,000	1,255	3,188	2,000
9	2,000	1,184	3,379	2,000
10	2,000	1,117	3,582	2,000

[a]See Equation 9-1

To illustrate the nature of this situation, let us consider two annuities. The first annuity is fixed (unresponsive to general price inflation) and yields $2,000 per year for 10 years. The second annuity is of the same duration and yields enough future dollars to be equivalent to $2,000 per year in real purchasing power. Assuming a general price inflation rate of 6% per year, pertinent values for the two annuities over a 10-year period are as shown in Table 9-3.

Thus, when the amounts are constant in actual dollars (unresponsive to general price inflation), their equivalent amounts in real dollars decline over the 10-year interval to $1,117 in the final year. When the future cash flow amounts are fixed in real dollars (responsive to general price inflation), their equivalent amounts in actual dollars rise to $3,582 by year 10.

> Included in engineering economy studies are certain quantities unresponsive to general price inflation, such as depreciation, or lease fees and interest charges based on an existing contract or loan agreement. For instance, depreciation write-offs, once determined, do not increase (with present accounting practices) to keep pace with general price inflation; lease fees and interest charges typically are contractually fixed for a given period of time. Thus, it is important when doing an actual dollar analysis to recognize the quantities that are unresponsive to general price inflation, and when doing a real dollar analysis to convert these A$ quantities to R$ quantities using Equation 9-2.
>
> If this is not done, all cash flows will not be in the same dollar domain (A$ or R$), and the analysis results will be distorted. Specifically, the equivalent worths of the cash flows for an A$ and a R$ analysis will not be the same in the base year, b, and the A$ IRR and the R$ IRR for the project will not have the proper relationship based on Equation 9-5; that is, $IRR_r = (IRR_c - f)/(1 + f)$.

9.2.5 The Impact of General Price Inflation on After-Tax Analysis

Engineering economy studies that include the effects of general price inflation present some difficulties because interest charges, depreciation deductions, and lease payments are actual dollar amounts based on past commitments. They are generally *unresponsive* to inflation. At the same time, many other types of cash flows (e.g., labor, materials) are *responsive* to inflation. Even though both types of amounts are usually present in a study, the effects of general price inflation can be readily included in an engineering economic analysis. In Example 9-5, an *after-tax analysis* that includes the effects of general price inflation is presented to show the correct handling of these considerations.

EXAMPLE 9-5

The cost of some newly installed and more efficient electrical circuit switching equipment is $180,000. It is estimated (in base year dollars, $b = 0$) that the equipment will reduce current net operating expenses by $36,000 per year for 10 years and will have a $30,000 market value at the end of the tenth year. These cash flows are estimated to increase at the general price inflation rate. Due to new computer control features on the equipment, it will be necessary to contract for some maintenance support during the first three years. The maintenance contract will cost $2,800 per year. This equipment will be depreciated under the MACRS (GDS) method, and it is in the five-year property class. The effective income tax rate (t) is 38%; the selected analysis period is 10 years; the projected general price inflation rate (f) is 8% per year; and the MARR (after taxes) is $i_c = 15\%$ per year.

(a) Based on an actual dollar, after-tax analysis, is this capital investment justified?

(b) Develop the after-tax cash flow in real dollars.

SOLUTION

(a) The actual dollar, after-tax economic analysis of the new equipment is shown in Table 9-4 (columns 1–9). The capital investment, savings in operating expenses, and market value (in the tenth year), which were estimated in real dollars with $b = 0$ (column 1), are adjusted to actual dollars (columns 2 and 3) using the general price inflation rate and Equation 9-1. The maintenance contract amounts for the first three years (column 4) are already in actual dollars (they are unresponsive to general price inflation). The algebraic sum of columns 3 and 4 equals the before-tax cash flow (BTCF) in actual dollars (column 5).

In columns 6, 7, and 8, the depreciation and income tax calculations are shown. The depreciation deductions in column 6 are based on the MACRS (GDS) method and, of course, are in actual dollars. The entries in columns 7 and 8 are calculated as discussed in Chapter 7. The effective income tax rate (t) is 38% as given. The entries in column 8 are equal to the entries in column 7 multiplied by $-t$. The algebraic sum of columns 5 and 8 equals the ATCF in actual dollars (column 9). The present worth of the actual dollar ATCF, using $i_c = 15\%$ per year, is

TABLE 9-4 Example 9-5 When the General Price Inflation Rate Is 8% Per Year

End of Year (k)	(1) R$ Cash Flow	(2) A$ Adjustment $(1+f)^{k-b}$	(3) A$ Cash Flows	(4) Contract (A$)	(5) BTCF (A$)	(6) Depreciation (A$)	(7) Taxable Income	(8) Income Taxes (t = 0.38)	(9) ATCF (A$)	(10) R$ Adjustment $[1/(1+f)]^{k-b}$	(11) ATCF (R$)
0	-$180,000	1.0000	-$180,000		-$180,000				-$180,000	1.0000	-$180,000
1	36,000	1.0800	38,880	-$2,800	36,080	$36,000	$80	-$ 30	36,050	0.9259	33,379
2	36,000	1.1664	41,990	- 2,800	39,190	57,600	- 18,410	+ 6,996	46,186	0.8573	39,595
3	36,000	1.2597	45,349	- 2,800	42,549	34,560	- 7,989	- 3,036	39,513	0.7938	31,366
4	36,000	1.3605	48,978		48,978	20,736	28,242	- 10,732	38,246	0.7350	28,111
5	36,000	1.4693	52,895		52,895	20,736	32,159	- 12,220	40,675	0.6806	27,683
6	36,000	1.5869	57,128		57,128	10,368	46,760	- 17,769	39,359	0.6302	24,804
7	36,000	1.7138	61,697		61,697		61,697	- 23,445	38,252	0.5835	22,320
8	36,000	1.8509	66,632		66,632		66,632	- 25,320	41,312	0.5403	22,320
9	36,000	1.9990	71,964		71,964		71,964	- 27,346	44,618	0.5003	22,320
10	36,000	2.1589	77,720		77,720		77,720	- 29,534	48,186	0.4632	22,320
10	30,000	2.1589	64,767		64,767		64,767	- 24,611	40,156	0.4632	18,600

$$\text{PW}(15\%) = -\$180,000 + \$36,050(P/F, 15\%, 1) + \cdots + \$40,156(P/F, 15\%, 10)$$
$$= \$33,790$$

Therefore, the project is economically justified.

(b) Next, Equation 9-1 is used to calculate the ATCF in real dollars from the entries in column 9. The real dollar ATCF (column 11) shows the estimated economic consequences of the new equipment in dollars that have the constant purchasing power of the base year. The actual dollar ATCF (column 9) is in dollars that have the purchasing power of the year in which the cost or saving occurs. The comparative information provided by the ATCF in both actual dollars and real dollars is helpful in interpreting the results of an economic analysis. Also, as illustrated in this example, the conversion between actual dollars and real dollars can easily be done. The PW of the real dollar ATCF (column 11) using $i_r = (i_c - f)/(1 + f) = (0.15 - 0.08)/1.08 = 0.06481$, or 6.48%, is

$$\text{PW}(6.48\%) = -\$180,000 + \$33,379(P/F, 6.48\%, 1) + \cdots + \$18,600(P/F, 6.48\%, 10)$$
$$= \$33,790$$

Thus, the PW (equivalent worth in the base year with $b = 0$) of the real dollar ATCF is the same as the present worth calculated previously for the actual dollar ATCF.

9.2.5.1 Variation in the General Price Inflation Rate

In Example 9-5, the general price inflation rate (f) was projected to be 8% each year during the 10-year analysis period. In the case where the estimated annual rates vary during the analysis period, these varying rates would be applied successively to the costs and revenues for the years involved. For example, assume that the annual rates for Example 9-5 were estimated to vary as shown in column 1 of Table 9-5. Then the operating cost savings and market value originally estimated in real dollars (column 2), would be adjusted to actual dollars (column 4) by successive application of the annual rates, as shown in column 3 (where the symbol $\prod$ means the product).

9.2.5.2 Calculating an Effective General Price Inflation Rate

In Table 9-5, the projected general price inflation rates varied during the 10-year analysis period. Suppose that these rates are the best estimates of future price inflation in your company. However, for a study of a small capital investment project, you consider the successive application of the varying annual rates in the analysis to be a refinement not justified by the situation. In this case, the analysis can be simplified by using an effective rate ($\bar{f}$) in the same way that $f = 8\%$ was used in the original solution to Example 9-5. Assume that the analysis period is 10 years for the small project. The calculation of $\bar{f}$ (based on the entries in column 1 of Table 9-5) would be

$$\bar{f} = \left[\prod_{k=1}^{N}(1 + f_k)\right]^{1/N} - 1 = \left[\prod_{k=1}^{10}(1 + f_k)\right]^{1/10} - 1 \tag{9-6}$$

$$= [(1.04)^1(1.055)^2(1.07)^3(1.08)^4]^{0.1} - 1 = (1.9292)^{0.1} - 1$$

$$= 0.0679, \text{ or } 6.79\%$$

TABLE 9-5 Variation in the General Price Inflation Rate

End of Year, k	(1) General Price Inflation Rate (f_k)	(2) R\$ Cash Flows	(3) A\$ Adjustment $\left[\prod_{l=1}^{k}(1+f_l)\right]$; for $b=0$	(4) A\$ Cash Flows
0	—	−\$180,000		−\$180,000
1	4.0	36,000	1.04	37,440
2	5.5	36,000	$(1.04)(1.055) = 1.0972$	39,499
3	5.5	36,000	$(1.04)(1.055)^2 = 1.1576$	41,672
4	7.0	36,000	$(1.04)(1.055)^2(1.07) = 1.2386$	44,589
5	7.0	36,000	$(1.04)(1.055)^2(1.07)^2 = 1.3253$	47,710
6	7.0	36,000	$(1.04)(1.055)^2(1.07)^3 = 1.4180$	51,050
7	8.0	36,000	$(1.04)(1.055)^2(1.07)^3(1.08) = 1.5315$	55,134
8	8.0	36,000	$(1.04)(1.055)^2(1.07)^3(1.08)^2 = 1.6540$	59,544
9	8.0	36,000	$(1.04)(1.055)^2(1.07)^3(1.08)^3 = 1.7893$	64,308
10⎤ 10⎦	8.0	36,000⎤ 30,000⎦	$(1.04)(1.055)^2(1.07)^3(1.08)^4 = 1.9292$	69,440 57,876

If this approach were applied to the original calculations in Table 9-5, the entries in column 4 would be slightly different for years one through nine. However, the operating cost savings in the tenth year would be $36,000(1.0679)^{10} = $69,440$, the same value calculated by successive application of the varying annual rates originally used in the table.

9.3 Differential Price Inflation or Deflation

The general price inflation rate (f) may not be the best estimate of future price changes for one or more cost and revenue cash flows in an engineering economy study. The variation between the general price inflation rate and the best estimate of future price changes for specific goods and services is called *differential price inflation* (or *deflation*), and it is caused by factors such as technological improvements, and changes in productivity, regulatory requirements, and so on. Also, a restriction in supply, an increase in demand, or a combination of both may change the value of a particular good or service relative to others. Price changes caused by some combination of general price inflation and differential price inflation can be represented by a *total price escalation* (or *de-escalation*) *rate*. These rates are further defined as follows:

1. Differential price inflation (or deflation) rate (e_j'): the increment (%) of price change (in the unit price, or cost for a fixed amount), above or below the general price inflation rate, during a time period (normally a year) for good or service j.
2. Total price escalation (or de-escalation) rate (e_j): The total rate (%) of price change (in the unit price, or cost for a fixed amount) during a time period

(normally a year), for good or service j. The total price escalation rate for a good or service includes the effects of both the general price inflation rate (f) and the differential price inflation rate (e_j') on price changes.

9.3.1 Relationship Among the Total Price Escalation Rate, the Differential Price Inflation Rate, and the General Inflation Rate

The differential price inflation rate (e_j') is a price change in good or service j in *real dollars* caused by various factors in the marketplace. Similarly, the total price escalation rate (e_j) is a price change in *actual dollars*. Consider the unit price, or cost for a fixed amount, of good or service j in the base time period, $(R\$)_{b,j} = (A\$)_{b,j}$. (Remember, in the base time period, the purchasing power of a real dollar and that of an actual dollar are equal. This is the only time period in which this is true if the general inflation rate is not equal to zero.) Then, the cost in real dollars in time period k for good or service j is

$$(R\$)_{k,j} = (R\$)_{b,j}(1 + e_j')^{k-b}$$

And, from Equation 9-1, the cost in actual dollars in time period k is

$$(A\$)_{k,j} = (R\$)_{k,j}(1 + f)^{k-b} = (R\$)_{b,j}(1 + e_j')^{k-b}(1 + f)^{k-b}$$

However, the cost in actual dollars in time period k is also a function of the total price escalation rate (e_j):

$$(A\$)_{k,j} = (A\$)_{b,j}(1 + e_j)^{k-b} = (R\$)_{b,j}(1 + e_j)^{k-b}$$

Since these two prices in actual dollars for good or service j in time period k have to equal each other, the following relationship among e_j, e_j', and f results:

$$(R\$)_{b,j}(1 + e_j)^{k-b} = (R\$)_{b,j}(1 + e_j')^{k-b}(1 + f)^{k-b}$$

$$(1 + e_j)^{k-b} = [(1 + e_j')(1 + f)]^{k-b}$$

$$1 + e_j = (1 + e_j')(1 + f) \tag{9-7}$$

$$e_j = e_j' + f + e_j'(f) \tag{9-8}$$

$$e_j' = \frac{e_j - f}{1 + f} \tag{9-9}$$

Thus, as shown in Equation 9-8, the total price escalation rate (e_j) for good or service j is the sum of the general price inflation rate and the differential price inflation rate plus their product. Also, as shown in Equation 9-9, the differential price inflation rate (e_j') can be calculated from the total price escalation rate and the general price inflation rate.

In practice, the general price inflation rate (f) and the total price escalation rate (e_j) for each good and service involved are usually estimated for the study period. For each of these rates, different values may be used for subsets of periods within the analysis period if justified by the available data. This was illustrated in Table 9-5 for the general price inflation rate. The differential price inflation rates (e_j'), when needed, normally are not estimated directly but are calculated using Equation 9-9.

In applying total price escalation rates in an actual dollar analysis to calculate future prices, remember that these rates are effective rates per period and have a compounding effect over time. Equation 9-10 relates the total price escalation rate (e_j) with the single payment compound amount factor (when the e_j value for a good or service does not vary by time period):

$$(A\$)_{k,j} = (A\$)_{b,j}(F/P, e_j\%, k - b) \tag{9-10}$$

where $(A\$)_{k,j}$ is the projected actual dollars in time period k for cash flow j, and $(A\$)_{b,j}$ is the actual dollars in the base time period, b. Similarly, in a real dollar analysis, we would use the differential price inflation rate (e_j') to calculate future amounts for a good or service:

$$(R\$)_{k,j} = (R\$)_{b,j}(F/P, e_j'\%, k - b) \tag{9-11}$$

Alternatively, we could convert projected actual dollar prices, if available, to real dollars using Equation 9-2. Also, if the e_j and e_j' values for a good or service vary during the study period, Equations 9-10 and 9-11 are used *within each subperiod* with the applicable values. When this is done, the F/P factor is applied to the last estimate in the previous subperiod.

EXAMPLE 9-6

The prospective maintenance expenses for a heating, ventilating, and air-conditioning system are estimated to be $12,200 per year in base year dollars (assume that $b = 0$). The total price escalation rate is estimated to be 7.6% for the next three years $(e_{1,2,3} = 7.6\%)$, and for years four and five it is estimated to be 9.3% $(e_{4,5} = 9.3\%)$. The general price inflation rate (f) for this five-year period is estimated to be 4.7% per year. Develop the maintenance expense estimates for years one through five in actual dollars and in real dollars, using e_j and e_j' values, respectively.

SOLUTION
The development of the annual maintenance expenses in actual dollars is shown in column 2 of Table 9-6. In this example, the general price inflation rate is not the best estimate of changes in future maintenance expenses. The five-year period is divided into two subperiods corresponding to the two different price escalation rates $(e_{1,2,3} = 7.6\%; e_{4,5} = 9.3\%)$. These rates are then used with the estimated expenses in the base year; $(A\$)_0 = (R\$)_0 = \$12,200$.

The development of the maintenance expenses in real dollars is shown in column 4. This development is the same as for actual dollars, except that the e_j' values (Equation 9-9) are used instead of the e_j values. The e_j' values in this example are as follows:

TABLE 9-6   Example 9-6 Calculations

(1) End of Year, k	(2) A\$ Adjustment (e_j)	(3) Maintenance Expenses, A\$	(4) R\$ Adjustment (e'_j)	(5) Maintenance Expenses, R\$
1	$-\$12,200(1.076)^1$	$-\$13,127$	$-\$12,200(1.0277)^1$	$-\$12,538$
2	$-12,200(1.076)^2$	$-14,125$	$-12,200(1.0277)^2$	$-12,885$
3	$-12,200(1.076)^3$	$-15,198$	$-12,200(1.0277)^3$	$-13,242$
4	$-12,200(1.076)^3(1.093)^1$	$-16,612$	$-12,200(1.0277)^3(1.0439)^1$	$-13,823$
5	$-12,200(1.076)^3(1.093)^2$	$-18,157$	$-12,200(1.0277)^3(1.0439)^2$	$-14,430$

$$e'_{1,2,3} = \frac{0.076 - 0.047}{1.047} = 0.0277, \text{ or } 2.77\%$$

$$e'_{4,5} = \frac{0.093 - 0.047}{1.047} = 0.0439, \text{ or } 4.39\%$$

This illustrates that differential inflation, or deflation, also results in market price changes in real dollars as well as in actual dollars.

9.3.2   Modeling Price Changes with Geometric Cash Flow Sequences

In Chapter 3 we discussed equivalence calculations involving projected cash flow patterns that are increasing at an effective rate, $\bar{f}$, per period. When total price escalation is included in an engineering economic analysis, projected prices of goods and services can be modeled as increasing at a constant rate per period. Thus, the resulting end-of-period cash flow pattern is often a geometric sequence.

In Section 9.2.2, the correct interest rate to use in an engineering economic analysis was shown to depend on the dollar terms of the cost and revenue cash flows; specifically, the combined interest rate (i_c) is used in an actual dollar analysis, and the real interest rate (i_r) is used in a real dollar analysis. An additional question is, what $\bar{f}$ value is used for each method of analysis when cost escalation is included and a geometric sequence cash flow model is appropriate? In the following table, we see that $\bar{f}$ is equal to e_j in an A\$ analysis and equal to e'_j in a R\$ analysis:

Method	Cash flows	Interest Rate (i)	Geometric Gradient ($\bar{f}$)
A	Actual dollars (A\$)	i_c	e_j
B	Real dollars (R\$)	i_r	e'_j

From this it follows that the "convenience rate" (Chapter 3) needed to calculate the PW of a geometric cash flow sequence involving price escalation would be as follows:

$$A\$ \ Analysis \qquad R\$ \ Analysis$$

$$i_{CR} = \frac{i_c - e_j}{1 + e_j} \qquad i_{CR} = \frac{i_r - e_j'}{1 + e_j'} \tag{9-12}$$

EXAMPLE 9-7

A water service public utility district is considering some new replacement pumping equipment to reduce operating expenses and improve service reliability. The base time period is the present, or year zero ($b = 0$). The estimated annual savings in year zero dollars is \$78,000. The utility district uses an eight-year study period for this type of equipment replacement study; the general price inflation rate is projected to be 4.6% per year; the total price escalation rate (e_j) for operating expenses is projected to be 6.2% per year; the utility district is using a MARR (which includes the effect of general price inflation) of 9.5% per year; the old equipment has no net market value; and no taxes are involved. Based on these estimates, calculate the maximum amount that could be paid for the equipment now (a) using an actual dollar analysis and (b) using a real dollar analysis.

SOLUTION
(a) Actual dollar analysis:

$$i_c = \text{MARR (as given)} = 9.5\%; \ \bar{f} = e_j = 6.2\%; \ N = 8$$

$$i_{CR} = \frac{i_c - e_j}{1 + e_j} = \frac{0.095 - 0.062}{1.062} = 0.03107, \text{ or } 3.11\%$$

From the data given, the annual savings of $\$78,000(1.062)^k$ for $1 \le k \le 8$ constitute a geometric cash flow sequence. By using Equation 3-27, we can write

$$PW(3.11\%) \text{ of savings} = \$78,000(P/A, 3.11\%, 8)$$

$$= \$78,000 \left[\frac{(1.0311)^8 - 1}{0.0311(1.0311)^8} \right] = \$545,000$$

which is the maximum amount that should be paid for the equipment.
(b) Real dollar analysis:

$$i_r = \text{MARR(Real)} = \frac{i_c - f}{1 + f} = \frac{0.095 - 0.046}{1.046} = 0.04685$$

$$\bar{f} = e_j' = \frac{e_j - f}{1 + f} = \frac{0.062 - 0.046}{1.046} = 0.01530$$

$$i_{CR} = \frac{i_r - e_j'}{1 + e_j'} = \frac{0.04685 - 0.0153}{1.0153} = 0.03107, \text{ or } 3.11\%$$

At this point in the real dollar analysis, we see that the convenience rate is the same value calculated for the actual dollar analysis, and the PW of the savings would thus be the same. This should make intuitive sense because we know that the equivalent worth of a cash flow is the same in the base time period using either actual dollar or real dollar analysis. Thus, the PW of the annual savings for the new pump would equal $545,000 when using method A or B since the base time period is the present ($b = 0$).

9.4 Application Strategy

In practice, should actual dollar or real dollar analysis be used, and when should price changes be included in an engineering economic analysis? There are no specific quantitative criteria that indicate when price changes are great enough to be included in an engineering economic analysis. Rather, judgment based on the prospective price change estimates, and sensitivity analysis, are used in practice. However, either the actual dollar or the real dollar analysis method may be used. Both methods, properly applied, result in the same equivalent worth for a cash flow in the base time period, require the same amount of information, and have no practical difference in application effort.

There is some difference, however, in the information available for interpreting the economic results. The results from an actual dollar analysis are in market purchasing power that varies with time, while the results from a real dollar analysis are in constant market purchasing power defined by the base time period (b). Thus, *real dollar analysis provides information in terms of a constant unit of measure, while actual dollar analysis provides information on the actual quantities that will occur during the study period.*

An analysis or *application strategy* that works well in engineering practice is to use actual dollar analysis for both before-tax and after-tax studies, and then, at the end of an analysis, to use Equation 9-1 or 9-2 to provide selected cash flows (particularly net cash flows) in real dollar terms. With little effort, this strategy provides additional information that is useful. In some organizations, a particular method of analysis may be specified; but even then, selected cash flows can be easily converted to the other dollar domain to assist with interpretation of results.

9.5 A Comprehensive Example

In most engineering economy studies of larger projects conducted in industry, price changes affecting revenues and costs have to be considered in addition to the applicable income tax provisions. To illustrate this situation, a fairly comprehensive and realistic project analysis is presented.

EXAMPLE 9-8

A company is considering an investment opportunity that requires the investment of $20,000 in production control equipment to increase the output of an assembly line. As a result, the revenue obtained from the modified assembly line is expected to increase. The following information applies to the investment opportunity.

Analysis period	10 years
Base time period	Present $(b = 0)$
Estimated useful life of the equipment	10 years
MACRS (GDS) property class	5 years
Effective income tax rate (t)	39%
Real after-tax MARR (i_r)	6%
General price inflation rate (f)	8% per year
Combined interest rate (i_c)	$14.48\% = [0.06 + 0.08 + (0.06)(0.08)]100$
Increased revenue (assume that revenue escalates at the general price inflation rate of 8% per year)	$15,000 per year in year zero dollars
Market value in 10 years	10% of the capital investment $(e_{MV} = 8\%)$

Annual Expenses

Category	Estimate (Year 0 Dollars)	Price Escalation Rate per Year (e_j)
Material	$1,200	10%
Labor	2,500	5.5%
Energy	2,500	15%
Other expenses	500	8%

Leased equipment is also required, which can be obtained for the first five years at a rate of $800 per year. The contract will be renegotiated at the beginning of the sixth year at an escalated value based on the general price inflation rate.

Perform an after-tax analysis of this project using the PW method and including the effects of total price escalation: (a) conduct an actual dollar analysis (and calculate the PW of the ATCF); (b) convert the actual dollar ATCF to a real dollar ATCF; and (c) calculate the PW of the real dollar ATCF and show that it is equivalent to the PW of the actual dollar ATCF.

SOLUTION

(a) Actual dollar analysis: The preliminary calculations required are described below:

1. *Revenue:* The estimated revenue of $15,000 per year in year zero dollars must be increased each year by the general price inflation rate.

$$\text{Revenue in year } k = \$15,000(1.08)^k$$

2. *Material, labor, energy, and other annual expenses:* These annual expenses, estimated in year zero dollars, are increased each year by the appropriate total

TABLE 9-7   Actual Dollar Cash Flow Analysis (with Conversion to Real Dollar ATCF) for Example 9-8

A$ After-Tax Analysis

End of Year, k	Revenue	Capital Investment	Material	Labor	Energy	Other Expenses	Leased Equipment
0		−$20,000					
1	$16,200		−$1,320	−$2,638	−$ 2,875	−$ 540	−$ 800
2	17,496		− 1,452	− 2,783	− 3,306	− 583	− 800
3	18,896		− 1,597	− 2,936	− 3,802	− 630	− 800
4	20,407		− 1,757	− 3,097	− 4,373	− 680	− 800
5	22,040		− 1,933	− 3,267	− 5,028	− 735	− 800
6	23,803		− 2,126	− 3,447	− 5,783	− 793	− 1,175
7	25,707		− 2,338	− 3,637	− 6,650	− 857	− 1,175
8	27,764		− 2,572	− 3,837	− 7,648	− 925	− 1,175
9	29,985		− 2,830	− 4,048	− 8,795	− 1,000	− 1,175
10	32,384		− 3,112	− 4,270	− 10,114	− 1,079	− 1,175
10	4,318[a]						

[a]Estimated MV.   (*Continued*)

price escalation rate (e_j):

$$\text{Material} = -\$1,200(1.1)^k$$
$$\text{Labor} = -\$2,500(1.055)^k$$
$$\text{Energy} = -\$2,500(1.15)^k$$
$$\text{Other expenses} = -\$500(1.08)^k$$

3. *Leased property:* The lease will be adjusted at the end of year five to account for five years of general price inflation at 8% per year:

$$\text{Lease (years 6–10)} = -\$800(1.08)^5 = -\$1,175$$

4. *Depreciation:* The MACRS depreciation amounts are as follows:

End of Year k	Cost Basis	MACRS (GDS) Recovery Rates	MACRS (GDS) Depreciation (A$)
1	$20,000	0.2000	$4,000
2	20,000	0.3200	6,400
3	20,000	0.1920	3,840
4	20,000	0.1152	2,304
5	20,000	0.1152	2,304
6	20,000	0.0576	1,152

TABLE 9-7	Actual Dollar Cash Flow Analysis (with Conversion to Real Dollar ATCF) for Example 9-8 (cont'd)

A$ After-Tax Analysis					R$ ATCF		
Before-Tax Cash Flow	Depre-ciation	Taxable Income	Income Taxes	A$ After-Tax Cash Flow	R$ Adjust-ment Factor $(1/1.08)^{k-0}$	R$ After-Tax Cash Flow
−$20,000				−$20,000	1.0	−$20,000
8,028	$4,000	$ 4,028	−$1,571	6,457	0.92593	5,979
8,572	6,400	2,172	− 847	7,725	0.85734	6,623
9,131	3,840	5,291	− 2,063	7,068	0.79383	5,611
9,700	2,304	7,396	− 2,884	6,816	0.73503	5,010
10,277	2,304	7,973	− 3,109	7,168	0.68058	4,878
10,479	1,152	9,327	− 3,638	6,841	0.63017	4,311
11,050		11,050	− 4,310	6,740	0.58349	3,933
11,607		11,607	− 4,527	7,080	0.54027	3,825
12,137		12,137	− 4,733	7,404	0.50025	3,704
12,634		12,634	− 4,927	7,707	0.46319	3,570
4,318		4,318[b]	− 1,684	2,634	0.46319	1,220
			PW(i_c =14.48%) = $16,780		PW(i_r = 6%) = $16,780	

[b]Recovery of depreciation (gain on disposal) — taxed like ordinary income.

5. *Market value:* The 10% MV, based on the capital investment, is a year zero amount and must be increased to account for the estimated 8% per year total price escalation rate ($e_{MV} = f$).

$$MV_{10} = 0.1(\$20,000)(1.08)^{10} = \$4,318$$

6. *Recaptured Depreciation:* The market value in actual dollars of $4,318 represents the recovery of depreciation and is taxed like ordinary income at the 39% rate.

The actual dollar, after-tax analysis is shown in Table 9-7. The PW of the actual dollar ATCF, using $i_c = 14.48\%$, is $16,780.

(b) Real dollar, ATCF: the conversion of the actual dollar ATCF to a real dollar ATCF is shown in the last two columns in Table 9-7. Equation 9-1, with $b = 0$, was used to make the conversion.

This solution method implements the strategy recommended in Section 9.4 of using an actual dollar analysis and then converting selected cash flows into real dollars. Reviewing the actual dollar ATCF in this case indicates an average annual positive cash flow during the analysis period of approximately $7,100 from the $20,000 investment in new equipment. However, the real dollar ATCF shows that in terms of dollars with constant purchasing power ($b = 0$), the net positive cash flow from the investment (except for year 2) decreases from $5,979 in year 1 to $3,569 in year 10.

(c) The PW of the real dollar ATCF, using $i_r = 6\%$, is $16,780. This is the same value as the PW of the actual dollar ATCF calculated in (a) using $i_c = 14.48\%$.

9.6 Foreign Exchange Rates and Purchasing Power Concepts

When domestic corporations make foreign investments, the resultant cash flows that occur over time are in a different currency from U.S. dollars. Typically, foreign investments are characterized by two (or more) translations of currencies: (1) when the initial investment is made and (2) when cash flows are returned to U.S.-based corporations. Exchange rates between currencies fluctuate, sometimes dramatically, over time, so a typical question that can be anticipated is "What return (profit) did we make on our investment in the synthetic fiber plant in Thatland?" For the engineer who is designing another plant in Thatland, the question might be "What is the PW (or IRR) that our firm will obtain by constructing and operating this new plant in Thatland?"

> Observe that changes in the exchange rate between two currencies over time are analogous to changes in the general inflation rate because the relative purchasing power between the two currencies is changing similar to the relative purchasing power between the actual dollar and the real dollar.

Assume the following:

i_{us} = rate of return in terms of a combined (market) interest rate relative to U.S. dollars.

i_{fc} = rate of return in terms of a combined (market) interest rate relative to the currency of a foreign country.

f_e = annual devaluation rate (rate of annual change in the exchange rate) between the currency of a foreign country and the U.S. dollar. In the following relationships, a positive f_e is used when the foreign currency is being devalued relative to the dollar, and a negative f_e is used when the dollar is being devalued relative to the foreign currency.

Then

$$1 + i_{us} = \frac{1 + i_{fc}}{1 + f_e}$$

or

$$i_{fc} = i_{us} + f_e + f_e(i_{us}) \tag{9-13}$$

and

$$i_{us} = \frac{i_{fc} - f_e}{1 + f_e} \tag{9-14}$$

EXAMPLE 9-9

The CMOS Electronics Company is considering a capital investment of 50,000,000 pesos in an assembly plant located in a foreign country. Currency is expressed in pesos, and the exchange rate is now 100 pesos per U.S. dollar.

The country has followed a policy of devaluing its currency against the dollar by 10% per year to build up its export business to the United States. This means that each year the number of pesos exchanged for a dollar increases by 10% (f_e = 10%), so in two years $(1.10)^2(100)$ = 121 pesos would be traded for one dollar. Labor is quite inexpensive in this country, so management of CMOS Electronics feels that the proposed plant will produce the following rather attractive after-tax cash flow, stated in pesos:

End of Year	0	1	2	3	4	5
ATCF (millions of pesos)	−50	+20	+20	+20	+30	+30

If CMOS Electronics requires a 15% rate of return per year, after taxes, in U.S. dollars (i_{us}) on its foreign investments, should this assembly plant be approved? Assume that there are no unusual risks of nationalization of foreign investments in this country.

SOLUTION

To earn a 15% annual rate of return in U.S. dollars, the foreign plant must earn, based on Equation 9-13, 0.15 + 0.10 + 0.15(0.10) = 0.265, which is 26.5% on its investment in pesos (i_{fc}). As shown below, the PW (at 26.5%) of the ATCF, in pesos, is 9,165,236, and its IRR is 34.6%. Therefore, investment in the plant appears to be economically justified. We can also convert pesos into dollars when evaluating the prospective investment:

End of Year	ATCF (Pesos)	Exchange Rate	ATCF (Dollars)
0	−50,000,000	100 pesos per $1	−500,000
1	20,000,000	110 pesos per $1	181,818
2	20,000,000	121 pesos per $1	165,289
3	20,000,000	133.1 pesos per $1	150,263
4	30,000,000	146.4 pesos per $1	204,918
5	30,000,000	161.1 pesos per $1	186,220
IRR:	34.6%	IRR:	22.4%
PW(26.5%):	9,165,236 pesos	PW(15%):	$91,632

The PW (at 15%) of the ATCF, in dollars, is $91,632, and its IRR is 22.4%. Therefore, the plant again appears to be a good investment in economic terms. Notice that the two IRRs can be reconciled by using Equation 9-14:

$$i_{us}(\text{ IRR in \$}) = \frac{i_{fc}(\text{ IRR in pesos}) - 0.10}{1.10}$$

$$= \frac{0.346 - 0.10}{1.10}$$

$$= 0.224, \text{ or } 22.4\%$$

Remember that devaluation of a foreign currency produces less expensive imports. Thus, the devaluation means that the U.S. dollar is strong relative to the foreign currency. That is, fewer dollars are needed to purchase a fixed amount (barrels, tons, items) of goods and services from the foreign source or, stated differently, more units of the foreign currency are required to purchase U.S. goods. This phenomenon was observed in Example 9-9.

Conversely, when exchange rates for foreign currencies drop against the U.S. dollar (f_e has a negative value), the prices for imported goods and services rise. In such a situation, U.S. products are less expensive in foreign markets. For example, in 1986 the U.S. dollar was exchanged for approximately 250 Japanese yen, but in late 1995 a weaker U.S. dollar was worth about 100 Japanese yen. As a consequence, U.S. prices on Japanese goods and services more than doubled in theory (but in actuality increased in the United States by lesser amounts). One explanation for this anomaly is that Japanese companies were willing to reduce their profit margins to retain their share of the U.S. market.

What does it mean to say that "the dollar is now *weaker* or *stronger* against a foreign currency"? For example, the dollar has weakened in each of these situations:

1. One U.S. dollar trades for 1.45 marks on Monday, January 22, and one dollar is worth 1.40 marks on Monday, April 23.
2. One dollar is worth 250 yen (average) in 1986, but in 1995 one dollar is worth approximately 100 yen.

A weaker dollar discourages U.S. purchases of foreign-made goods and favors the sale of our products in foreign markets. A weaker dollar signifies a loss in the value of our currency relative to the buying power of other currencies (f_e is negative). The opposite is true for a strong U.S. dollar (f_e is positive).

9.7 Spreadsheet Applications

The following example illustrates the use of spreadsheets to convert actual dollars to real dollars and vice versa.

EXAMPLE 9-10

Sara B. Goode wishes to retire in the year 2022 with personal savings of $500,000 (1997 spending power). Assume that the expected inflation rate in the economy will average 3.75% per year during this time period. Sara plans to invest in a 7.5% per year savings account, and her salary is expected to increase by 8.0% per year between 1997 and 2022. Assume that Sara's 1997 salary is $60,000 and that the first deposit takes place at the end of 1997. What percent of her yearly salary must Sara put aside for retirement purposes to make her retirement plan a reality?

This example demonstrates the flexibility of a spreadsheet, even in instances where all of the calculations are based on a piece of information (% of salary to be saved) that we do not yet know. If we deal in actual dollars, the cash flow relationships are straightforward. The formula in cell F7 converts the desired

	A	B	C	D	E	F
1	Base Rate	Figure Entry		Starting Salary in 1997		$ 60,000
2	10.000%	1		Annual Salary Increase		8.00%
3	1.000%	2		Savings Interest Rate		7.50%
4	0.100%	4		Average Inflation Rate		3.75%
5	0.010%	5		Desired 2022 Amount (R$)		$ 500,000
6						
7	Savings Rate	12.45%		Desired 2022 Amount (A$)		$ 1,255,084
8				Year 2022 Bank Balance		$ 1,255,913
9						
10					Bank	
11			Salary	Savings	Balance	
12		Year	(A$)	(A$)	(A$)	
13		1997	$ 60,000	$ 7,470	$ 7,470	
14		1998	64,800	8,068	16,098	
15		1999	69,984	8,713	26,018	
16		2000	75,583	9,410	37,380	
17		2001	81,629	10,163	50,346	
18		2002	88,160	10,976	65,098	
19		2003	95,212	11,854	81,834	
20		2004	102,829	12,802	100,774	
21		2005	111,056	13,826	122,158	
22		2006	119,940	14,933	146,253	
23		2007	129,535	16,127	173,349	
24		2008	139,898	17,417	203,767	
25		2009	151,090	18,811	237,861	
26		2010	163,177	20,316	276,016	
27		2011	176,232	21,941	318,658	
28		2012	190,330	23,696	366,253	
29		2013	205,557	25,592	419,314	
30		2014	222,001	27,639	478,402	
31		2015	239,761	29,850	544,132	
32		2016	258,942	32,238	617,180	
33		2017	279,657	34,817	698,286	
34		2018	302,030	37,603	788,261	
35		2019	326,192	40,611	887,991	
36		2020	352,288	43,860	998,450	
37		2021	380,471	47,369	1,120,703	
38		2022	$ 410,909	$ 51,158	$ 1,255,913	

FIGURE 9-2 Spreadsheet for Example 9-10

ending balance into actual dollars. The salary is paid at the end of the year, at which point some percentage is placed in a bank account. The interest calculation is based on the cumulative deposits and interest in the account at the beginning of the year, but not on the deposits made at the end of the year. The salary is increased and the cycle repeats.

The spreadsheet model is shown in Figure 9-2. We can enter the formulas for the geometric gradient representing the salary increase (column C), the percentage of the salary (cell B7) that goes into savings (column D), and the bank balance at the end of the year (column E) without knowing the percent of salary being saved.

Most spreadsheets have a *solver* feature that will automatically determine the desired savings strategy. This example illustrates an approach that is not as elegant, but is nonetheless fast and will work for software that does not have this solver feature.

The approach is to revise the base savings rate systematically and compare the ending bank balance (copied to cell F8 for ease of viewing on the screen) with the desired year 2022 balance. To save keystrokes, the base rate is broken down by powers of 10 into separate cells, in the range B2:B5. The base rate is recombined with the formula in cell B7. Starting with the highest power of 10 (cell B2), we bracket the savings rate that will set cells F7 and F8 equal (or nearly equal). The highlighted cell formulas in Figure 9-2 are as follows.

Cell	Contents
B7	= A2 * B2 + A3 * B3 + A4 * B4 + A5 * B5
F7	= F5 * (1 + F4)^25
F8	= E38
C13	= F1
C16	= C15 * (1 + $F$2)
D16	= C16 * $B$7
E16	= E15 * (1 + $F$3) + D16

Once the problem has been formulated in a spreadsheet, we can determine the impact of different interest rates, inflation rates, etc., on the retirement plan with minimal changes and effort.

9.8 Summary

Inflation and price changes are an economic and business reality that can affect the comparison of alternatives and the quality of project decision making in an organization. Much of this chapter has dealt with incorporating price changes into before-tax and after-tax engineering economy studies.

In this regard, it must be ascertained whether cash flows have been estimated in actual dollars or real dollars. The appropriate interest rate to use when discounting or compounding actual dollar amounts is a combined, or marketplace, rate, while the corresponding rate to apply in real dollar analysis is the firm's real interest rate. A common error that must be avoided is estimating cash flows in real dollars and then using the combined interest rate (which includes the effect of general price inflation) in subsequent equivalency calculations.

Engineering economy studies often involve quantities that do not respond to inflation, such as depreciation, interest charges, and lease fees and other amounts established by contract. Identifying these quantities and handling them properly in an analysis are necessary to avoid erroneous economic results.

The usefulness of doing an actual dollar and a real dollar analysis has been discussed in the chapter. A suggested application strategy is to do an actual dollar analysis and then convert the before-tax and after-tax net cash flows to real dollars to aid the interpretation of results. The use of the chapter's basic concepts in dealing with foreign exchange rates has also been demonstrated.

9.9 References

FREIDENFELDS, J., and M. KENNEDY, "Price Inflation and Long-Term Present Worth Studies," *The Engineering Economist,* vol. 24, no. 3, Spring 1979, pp. 143–160.

Industrial Engineering, vol. 12, no. 3, March 1980. The entire issue is devoted to "The Industrial Engineer and Inflation." Of particular interest are the following articles:

    a. ESTES, C. B., W. C. TURNER, and K. E. CASE, "Inflation—Its Role in Engineering-Economic Analysis," pp. 18–22.

    b. SULLIVAN, W. G., and J. A. BONTADELLI, "How an IE Can Account for Inflation in Decision-Making," pp. 24–33.

    c. WARD, T. L., "Leasing During Inflation: A Two-Edged Sword," pp. 34–37.

JONES, B. W., *Inflation in Engineering Economic Analysis.* New York: John Wiley & Sons, 1982.

LEE, P. M., and W. G. SULLIVAN, "Considering Exchange Rate Movements in Economic Evaluation of Foreign Direct Investments," *The Engineering Economist,* vol. 40, no. 2, Winter 1995, pp. 171–199.

WATSON, F. A., and F. A. HOLLAND, "Profitability Assessment of Projects Under Inflation," *Engineering and Process Economics,* vol. 2, no. 3, 1976, pp. 207–221.

9.10 Problems

The number in parentheses ( ) that follows each problem refers to the section from which the problem is taken.

9-1. Your rich aunt is going to give you an end-of-year gift of $1,000 for each of the next 10 years.

    a. If general price inflation is expected to average 6% per year during the next 10 years, what is the equivalent value of these gifts at the present time? The real interest rate is 4% per year.

    b. Suppose that your aunt specified that the annual gifts of $1,000 are to be increased by 6% each year to keep pace with inflation. With a real interest rate of 4% per year, what is the current PW of the gifts? (9.2)

9-2. Because of general price inflation in our economy, the purchasing power of the dollar shrinks with the passage of time. If the average general price inflation rate is expected to be 8% per year into the foreseeable future, how many years will it take for the dollar's purchasing power to be one-half of what it is now? (That is, at what future point in time will it take $2 to buy what can be purchased today for $1?) (9.2)

9-3. Which of these situations would you prefer? (9.2)

    1. You invest $2,500 in a certificate of deposit that earns an effective interest rate of 8% per year. You plan to leave the money alone for five years, and the general price inflation rate is expected to average 5% per year. Taxes are ignored.

    2. You spend $2,500 on a piece of antique furniture. You believe that in five years the furniture can be sold for $4,000. Assume that the average general price inflation rate is 5% per year. Again, taxes are ignored.

9-4. Annual expenses for two alternatives have been estimated on different bases as follows:

End of Year	Alternative A Annual Expenses Estimated in Actual (Inflated) Dollars	Alternative B Annual Expenses Estimated in Real (Constant) Dollars with $b = 0$
1	−$120,000	−$100,000
2	− 132,000	− 110,000
3	− 148,000	− 120,000
4	− 160,000	− 130,000

If the average general price inflation rate is expected to be 6% per year and the real rate of interest is 9% per year, show which alternative has the least negative PW at time 0. (9.2)

9-5. Suppose that you deposit $1,000 in a Swiss bank account that earns an effective interest rate of 18% per year, and you withdraw the principal after six years. You receive the interest each year and spend it on your favorite hobby. What is the real annual rate of return on your investment if the general price inflation rate is 10% per year? Be exact! (9.2)

9-6. A recent engineering graduate has received the annual salaries shown in the following table over the past four years. During this time, the CPI has performed as indicated. Determine the engineer's annual salaries in *year 0 dollars* ($b = 0$) using the CPI as the indicator of general price inflation. (9.2)

End of Year	Salary (A$)	CPI
1	$34,000	7.1%
2	36,200	5.4%
3	38,800	8.9%
4	41,500	11.2%

9-7. Your company has just issued bonds, each with a face value of $1,000. They mature in 10 years and pay annual dividends of $100. At present they are being sold for $887. If the average annual general price inflation rate over the next 10 years is expected to be 6%, what is the real rate of return per year on this investment? (9.2)

9-8. A reactor vessel cost $375,000 ten years ago. The reactor had the capacity of producing 500 pounds of product per hour. Today, it is desired to build a vessel of 1,000 pounds per hour capacity. With a general price inflation rate of 5% per year and assuming a cost-capacity factor to reflect economies of scale, X, to be 0.75, what is the approximate cost of the new reactor now? (Chap. 8 and Sec. 9.2)

9-9. A commercial building design cost $89/ft² to construct eight years ago (for an 80,000-ft² building). This construction cost has escalated 5.4% per year since then. Presently, your company is considering construction of a 125,000-ft² building of the same design. The cost capacity factor is $X = 0.92$. In addition, it is estimated that working capital will be 5% of construction costs, and that project management, engineering services, and overhead will be 4.2%, 8%, and 31%, respectively, of construction costs. Also, it is estimated that annual expenses in the first year of operation will be $5/ft², and these are estimated to increase 5.66% per year thereafter. The future general inflation rate is estimated to be 7.69% per year, and the market-based $MARR_c = 12$% per year. (9.2)

a. What is the estimated capital investment for the 125,000-ft² building?

b. Based on a before-tax analysis, what is the PW for the first 10 years of ownership of the building?

c. What is the AW for the first 10 years of ownership in real dollars (R$)?

9-10. The operating budget estimate for an engineering staff for fiscal year 1997 is $1,780,000. The actual budget expenditures of the staff for the previous two fiscal years, as well as estimates for the next two years, are as shown in the following table. These are actual dollar amounts. Management, however, also wants annual budget amounts for these years, using a constant dollar perspective. The 1997 fiscal year is to be used for this purpose ($b = 1997$). The estimated annual

Fiscal Year	Budget Amount (A$)
1995	$1,615,000
1996	1,728,000
1997	1,780,000
1998	1,858,300
1999	1,912,200

general price inflation rate is 5.6%. What are the annual constant (real) dollar budget amounts? (9.2)

9-11. An individual wishes to have a preplanned amount in a savings account for retirement in 20 years. This amount is to be equivalent to $30,000 in today's purchasing power. If the expected average inflation rate is 7% per year and the savings account earns 5% interest, what lump sum should be deposited now in the savings account? (9.2)

9-12. The AZROC Corporation needs to acquire a computer system for one of its regional engineering offices. The purchase price of the system has been quoted at $50,000, and the system will reduce annual expenses by $18,000 per year in real dollars. Historically, these annual expenses have escalated at an average rate of 8% per year, and this is expected to continue into the future. Maintenance services will be contracted for, and their cost per year (in actual dollars) is constant at $3,000.

What is the minimum (integer-valued) life of the system such that the new computer can be economically justified? Assume that the computer's market value is zero at all times. The firm's MARR is 25% per year (which includes an adjustment for anticipated inflation in the economy). Show all calculations. (9.2)

9-13. An investor lends $10,000 today, to be repaid in a lump sum at the end of 10 years with interest at 10% ($= i_c$) compounded annually. What is the real rate of return, assuming that the general price inflation rate is 8% annually? (9.2)

9-14. An investor established an individual savings account in 1991 that involves a *series* of 20 deposits as shown in the following diagram.

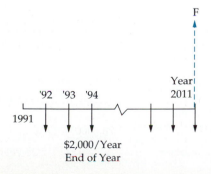

$2,000/Year
End of Year

The account is expected to compound at an average interest rate of 12% per year through the year 2011. General price inflation is expected to average 6% per year during this time. (9.2)

a. What is the FW of the savings account at the end of year 2011?

b. What is the FW of the savings account in 1991 (base time period) spending power?

9-15. Consider a project requiring an investment of $20,000 that is expected to return, in actual dollars, $6,000 at the end of the first year, $8,000 at the end of the second year, and $12,000 at the end of the third year. The general price inflation rate is 5% per year, and the real interest rate is 10% per year. Compare the PW of this project using before-tax actual dollar and real dollar analysis. (9.2)

9-16. Listed in the following table are the estimated annual price changes, in percentages, for two products over the next seven years. You want to simplify the modeling of these price change effects in some cost analyses being done. What single annual rate for each product would you use in your simplified model for this seven-year period? (9.2)

	Price Change (%)	
Year	Product A	Product B
1	4.6	8.3
2	4.8	7.5
3	6.1	9.0
4	6.9	8.0
5	5.8	7.0
6	7.2	9.0
7	6.6	9.5

9-17. Suppose that the general price inflation rate (f) is expected to average 8% per year during the next four years. For a certain commodity, the differential price inflation rate (e_j') has been estimated directly and is projected to be +2%. What will the marketplace prices be for this commodity over the next four years if it is now selling for $6.30 per unit ($b = 0$)? (9.3)

9-18. In Problem 9-17, determine what the real dollar prices will be for the commodity over the next four years. (9.3)

9-19. The capital investment for a new highway paving machine is $838,000. The estimated annual expense, in year zero dollars, is $92,600.

This expense is estimated to increase at the rate of 6.3% per year. Assume that $f = 4.5\%$, $N = 7$, SV at the end of year seven is 15% of the capital investment, and the MARR (in real terms) is 10.05% per year. What uniform annual revenue (before taxes), in actual dollars, would the machine need to generate to break even? (9.3)

9-20. A gas-fired heating unit is expected to meet an annual demand for thermal energy of 500 million Btu, and the unit is 80% efficient. Assume that each 1,000 cubic feet of natural gas, if burned at 100% efficiency, can deliver one million Btu. Suppose further that natural gas is now selling for $2.50 per 1,000 cubic feet. What is the PW of fuel cost for this heating unit over a 12-year period if natural gas prices are expected to escalate at an average rate of 10% per year? The firm's MARR is 18% per year. (9.3)

9-21. Your company uses the same large electric drive motor at several locations in its electric power generating plants. A new motor, which is much more efficient, is available. The market price of the new model is $71,000. Assume the following:

- Analysis period is 10 years.
- General inflation rate is 3.2%.
- The total price escalation rate on annual savings in operating expenses is 5.7% per year. Assume any savings in year one would escalate at this rate thereafter.
- Before-tax MARR = 12% per year (does not include inflation component).
- Base time period is year zero ($b = 0$).

If you ignore any market value or income taxes, what would the *annual savings in year one* need to be per motor to break even, if purchased at the $71,000 market price? (Use an actual dollar analysis). (9.3)

9-22. A small heat pump, including the duct system, now costs $2,500 to purchase and install. It has a useful life of 15 years and incurs annual maintenance expenses of $100 per year in real (year zero) dollars over its useful life. A compressor replacement is required at the end of the eighth year at a cost of $500 in real dollars. The annual cost of electricity for the heat pump is $680 based on present prices. Electricity prices are projected to escalate at an annual rate of 10%. All other costs are expected to escalate at

6%, which is the projected general price inflation rate. The firm's MARR, which includes an allowance for general price inflation, is 15% per year. No market value is expected from the heat pump at the end of 15 years. (9.3)

a. What is the AW, expressed in actual dollars, of owning and operating the heat pump?

b. What is the AW in real dollars of owning and operating the heat pump?

9-23. Suppose that you have just graduated with a BS degree in engineering, and the Omega Corporation offers you $30,000 per year as a starting salary. The company is located in New York City, and your guaranteed raises over the next five years will be 15% per year, starting in the second year of employment. Another company offers you a position in Jonesville but says that it is willing to negotiate a starting salary with you. Guaranteed raises with the Jonesville-based company will be 8% per year over a five-year period. What starting salary should you request of the Jonesville company if you are indifferent about where you would like to work? Ignore the effects of income taxes. Other data are the following: (9.2, 9.3)

1. Your opportunity cost of capital (i_c) is 15% per year.

2. The average annual cost of living index for the next five years is 108 in New York City and 100 in Jonesville (based on actual dollar spending power differences).

3. The general price inflation rate (f) of the economy is projected to be 8% over the next five years.

9-24. An electric utility company in the Northeast is trying to decide whether to switch from oil to coal at one of its generating stations. After much investigation, the problem has been reduced to these economic trade-offs:

	Oil	Coal
Cost of retrofitting boilers to burn coal	—	?
Annual fuel expense (year 0 dollars)	$-\$25 \times 10^6$	$-\$17 \times 10^6$
Escalation rate (e_j)	10%/yr	6%/yr
Life of plant	25 years	25 years

	Alternative			(Pr. 9-25)
			Price Escalation Rate (e_j)	
	Domestic	Import	(% per year)	
Negotiated purchase price (including tax)	−$8,500	−$9,200	—	
Fuel/year	− 1,050	− 700	8	
Maintenance/year	− 200	− 400	5	
Insurance/year	− 400	− 440	5	
Miscellaneous/year	− 100	− 125	7	
Ownership period, N	5 years	5 years	—	
Trade-in value at end of year N	$1,000	$2,000	6	

Determine the cost of retrofitting the boilers (to burn coal) that could be justified at this generating station. The utility's real MARR is 3% per year, and the general price inflation rate in the economy will average 6% per year over the next 25 years. (9.2, 9.3)

a. Solve using an actual dollar analysis.

b. Solve using a real dollar analysis.

9-25. Suppose that you are faced with the problem of deciding between two automobiles (see table at the top of this page). All cost estimates are based on 15,000 driving miles per year and are expressed in year zero dollars. Your personal MARR will average 12% per year over the next five years; and this is a marketplace rate of interest that includes an allowance for general price inflation ($f = 7\%$). Which automobile would you select based solely on monetary considerations? Solve using an actual dollar analysis. Then, for the selected alternative convert its net before-tax cash flow into real dollars. (9.2, 9.3)

9-26. Because of tighter safety regulations, an improved air filtration system must be installed at a plant that produces a highly corrosive chemical compound. The capital investment in the system is $260,000 in present-day dollars. The system has a useful life of 10 years, and is in the MACRS (GDS) five-year property class. It is expected that the MV of the system at the end of its 10-year life will be $50,000 in present-day dollars. Annual expenses, estimated in present-day dollars, are expected to be $6,000 per year, not including an annual property tax of 4% of the investment cost (*does not inflate*). Assume that the *plant* has a remaining life of 20 years, and that replacement costs, annual expenses, and MV escalate at 6% per year.

If the effective income tax rate is 40%, set up a table to determine the ATCF for the system over a 20-year period. The after-tax market rate of return desired on investment capital is 12% per year. What is the PW of costs of this system after income taxes have been taken into account? Develop the real dollar ATCF. (Assume that the annual general price inflation rate is 4.5% over the 20-year period.) (Chap. 7 and Secs. 9.4, 9.5)

9-27. Referring to the comprehensive example in Section 9.5 (Example 9-8), solve the problem with the following changes: (Chap. 7 and Secs. 9.4, 9.5)

1. Assume that the asset involved in the problem is one that is in the MACRS (GDS) seven-year property class instead of the five-year class.
2. The effective income tax rate (t) is 46% instead of 39%.
3. The real after-tax MARR (i_r) is 10% instead of 6%.
4. The price escalation rate (e_j) for labor is 9% instead of 5.5%.

9-28. Your company manufactures circuit boards and other electronic parts for various commercial products. Design changes in part of the product line, which are expected to increase sales, will require changes in the manufacturing operation. The cost basis of new equipment required is $220,000 (MACRS five-year property class assets). Increased annual revenues, in year zero dollars, are estimated to be $360,000. Increased annual expenses, in year zero dollars, are estimated to be $239,000. The estimated market value of equipment at the end of the six-year analysis period is $40,000. General price inflation is estimated at 4.9% per year; the total escalation rate on annual revenues is 2.5%, and for annual

Cash Flow Item	Estimate in Year 0 Dollars		Best Estimate of Price Change (% per year, e_j)	(Pr. 9-29)
	Purchase	Lease		
Capital investment	−$600,000	—	—	
MV at end of six years	90,000	—	2%	
Annual operating, insurance and other expenses	− 26,000	−$26,000	6	
Annual maintenance expenses	− 32,000	− 32,000	9	

expenses it is 5.6%; the after-tax MARR (in market terms) is 10% per year; and $t = 39\%$. (Chap. 7 and Secs. 9.4, 9.5)

a. Based on an after-tax, actual dollar analysis, what is the maximum amount that your company can afford to spend on the total project (i.e., changing the manufacturing operations)? Use the PW method of analysis.

b. Develop (show) your ATCF in real dollars.

9-29. You have been assigned the task of analyzing for your company whether to purchase or lease some transportation equipment. The analysis period is six years, and the base year is year zero ($b = 0$). Other pertinent information is given in the table at the top of this page. Also:

1. The contract terms for the lease specify a cost of $300,000 in the first year and $200,000 annually in years two through six (the contract, i.e., these rates, do not cover the annual expense items).

2. The after-tax MARR (*not* including inflation) is 13.208% per year.

3. The general inflation rate (f) is 6%.

4. The effective income tax rate (t) is 34%.

5. Assume the equipment is in the MACRS (GDS) five-year property class.

Which alternative is preferred (use an after-tax, actual dollar analysis and the FW criterion)? (Chap. 7 and Secs. 9.4, 9.5)

9-30. A company requires a 26% rate of return (before taxes) in U.S. dollars on project investments in foreign countries. (9.6)

a. If the currency of Country A is projected to average an 8% annual devaluation relative to the dollar, what rate of return (in terms of the currency there) would be required for a project?

b. If the dollar is projected to devaluate 6% annually relative to the currency of Country B,

what rate of return (in terms of the currency there) would be required for a project?

9-31. A U.S. company is considering a high-technology project in a foreign country. The estimated economic results for the project (after taxes) in the foreign currency (T-marks), is shown in the following table for the seven-year analysis period being used. The company requires an 18% rate of return in U.S. dollars (after taxes) on any investments in this foreign country. (9.6)

a. Should the project be approved, based on a PW analysis in U.S. dollars, if the devaluation of the T-mark, relative to the U.S. dollar, is estimated to average 12% per year and the present exchange rate is 20 T-marks per dollar?

b. What is the IRR of the project in T-marks?

c. Based on your answer in (b), what is the IRR in U.S. dollars?

End of year	NCF (T-marks after Taxes)
0	−3,600,000
1	450,000
2	1,500,000
3	1,500,000
4	1,500,000
5	1,500,000
6	1,500,000
7	1,500,000

9-32. An automobile manufacturing company in Country X is considering the construction and operation of a large plant on the eastern seaboard of the United States. Their MARR = 20% per year on a before-tax basis (*This is a market rate relative to their currency in Country X*). The study period used by the company for this type of investment is 10 years. Additional information is provided as follows:

- The currency in Country X is the Z-Kron.
- It is estimated that the U.S. dollar will become weaker relative to the Z-Kron during the next 10 years. Specifically, the dollar is estimated to be devalued at a rate of 2.2% per year.
- The present exchange rate is 92 Z-Krons per U.S. dollar.
- The estimated before-tax net cash flow (in U.S. dollars) is shown below:

EOY	Net Cash Flow (U.S. Dollars)
0	−$168,000,000
1	− 32,000,000
2	69,000,000
⋮	⋮
10	69,000,000

Based on a before-tax analysis, will this project meet the company's economic decision criterion? (9.6)

9-33. Refer to Section 9.1. Hypothesize a weighted mix of goods and services used in the operation of any type of company you select. Then obtain historical and forecasted inflation/deflation data for these goods and services from the types of sources referred to in Section 9.1. Use this data to develop a "tailored" general inflation rate (f) for the company for each of the next five years. What average effective rate per year might you consider using for this period instead of the varying yearly rates? (9.1)

9-34. *Brain Teaser*
This case study is a justification of a minicomputer system for a theoretical firm, the ABC Manufacturing Company. The following are known data:

- The combined hardware and software cost is $80,000.

- Contingency costs have been set at $15,000 (these are not necessarily incurred).
- A service contract on the hardware costs $500 per month.
- The effective income tax rate (t) is 38%.
- Company management has established a 15% per year hurdle rate (MARR).

In addition, the following assumptions/projections have been made:

- In order to support the system on an ongoing basis, a programmer/analyst will be required. The starting salary is $28,000 per year, and fringe benefits amount to 30% of the base salary. Salaries are expected to increase by 6% each year.
- The system is expected to yield a staff savings of three persons (to be reduced through normal attrition) at an average salary of $16,200 per year per person (base plus fringes) at the current year (i.e., year zero) base. It is anticipated that one person will retire during the first year, another in the second year, and a third in the third year.
- A 3% reduction in purchased material costs is expected; current year purchases are $1,000,000 and are expected to grow at a compounded rate of 10% per year.
- The project life is expected to be six years, and the minicomputer capital expenditure will be fully depreciated over that time period [MACRS (GDS) five-year property class].

Based on this information, perform an ATCF analysis. Is this investment acceptable based on economic factors alone? (9.2, 9.5)

10

Dealing with Uncertainty

*T*he objective of this chapter is to present and discuss several practical non-probabilistic methods that are helpful in analyzing investment situations where uncertainty exists.

The following topics are discussed in this chapter:

The nature of risk, uncertainty, and sensitivity

Sources of uncertainty

Common nonprobabilistic methods for dealing with uncertainty

Breakeven analysis

Sensitivity analysis

Optimistic-pessimistic estimates

Graphical sensitivity displays

Risk-adjusted MARR

Reduction in useful life

▼ 10.1 Introduction

In previous chapters we stated specific assumptions concerning applicable revenues, costs, and other quantities important to an engineering economic analysis. It was assumed that a high degree of confidence could be placed in all estimated values. That degree of confidence is sometimes called *assumed certainty*. Decisions made solely on the basis of this kind of analysis are sometimes called *decisions under certainty*. The term is rather misleading in that there rarely is a case in which estimated quantities can be assumed as certain.

In virtually all situations there is doubt as to the ultimate results that will be obtained from an investment. We now examine techniques applicable to Step 5 of the seven-step procedure for conducting engineering economy studies (Chapter 1). The motivation for dealing with risk and uncertainty is to establish the bounds of error in our estimates such that another alternative being considered may turn out to be a better choice than the one we recommended under assumed certainty.

10.2 What Are Risk, Uncertainty, and Sensitivity?

Both *risk* and *uncertainty* in decision-making activities are caused by lack of precise knowledge regarding future business conditions, technological developments, synergies among funded projects, and so on. *Decisions under risk* are decisions in which the analyst models the decision problem in terms of assumed possible future outcomes, or scenarios, whose probabilities of occurrence can be estimated. A *decision under uncertainty*, by contrast, is a decision problem characterized by several unknown futures for which probabilities of occurrence cannot be estimated.

In reality the difference between risk and uncertainty is somewhat arbitrary. One contemporary school of thought posits that representative and likely future outcomes and their probabilities can always be subjectively developed.* Hence, it is not unreasonable to suggest that decision making under risk is the more plausible and tractable framework for dealing with lack of perfect knowledge concerning the future. Although we may make a technical distinction between risk and uncertainty, both can cause study results to vary from predictions, and there seldom is anything significant to be gained by attempting to treat them separately. Therefore, in the remainder of this book, the terms *risk* and *uncertainty* are used interchangeably.

P R I N C I P L E 6 —MAKE UNCERTAINTY EXPLICIT (PAGE 7)

In dealing with uncertainty, it is often helpful to determine to what degree changes in an estimate would affect a capital investment decision, that is, how *sensitive* a given investment is to changes in particular factors (parameters) that are not known with certainty. If a parameter such as project life or annual revenue can be varied over a wide range without causing much effect on the investment decision, the decision under consideration is said *not* to be sensitive to that particular factor. Conversely, if a small change in the relative magnitude of a parameter will reverse an investment decision, the decision is highly sensitive to that parameter.

In this chapter, we discuss several nonprobabilistic techniques for considering uncertainty in engineering economic analysis. The use of probability models is introduced in Chapter 14.

---

*R. Schlaifer, *Analysis of Decisions Under Uncertainty* (New York: McGraw-Hill, 1969).

▼ 10.3 Sources of Uncertainty

It is useful to consider the factors that affect the uncertainty involved in a capital investment, so that they may be related to the economic measure of merit that must be met or exceeded in order for an investment to be justified. It would be almost impossible to list and discuss all of these factors. There are four major sources of uncertainty, however, which are nearly always present in engineering economy studies.

The first source that is always present is the *possible inaccuracy of the estimates used in the study.* If exact information is available regarding the items of revenue and expense, the resulting accuracy should be good. If, on the other hand, little factual information is available and nearly all the values have to be estimated, the accuracy may be high or low, depending on the manner in which the estimated values are obtained. Are they sound estimates based on good information or merely guesses?

The accuracy of the cash inflow estimates is often difficult to determine. If they are based on a considerable amount of past experience or have been determined by adequate market surveys, a fair degree of reliance may be placed on them. On the other hand, if they are merely the result of guesswork, with a considerable element of hope thrown in, they must of course be considered to contain a sizable element of uncertainty.

A saving in existing operating expenses should involve less uncertainty. It is usually easier to determine what the saving will be because there is considerable experience and past history on which to base the estimates. Similarly, there should be no large error in estimates of capital required. Uncertainty in capital investment requirements is often reflected as a *contingency* above the actual cost of plant and equipment.

The second major source affecting uncertainty is the *type of business involved in relation to the future health of the economy.* Some lines of business are notoriously less stable than others. For example, most mining enterprises are more risky than large retail food stores. However, we cannot arbitrarily say that an investment in any retail food store always involves less uncertainty than investment in mining property. Whenever capital is to be invested in an enterprise, the nature and history of the business as well as expectations of future economic conditions (e.g., interest rates) should be considered in deciding what risk is present.

A third source affecting uncertainty is the *type of physical plant and equipment involved.* Some types of structures have rather definite economic lives and market values. Little is known of the physical or economic lives of others, and they have almost no resale value. A good engine lathe generally can be used for many purposes in nearly any fabrication shop. Quite different would be a special type of lathe that was built to do only one unusual job. Its value would be dependent almost entirely upon the demand for the special task that it can perform. Thus, the type of physical property involved will have a direct bearing upon the accuracy of the estimated income and expenditure patterns. Where money is to be invested in specialized plant and equipment, this factor should be considered carefully.

The fourth very important source of uncertainty that must always be considered in evaluating uncertainty is the *length of the assumed study period*. The conditions that have been assumed in regard to revenue and expense must exist throughout the study period in order to obtain a satisfactory return on the capital investment. A long study period naturally decreases the probability of all the factors turning out as estimated. Therefore, a long study period, all else being equal, always increases the uncertainty of a capital investment.

10.4 Common Nonprobabilistic Methods for Dealing with Uncertainty

There are different methods for taking uncertainty resulting from the four major sources just described into account. This chapter discusses and illustrates each of the following five popular *nonprobabilistic* methods.

1. *Breakeven analysis* is commonly used when the selection among alternatives is heavily dependent on a single factor, such as capacity utilization, that is uncertain. A breakeven point for the factor is determined such that two alternatives are equally desirable from an economic standpoint. It is then possible to choose between the alternatives by estimating the most likely value of the uncertain factor and comparing this estimate to the breakeven value.

2. *Sensitivity analysis* is a basic and often employed technique when one or more factors are subject to uncertainty. The questions that sensitivity analysis attempts to resolve are these: (a) What is the behavior of the measure of merit (e.g., PW) to $\pm x\%$ changes in each individual factor? (b) What is the amount of change in a particular factor that will cause a reversal in preference for an alternative? (c) What is the change in the measure of merit to selected combinations of changes in two or more factors?

3. *Optimistic–pessimistic estimation* of parameters included in an engineering economy study has been used to establish a range of values for the economic measure of merit. If the range falls *entirely* in the acceptable region (e.g., annual worth ≥ 0), the alternative is desirable. When all factors are estimated under conservative (pessimistic) conditions, an unacceptable alternative often is the result. This method directs attention to the best and worst outcomes of going ahead with an alternative and requires managerial judgment to make the final go or no-go determination.

4. *Risk-adjusted MARRs* are sometimes used to deal with estimation uncertainties. This method involves the use of higher MARRs for alternatives that are classified as highly uncertain and lower MARRs for projects for which there appear to be fewer uncertainties.

5. *Reduction of the useful life* of an alternative is another means for attempting to include the effects of uncertainty. Here the estimated project life is reduced by a fixed percentage, for instance 50%, and each alternative is evaluated regarding its acceptability over only this reduced life span.

10.5 Breakeven Analysis

Essentially all data employed in an engineering economy study are uncertain simply because they represent estimates of the future. Breakeven analysis is useful when we must make a decision between alternatives that are highly sensitive to a single factor that is difficult to estimate. Through breakeven analysis, we can solve for the value of that factor at which the conclusion is a standoff. That value is known as the *breakeven point*. (The use of breakeven points with respect to production and sales volumes was discussed in Chapter 2.) If we can then estimate whether the actual outcome of the common factor will be higher or lower than the breakeven point, the best alternative becomes apparent.

In mathematical terms, we have

$$EW_A = f_1(y) \text{ and } EW_B = f_2(y)$$

where EW_A = an equivalent worth value for the net cash flow of Alternative A
   EW_B = an equivalent worth value for the net cash flow of Alternative B
   y = a common factor of interest affecting the equivalent worth of Alternative A and Alternative B

Therefore, the breakeven point between Alternative A and Alternative B is the value of factor y for which the two equivalent worths are equal. That is, $EW_A = EW_B$, or $f_1(y) = f_2(y)$, which may be solved for y.

The following are examples of common factors for which breakeven analyses might provide useful insights into the decision problem:

1. *Annual revenue and expenses:* Solve for the annual revenue required to equal (break even with) annual expenses. Breakeven annual expenses of an alternative can also be determined in a pairwise comparison when revenues are identical for both alternatives being considered.
2. *Rate of return:* Solve for the rate of return on the increment of invested capital at which two given alternatives are equally desirable.
3. *Market (or salvage) value:* Solve for the future resale value that would result in indifference as to preference for an alternative.
4. *Equipment life:* Solve for the useful life required for an alternative to be justified.
5. *Capacity utilization:* Solve for the hours of utilization per year, for example, at which an alternative is justified or at which two alternatives are equally desirable.

> The usual breakeven problem involving two alternatives can be most easily approached mathematically by equating an equivalent worth of the two alternatives expressed as a function of the factor of interest. In breakeven studies, project lives may or may not be equal, so care should be taken to determine whether the coterminated or repeatability assumption best fits the situation.

The following examples illustrate both mathematical and graphical solutions to typical breakeven problems.

EXAMPLE 10-1

Suppose that there are two alternative electric motors that provide 100 hp output. An Alpha motor can be purchased for $12,500 and has an efficiency of 74%, an estimated useful life of 10 years, and estimated maintenance expenses of $500 per year. A Beta motor will cost $16,000 and has an efficiency of 92%, a life of 10 years, and annual maintenance expenses of $250. Annual taxes and insurance expenses on either motor will be $1\frac{1}{2}$ % of the investment. If the MARR is 15%, how many hours per year would the motors have to be operated at full load for the annual expenses to be equal? Assume that market values for both are negligible and that electricity costs $0.05 per kilowatt-hour.

SOLUTION BY MATHEMATICS

Note: 1 hp = 0.746 kW and input = output/efficiency. If X = number of hours of operation per year, components of the equivalent annual worth, AW_α, for the Alpha motor would be as follows:

Capital recovery amount (depreciation and minimum profit):

$$-\$12,500(A/P, 15\%, 10) = -\$12,500(0.1993) = -\$2,490 \text{ per year}$$

Operating expense for power:

$$-(100)(0.746)(\$0.05)X/0.74 = -\$5.04X \text{ per year}$$

Maintenance expense:

$$-\$500 \text{ per year}$$

Taxes and insurance:

$$-\$12,500(0.015) = -\$187 \text{ per year}$$

Similarly, components of the equivalent annual worth, AW_β, for the Beta motor would be as follows:

Capital recovery amount:

$$-\$16,000(A/P, 15\%, 10) = -\$16,000(0.1993) = -\$3,190 \text{ per year}$$

Operating expense for power:

$$-(100)(0.746)(\$0.05)X/0.92 = -\$4.05X \text{ per year}$$

Maintenance expense:

$$-\$250 \text{ per year}$$

Taxes and insurance:

$$-\$16,000(0.015) = -\$240 \text{ per year}$$

At the breakeven point, $AW_\alpha = AW_\beta$. Thus,

$$-\$2,490 - \$5.04X - \$500 - \$187 = -\$3,190 - \$4.05X - \$250 - \$240$$
$$-\$5.04X - \$3,177 = -\$4.05X - \$3,680$$
$$\hat{X} \approx 508 \text{ hours/year}$$

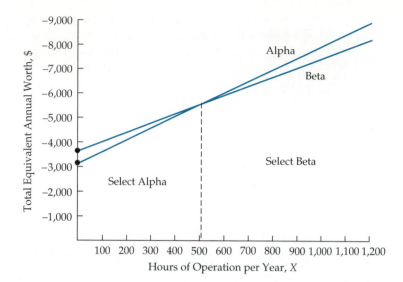

FIGURE 10-1

Graphical Solution
of the Breakeven Point
for Example 10-1

SOLUTION BY GRAPHICS

Figure 10-1 shows a plot of the total equivalent annual worths of each motor as a function of the number of hours of operation per year. The constant annual costs (AW intercepts) are −$3,177 and −$3,680 for Alpha and Beta, respectively, and the expenses that vary directly with hours of operation per year (slopes of lines) are −$5.04 and −$4.05 for Alpha and Beta, respectively. Of course, the breakeven point is the value of the independent variable, X, at which the linear annual worth functions for the two alternatives intersect.

Hours of operation per year was the measure of business activity in Example 10-1 and was used as the variable for which a breakeven value was desired. If hours of operation had been independently estimated to be, say, 1,100 hours per year, the choice of the Beta motor would be apparent from Figure 10-1.

In Example 10-1, breakeven analysis was applied where only two alternatives were involved. However, breakeven analysis can be extended to multiple alternatives, which is demonstrated in Example 10-2.

EXAMPLE 10-2

The Universal Postal Service is considering the possibility of putting wind deflectors on the tops of 500 of their long-haul tractors. Three types of deflectors, with the following characteristics, are being considered (MARR = 10% per year):

	Windshear	Blowby	Air-vantage
Capital investment	−$1,000	−$400	−$1,200
Drag reduction	20%	10%	25%
Maintenance/year	−$10	−$5	−$5
Useful life	10 years	10 years	5 years

If 5% in drag reduction means 2% in fuel savings per mile, how many miles do tractors have to be driven per year before the Windshear deflector is favored over the other deflectors? Over what range of miles driven per year is Air-vantage the best choice? (*Note:* Fuel cost is expected to be $1.00 per gallon and average fuel consumption is five miles per gallon without the deflectors). State any assumptions you make.

SOLUTION

The annual operating expenses of long-haul tractors equipped with the various deflectors are calculated as a function of mileage driven per year, X:

$$\text{Windshear: } -[(X \text{ miles/yr})(0.92)(0.2 \text{ gal/mi})(\$1.00/\text{gal})] = -\$0.184X/\text{yr}$$

$$\text{Blowby: } -[(X \text{ miles/yr})(0.96)(0.2 \text{ gal/mi})(\$1.00/\text{gal})] = -\$0.192X/\text{yr}$$

$$\text{Air-vantage: } -[(X \text{ miles/yr})(0.90)(0.2 \text{ gal/mi})(\$1.00/\text{gal})] = -\$0.180X/\text{yr}$$

Plotting equivalent annual worths (AW) of the deflectors yields the breakeven values of X shown in Figure 10-2. In summary, when $X \leq 12,831$, Blowby would be selected. If $X \geq 37,203$, the Air-vantage deflector would be chosen; otherwise, Windshear is the preferred alternative. Mathematically, the exact values can be calculated for each pair of AW equations (Windshear versus Blowby, Windshear

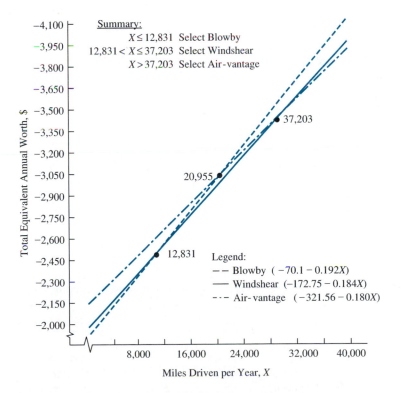

FIGURE 10-2 Graphical Breakeven Analysis of Example 10-2

versus Air-vantage, and Blowby versus Air-vantage). For example, the breakeven value between the Windshear deflector and the Blowby deflector is

$$-\$1,000(A/P, 10\%, 10) - \$10 - \$0.184X = -\$400(A/P, 10\%, 10) - \$5 - \$0.192X$$

$$-\$172.75 - \$0.184X = -\$70.1 - \$0.192X$$

$$X = \frac{102.65}{0.008} = 12{,}831 \text{ miles/year}$$

The repeatability assumption, which is appropriate in this situation, allows the annual worths to be compared over different periods of time.

It is often helpful to know at what future date a deferred investment will be needed so that an alternative permitting deferred investment will break even with one that provides immediately for all future requirements. Where only the costs of acquiring the assets by the two alternatives need to be considered or where the annual expenses throughout the entire life are not affected by the date of acquisition of the deferred asset, the breakeven point may be determined very easily and may be helpful in arriving at a decision between alternatives. Example 10-3 illustrates this type of breakeven study.

EXAMPLE 10-3

In planning a small two-story office building, the architect has submitted two designs. The first provides foundation and structural details so that two additional stories can be added to the required initial two stories at a later date and without modifications to the original structure. This building would cost $1,400,000. The second design, without such provisions, would cost only $1,250,000. If the first plan is adopted, an additional two stories could be added at a later date at a cost of $850,000. If the second plan is adopted, however, considerable strengthening and reconstruction would be required, which would add $300,000 to the cost of a two-story addition. Assuming that the building is expected to be needed for 75 years, by what time would the additional two stories have to be built to make the adoption of the first design justified? (The MARR is 10% per year.)

SOLUTION

The breakeven deferment period, $\hat{T}$, is determined as follows:

	Provide Now	No Provision
PW cost:		
First unit	−$1,400,000	−$1,250,000
Second unit	−$850,000$(P/F, 10\%, \hat{T})$	−$1,150,000$(P/F, 10\%, \hat{T})$
Equating total PW costs:		

$$-\$1{,}400{,}000 - \$850{,}000(P/F, 10\%, \hat{T}) = -\$1{,}250{,}000 - \$1{,}150{,}000(P/F, 10\%, \hat{T})$$

Solving, we have

$$(P/F, 10\%, \hat{T}) = 0.5$$

From the 10% interest table in Appendix C, $\hat{T}$ = seven years (approximately). Thus, if the additional space will be required in less than seven years, it would be more economical to make immediate provision in the foundation and structural details. If the addition would not likely be needed until after seven years, greater economy would be achieved by making no such provision in the first structure.

There are many situations in which the relationship between the dependent and independent variables is not continuous and therefore cannot be expressed readily in mathematical terms. In other cases the relationship may be very complex, so that the time required to develop a mathematical formula would be so great that it would be uneconomical to do so. In such cases, a graphical solution may be used to determine the breakeven point. The following is a case in which a graphical solution is used to advantage.

Products ordered from a wholesale drug company are collected in large wire-mesh baskets and taken to the shipping department for packing. Two methods of packing are used—one with heavy plastic bags and the other with cartons. Figure 10-3 shows the packing costs for the two methods as a function of the volume of the merchandise. The relationship of volume and cost is not a simple one; for shipments whose volume is less than approximately 3,500 cubic inches the bag method should be used, and shipments whose volume is larger should be packed in cartons. In this instance, because the sides of the wire-mesh baskets are nearly vertical, the packers are able to determine which packing method to use (and the correct size of bag or carton) merely by checking the height of the merchandise in the basket against a colored vertical scale on the end of the packing table.

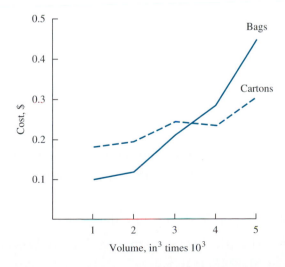

FIGURE 10-3

Packing Costs for Plastic Bags and Cartons

▼
10.6 Sensitivity Analysis

In many cases, a simple breakeven analysis does not provide adequate information about the potential impact of uncertainty in estimated factor values on study results. In such instances, and in fact in most cases, it is helpful to determine how sensitive the situation is to the several factors of concern so that proper weight and consideration may be assigned to them. *Sensitivity*, in general, means the relative magnitude of change in the measure of merit (such as PW) caused by one or more changes in estimated study factor values. Sometimes sensitivity is more specifically defined to mean the relative magnitude of the change in one or more factors that will reverse a decision among alternatives.

> In engineering economy studies, sensitivity analysis is a basic nonprobabilistic technique, readily available, to provide information about the potential impact of uncertainty in selected factor estimates. Its routine use is fundamental to achieving sound results useful in the decision process.

Example 10-4 demonstrates sensitivity analysis for several factors taken one at a time.

EXAMPLE 10-4

A machine for which most likely cash flow estimates are given in the following list is being considered for immediate installation. Because of the new technology built into this machine, it is desired to investigate its PW over a range of ±40% in (a) capital investment, (b) annual net cash flow, (c) market value, and (d) useful life. Based on these estimates, how much can the initial investment increase without making the machine an unattractive venture?

Capital investment, I	−$11,500
Revenues/yr $\Big\}\,A$	5,000
Expenses/yr	−2,000
Market value, MV	1,000
Useful life, N	6 years

Draw a diagram that summarizes the sensitivity of present worth to changes in each separate parameter when the MARR $=10\%$ per year.

SOLUTION
The PW of this capital investment for the estimates of parameters given above is

$$PW(10\%) = -\$11,500 + \$3,000(P/A, 10\%, 6) + \$1,000(P/F, 10\%, 6)$$
$$= \$2,130$$

(a) When the capital investment varies by $\pm p\%$ the PW is

$$PW(10\%) = -(1 \pm p\%/100)(\$11,500) + \$3,000(P/A, 10\%, 6)$$
$$+ \$1,000(P/F, 10\%, 6)$$

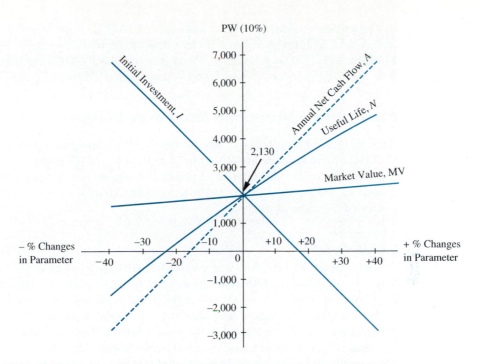

FIGURE 10-4 Sensitivity Analysis of Four Parameters in Example 10-4

If we let $p\%$ vary in increments of 10% to $\pm40\%$, the resultant calculations of PW(10%) can be plotted as shown in Figure 10-4.

(b) The equation for PW can be modified to reflect $\pm a\%$ changes in net annual cash flow, A:

$$PW(10\%) = -\$11,500 + (1 \pm a\%/100)(\$3,000)(P/A, 10\%, 6)$$
$$+ \$1,000(P/F, 10\%, 6)$$

Results are plotted in Figure 10-4 for 10% increments in A within the prescribed $\pm40\%$ interval.

(c) When market value varies by $\pm s\%$, the PW is

$$PW(10\%) = -\$11,500 + \$3,000(P/A, 10\%, 6)$$
$$+ (1 \pm s\%/100)(\$1,000)(P/F, 10\%, 6)$$

Results are shown in Figure 10-4 for $\pm40\%$ changes in MV.

(d) Plus and minus $n\%$ changes in useful life, as they affect PW(10%), can be represented by this equation:

$$PW(10\%) = -\$11,500 + \$3,000[P/A, 10\%, 6(1 \pm n\%/100)]$$
$$+ \$1,000[P/F, 10\%, 6(1 \pm n\%/100)]$$

When $n\%$ varies by 10% increments within the desired $\pm40\%$ interval, resultant changes in PW(10%) can be calculated and plotted as shown in Figure 10-4.

The measure of merit, PW, is insensitive to MV, but quite sensitive to changes in I, A, and N. From part (a), it is clear that the capital investment can increase by as much as $2,130 (to $13,630) without causing the PW(10%) to become negative. This is an 18.5% increase, which can be approximated from Figure 10-4.

10.7 Analyzing a Proposed Business Venture

A more detailed illustration of the use of sensitivity analysis is provided in Example 10-5, in which a new business venture is analyzed. This example extends sensitivity explorations to several factors in the study whose outcomes are believed to be crucial to the success of the venture. Tabular displays are used to summarize results of the various analyses.

EXAMPLE 10-5

A small group of investors is considering starting a premixed-concrete plant in a rapidly developing suburban area about 15 miles from a large city. The group believes that there will be a good market for premixed concrete in this area for at least the next 10 years, and that if they establish such a local plant, it will be unlikely that another local plant would be established. Existing plants in the adjacent large city would, of course, continue to serve this new area. The investors believe that the plant could operate at about 75% of capacity 250 days per year, because it is located in an area where the weather is mild throughout the year.

The plant will cost $100,000 and will have a maximum capacity of 72 cubic yards of concrete per day. Its market value at the end of 10 years is estimated to be $20,000, which is the value of the land. To deliver the concrete, four secondhand trucks would be acquired, costing $8,000 each, having an estimated life of 5 years and a trade-in value of $500 each at the end of that time. In addition to the four truck drivers, who would be paid $50.00 per day each, four people would be required to operate the plant and office, at a cost of $175.00 per day. Annual operating and maintenance expenses for the plant and office are estimated at $7,000 and for each truck at $2,250, both in view of 75% capacity utilization. Raw material costs are estimated to be $27.00 per cubic yard of concrete. Payroll taxes, vacations, and other fringe benefits would amount to 25% of the annual payroll. Annual taxes and insurance on each truck would be $500, and taxes and insurance on the plant would be $1,000 per year. The investors would not contribute any labor to the business, but a manager would be employed at an annual salary of $20,000.

Delivered, premixed concrete currently is selling for an average of $45 per cubic yard. A useful plant life of 10 years is expected, and capital invested elsewhere by these investors is earning about 15% per year before income taxes. It is desired to find the AW for the expected conditions described and to perform sensitivity analyses for certain factors.

SOLUTION BY THE AW METHOD
Annual revenue:

$$72 \times 250 \times \$45 \times 0.75 = \$607,500$$

Annual expenses:
1. Capital recovery amount
    Plant: $-\$100,000(A/P, 15\%, 10)$
    $+\$20,000(A/F, 15\%, 10)$ $=$ $-\$18,940$
    Trucks: $4[-\$8,000(A/P, 15\%, 5)$
    $+\$500(A/F, 15\%, 5)]$ $=$ $-9,250$
    $-\$28,190$

2. Labor:
    Plant and office: $-\$175 \times 250$ $=$ $-43,750$
    Truck drivers: $4 \times -\$50 \times 250$ $=$ $-50,000$
    Manager $=$ $-20,000$
    $-113,750$

3. Payroll taxes, fringe benefits, etc.:
    $-\$113,750 \times 0.25$ $-28,438$

4. Taxes and insurance:
    Plant $=$ $-1,000$
    Trucks: $-\$500 \times 4$ $=$ $-2,000$
    $-3,000$

5. Operations and maintenance at 75% capacity:
    Plant and office $=$ $-7,000$
    Trucks: $-\$2,250 \times 4$ $=$ $-9,000$
    $-16,000$

6. Materials: $72 \times 0.75 \times 250 \times -\$27.00$ $-364,500$

         Total $-\$553,878$

The net AW for these most likely estimates is $\$607,500 - \$553,878 = \$53,622$. Apparently, the project is an attractive investment opportunity.

In Example 10-5 there are three factors that are of great importance and that must be estimated: *capacity utilization*, the *selling price of the product*, and the *useful life of the plant*. A fourth factor—raw material costs—is important, but any significant change in this factor would probably be equally beneficial to competitors and probably would be reflected in a corresponding change in the selling price of mixed concrete. The other cost elements should be determinable with considerable accuracy. Therefore, we need to investigate the effect of variations in the plant utilization, selling price, and useful life. Sensitivity analysis is a good method for doing this.

10.7.1 Sensitivity to Capacity Utilization

As a first step, we must determine how expenses would vary, if at all, as capacity utilization is varied. In this case, it is probable that the annual expense items listed under groups 1, 2, 3, and 4 in the previous tabulation would be virtually unaffected if capacity utilization should vary over a quite wide range—from 50% to 90%, for example. To meet peak demands, the same amount of plant, trucks, and personnel probably would be required. Operations and maintenance expenses (group 5) would be affected somewhat. For this factor we must try to determine what the variation would be or make a reasonable assumption as to the probable variation. For this case it will be *assumed* that one half of these expenses would be fixed and the other half would vary with capacity utilization by a straight-line relationship. Certain other factors, such as the cost of materials in this case, will vary in direct proportion to capacity utilization.

Using these assumptions, Table 10-1 shows how the revenue, expenses, and net AW would change with different capacity utilizations. It will be noted that the AW is moderately sensitive to capacity utilization. The plant could be operated at a little less than 65% of capacity, instead of the assumed 75%, and still produce an AW greater than zero. Also, quite clearly, if they should be able to operate above the assumed 75% of capacity, the AW would be very good. This type of analysis gives the analyst a good idea of how much leeway the company can have in capacity utilization and still have an acceptable venture.

10.7.2 Sensitivity to Selling Price

Examination of the sensitivity of the project to the selling price of the concrete reveals the situation shown in Table 10-2. The values in this table assume that the plant would operate at 75% of capacity; the expenses would thus remain

TABLE 10-1 AW at $i = 15\%$ per year for a Premixed-Concrete Plant for Various Capacity Utilizations (Average Selling Price Is $45 per Cubic Yard)

	50% Capacity	65% Capacity	90% Capacity
Annual revenue	$405,000	$526,500	$729,000
Annual expenses:			
Capital recovery	−28,190	−28,190	−28,190
Labor	−113,750	−113,750	−113,750
Payroll taxes and similar items	−28,438	−28,438	−28,438
Taxes and insurance	−3,000	−3,000	−3,000
Operations and maintenance[a]	−13,715	−15,086	−17,372
Materials	−243,000	−315,900	−437,400
Total expenses	−$430,093	−$504,364	−$628,150
AW	−$ 25,093	+$ 22,136	+$100,850

[a]Let $x =$ annual operations and maintenance expenses and assume that 50% of the cost varies directly with capacity utilization. At 75% capacity utilization, $x/2 + (x/2)(0.75) = \$16,000$, so that $x = \$18,286$ at 100% capacity utilization. Therefore, at 50% utilization, the operations and maintenance expense would be $\$9,143 + 0.5(\$9,143) = \$13,715$.

TABLE 10-2 Effect of Various Selling Prices on the AW for the Premixed-Concrete Plant Operating at 75% of Capacity

	Selling Price			
	$45.00	$43.65(3%)[a]	$42.75(5%)[a]	$40.50(10%)[a]
Annual revenue	$607,500	$589,275	$577,125	$546,750
Annual expenses	−553,878	−553,878	−553,878	−553,878
AW	$ 53,622	$ 35,397	$ 23,247	−$ 7,128

[a]Percentage values shown in parentheses are reductions in price below $45.

constant, with only the selling price varying. Here it will be noted that the project is quite sensitive to price. A decrease in price of 10% would reduce the IRR to less than 15% (i.e., AW < 0). Since a decrease of 10% is not very large, the investors would want to make a thorough study of the price structure of concrete in the area of the proposed plant, particularly with respect to the possible effect of the increased competition that the new plant would create. If such a study reveals price instability in the market for concrete, the plant could be a risky investment.

10.7.3   Sensitivity to Useful Life

The effect of the third factor, assumed useful life of the plant, can be investigated readily. If a life of five years were assumed for the plant, instead of the assumed value of ten years, the only factor in the study that would be changed would be the cost of capital recovery. If the market value is assumed to remain constant, the capital recovery amount over a five-year period is

$$-\$100{,}000(A/P, 15\%, 5) + \$20{,}000(A/F, 15\%, 5) = -\$26{,}866 \text{ per year}$$

which is $7,926 more expensive than the initial value of −$18,940. In this case the AW would be reduced to $45,696—a decline of 14.8%. Hence, a 50% reduction in useful life causes only a 14.8% reduction in AW. Clearly, the venture is rather insensitive to the assumed useful life.

With the added information supplied by the sensitivity analyses that have just been described, those who make the investment decision concerning the proposed concrete plant would be in a much better position than if they had only the initial study results, based on an assumed utilization of 75% of capacity, available to them. They would know which factors were critical and thus could seek more information about these particular items if desired.

10.8   Optimistic–Pessimistic Estimates

A useful method for exploring sensitivity is to estimate one or more factors in a favorable (optimistic) direction and in an unfavorable (pessimistic) direction to investigate the effect of these changes on study results. This is a simple and effective method for including uncertainty in the analysis.

In applications of this method, the optimistic condition for a factor is often specified as a value that has 1 chance in 20 of being exceeded by the actual outcome. Similarly, the pessimistic condition has 19 chances out of 20 of being exceeded by the actual outcome. In operational terms, the optimistic condition for a factor is the value when things occur as well as can be reasonably expected, and the pessimistic estimate is the value when things occur as detrimentally as can be reasonably expected.

As an example, consider a proposed ultrasound inspection device for which the optimistic, pessimistic, and most likely (or best) estimates are given in Table 10-3. The MARR is 8% per year. Also shown at the end of Table 10-3 are the AWs for all three estimation conditions. Note that the AW for optimistic conditions is highly favorable (+$73,995), while for pessimistic conditions it is quite unfavorable

TABLE 10-3   Optimistic, Most Likely, and Pessimistic Estimates and AWs for Proposed Ultrasound Device

	Estimation Condition		
	Optimistic (O)	Most Likely (M)	Pessimistic (P)
Capital investment	−$150,000	−$150,000	−$150,000
Life	18 years	10 years	8 years
Market value	0	0	0
Annual revenues	$110,000	$ 70,000	$ 50,000
Annual expenses	− 20,000	− 43,000	− 57,000
AW	+$ 73,995	+$ 4,650	−$ 33,100

(−$33,100). After obtaining this information, the decision maker may be willing to make a go–no-go decision on the proposed device. However, he or she should recognize that these are extreme outcomes; the optimistic AW assumes that *all* estimated factors turn out according to the optimistic estimates, and the pessimistic AW assumes that *all* estimated factors turn out per the pessimistic estimates. It is reasonable to assume that this will not happen, but instead that different factors may turn out to have a mixture of optimistic, most likely, and pessimistic outcomes. One good way to reflect such results is shown in Table 10-4, which shows AWs for all combinations of estimated outcomes for three of the key factors being estimated. This could also have been done for four or more factors if they had been subject to significant variation, but the size of the matrix display would grow enormously. Even the $3 \times 3 \times 3$ matrix shown in Table 10-4 quickly becomes cumbersome because of the proliferation of numbers.

10.8.1   Making Matrices Easier to Interpret

It should be recognized that the AW numbers in Table 10-4 result from estimates subject to varying degrees of uncertainty. Hence, little information of value would

TABLE 10-4   AWs ($) for All Combinations of Estimated Outcomes[a] for Annual Revenues, Annual Expenses, and Life: Proposed Ultrasound Device

	Annual Expenses								
	O			M			P		
	Life			Life			Life		
Annual Revenues	O	M	P	O	M	P	O	M	P
O	73,995	67,650	63,900	50,995	44,650	40,900	36,995	30,650	26,900
M	34,000	27,650	23,900	10,995	4,650	900	−3,005	−9,350	−13,100
P	14,000	7,650	3,900	−9,005	−15,350	−19,100	−23,005	−29,350	−33,100

[a]Estimates: O, optimistic; M, most likely; P, pessimistic.

TABLE 10-5 Results in Table 10-4 Made Easier to Interpret (AWs in $000s)[a,b]

Annual Revenues	Annual Expenses								
	O			M			P		
	Life			Life			Life		
	O	M	P	O	M	P	O	M	P
O	(74)	(68)	(64)	(51)	45	41	37	31	27
M	34	28	24	11	5	1	− 3	− 9	−13
P	14	8	4	− 9	−15	−19	−23	−29	−33

[a] Estimates: O, optimistic; M, most likely; P, pessimistic.
[b] Circled entries, AW > $50,000 (4 out of 27 combinations); underscored entries, AW < $0 (9 out of 27 combinations).

be lost if the numbers were rounded to the nearest thousand dollars. Further, suppose that management is most interested in the number of combinations of conditions in which the AW is, say, (1) more than $50,000 and (2) less than $0. Table 10-5 shows how Table 10-4 might be changed to make it easier to interpret and use in communicating study results to management.

From Table 10-5 it is apparent that four combinations result in AW > $50,000, while nine produce AW < $0. Each combination of conditions is not necessarily equally likely. Therefore, statements such as "There are 9 chances out of 27 that we will lose money on this project" are not appropriate in this example.

10.8.2 Graphical Sensitivity Displays

As was demonstrated in Figure 10-4 (for Example 10-4), an effective way of displaying and examining sensitivity is to graph the measure of merit for independent variation of all factors of interest by expressing variation for each on a common abscissa in terms of percent deviation from its most likely value. This is shown in Figure 10-5 for the proposed ultrasound inspection device, for which the most likely estimates were given in Table 10-3. Figure 10-5 shows, among other things, that AW is relatively sensitive to changes in annual revenues, annual expenses, and reductions in project life. It also shows that AW is relatively insensitive to changes in the MARR and to increases in the project life.

Another type of sensitivity test that is often quite valuable is to determine the relative (or absolute) change in one or more factors that will just reverse the decision. Applied to the example in Table 10-3, this means determining the relative change in each factor that will decrease the AW by $4,650 so that it reaches $0. Table 10-6 shows this application by using a table and bars of varying lengths to emphasize that the AW of the device is (1) most sensitive to changes in the estimated annual revenues and (2) least sensitive to changes in the market value.

It is clear that even with a few factors, the number of possible combinations of conditions in a sensitivity analysis can become quite large, and the task of investigating all of them might be quite time-consuming. This was made obvious in Example 10-4. Ordinarily, sensitivity analysis involves eliminating from detailed

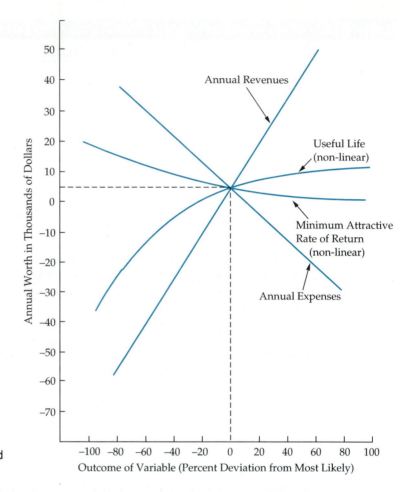

FIGURE 10-5

Graph of Sensitivity to Multiple Variables, Each Independently Deviating from the Most Likely Estimate (Proposed Ultrasound Device)

TABLE 10-6   Sensitivity of Decision Reversal to Changes in Selected Estimtates

	Most Likely Estimate	Required Outcome[a]	Amount of Change	Change Amount as Percentage of Most Likely Estimate	
Capital investment	−$150,000	−$181,000	−$31,200	+20.8%	_____
Useful life	10 years	7.3 years	−2.7 years	−27.0%	_____
Market value	0	− 67,890	− 67,890	∞	⟶
Annual revenues	70,000	65,350	− 4,650	− 6.6%	__
Annual expenses	−43,000	−47,650	− 4,650	+10.8%	___
MARR	8%	12.5%	+ 4.5%	+ 56%	_____

[a]To reverse decision (decrease AW to $0). Notice that reversal of AW is most sensitive to change in annual revenues.

consideration those factors for which the measure of merit is quite insensitive and highlighting the conditions for other factors to be studied further in accordance with the degree of sensitivity of each. Thus, the number of combinations of conditions included in the analysis can perhaps be kept to a manageable size.

10.9 Risk-Adjusted Minimum Attractive Rates of Return

Uncertainty causes factors inherent to engineering economy studies, such as cash flows and project life, to become random variables in the analysis. (Simply stated, a random variable is a function that assigns a unique numerical value to each possible outcome of a probabilistic quantity.) A widely used industrial practice for including some consideration of uncertainty is to increase the MARR when a project is thought to be relatively uncertain. Most likely estimates of other factors are then utilized in the study. Hence, a procedure has emerged that employs *risk-adjusted* interest rates. It should be noted, however, that many pitfalls of performing studies of financial profitability with risk-adjusted MARRs have been identified.* Also, the procedure does not make the uncertainty in the project estimates explicit.

In general, the preferred practice to account for uncertainty in estimates (cash flows, project life, etc.) is to deal directly (explicitly) with their suspected variations in terms of probability assessments (Chapter 14) rather than to manipulate the MARR as a means of reflecting the virtually certain versus highly uncertain status of a project. Intuitively, the risk-adjusted interest rate procedure can be defended because more certainty regarding the overall profitability of a project exists in the early years compared to, say, the last two years of its life. Increasing the MARR places emphasis on early cash flows rather than on longer-term benefits, and this would appear to help compensate for time-related project uncertainties. But the question of uncertainty in cash flow amounts is not directly addressed. The following example illustrates how this method of dealing with uncertainty can lead to an illogical recommendation.

EXAMPLE 10-6

The Atlas Corporation is considering two alternatives, both affected by uncertainty to different degrees, for increasing the recovery of a precious metal from its smelting process. The following data concern capital investment requirements and estimated annual savings of both alternatives. The firm's MARR for its risk-free investments is 10% per year.

	Alternative	
End-of-Year, k	P	Q
0	−$160,000	−$160,000
1	120,000	20,827
2	60,000	60,000
3	0	120,000
4	60,000	60,000

---

*A. A. Robichek, and S. C. Myers, "Conceptual Problems in the Use of Risk-Adjusted Discount Rates," *Journal of Finance*, vol. 21 (December 1966), 727–730.

Because of the technical considerations involved, alternative P is thought to be *more uncertain* than alternative Q. Therefore, according to the Atlas Corporation's engineering economy handbook, the risk-adjusted MARR applied to P will be 20% per year and the risk-adjusted MARR for Q has been set at 17% per year. Which alternative should be recommended?

SOLUTION

At the risk-free MARR of 10%, both alternatives have the same PW of $39,659. All else being equal, alternative Q would be chosen because it is less uncertain than alternative P. Now a PW analysis is performed for the Atlas Corporation, using its prescribed risk-adjusted MARRs for the two options.

$$PW_P(20\%) = -\$160,000$$
$$+ \$120,000(P/F, 20\%, 1) + \$60,000(P/F, 20\%, 2)$$
$$+ \$60,000(P/F, 20\%, 4) = \$10,602$$
$$PW_Q(17\%) = -\$160,000 + \$20,827(P/F, 17\%, 1)$$
$$+ \$60,000(P/F, 17\%, 2)$$
$$+ \$120,000(P/F, 17\%, 3)$$
$$+ \$60,000(P/F, 17\%, 4) = \$8,575$$

Without considering uncertainty (i.e., MARR = 10% per year), the selection was seen to be alternative Q. But when the more uncertain alternative P is "penalized" by applying a higher risk-adjusted MARR to compute its PW, the comparison of alternatives favors alternative P. One would expect to see alternative Q recommended with this procedure. This contradictory result can be seen clearly in Figure 10-6, which demonstrates the general situation in which contradictory results might be expected.

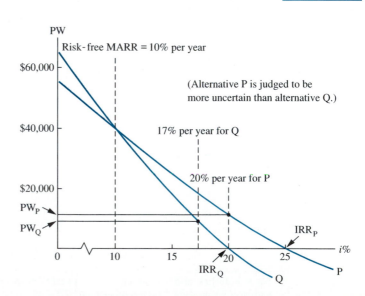

FIGURE 10-6
Graphical Portrayal of Risk-Adjusted
Interest Rates (Example 10-6)

Even though the intent of the risk-adjusted MARR is to make more uncertain projects appear less economically attractive, the opposite was shown to be true in Example 10-6. Furthermore, a related shortcoming of the risk-adjusted MARR procedure is that cost-only projects are made to appear more desirable (to have a less negative PW, for example) as the interest rate is adjusted upward to account for uncertainty. At extremely high interest rates, the alternative having the lowest investment requirement would be favored, regardless of subsequent cost cash flows. Because of difficulties such as those illustrated, this procedure is not generally recommended as an acceptable means of dealing with uncertainty.

10.10 Reduction of Useful Life

Many of the methods for dealing with uncertainty that have been discussed to this point have attempted to compensate for potential losses that could be incurred if conservative decision-making practices are not followed. Thus, dealing with uncertainty in an engineering economy study tends to lead to the adoption of conservative (pessimistic) estimates of factors so as to reduce downside risks of making a wrong decision.

The method considered in this section makes use of a truncated project life that is often considerably less than the estimated useful life. By dropping from consideration those revenues (savings) and expenses that may occur after the reduced study period, heavy emphasis is placed on rapid recovery of investment capital in the early years of a project's life. Consequently, this method is closely related to the discounted payback technique discussed in Chapter 4, and it suffers from most of the same deficiencies that beset the payback method.

EXAMPLE 10-7

Suppose that the Atlas Corporation referred to in Example 10-6 decided not to utilize risk-adjusted interest rates as a means of recognizing uncertainty in their engineering economy studies. Instead, they have decided to truncate the study period at 75% of the most likely estimate of useful life. Hence, all cash flows past the third year would be ignored in the analysis of alternatives. By using this method, should alternative P or Q be selected when MARR $= 10\%$ per year?

SOLUTION

Based on the PW criterion, it is apparent that neither alternative would be the choice with this procedure for recognizing uncertainty.

$$PW_P(10\%) = -\$160,000 + \$120,000(P/F, 10\%, 1)$$
$$+ \$60,000(P/F, 10\%, 2) = -\$1,322$$

$$PW_Q(10\%) = -\$160{,}000 + \$20{,}827(P/F, 10\%, 1)$$
$$+ \$60{,}000(P/F, 10\%, 2)$$
$$+ \$120{,}000(P/F, 10\%, 3) = -\$1{,}322$$

EXAMPLE 10-8

A proposed new product line requires $2,000,000 in capital investment over a two-year construction period. Projected revenues and expenses over this product's anticipated eight-year commercial life are shown, along with its capital requirements.

Type of Cash Flow	End of Year (Millions of $)									
	−1	0	1	2	3	4	5	6	7	8
Capital investment	−0.9	−1.1	0	0	0	0	0	0	0	0
Revenues	0	0	1.8	2.0	2.1	1.9	1.8	1.8	1.7	1.5
Expenses	0	0	−0.8	−0.9	−0.9	−0.9	−0.8	−0.8	−0.8	−0.7

The company's maximum simple payback period is four years (after taxes), and its after-tax MARR is 15% per year. This investment will be depreciated using the MACRS (GDS) method and a five-year property class. (GDS recovery rates are given in Table 7-3.) An effective income tax rate of 40% applies to taxable income produced by this new product.

The company's management is quite concerned about the financial attractiveness of this venture if unforeseen circumstances (e.g., loss of market and/or technological breakthroughs) occur. They are leery of investing a large sum of capital in this product because competition is quite keen and companies that wait to enter the market may be able to purchase more cost-efficient technology. You have been given the assignment of assessing the downside profitability of the product when the primary concern is its staying power (life) in the marketplace. That is, they must determine the minimum life of the product that will produce an acceptable after-tax IRR. Draw a graph of your results and list all appropriate assumptions.

SOLUTION

An analysis of after-tax cash flows is shown in Table 10-7 for the most likely product life of eight years.

It has been assumed that the residual (market) value of the investment is zero. Moreover, the MACRS depreciation deductions are assumed to be unaffected by the useful life of this product, and they begin in the first year of commercial operation (year one). A plot of after-tax IRR versus actual life of the product line is shown in Figure 10-7. To make at least 15% per year after taxes on this venture, the product's life must be four years or more. It can be quickly determined from Table 10-7 that the *simple* payback period is three years. Consequently, this new product would appear to be a judicious investment as long as its actual life turns out to be four years or greater.

TABLE 10-7   After-Tax Analysis of Example 10-8

End of Year, k	(A) BTCF	(B) Depreciation Deduction	(C) = (A) − (B) Taxable Income	(D) = −0.4(C) Cash Flow for Income Taxes	(E) = (A) + (D) ATCF
−1	−$ 900,000	—	—	—	−$ 900,000
0	− 1,100,000	—	—	—	− 1,100,000
1	1,000,000	$400,000	$600,000	−$240,000	760,000
2	1,100,000	640,000	460,000	− 184,000	916,000
3	1,200,000	384,000	816,000	− 326,400	873,600
4	1,000,000	230,400	769,600	− 307,840	692,160
5	1,000,000	230,400	769,600	− 307,840	692,160
6	1,000,000	115,200	884,800	− 353,920	646,080
7	900,000	0	900,000	− 360,000	540,000
8	800,000	0	800,000	− 320,000	480,000

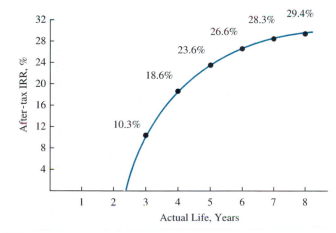

FIGURE 10-7

IRR for Different Product Lives in Example 10-8

10.11   Spreadsheet Applications

Spreadsheet applications provide an excellent capability to answer "what-if" questions. In the following example, a spreadsheet is used to determine the sensitivity of a project's present worth to several factors.

EXAMPLE 10-9

In this example we explore the impact on the present worth of an engineering project relative to changes in the capital investment, annual savings, market value, study period, and MARR.

Figure 10-8 shows the resulting table of present worth values as each parameter of the present worth calculation is varied over a range of ±50% from the most likely estimate. Each column has a unique formula that refers to the parameters located in the range C2:C6 to determine present worth. The particular parameter

	A	B	C	D	E	F
1	Most Likely Estimates					
2	Capital Investment (I):		($50,000)			
3	Annual Savings (A):		$12,000			
4	Market Value (MV):		$5,000			
5	Study Period (N):		8			
6	MARR (i):		10%			
7						
8	% Change	I	A	MV	N	i
9						
10	-50%	$ 41,352	$ (15,658)	$ 15,185	$ (8,547)	$ 30,943
11	-40%	$ 36,352	$ (9,256)	$ 15,419	$ (2,780)	$ 27,655
12	-30%	$ 31,352	$ (2,854)	$ 15,652	$ 2,563	$ 24,566
13	-20%	$ 26,352	$ 3,548	$ 15,885	$ 7,514	$ 21,661
14	-10%	$ 21,352	$ 9,950	$ 16,118	$ 12,101	$ 18,927
15	0%	$ 16,352	$ 16,352	$ 16,352	$ 16,352	$ 16,352
16	10%	$ 11,352	$ 22,754	$ 16,585	$ 20,290	$ 13,923
17	20%	$ 6,352	$ 29,155	$ 16,818	$ 23,940	$ 11,631
18	30%	$ 1,352	$ 35,557	$ 17,051	$ 27,321	$ 9,466
19	40%	$ (3,648)	$ 41,959	$ 17,285	$ 30,454	$ 7,419
20	50%	$ (8,648)	$ 48,361	$ 17,518	$ 33,357	$ 5,482

FIGURE 10-8
Spreadsheet for Performing
a Sensitivity Analysis

of interest, for example the study period in column E, is multiplied by the factor
(1 + % change) to create the table. You can verify your formulas by noting that all
columns are equal at the most likely value (% Change = 0). The formulas in the
highlighted cells in Figure 10-8 are as follows.

Cell	Contents
B10	= $C$2 * (1 + A10) + PV($C$6, $C$5, −$C$3) + $C$4/(1 + $C$6)^$C$5
C10	= $C$2 + PV($C$6, $C$5, −$C$3 * (1 + A10)) + $C$4/(1 + $C$6)^$C$5
D10	= $C$2 + PV($C$6, $C$5, −$C$3) + $C$4 * (1 + A10)/(1 + $C$6)^$C$5
E10	= $C$2 + PV($C$6, $C$5 * (1 + A10), −$C$3) + $C$4/(1 + $C$6)^($C$5 * (1 + A10))
F10	= $C$2 + PV($C$6 * (1 + A10), $C$5, −$C$3) + $C$4/(1 + $C$6 * (1 + A10))^$C$5

For ease of interpretation, it is helpful to graph the results of the sensitivity
analysis, which is easily accomplished using the charting feature that exists in most
spreadsheet packages. The graphed results of this analysis are shown in Figure
10-9. The graph uses the "% Change" column as the independent (X) axis and
columns B through F.

This graph indicates that present worth is most sensitive to annual savings.
The next most influential parameter is the capital investment. The least sensitive
parameter is the market value, which is as expected because it is a small dollar
amount and is more heavily discounted because it occurs at the end of the study
period.

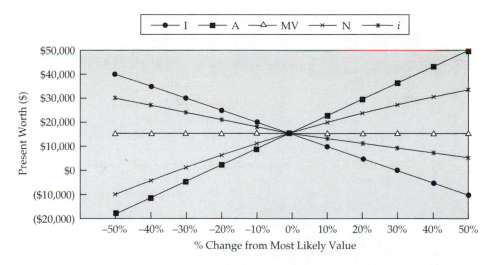

FIGURE 10-9   Sensitivity Analysis of Five Parameters in Example 10-9

10.12   Summary

Engineering economy involves decision making among competing uses of scarce capital resources. The consequences of resultant decisions usually extend far into the future. In this chapter, we have used nonprobabilistic techniques to deal with the realization that the consequences (cash flows, useful lives, etc.) of engineering projects can never be known with absolute certainty. This situation is generally referred to as *decision making under uncertainty*.

Several of the most commonly applied and useful nonprobabilistic procedures for dealing with uncertainty in engineering economy studies have been presented in this chapter: (1) breakeven analysis, (2) sensitivity analysis, (3) optimistic-pessimistic estimates, (4) risk-adjusted MARRs, and (5) reduction in useful life. Breakeven analysis determines the value of a key common factor, such as utilization of capacity, at which the economic desirability of two alternatives is equal. This breakeven point is then compared to an independent estimate of the factor's most likely value to assist with the selection between alternatives. Similarly, sensitivity analysis typically determines the range of values that a key parameter may have without reversing the superiority that the best alternative has over others being considered. The remaining three procedures for dealing with uncertainty are aimed at selecting the best course of action when one or more consequences of alternatives being evaluated lack estimation precision.

Regrettably, there is no quick and easy answer to the question "How should uncertainty best be considered in an engineering economic analysis?" Generally, simple procedures (e.g., breakeven analysis and sensitivity analysis) allow reasonable discrimination among alternatives to be made on the basis of the uncertainties present, and they are relatively inexpensive to apply. Additional discrimination among alternatives is possible with more complex procedures that utilize probabilistic concepts (Chapter 14), but their difficulty of application and expense may be prohibitive.

10.13 References

CANADA, J. R., W. G. SULLIVAN, and J. A. WHITE. *Capital Investment Decision Analysis for Engineering and Management*, 2nd ed. Englewood Cliffs, N.J.: Prentice-Hall, Inc., 1996.

CHURCHMAN, C. W., R. L. ACKOFF, and E. L. ARNOFF. *Introduction to Operations Research*. New York: John Wiley & Sons, 1957.

FLEISCHER, G.A. *Engineering Economy*. Monterey, Calif.: Brooks/Cole Engineering Division of Wadsworth, Inc., 1984.

GRANT, E. L., W. G. IRESON, and R. S. LEAVENWORTH. *Principles of Engineering Economy*. New York: John Wiley & Sons, 1990.

MORRIS, W. T. *The Analysis of Management Decisions*. Homewood, Ill.: Richard D. Irwin Co., 1964.

10.14 Problems

The number in parentheses ( ) that follows each problem refers to the section from which the problem is taken.

10-1. Explain why the terms *risk* and *uncertainty* can be used interchangeably throughout this book. (10.2)

10-2. Why should the effects of uncertainty be considered in engineering economy studies? What are some likely sources of uncertainty in these studies? (10.3)

10-3. Construct your own *nonlinear* breakeven analysis problem, develop a solution for it, and bring a one-page summary of your problem and solution to class for discussion. (10.5)

10-4. How are the optimistic and pessimistic values of parameters determined when using the optimistic-pessimistic estimation approach to dealing with uncertainty? (10.8)

10-5. Consider these two alternatives:

	Alternative 1	Alternative 2
Capital investment	−$4,500	−$6,000
Annual revenues	$1,600	$1,850
Annual expenses	−$400	−$500
Estimated market value	$800	$1,200
Useful life	8 years	10 years

a. Suppose that the market value of Alternative 1 is known with certainty. By how much would the estimate of market value for Alternative 2 have to vary so that the *initial* decision based on these data would be reversed? The annual MARR is 15% per year. (10.5)

b. Determine the life of Alternative 1 for which the AWs are equal. (10.5)

10-6. Two 100 horsepower motors are being considered for use (See the top of page 429):

a. If power cost is $0.10 per kWh, and the interest rate is 12% per year, how many hours of operation per year are required to justify the purchase of XYZ brand motor? (1 hp = 0.746 kW) (10.5)

b. Given your answer in part (a), which motor would you select if the motor is expected to operate 2000 hours per year? Explain why. (10.5)

10-7. A set of four "100-year," sixty (60) watt light bulbs costs $24.95. If *H* equals the annual number of hours the bulb is used, each bulb provides $100H$ hours of light, and the efficiency of each bulb is 80%. A set of four, sixty (60) watt *standard* bulbs costs $2.60. Each standard bulb provides *H* hours of light and the efficiency of each bulb is 95%. If the cost of electricity is $0.0574/kWh, how many hours of operation each year, *H*, will make the costs of the two types of bulbs *breakeven*? Your personal MARR is 5% per

	ABC Brand	XYZ Brand
Purchase price	−$1,900	−$6,200
Useful life in years	10	10
Market value	none	none
Annual maintenance expense	−$170	−$310
Efficiency	80%	90%

(Pr. 10-6)

year. Assume repeatability. State other realistic assumptions you must make. (10.5)

10-8. Two electric motors are being considered to power an industrial hoist. Each is capable of providing 90 hp. Pertinent data for each motor are as follows:

	Motor	
	D-R	**Westhouse**
Capital investment	−$2,500	−$3,200
Electrical efficiency	0.74	0.89
Maintenance/ year	−$40	−$60
Useful life	10 years	10 years

If the expected usage of the hoist is 500 hours per year, what would the cost of electrical energy have to be (in cents per kilowatt-hour) before the D-R motor is favored over the Westhouse motor? The MARR is 12% per year. [*Note:* 1 hp = 0.746 kW] (10.5)

10-9. Your company operates a fleet of light trucks that are used to provide contract delivery services. As the engineering and technical manager, you are analyzing the purchase of 55 new trucks as an addition to the fleet. These trucks would be used for a new contract the sales staff is trying to obtain. If purchased, the trucks would cost $21,200 each; estimated use is 20,000 miles per year per truck; estimated operation and maintenance and other related expenses (year zero dollars) are $0.45 per mile, which is forecasted to escalate (increase) at the rate of 5% per year; and the trucks are MACRS (GDS) three-year property class assets. The analysis period is four years; $t = 38\%$; MARR = 15% per year (after taxes; includes an inflation component); and the estimated MV at the end of four years (in year zero dollars) is 35% of the purchase price of the vehicles. This estimate is expected to escalate at the rate of 2% per year.

Based on an after-tax, actual dollar analysis, what is the annual revenue required by your company from the contract to justify these expenditures before any profit is considered? This calculated amount for annual revenue is the breakeven point between purchasing the trucks and which other alternative? (10.5)

10-10. A nationwide motel chain is considering locating a new motel in Bigtown, USA. The cost of building a 150-room motel (excluding furnishings) is $5 million. The firm uses a 15 year planning horizon to evaluate investments of this type. The furnishings for this motel must be replaced every 5 years at an estimated cost of $1,875,000 (at $k = 0$, 5, and 10). The old furnishings have no market value. Annual operating and maintenance expenses for the facility are estimated to be $125,000. The market value of the motel after 15 years is estimated to be 20% of the original building cost.

Rooms at the motel are projected to be rented at an average rate of $45 per night. On the average, the motel will rent 60% of its rooms each night. Assume the motel will be open 365 days per year. The MARR is 10% per year. (10.6)
a. Using an annual worth measure of merit, is the project economically attractive?
b. Investigate sensitivity to decision reversal for the following three factors: (1) capital investment, (2) MARR, and (3) occupancy rate (average percent of rooms rented per night). To which of these factors is the decision most sensitive?
c. Graphically investigate the sensitivity of the annual worth to changes in these three factors. Investigate changes over the interval ±40%. On your graph, use percent change as the x-axis and annual worth as the y-axis.

10-11. Your engineering group is analyzing a new model of a large specialty pump to replace an existing pump in a chemical plant operation. The analysis period is six years; assume MACRS (GDS) depreciation with a five-year recovery period; the base year is year zero ($b = 0$). Estimates of costs and savings, and other information are as follows:

Cash Flow Item	Estimate in Year 0 Dollars	Annual Price Change (e_j)
Capital investment	$-I = ?$	—
Annual savings	60,000	5%
MV at end of 6 years	20,000	3%

NOTES:

1. The general inflation rate (f) is 5%.
2. Assume that your company is in the 34% federal income tax bracket.
3. The state income tax rate is 3.8%.
4. The after-tax MARR (*not* including inflation) is 9.524% per year.

What is the maximum amount justified for the capital investment of the pump? Use an after-tax, actual dollar analysis and the PW method as applicable. This calculated capital investment amount is the breakeven point between the new pump and which other alternative? (10.5)

10-12. Consider the following cash flow diagram. Plot changes in PW to ±20% and ±40% changes in the project's life, N. Let $i = 10\%$ per year and assume that MV = 0. State any other assumptions you make. (10.6)

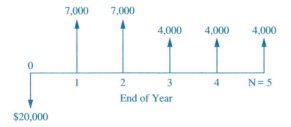

10-13. A new steam flow monitoring device must be purchased immediately by a local municipality. These most likely estimates have been developed by a group of engineers:

Capital investment	$-\$140,000$
Annual savings	$\$ 25,000$
Useful life	12 years
Market value (end of year 12)	$\$ 40,000$
MARR	10%/year

Because considerable uncertainty surrounds these estimates, it is desired to evaluate the sensitivity of PW to ±50% changes in the most likely estimates of (a) annual savings, (b) useful life, and (c) interest rate (MARR). Graph the results and determine to which factor the decision is most sensitive. (10.6)

10-14. It is desired to determine the most economic thickness of insulation for a large cold-storage room. Insulation is expected to cost $150 per 1,000 sq ft of wall area per inch of thickness installed and to require annual property taxes and insurance of 5% of the capital investment. It is expected to have $0 market value after a 20-year life. The following are estimates of the heat loss per 1,000 sq ft of wall area for several thicknesses:

Insulation, in.	Heat loss, Btu per hr
3	4,400
4	3,400
5	2,800
6	2,400
7	2,000
8	1,800

The cost of heat removal is estimated at $0.02 per 1,000 Btu per hr. The MARR is 20% per year. Assuming continuous operation throughout the year, analyze the sensitivity of the optimal thickness to errors in estimating the cost of heat removal. Use the AW technique (a computer spreadsheet should be considered here). (10.6)

10-15. An industrial machine costing $10,000 will produce net cash savings of $4,000 per year. The machine has a five-year economic life but must be returned to the factory for major repairs after three years of operation. These repairs cost $5,000. The company's MARR is 10% per year. What internal rate of return will be earned on purchase of this machine? Analyze the sensitivity of the internal rate of return to ±$2,000 changes in the repair cost. (10.6)

	Number of floors				(Pr. 10-16)
	2	3	4	5	
Capital investment	-$200,000	-$250,000	-$320,000	-$400,000	
Annual revenue	40,000	60,000	85,000	100,000	
Annual expenses	-15,000	-25,000	-25,000	-45,000	

10-16. It is desired to determine the optimal height for a proposed building that is expected to last 40 years and then be demolished at zero net market value. Pertinent data are shown at the top of this page. In addition to the building capital investment, the land requires an investment of $50,000 and is expected to retain that value throughout the useful life period. Analyze the sensitivity of the decision due to changes in estimates of the MARR between 10%, 15%, and 20%. Use the PW method and ignore income taxes. (10.6)

10-17. An office building is considering converting from a coal burning furnace to one that burns either fuel oil or natural gas. The cost of converting to fuel oil is estimated to be $80,000 initially; annual operating expenses are estimated to be $4,000 less than that experienced using the coal furnace. Approximately 140,000 Btus are produced per gallon of fuel oil; fuel oil is anticipated to cost $1.10 per gallon.

The cost of converting to natural gas is estimated to be $60,000 initially; additionally, annual operating and maintenance expenses are estimated to be $6,000 less than that for the coal-burning furnace. Approximately 1,000 Btus are produced per cubic foot of natural gas; it is estimated natural gas will cost $0.02 per cu ft.

A planning horizon of twenty years is to be used. Zero market values and a MARR of 10% per year are appropriate. Perform a sensitivity analysis for the annual Btu requirement for the heating system. (*Hint:* First calculate the break-even number of Btus (in thousands). Then determine AWs if Btu requirement varies over ±30% of the break-even amount.) (10.6)

10-18. Suppose that for an engineering project the optimistic, most likely, and pessimistic estimates are as shown below. (10.8)
  a. What is the AW for each of the three estimation conditions?
  b. It is thought that the most critical elements are useful life and net annual cash flow. Develop a table showing the AW for all combinations of estimates for these two factors, assuming that all other factors remain at their most likely values.

10-19. Suppose that for a certain potential investment project the optimistic-pessimistic estimates are shown at the top of page 432.
  a. What is the annual worth for each of the three estimation conditions?
  b. It is thought that the most critical elements are useful life and net annual cash flow. Develop a table showing the net annual worth for all combinations of estimates for those two elements assuming all other elements to be at their "most likely" values. Also, show results with histogram bars for entries in your table. (10.8)

	Optimistic	Most Likely	Pessimistic	(Pr. 10-18)
Capital investment	-$80,000	-$95,000	-$120,000	
Useful life	12 years	10 years	6 years	
Market value	$30,000	$20,000	0	
Net annual cash flow	$35,000	$30,000	$20,000	
MARR	12%/yr	12%/yr	12%/yr	

	Optimistic	Most Likely	Pessimistic	(Pr. 10-19)
Capital investment	−$90,000	−$100,000	−$120,000	
Useful life	12 years	10 years	6 years	
Market value	$30,000	$20,000	0	
Net annual cash flow	$35,000	$30,000	$20,000	
MARR(per year)	10%	10%	10%	

10-20. A bridge is to be constructed now as part of a new road. Engineers have determined that traffic density on the new road will justify a two-lane road and a bridge at the present time. Because of uncertainty regarding future use of the road, the time at which an extra two lanes will be required is currently being studied.

The two-lane bridge will cost $200,000 and the four-lane bridge, if built at one time, will cost $350,000. The cost of widening a two-lane bridge to four lanes will be an extra $200,000 plus $25,000 for every year that construction is delayed. The MARR used by the highway department is 12% per year. The following estimates have been made of the times at which the four-lane bridge will be required:

Pessimistic estimate	4 years
Most likely estimate	5 years
Optimistic estimate	7 years

In view of these estimates, what would you recommend? What difficulty, if any, do you have in interpreting your results? List some advantages and disadvantages of this method of preparing estimates. (10.8)

10-21. A small-town cable TV company wishes to expand its coverage into a rural area just east of town. A study indicates that 3,000 homes will be within connecting range of the proposed cable system, and the company knows that the typical subscriber rate is 43% (i.e., 43% of the homes with the option to get cable TV actually do so). The basic charge is $20 per month, plus a one-time $50 installation fee. The company

must pay 3% of its gross revenue to the city government and 20% of its gross revenue to the program sources (e.g., HBO, Showtime). The company uses a 10-year planning horizon for this type of project (zero salvage value is assumed). The company's present personnel can easily service 2,000 new subscribers without having to use overtime or hire new people. However, some additional overtime maintenance expenses would be incurred. There is a choice between (a) using company people to install and maintain the new lines, taps, and distribution amplifiers and (b) letting CABLEX, an outside company that specializes in CATV installations and maintenance, do the job for a fee (see the table at the end of this page). The company's MARR is 15% per year. (10.5)

a. Which alternative is more attractive?

b. What is the breakeven point of the project in percentage of homes subscribing?

10-22. Motor XYZ in Problem 10-6 is manufactured in a foreign country and is believed to be less reliable than motor ABC. To cope with this uncertainty, a risk-adjusted MARR of 20% is utilized to calculate its AW. When hours of operation per year total 1,000, which motor would be selected? What difficulty is encountered with this method? (10.9)

10-23. Refer to Example 10-5, Section 10.7. For this premixed-concrete plant example, do the following: (10.7)

a. Graph the results of the sensitivity analysis for the three factors (capacity utilization, selling price, and useful life of plant) and include

	In-House	Cablex	(Pr. 10-21)
Materials and equipment	−$675,000	−$615,000	
Labor (one-time; installation)	− 110,000	− 145,000	
Maintenance	− 30,000/year	− 42,000/year	

any additional values of these factors you consider necessary. Also, include raw material costs as an additional factor in the sensitivity analysis under the assumption that all competitors may not respond to changes in these costs the same way.

b. In addition to graphing the sensitivity analysis results for each factor (a), select several combinations of factor values that you consider particularly insightful and develop the AW results for each.

c. Prepare information from (a) and (b) in a form you consider effective for communicating with the investor group considering the plant.

10-24. *Brain Teaser* (10.5, 10.6, 10.10)
Consider these two alternatives for solid waste removal:

Alternative A: Build a solid waste processing facility.

Capital investment	−$108 million in 1996 (commercial operation starts in 1996)
Expected life of facility	20 years
Annual operating expenses	−$3.46 million (expressed in 1996 dollars)
Estimated market value	40% of initial capital cost at all times

Alternative B: Contract with vendors for solid waste disposal after intermediate recovery.

Capital investment	−$17 million in 1996 (this is for *intermediate* recovery from the solid waste stream)
Expected contract period	20 years

Annual operating expenses	−$2.10 million (in 1996 dollars)
Repairs to intermediate recovery system every 5 years	−$3.0 million (in 1996 dollars)
Annual fee to vendors	−$10.3 million (in 1996 dollars)
Estimated market value at all times	$0

Related Data

MACRS (GDS) property class	15 years (see Table 7-4)
Study period	20 years
Effective income tax rate	40%
Company MARR (after-tax)	10% per year
Inflation rate	0% (ignore inflation)

a. How much more expensive (in terms of capital investment only) could Alternative B be in order to break even with Alternative A?

b. How sensitive is the after-tax PW of Alternative B to cotermination of both alternatives at the end of *year 10*?

c. Is the initial decision to adopt Alternative B in (a) reversed if our company's annual operating expenses for Alternative B only ($2.10 million per year) unexpectedly double? Explain why (or why not).

d. Use a computer spreadsheet available to you to solve this problem.

*S*pecial Topics in Engineering Economy

Money is the seed of money, and the first guinea is sometimes more difficult to acquire than the second million.

—Jean Jacques Rousseau, "A Discourse on Political Economy" in *The Social Contract*, 1762

*R*eplacement Analysis

*R*eplacement decisions are critically important to an operating organization. The objectives of this chapter are (1) to discuss the considerations involved in replacement studies and (2) to address the key question of whether an asset should be kept one or more years or immediately replaced with the best available challenger.

The following topics are discussed in this chapter:

Reasons for the replacement of assets

Factors that must be considered in replacement studies

A typical replacement problem

A decision roadmap for replacement analysis

Determining the economic life of the challenger(s)

Determining the economic life of a defender

Comparisons when the useful life of the defender and the challenger differ

Retirement without replacement (abandonment)

A comprehensive example

11.1 Introduction

A decision situation often encountered in business firms and government organizations, as well as by individuals, is whether an existing asset should be retired from use, continued in service, or replaced with a new asset. As the pressures of worldwide competition continue to increase, requiring better quality in goods and services, shorter response times, and other changes, this type of decision is occurring more frequently. Thus, the *replacement problem*, as it is commonly called, requires careful engineering economy studies to provide the information needed for sound decisions that improve the operating efficiency and the competitive position of an enterprise.

> Engineering economy studies of replacement situations are performed using the same basic methods as other economic studies involving two or more alternatives. The specific decision situation, however, occurs in different forms. Sometimes it may be whether to retire an asset without replacement (*abandonment*) or whether to retain the asset for back-up rather than primary use. Also, the decision may be whether the changed requirements can be met by *augmenting* the capacity or capability of the existing asset(s). Often, however, the decision is whether to replace an existing (old) asset, descriptively called the *defender*, with a new asset. The one or more alternative replacement (new) assets are then called *challengers*.

11.2 Reasons for Replacement Analysis

The need to evaluate the replacement, retirement, or augmentation of assets results from changes in the economics of their use in an operating environment. Various reasons can underlie these changes, and unfortunately they are sometimes accompanied by unpleasant financial facts. The four major reasons that summarize most of the factors involved are:

1. *Physical impairment (deterioration):* These are changes that occur in the physical condition of the asset. Normally, continuing use (aging) results in the operation of an asset becoming less efficient. Routine maintenance as well as breakdown repair costs increase, energy use may increase per unit of output, more operator time is required, and so forth. Or, some unexpected incident such as an accident occurs that affects the physical condition and the economics of ownership and use of the asset.

2. *Altered Requirements:* Capital assets are used to produce goods and services that satisfy human wants. When the demand for a good or service either increases or decreases or the design of a good or service changes, the related asset(s) may have the economics of its use affected.

3. *Technology:* The impact of changes in technology varies among different types of assets. For example, the relative efficiency of heavy highway construction equipment is impacted less rapidly by technological changes than automated manufacturing equipment. In general, the costs per unit of production, as well as quality and other factors, are favorably impacted by changes in technology, which results in more frequent replacement of existing assets with new and better challengers.

4. *Financing:* Financial factors involve economic opportunity changes external to the physical operation or use of assets and may involve income tax consid-erations.* For example, the rental (lease) of assets may become more attractive

---

*In this chapter, we often refer to Chapter 7 for details on depreciation methods and after-tax analysis.

than ownership. Regardless of the financial factors involved, the replacement asset does not necessarily have to be different from the existing asset.

Reason 2 (altered requirements) and Reason 3 (technology) are sometimes referred to as different categories of *obsolescence*. Even financial changes (Reason 4) could be considered a form of obsolescence. In any replacement problem, however, factors from more than one of these four major areas may be involved. Regardless of the specific considerations and even though there is a tendency to regard it with some apprehension, the replacement of assets often represents economic opportunity for the firm.

For the purposes of our discussion of replacement studies, the following is a distinction between various types of lives for typical assets.

Economic life is the period of time (years) that results in the minimum equivalent uniform annual cost (EUAC) of owning and operating an asset.* If we assume good asset management, economic life should coincide with the period of time extending from the date of acquisition to the date of abandonment, demotion in use, or replacement from the primary intended service.

Ownership life is the period between the date of acquisition and the date of disposal by a specific owner. A given asset may have different categories of use by the owner during this period. For example, a car may serve as the primary family car for several years and then serve only for local commuting for several more years.

Physical life is the period between original acquisition and final disposal of an asset over its succession of owners. For example, the car just described may have several owners over its existence.

Useful life is the time period (years) that an asset is kept in productive service (either primary or backup). It is an estimate of how long an asset is expected to be used in a trade or business to produce income.

11.3 Factors That Must Be Considered in Replacement Studies

There are several factors that must be considered in replacement studies. If they are not properly included in the analysis, erroneous study results may jeopardize reaching a sound decision. Once a proper perspective has been established regarding these factors, however, little difficulty should be experienced in making replacement studies. Six factors and related concepts are discussed in this section:

1. Recognition and acceptance of past errors
2. Sunk costs
3. Existing asset value and the *outsider viewpoint*

---

*The AW of a primarily cost cash flow pattern is sometimes called the equivalent uniform annual cost (EUAC). Because this term is commonly used in the definition of the economic life of an asset, we will often use EUAC in this chapter.

4. Income tax considerations
5. Economic life of the proposed replacement asset (challenger)
6. Remaining (economic) life of the old asset (defender)

11.3.1 Past Estimation Errors

The economic focus in a replacement study is the future. Any *estimation errors* made in a previous study related to the defender are not relevant (unless there are income tax implications). For example, when an asset's BV is greater than its current MV, the difference frequently has been designated as an estimation error. Such "errors" also arise when capacity is inadequate, maintenance expenses are higher than anticipated, and so forth.

This implication is unfortunate because in most cases these differences are not the result of errors but of honest inability to foresee future conditions. Acceptance of unfavorable economic realities may be made easier by posing a hypothetical question: "What will be the costs of my competitor, who has no past errors to consider?" In other words, we must decide whether we wish to live in the *past*, with its errors and discrepancies, or to be in a sound competitive position in the *future*. A common reaction is "I can't afford to take the loss in value of the existing asset that will result if the replacement is made." The fact is that the loss already has occurred, whether or not it could be afforded, and it exists whether or not the replacement is made.

11.3.2 The Sunk Cost Trap

Only present and future cash flows should be considered in replacement studies. Any unamortized values (i.e., unallocated value of an asset's capital investment) of an existing asset under consideration for replacement are strictly the result of *past* decisions—the initial decision to invest in that asset and decisions as to the method and number of years to be used for depreciation purposes. For purposes of this chapter, we define a *sunk cost* to be the difference between an asset's BV and its MV at a particular point in time. Sunk costs have no relevance to the replacement decisions that must be made (*except to the extent that they affect income taxes*). When tax considerations are involved, we must include them in an engineering economy study. Clearly, serious errors can be made in practice when sunk costs are incorrectly handled in replacement studies.

11.3.3 Investment Value of Existing Assets and the Outsider Viewpoint

Recognition of the nonrelevance of BVs and sunk costs leads to the proper viewpoint to use in placing value on existing assets for replacement study purposes. In this chapter we use the so-called *"outsider viewpoint"* for approximating the investment amount of an existing asset (defender). In particular, the outsider viewpoint is the perspective that would be taken by an impartial third party to establish the fair MV of a used (secondhand) asset. This viewpoint forces the analyst to focus on present and future cash flows in a replacement study, thus avoiding the temptation to dwell on past (sunk) costs.

The *present realizable* MV (modified by any income tax effects) is the correct investment amount to be assigned to an existing asset in replacement studies. A good way to reason that this is true is to use the *opportunity cost* or *opportunity foregone principle*. That is, if it should be decided to keep the existing asset, we are giving up the opportunity to obtain the net realizable MV at that time. Thus this represents the *opportunity cost* of keeping the defender.

There is one addendum to this rationale: If any new investment expenditure (such as for overhaul) is needed to upgrade the existing asset so that it will be competitive in level of service with the challenger, the extra amount should be added to the present realizable MV to determine the total investment in the existing asset for replacement study purposes.

> When using the outsider viewpoint, the total investment in the defender is the opportunity cost of not selling the existing asset for its current MV, *plus* the cost of upgrading it to be competitive with the best available challenger (all feasible challengers are to be considered).

Clearly, the MV of the defender must not also be claimed as a reduction in the challenger's capital investment because doing so would provide an unfair advantage to the challenger due to double counting the defender's selling price.

EXAMPLE 11-1

The capital investment of a machine purchased two years ago was $20,000. It has been depreciated by the MACRS (GDS) method, and its current BV is $9,600. The MV of the machine, if sold now, is $5,000, and it would cost $2,000 to overhaul the machine to make it serviceable for another five years. What are (a) the total investment in the defender and (b) the unamortized value?

SOLUTION

The investment in an existing asset is its present realizable MV plus any required expenditures to make it serviceable (and comparable) relative to new machines that may be available. Therefore, the investment associated with keeping the present machine is $5,000 + $2,000 = $7,000. If this machine were sold for $5,000, the unamortized value would be $9,600 − $5,000 = $4,600.

11.3.4   The Importance of Income Tax Consequences

The replacement of assets often results in capital gains or losses, or gains or losses from the sale of land or *depreciable property*, as discussed in Chapter 7. Consequently, to obtain an accurate economic analysis in such cases, the studies must be made on an *after-tax basis*. It is evident that the existence of a taxable gain or loss, in connection with replacement, can have a considerable effect on the results of an engineering economy study. A prospective gain from the disposal of assets can be reduced by as much as 40% or 50%, depending on the effective income tax rate used in a particular problem. Consequently, the normal propensity for disposal or retention of existing assets can be influenced considerably by income tax considerations.

11.3.5 Economic Life of the Challenger

The economic life of an asset minimizes the equivalent uniform annual cost of owning and operating an asset, and it is often shorter than the useful or physical life. It is essential to know a challenger's economic life in view of the principle that new and existing assets should be compared over their economic (optimum) lives.

11.3.6 Economic Life of the Defender

As we shall see later in this chapter, the economic life of the defender is often one year. Consequently, care must be taken when comparing the defender with a challenger asset because *different lives* are involved in the analysis. We shall see that the defender should be kept longer than its apparent economic life as long as its *marginal cost* is less than the minimum equivalent uniform annual cost of the challenger. What *assumptions* are involved when two assets having different apparent economic lives are compared, knowing that the defender is a nonrepeating asset? These concepts will be discussed in Sections 11.7 and 11.8.

11.4 A Typical Replacement Problem

The following is a typical replacement situation used to illustrate several of the factors that must be considered in replacement studies. The example is first solved on a before-tax basis. Then an after-tax analysis is performed using the MACRS general depreciation system (GDS) for the new pump (challenger), and depreciation of the present pump is continued under the MACRS alternative depreciation system (ADS), which was elected at the time it was put into service. Both analyses use the outsider viewpoint to determine the investment in the present pump if it is not replaced by the challenger.

EXAMPLE 11-2

The manager of a carpet manufacturing plant became concerned about the operation of a critical pump in one of the liquid flow processes. After discussing it with the supervisor of plant engineering, they decided that a replacement study should be done, and that a nine-year study period would be appropriate for this situation. The company that owns the plant is using a MARR of 10% per year for its capital investment projects before taxes, and 6% per year after taxes. The effective income tax rate is 40%.

The present pump, *pump A*, including driving motor with integrated controls, cost $17,000 five years ago. The ADS recovery period for this pump, when used in a carpet manufacturing facility, is nine years. The accounting records show depreciation has been occurring on this basis using the straight-line method with half-year convention.

Some reliability problems have been experienced with pump A, including annual replacement of the impeller and bearings at a cost of $1,750. Annual operating and maintenance (O&M) expenses have been averaging $3,250. Annual insurance and property tax expenses are 2% of the initial capital investment. It appears that the pump will provide adequate service for another nine years if the present maintenance and repair practice is continued. It is estimated that if this pump is continued in service, its final MV after nine more years will be about $200.

An alternative to keeping the existing pump in service is to sell it immediately and to purchase a new type of replacement pump, *pump B*, for $16,000. An estimated market value of $750 could be obtained for the existing pump. A nine-year class life (MACRS five-year property class) would be applicable to the new pump under the GDS because it would be in service in a carpet manufacturing facility. An estimated market value would be 20% of the initial capital investment at the end of year nine. O&M expenses for the new pump are estimated to be $3,000 per year. Annual taxes and insurance would total 2% of initial capital investment. The data for Example 11-2 are summarized in Table 11-1.

TABLE 11-1   Summary of Information for Example 11-2

Existing pump A (defender)
Capital investment when purchased 5 years ago		−$17,000
Class life (and ADS recovery period)		9 years
Annual expenses		
Replacement of impeller and bearings	−$1,750	
Operating and maintenance	−3,250	
Taxes and insurance: $17,000 × 2%	−340	
		−$ 5,340
Present market value		$ 750
Estimated market value at the end of 9 additional years		$ 200
Current book value		$ 8,500

Replacement pump B (challenger)
Capital investment		−$16,000
Class life		9 years
MACRS property class		5 years
Estimated market value at the end of 9 years		$ 3,200
Annual expenses:		
Operating and maintenance	−$3,000	
Taxes and insurance: $16,000 × 2%	−320	
		−$ 3,320

Effective income tax rate = 40%
MARR (before taxes) = 10%
  and MARR (after taxes) = 6%

SOLUTION

In our *before-tax analysis* of the defender and challenger, care must be taken to identify correctly the investment amount in the existing pump. Based on the outsider viewpoint, this would be the current MV of $750; that is, the *opportunity cost* of keeping the defender. Note that the investment amount of pump A ignores the original purchase price of $17,000 and the present book value of

$$BV = \$17,000[1 - 0.0556 - 4(0.1111)] = \$8,500$$

By using the principles discussed thus far, a before-tax analysis of *equivalent annual worth* (AW) of pump A and pump B can now be made.

The solution of Example 11-2 by the AW method (before taxes) follows:

Study Period = 9 Years	Keep Old Pump A	Replacement Pump B
AW(10%):		
Expenses	−$5,340	−$3,320
Capital recovery cost (Equation 4-7):		
−($750 − $200)($A/P$, 10%, 9) − $200(0.10)	− 115	
−($16,000 − $3,200)($A/P$, 10%, 9) − $3,200(0.10)		− 2,542
Total AW(10%)	−$5,455	−$5,862

Because Pump A has the least negative AW value (−$5,455 > −$5,862), the replacement pump apparently is not justified and the defender should be kept at least one more year. We could also make the analysis using other methods (e.g., PW), and the indicated choice would be the same.

It must now be acknowledged that a before-tax analysis often is not a valid basis for a replacement decision because of the effect of depreciation and any significant gain or loss on disposal on income tax consequences of the alternatives. An *after-tax analysis* is a valid basis for such decisions.

To evaluate correctly the defender's ATCFs, we use the outsider viewpoint as follows. First, suppose that an independent secondhand equipment dealer has the *option* of selling the defender on your company's behalf. In Example 11-2, the year zero transactions if pump A *is sold* now would be as follows:

End of Year, k	BTCF	Depreciation	Taxable Income	Income Taxes at 40%	ATCF
0	$750 MV	None	$750 − $8,500* MV − BV	$3,100 −0.4(MV − BV)	$3,850 0.6MV + 0.4BV

*BV = $17,000[1 − 0.0556 − 4(0.1111)] = $8,500

The investment amount of the defender (which in general may include overhaul costs) is next determined by further supposing that the equipment dealer changes her mind and decides *to keep* the asset for your company. The preceding entries now become the opportunity costs of retaining the defender:

End of Year, k	BTCF	Depreciation	Taxable Income	Income Taxes at 40%	ATCF
0	−$750	None	−$750 + $8,500	−$3,100	−$3,850

Finally, the complete ATCFs of the existing and replacement pumps in Example 11-2 are computed as shown in Tables 11-2 and 11-3, respectively.

Note in Table 11-2 that for the amounts for the defender alternative on the year zero line, we have placed reversed signs in parentheses because, as explained previously, the amounts first were determined by assuming the sale and subsequent replacement of the existing pump A. The signs are then *reversed* to indicate retention of the asset. Also shown in Table 11-3 are the after-tax capital investment in the replacement pump B and its after-tax MV.

After the one-time and annual effects have been determined in Tables 11-2 and 11-3, the next step in an after-tax replacement study involves the equivalence calculations using an after-tax MARR. The following is the after-tax AW analysis for Example 11-2.

TABLE 11-2   ATCF Computations for the Defender (Existing Pump A) in Example 11-2

End of Year, k	(A) BTCF	(B) MACRS (ADS) Depreciation	(C) = (A) − (B) Taxable Income	(D) = −0.4(C) Income Taxes at 40%	(E) = (A) + (D) ATCF
0	(−) $ 750		(+)−$7,750	(−)+$3,100	(−) $3,850
1–4	− 5,340	$1,889	− 7,229	+ 2,892	− 2,448
5	− 5,340	944	− 6,284	+ 2,514	− 2,826
6–9	− 5,340	0	− 5,340	+ 2,136	− 3,204
9	200		200[a]	− 80	+ 120

[a]Gain on disposal (taxable at the 40% rate)

TABLE 11-3   ATCF Computations for the Challenger (Replacement Pump B) in Example 11-2

End of Year, k	(A) BTCF	(B) MACRS (GDS) Depreciation	(C) = (A) − (B) Taxable Income	(D) = −0.4(C) Income Taxes at 40%	(E) = (A) + (D) ATCF
0	−$16,000				−$16,000
1	− 3,320	$3,200	−$6,520	+$2,608	− 712
2	− 3,320	5,120	− 8,440	+ 3,376	+ 56
3	− 3,320	3,072	− 6,392	+ 2,557	− 763
4	− 3,320	1,843	− 5,163	+ 2,065	− 1,255
5	− 3,320	1,843	− 5,163	+ 2,065	− 1,255
6	− 3,320	922	− 4,242	+ 1,697	− 1,623
7–9	− 3,320	0	− 3,320	+ 1,328	− 1,992
9	+ 3,200		+ 3,200[a]	− 1,280	+ 1,920

[a]Gain on disposal (taxable at the 40% rate)

$$\text{AW (6\%) of pump A} = -\$3{,}850(A/P, 6\%, 9)$$
$$-\$2{,}448(P/A, 6\%, 4)(A/P, 6\%, 9)$$
$$-[\$2{,}826(F/P, 6\%, 4) + \$3{,}204(F/A, 6\%, 4) - 120](A/F, 6\%, 9)$$
$$= -\$3{,}333$$

$$\text{AW (6\%) of pump B} = -\$16{,}000(A/P, 6\%, 9)$$
$$-[\$712(P/F, 6\%, 1) - \$56(P/F, 6\%, 2)$$
$$+ \$763(P/F, 6\%, 3)$$
$$+ \cdots + \$1{,}992(P/F, 6\%, 9)](A/P, 6\%, 9)$$
$$+ \$1{,}920(A/F, 6\%, 9)$$
$$= -\$3{,}375$$

Because the AWs of both pumps are very close, other considerations, such as the improved reliability of the new pump, could detract from the slight economic preference for pump A. The after-tax annual costs of both alternatives are considerably less than their before-tax annual costs.

Also note that an after-tax analysis may *reverse* the results of the before-tax analysis for the same problem. Identical recommendations should not necessarily be expected, due to the income tax considerations.

11.5 A Decision Roadmap for Replacement Analysis

A sound methodology for the management of assets should include the *periodic evaluation* of all factors related to the ownership and use of assets in their operating environment. The proper consideration of all factors will help identify the specific changes in requirements and status of assets that need to be included in replacement studies. For example, decreasing demand for a product may require the future abandonment of an asset. Similarly, increasing demand and technology changes may result in the immediate replacement of an asset with a new challenger.

A decision roadmap for replacement studies, which begins with a periodic evaluation of assets, is shown in Figure 11-1. This logical process will be helpful in engineering practice, as well as aiding in the study of the remainder of this chapter. Based on the results of the initial evaluation step, one or more assets may need to be analyzed for potential replacement. The type of study done in each case will depend upon the specific changes involved. After completing the studies required (or if no replacement studies are needed now), a time should be set for the next evaluation of changes in the operating situation. This time is normally within one year.

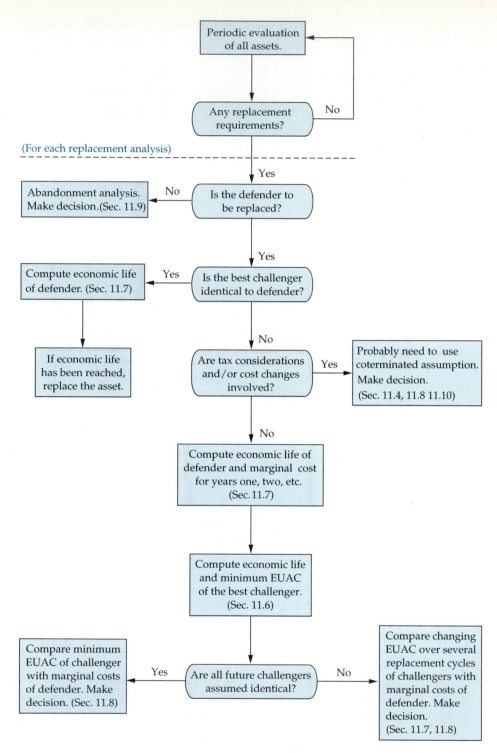

FIGURE 11-1   Asset Management and a Decision Roadmap for Replacement Analysis

11.6 Determining the Economic Life of a New Asset (Challenger)*

> *Sometimes in practice the useful lives of the defender and the challenger(s) are not known and cannot be reasonably estimated.* The period of time an asset is kept in productive service might be extended indefinitely with adequate maintenance and other actions, or it might be suddenly jeopardized by an external factor such as technological change. Under this situation, it is important to know the *economic life, minimum EUAC, and total year-by-year or marginal costs* for both the best challenger and the defender *so they can be compared based on an evaluation of their economic lives and the costs most favorable to each.*

The economic life of an asset was defined in Section 11.2 as the period of time that results in the minimum EUAC of owning and operating the asset. Also, the economic life is sometimes called the minimum-cost life or optimum replacement interval. For a new asset, it can be computed if the capital investment, annual expenses, and year-by-year market values are known or can be estimated. The apparent difficulties of estimating these values in practice may discourage doing the economic life and equivalent cost calculations. Similar difficulties, however, are encountered in most engineering economy studies when estimating the *future* economic consequences of alternative courses of action. Therefore, the estimating problems in replacement analysis are not unique and can be overcome in most application studies.

11.6.1 Before-Tax Analysis

The estimated initial capital investment, as well as the annual expense and market value estimates, may be used to determine the PW through year k of total costs, PW_k. That is, on a *before-tax* basis,

$$PW_k(i\%) = I - MV_k(P/F, i\%, k) + \sum_{j=1}^{k} E_j(P/F, i\%, j) \qquad \text{(11-1)}$$

which is the sum of the initial capital investment (PW of the initial investment amounts if any occur after time zero) adjusted by the PW of the MV at end of year k, and of the PW of annual expenses (E_j) through year k. The *total marginal cost* for each year k, TC_k, is calculated using Equation 11-1 to find the increase in the PW of total cost from year $k - 1$ to year k, and then to determine the equivalent worth of this increase at the end of year k. That is, $TC_k = (PW_k - PW_{k-1})(F/P, i\%, k)$. The algebraic simplification of this relationship results in

$$TC_k(i\%) = MV_{k-1} - MV_k + iMV_{k-1} + E_k \qquad \text{(11-2)}$$

---

*For convenience, an opposite sign convention will be used in this section (11.6) and in Section 11.7 with cash outflows being positive and cash inflows being negative.

which is the sum of the loss in MV during the year of extended service, the opportunity cost of capital invested in the asset at the beginning of year k, and the annual expenses incurred in year k. These total marginal (or year-by-year) costs, based on Equation 11-2, are then used to find the EUAC through each year k. The minimum $EUAC_k$ value during the useful life of the asset identifies its before-tax economic life, N^*. This procedure is illustrated in Example 11-3.

EXAMPLE 11-3

A new forklift truck will require an investment of $20,000 and is expected to have year-end market values and annual expenses as shown in columns 2 and 5, respectively, of Table 11-4. If the before-tax MARR is 10% per year, how long should the asset be retained in service?

SOLUTION

The solution to this problem is obtained by using Equation 11-2 and completing Columns 3, 4, 6, and 7 of Table 11-4. In the solution, the customary year-end occurrence of all cash flows is assumed. The *actual* depreciation for any year is the difference between the beginning and year-end market values. In this before-tax example, depreciation is not computed according to any formal method but rather results from expected forces in the marketplace. The opportunity cost of capital in year k is 10% of the capital unrecovered (invested in the asset) at the beginning of each year. The values in column 7 are the equivalent uniform annual costs that would be incurred each year if the asset were retained in service through year k and then replaced at the end of the year. The minimum EUAC occurs at the end of year N^*.

It is apparent from the values shown in column 7 that the asset will have a minimum EUAC if it is kept in service for only three years (i.e., $N^* = 3$).

The computational approach in the preceding example, as shown in Table 11-4, was to determine the total marginal (or year-by-year) cost and then to convert these into an EUAC. The EUAC for any life can also be calculated using the more familiar capital recovery formulas in Chapter 4. For example, for a life of two years, the EUAC can be calculated as follows:

$$EUAC_2(10\%) = \$20,000(A/P, 10\%, 2) - \$11,250(A/F, 10\%, 2)$$
$$+ [\$2,000(P/F, 10\%, 1)$$
$$+ \$3,000(P/F, 10\%, 2)](A/P, 10\%, 2)$$
$$= \$8,643$$

which checks with the corresponding row in column 7 of Table 11-4.

TABLE 11-4 Determination of the Before-Tax Economic Life N* of a New Asset (Example 11-3)

(1)	(2)	Cost of Service for kth Year				EUAC Through Year k
		(3)	(4)	(5)	(6)	(7)
End of Year, k	MV, End of Year k	Actual Depreciation During Year k	Cost of Capital = 10% of Beginning of Year MV	Annual Expenses (E_k)	[= (3) + (4) + (5)] Total (Marginal) Cost for Year k (TC_k)	$EUAC_k = [\sum_{j=1}^{k} TC_j(P/F, 10\%, j)](A/P, 10\%, k)$
0	$20,000	—	—	—	—	—
1	15,000	$5,000	$2,000	$2,000	$ 9,000	$9,000
2	11,250	3,750	1,500	3,000	8,250	8,643
3	8,500	2,750	1,125	4,620	8,495	8,600 ← minimum EUAC (N^* = 3)
4	6,500	2,000	850	8,000	10,850	9,082
5	4,750	1,750	650	12,000	14,400	9,965

11.6.2 After-Tax Analysis

In Example 11-3, the economic life (N^*) of a new asset was determined on a before-tax basis. However, an *after-tax* analysis can also be used to determine the economic life of an asset by extending Equation 11-1 to account for income tax effects, which is shown in Equation 11-3.

$$PW_k(i\%) = I + \sum_{j=1}^{k}[(1-t)E_j - td_j](P/F,i\%,j) - [(1-t)MV_k + t(BV_k)](P/F,i\%,k)$$

(11-3)

This computation finds the PW of the ATCF through year k, PW_k, on an after-tax basis for the asset by (1) adding the initial capital investment (PW of investment amounts if any occur after time zero) and the sum of the after-tax PW of annual expenses through year k, including adjustments for annual depreciation amounts (d_j), and then (2) adjusting this total after-tax PW of costs by the after-tax consequences of gain or loss on disposal of the asset at the end of year k. Similar to the previous before-tax analysis using Equation 11-1, Equation 11-3 is used to determine the after-tax total marginal cost for each year k, TC_k. That is, $TC_k = (PW_k - PW_{k-1})(F/P, i\%, k)$. The algebraic simplification of this relationship results in Equation 11-4:

$$TC_k(i\%) = (1-t)(MV_{k-1} - MV_k + iMV_{k-1} + E_k) + i(t)(BV_{k-1}) \qquad \textbf{(11-4)}$$

which is $1 - t$ times Equation 11-2 plus interest on the tax credit for the book value of the asset at the beginning of year k. A tabular format incorporating Equation 11-4 is used in the solution of the next example to find the economic life of a new asset on an after-tax basis (N_{AT}^*).

TABLE 11-5   Determination of the After-Tax Economic Life for the Asset in Example 11-3

(1) End of Year, k	(2) MV, End of Year k	(3) Actual Depreciation during Year k	(4) Cost of Capital = 6% of BOY MV in Col. 2	(5) Annual Expenses	(6) Approximate After-Tax Total (Marginal) Cost for Year k $(1-t)\cdot$(Col. 3 + 4 + 5)
0	$20,000	0	0	0	0
1	15,000	$5,000	$1,200	$2,000	$4,920
2	11,250	3,750	900	3,000	4,590
3	8,500	2,750	675	4,620	4,827
4	6,500	2,000	510	8,000	6,306
5	4,750	1,750	390	12,000	8,484

(continued)

EXAMPLE 11-4

Find the economic life on an after-tax basis for the new fork lift truck (challenger) described in Example 11-3. Assume that the new fork lift is depreciated as a MACRS (GDS) three-year property class asset, the effective income tax rate is 40%, and the after-tax MARR is 6%.

SOLUTION

The calculations using Equation 11-4 are shown in Table 11-5, which starts at the bottom of page 450. The expected year-by-year market values and annual expenses are repeated from Example 11-3 in columns 2 and 5, respectively. In column 6, the *sum* of actual depreciation in year k, cost of capital based on the MV at the beginning-of-year (BOY) k, and annual expenses in year k are multiplied by $1 - t$ to determine an *approximate* after-tax total marginal cost in year k.

The BV amounts at the end of each year, based on the new fork lift truck being a MACRS (GDS) three-year property class asset, are shown in column 7. These amounts are then used in column 8 to determine an annual tax credit adjustment (last term in Equation 11-4), based on the BOY book values (BV_{k-1}). This annual tax credit adjustment is algebraically added to the entry in column 6 to obtain an *adjusted* after-tax total marginal cost in year k, TC_k. The total marginal cost amounts are used in column 10 to calculate, successively, the equivalent uniform annual cost, $EUAC_k$, of retirement of the asset at the end of year k. In this case, the after-tax economic life (N_{AT}^*) is three years, the same result obtained on a before-tax basis in Example 11-3.

It is not uncommon for the before-tax and after-tax economic lives of a new asset to be the same (as occurred in Examples 11-3 and 11-4). For this reason, we shall confine our attention to before-tax economic life in the sections that follow.

TABLE 11-5 *(cont'd)*

End of Year, k	(7) MACRS BV at End of Year k	(8) Interest on Tax Credit = 6% · t · BOY BV in Col. 7	(9) Adjusted After-Tax Total (Marginal) Cost (TC_k) (Col. 6 + Col. 8)	(10) EUAC (After-Tax) Through Year k $EUAC_k = \sum_{j=1}^{k}[(Col.\ 9)_j \cdot (P/F, 6\%, j)](A/P, 6\%, k)$
0	$20,000	0	0	0
1	13,334	$480	$5,400	$5,400
2	4,444	320	4,910	5,162
3	1,482	107	4,934	5,090 ← minimum EUAC
4	0	36	6,342	5,376 ($N_{AT}^* = 3$)
5	0	0	8,484	5,928

11.7 Determining the Economic Life of a Defender*

In replacement analyses, we must also determine the economic life that is most favorable to the defender. When a major outlay for defender alteration or overhaul is needed, the life that will yield the least EUAC is likely to be the period that will elapse before the next major alteration or overhaul will be needed. Alternatively, *when there is no defender MV now or later (and no outlay for alteration or overhaul) and when defender operating expenses are expected to increase annually, the remaining life that will yield the least EUAC will be one year.*

When MVs are greater than zero and expected to decline from year to year, it is necessary to calculate the apparent remaining economic life, which is done in the same manner as in Example 11-3 for a new asset using Equation 11-2 for a before-tax analysis. By using the outsider viewpoint, the present realizable MV of the defender is considered to be its investment value.

> Regardless of how the remaining economic life for the defender is determined, a decision to keep the defender does not mean that it should be kept only for this period of time. Indeed, the defender should be kept longer than the apparent economic life as long as its *marginal* cost (total cost for an additional year of service) is less than the minimum EUAC for the best challenger.

This important principle of replacement analysis is illustrated in Example 11-5.

EXAMPLE 11-5

Suppose that it is desired to determine how much longer a forklift truck should remain in service before it is replaced by the new truck (challenger) for which data were given in Example 11-3 and Table 11-4. The defender in this case is two years old, originally cost $13,000, and has a present realizable MV of $5,000. If kept, its market values and annual expenses are expected to be as follows:

End of Year, k	MV, End of Year k	Annual Expenses, E_k
1	$4,000	$5,500
2	3,000	6,600
3	2,000	7,800
4	1,000	8,800

Determine the most economical period to keep the defender *before* replacing it (if at all) with the present challenger of Example 11-3. The cost of capital is 10% per year.

---

*For convenience, the opposite sign convention used in Section 11.6 is continued in this section.

TABLE 11-6 Determination of the Economic Life of an Old Asset (Example 11-5)

(1) End of Year, k	(2) Actual Depreciation During Year k	(3) Cost of Capital = 10% of Beginning of Year MV[a]	(4) Annual Expenses (E_k)	(5) Total (Marginal) Cost for Year, (TC_k)	(6) EUAC Through Year k
1	$1,000	$500	$5,500	$ 7,000	$7,000
2	1,000	400	6,600	8,000	7,475
3	1,000	300	7,800	9,100	7,966
4	1,000	200	8,800	10,000	8,406

[a]Year one is based on a present realizable MV of $5,000.

SOLUTION

Table 11-6 shows the calculation of total cost for each year (marginal cost) and the EUAC at the end of each year for the defender based on the format used in Table 11-4. Note that the minimum EUAC of $7,000 corresponds to keeping the defender for one more year. However, the marginal cost of keeping the truck for the second year is $8,000, which is still less than the minimum EUAC for the challenger (i.e., $8,600, from Example 11-3). The *marginal cost* for keeping the defender the third year and beyond is greater than the $8,600 minimum EUAC for the challenger. Based on the currently available data shown, it would be most economical to keep the defender for two more years and then to replace it with the challenger. This situation is portrayed graphically in Figure 11-2.

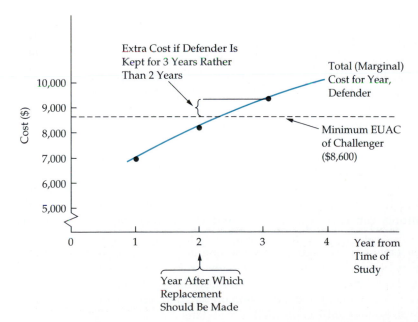

FIGURE 11-2

Defender Versus Challenger Forklift Trucks (Based on Examples 11-3 and 11-5)

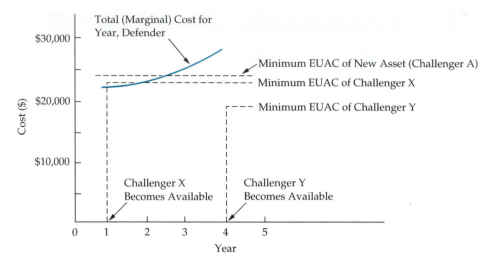

FIGURE 11-3   Old Versus New Asset Costs with Improved Challengers Becoming Available in the Future

Example 11-5 assumes that a comparison is being made with the best challenger alternative available. In this situation, if the defender is retained beyond the point where its marginal (year-by-year) costs exceed the minimum EUAC for the challenger, the difference in costs continues to grow, and replacement becomes more urgent. This is illustrated to the right of the intersection in Figure 11-2.

Figure 11-3 illustrates the effect of improved new challengers in the future. If an improved challenger X becomes available before replacement with the new asset of Figure 11-2, then a new replacement study probably should take place to consider the improved challenger. If there is a possibility of a further-improved challenger Y as of, say, four years later, it may be better still to postpone replacement until that challenger becomes available. Although retention of the old asset beyond its breakeven point with the best available challenger has a cost that may well grow with time, this cost of waiting can, in some instances, be worthwhile if it permits the purchase of an improved asset having economies that offset the cost of waiting. Of course, a decision to postpone a replacement may also "buy time and information." Because technological change tends to be sudden and dramatic rather than uniform and gradual, new challengers with significantly improved features can arise sporadically and can change replacement plans substantially.

When replacement is not signaled by the engineering economy study, more information may become available before the next analysis of the defender. Hence, the next study should include that additional information. *Postponement* generally should mean a postponement of the decision on when to replace, not the decision to postpone replacement until a specified future date.

▼ 11.8   Comparisons in Which the Defender's Useful Life Differs from That of the Challenger

In Section 11.4, we discussed a typical replacement situation in which the useful lives of the defender and the challenger were known and the same, as well as equal to the study period. When this situation occurs, any of the correct analysis methods, properly applied, may be used.

In the previous two sections (11.6 and 11.7), we discussed the economic lives of a new asset and of a defender and how these results (along with the related cost information) are used in replacement analysis when the useful lives of the assets may or may not be known.

A third situation occurs when the useful lives of the best challenger and the defender are known, or can be estimated, but are not the same. The comparison of the challenger and the defender under these circumstances is the topic of this section.

In Chapter 5, two assumptions were described that are used for the economic comparisons of alternatives, including those having different useful lives: (1) *repeatability* and (2) *cotermination*. Under either assumption, the analysis period used is the same for all alternatives in the study. The repeatability assumption, however, involves two main stipulations:

1. The period of needed service for which the alternatives are being compared is either indefinitely long or a length of time equal to a common multiple of the useful lives of the alternatives.
2. What is estimated to happen in the first useful life span will happen in all succeeding life spans, if any, for each alternative.

For replacement analyses, the first of these conditions may be acceptable, but normally the second is not reasonable for the defender. The defender is typically an older and used piece of equipment. An identical replacement, even if it could be found, probably would have an installed cost in excess of the current MV of the defender.

Failure to meet the second stipulation can be circumvented if the period of needed service is assumed to be indefinitely long and *if we recognize that the analysis is really to determine if* now *is the time to replace the defender*. When the defender is replaced, it will be by the challenger—the best available replacement.

Example 11-5, involving a before-tax analysis of the defender versus challenger forklift trucks, made use of the *repeatability* assumption. That is, it was assumed that the particular challenger analyzed in Table 11-4 would have a minimum EUAC regardless of when it replaces the defender. Figure 11-4 shows time diagrams of the cost consequences of keeping the defender for two more years versus adopting the challenger now, with the challenger costs to be repeated into the indefinite future. Recall that the economic life of the challenger is three years. *It can be seen in Figure 11-4 that the only difference between the alternatives is in years one and two.*

The repeatability assumption, applied to replacement problems involving assets with different useful or economic lives, often simplifies the economic comparison

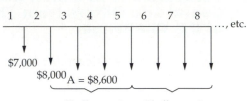

A: Keep Defender for Two More Years

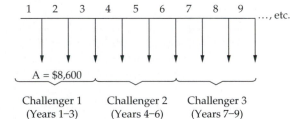

B: Replace with Challenger Now

FIGURE 11-4

Effect of the Repeatability Assumption
Applied to Alternatives in Example 11-5

of the alternatives. For example, the comparison of the PW values of the alternatives in Figure 11-4 *over an infinite analysis period* will confirm our previous answer to Example 11-5 that Alternative A (keep defender for two more years) is preferred to Alternative B (replace with challenger now). Using MARR = 10% per year, we have

$$PW_A(10\%) = -\$7,000(P/F, 10\%, 1) - \$8,000(P/F, 10\%, 2) - \frac{\$8,600}{0.10}(P/F, 10\%, 2)$$

$$= -\$84,050$$

$$PW_B(10\%) = -\frac{\$8,600}{0.10} = -\$86,000$$

The difference ($PW_B - PW_A$) is −$1,950, which confirms that the additional cost of the challenger over the next two years is not justified and it is best to keep the defender for two more years before replacing it with the challenger.

> Whenever the *repeatability* assumption is not applicable, the *coterminated* assumption may be used; it involves using a finite study period for all alternatives. As described in Chapter 5, use of the *coterminated* assumption requires detailing what and when cash flows are expected to occur for each alternative and then determining which is most economical, using any of the correct economic analysis methods. *When the effects of inflation are to be considered in replacement studies, it is recommended that the coterminated assumption be used.*

EXAMPLE 11-6

Suppose that we are faced with the same replacement problem as in Example 11-5, except that the period of needed service is (a) three years or (b) four years. That is, a finite analysis period under the coterminated assumption is used. In each case, which alternative should be selected?

SOLUTION

(a) For a planning horizon of three years, we might intuitively think that either the defender should be kept for three years or it should be replaced immediately by the challenger to serve for the next three years. From Table 11-6 the EUAC for the defender for three years is $7,966, and from Table 11-4 the EUAC for the challenger for three years is $8,600. Thus, following this reasoning, the defender would be kept for three years. However, this is not quite right. Focusing on the "Total (Marginal) Cost for Year" columns, we can see that the defender has the lowest cost in the first two years, but in the third year its cost is $9,100; the cost of the first year of service for the challenger is only $9,000. Hence, it would be economical to replace the defender after the second year. *This conclusion can be confirmed by enumerating all replacement possibilities and their respective costs and then computing the EUAC for each*, as will be done for the four-year planning horizon in part (b).

(b) For a planning horizon of four years, the alternatives and their respective costs for each year and the EUAC of each are given in Table 11-7. Thus, the most economical alternative is to keep the defender for two years and then replace with the challenger, to be kept for the next two years. The decision to keep the defender for two years happens to be the same as when the repeatability assumption was used, which, of course, would not necessarily be true in general.

When a replacement analysis involves a defender that cannot be continued in service because of changes in technology, service requirements, and so on, a choice among two or more new challengers must be made. Under this situation,

TABLE 11-7   Determination of When to Replace the Defender for a Planning Horizon of Four Years (Example 11-6, Part b)

Keep Defender for	Keep Challenger for	Total (Marginal) Costs for Each Year				EUAC at 10% for 4 Years
		1	2	3	4	
0 years	4 years	−$9,000[a]	−$8,250[a]	−$8,495[a]	−$10,850[a]	−$9,082
1	3	− 7,000	− 9,000	− 8,250	− 8,495	− 8,301
2	2	− 7,000	− 8,000	− 9,000	− 8,250	− 8,005 ← Least
3	1	− 7,000	− 8,000	− 9,100	− 9,000	− 8,190 negative
4	0	− 7,000[b]	− 8,000[b]	− 9,100[b]	− 10,000[b]	− 8,406

[a]Column 6 of Table 11-4.
[b]Column 5 of Table 11-6.

the repeatability assumption may be a convenient economic modeling approach for comparing the alternatives and making a present decision.

EXAMPLE 11-7

An existing robot is used in a commercial materials laboratory to handle ceramic samples in the high-temperature environment that is part of a test procedure. Due to changing customer needs, the robot will not meet future service requirements. The handling of larger test samples of ceramic material, as well as increased temperature testing, will be required in the future. Both of these changes will exceed the operating capabilities of the existing robot.

Because of this situation, two new advanced technology robots have been selected for economic analysis and comparison. The following estimates have been developed from information provided by some current users of each robot and data obtained from the manufacturers. The firm's before-tax MARR is 25% per year. Based on this information, which robot is economically preferable?

	Robot	
	R1	**R2**
Capital investment		
Purchase price	−$38,200	−$51,000
Installation cost	−2,000	−5,500
Annual expenses	−1,400 in year one, and increasing at the rate of 8%/yr thereafter.	−1,000 in year one, and increasing $150/yr thereafter.
Useful life (years)	6	10
Market value	−$1,500	+$7,000

SOLUTION

The repeatability assumption and the AW method are used in the comparison of the two robots. The use of repeatability as a simplified modeling approach can be supported in this case. Either robot, if selected, is expected to provide the required service for its total useful life period. Also, either robot most likely would be replaced at the end of its useful life with a new and better challenger. The annual equivalent cost of a new challenger at that time should be slightly less than for either Model R1 or Model R2 and should provide equal or better service because of continuing technological advances and competition among the robot manufacturers.

The estimated annual expenses for R1 are a geometric cash flow sequence (Chapter 3) beginning in year one. The convenience rate (Chapter 3) needed to calculate the PW of this sequence is $i_{cr} = (0.25 - 0.08)/1.08 = 0.1574$, or 15.74%. The negative estimated salvage value (−$1,500) indicates an expected net cost for disposal of the asset at the end of six years.

$$AW_{R1}(25\%) = -(\$38,200 + \$2,000)(A/P, 25\%, 6)$$

$$- \frac{\$1,400}{1.08}(P/A, 15.74\%, 6)(A/P, 25\%, 6) - \$1,500(A/F, 25\%, 6)$$

$$= -\$15,382$$

and, for Model R2, the AW over its useful life is

$$AW_{R2}(25\%) = -(\$51,000 + \$5,500)(A/P, 25\%, 10)$$
$$- [\$1,000(P/A, 25\%, 10) + \$150(P/G, 25\%, 10)](A/P, 25\%, 10)$$
$$+ \$7,000(A/F, 25\%, 10)$$
$$= -\$17,035$$

The R1 replacement robot is economically preferable because the AW over its useful life has the least negative value ($-\$15,382$).

11.9 Retirement Without Replacement (Abandonment)

Consider a project for which the period of required service is finite and that has *positive* net cash flows following an initial capital investment. Market values, or abandonment values, are estimated for the end of each remaining year in the project's life. In view of an opportunity cost of capital of $i\%$ per year, should the project be undertaken? What is the best year to abandon the project, given that we have decided to implement it?

For this type of problem, the following assumptions are applicable:

1. Once a capital investment has been made, the firm desires to postpone the decision to abandon a project as long as its equivalent worth is not decreasing.
2. The existing project will be terminated at the best abandonment time and will not be replaced by the firm.

Solving the abandonment problem is similar to determining the economic life of an asset. In abandonment problems, however, annual benefits (cash inflows) are present but in economic life analysis, costs (cash outflows) are dominant.

EXAMPLE 11-8

A baling machine for recycled paper is being considered by the XYZ Company. After-tax annual revenues less expenses have been estimated for the project in addition to the after-tax abandonment value of the machine at the end of each year.

	End of Year							
	0	1	2	3	4	5	6	7
Annual (after-tax) revenues less expenses	−$50,000[a]	$10,000	$15,000	$18,000	$13,000	$ 9,000	$ 6,000	$ 5,000
Abandonment value of machine[b]	—	40,000	32,000	25,000	21,000	18,000	17,000	15,000

[a]Capital investment
[b]Estimated market value

The firm's after-tax MARR is 12% per year. When is the best time to abandon the project if the firm has already decided to acquire the baling machine and use it for no longer than seven years?

SOLUTION

The AWs that result from deciding now to keep the machine exactly one, two, three, four, five, six, and seven years are shown here. As you can see, AW is maximized by retaining the machine for six years, which would be the best abandonment time.

Keep for one year:

$$AW_1(12\%) = -\$50,000(A/P, 12\%, 1) + (\$40,000 + \$10,000)(A/F, 12\%, 1)$$
$$= -\$6,000$$

Keep for two years:

$$AW_2(12\%) = [-\$50,000 + \$10,000(P/F, 12\%, 1)$$
$$+ (\$15,000 + \$32,000)(P/F, 12\%, 2)](A/P, 12\%, 2)$$
$$= -\$2,132$$

In the same manner, the net AW for years three through seven can be computed as follows:

Keep for three years: $AW_3(12\%) = \$622$
Keep for four years: $AW_4(12\%) = \$1,747$
Keep for five years: $AW_5(12\%) = \$2,020$
Keep for six years: $AW_6(12\%) = \$2,121$
Keep for seven years: $AW_7(12\%) = \$2,006$

In some cases, management may decide that although an existing asset is to be retired from its current use, it will not be replaced or removed from all service. Although the existing asset may not be able to compete economically at the moment, it may be desirable and even economical to retain the asset as a standby unit or for some different use. The cost to retain the defender under such conditions may be quite low, owing to its relatively low realizable MV and perhaps low annual expenses. Often income tax considerations also bear on the true cost of retaining the defender, as discussed in Section 11.4.

▼11.10 A Comprehensive Example

Sometimes in engineering practice a replacement analysis involves an existing asset that cannot meet future service requirements without *augmentation* of its capabilities. When this is the case, the defender with increased capability should be competitive with the best available challenger. The analysis of this situation is included in the following comprehensive example.

EXAMPLE 11-9

The emergency electrical supply system of a hospital owned by a medical service corporation is presently supported by an 80-kW diesel-powered electrical generator (defender) that was put into service five years ago [capital investment = $210,000; MACRS (GDS) seven-year property class]. An engineering firm is designing modifications to the electrical and mechanical systems of the hospital as part of an expansion project. The redesigned emergency electrical supply system will require 120 kW of generating capacity to serve the increased demand. Two preliminary designs for the system are being considered. The first involves the augmentation of the existing 80-kW generator with a new 40-kW diesel-powered unit (GDS seven-year property class). The second design alternative includes replacement of the existing generator with the best available alternative, a new turbine-powered unit with 120 kW of generating capacity (challenger). Both alternatives will provide the same level of service to the operation of the emergency electrical supply system.

The challenger, if selected, will be leased by the hospital for a ten-year period. At that time, the lease contract would be renegotiated for the original piece of equipment or for a replacement generator with the same capacity. The following additional information has been estimated for use in the replacement analysis.

	Alternative		
	Defender		Challenger
	80 kW	40 kW	
Capital investment	−$90,000[a]	−$140,000	−$10,000[b]
Annual lease amount	0	0	−$39,200
Operating hours/year	260	260	260
Annual expenses (year zero $):			
Operating and maintenance (O&M) costs per hour	−$80	−$35	−$85
Other expenses	−$3,200	−$1,000	−$2,400
Useful life (years)	10	15	15

[a]Opportunity cost based on present market value of the defender (outsider viewpoint).
[b]Deposit required by the terms of the contract to lease the challenger. It is refundable at the end of the ten-year period.

The annual lease amount for the challenger will not change over the ten-year contract period. The O&M costs per hour of operation and the other annual expense amounts are estimated in year zero dollars, and expected to escalate at the rate of 4% per year (assume base year, b, is year zero; see Chapter 9 for dealing with price changes).

The present estimated market value of the defender is $90,000, and its estimated market value at the end of an additional ten years, in year zero dollars, is $30,000. The estimated market value of the new 40-kW generator, ten years from now in year zero dollars, is $38,000. Both future market values are estimated to escalate at the rate of 2% per year.

The corporation's after-tax, market-based MARR (i_c) is 12% per year, and its effective income tax rate is 40%. A ten-year planning horizon (study period) is considered appropriate for this decision situation (note that with tax considerations and price changes in the analysis, a selected study period based on the coterminated assumption is being used).

Based on an after-tax, actual dollar analysis, which alternative (augmentation of the defender or lease of the new challenger) should be selected as part of the design of the modified emergency electrical power system?

SOLUTION

The after-tax analysis of the first alternative, keeping the defender and augmenting its capacity with a new 40-kW generator, is shown in Table 11-8. The initial $230,000 capital investment amount is the sum of (1) the present $90,000 market value of the defender, which is an opportunity cost based on an outsider viewpoint, and (2) the $140,000 capital investment for the new 40-kW generator. The −$33,772 of taxable income at time zero is due to the gain on disposal, *which is not incurred when the defender is kept instead of sold*. The after-tax PW (MARR = 12% per year) of keeping the defender and augmenting its capacity is

$$PW_D(12\%) = -\$216{,}491 - \$5{,}783(P/F, 12\%, 1) - \cdots$$
$$+ (\$49{,}735 - \$30{,}286)(P/F, 12\%, 10)$$
$$= -\$277{,}098$$

Under the contract terms for leasing the challenger, there is an initial $10,000 deposit, which is fully refundable at the end of the ten-year period. There are no tax consequences associated with the deposit transaction. The annual before-tax cash

TABLE 11-8   Defender Augmented with a New 40-kW Generator (Example 11-9)

End of Year, k	BTCF	Depreciation 80 kW	Depreciation 40 kW	Taxable Income	Income Taxes at 40%	ATCF
0	−$230,000			−$33,772[c]	$13,509	−$216,491
1	− 35,464	$18,732	$20,006	− 74,202	29,681	− 5,783
2	− 36,883[a]	18,753	34,286	− 89,922	35,969	− 914
3	− 38,358	9,366	24,486	− 72,210	28,884	− 9,474
4	− 39,892		17,486	− 57,378	22,951	− 16,941
5	− 41,488		12,502	− 53,990	21,596	− 19,892
6	− 43,147		12,488	− 55,635	22,254	− 20,893
7	− 44,873		12,502	− 57,375	22,950	− 21,923
8	− 46,668		6,244	− 52,912	21,165	− 25,503
9	− 48,535			− 48,535	19,414	− 29,121
10	− 50,476			− 50,476	20,190	− 30,286
10 (MV)	82,892[b]			82,892	− 33,157	49,735

[a] −[260($80 + $35) + ($3,200 + $1,000)](1.04)² = −$36,883
[b] MV_{10} = ($30,000 + $38,000)(1.02)¹⁰ = $82,892
[c] If the defender was sold now, gain on disposal = $90,000 − $56,228 = $33,772; where BV now, if sold, is $56,228.

flow (BTCF) for the challenger is the sum of (1) the annual lease amount, which stays constant over the ten-year contract period, and (2) the annual O&M and other expenses, which escalate at the rate of 4% per year. For example, the BTCF for the challenger in year one is $-\$39{,}200 - [\$85(260) + \$2{,}400](1.04) = -\$64{,}680$. These annual BTCF amounts for years one through ten are fully deductible from taxable income by the corporation, and they are also the taxable income amounts for the alternative (the corporation cannot claim any depreciation on the challenger because it does not own the equipment). Hence, the after-tax PW (MARR = 12% per year) of selecting the challenger, assuming it is leased under these contract terms, is

$$
\begin{aligned}
PW_C(12\%) = & -\$10{,}000 + \$10{,}000(P/F, 12\%, 10) \\
& - (1 - 0.4)(\$39{,}200)(P/A, 12\%, 10) \\
& - (1 - 0.4)[\$85(260) + \$2{,}400](P/A, i_{cr} = 7.69\%, 10) \\
= & -\$239{,}705
\end{aligned}
$$

where $i_{cr} = (0.12 - 0.04)/(1.04) = 0.0769$, and $(P/A, 7.69\%, 10) = 6.8049$.

Based on an after-tax analysis, the challenger is economically preferable for use in the emergency electrical supply system because its PW has the least negative value ($-\$239{,}705$).

11.11 Spreadsheet Applications

A vital ingredient of replacement studies is an asset's economic life. Similarly, when an asset is not to be replaced, the best time to abandon the project is of interest. The first example in this section provides a general spreadsheet model that can be used to determine the economic life of an asset given the capital investment, year-by-year market value, and the annual operating expenses. The second example uses a similar spreadsheet to compute the best abandonment time for a project.

EXAMPLE 11-10

The estimated year-by-year market values and operating expenses for a replacement piece of equipment are shown in Figure 11-5 (column B and column E, respectively). The market values are used to compute actual depreciaton (column C) and the cost of the capital (column D). The resulting capital recovery amount is combined with the expenses for the year (shown as a net cash flow in column E) to determine the total marginal cost for the year (column F). Column G calculates the equivalent annual worth of the cash flows in column F up to the period in question. Column H contains an IF( ) function that places the label "Economic Life" next to the maximum equivalent annual worth (which corresponds to the minimum equivalent uniform annual cost) in column G. The formulas from the highlighted cells in Figure 11-5 are given in the following table.

	A	B	C	D	E	F	G	H
1	MARR	15%						
2								
3								
4			Actual		Net	Total (Marginal)	Equivalent	
5	End of	MV at End	Depreciation During	Cost of	Cash Flow for Year	Cash Flow for Year	Annual Worth	
6	Year	of Year	Year	Capital	(R - E)	(R - E - CR)	Through Year	
7	0	$ 15,000						
8	1	$ 12,000	$ 3,000	$ 2,250	$ (1,000)	$ (6,250)	($6,250)	
9	2	$ 10,000	$ 2,000	$ 1,800	$ (1,100)	$ (4,900)	($5,622)	Economic Life
10	3	$ 7,000	$ 3,000	$ 1,500	$ (1,300)	$ (5,800)	($5,673)	
11	4	$ 3,000	$ 4,000	$ 1,050	$ (2,000)	$ (7,050)	($5,949)	
12	5	$ 500	$ 2,500	$ 450	$ (2,500)	$ (5,450)	($5,875)	

FIGURE 11-5   Spreadsheet for Determining Economic Life in Example 11-10

Cell	Contents
C10	= B9 − B10
D10	= B9 * $B$1
F10	= E10 − (C10 + D10)
G10	= −PMT($B$1, A10, NPV(F$8:F10))
H10	= IF(G10 = MAX(G8:G12), "Economic Life", "")

EXAMPLE 11-11

In this example, we use the same spreadsheet to compute the best abandonment time for the baling machine project of Example 11-8. Figure 11-6 shows the spreadsheet solution to this example. Solving for the abandonment time is essentially the same as solving for the economic life, except that benefits are present in abandonment analyses, whereas costs are the primary component in economic life analyses. The spreadsheet model presented in Figures 11-5 and 11-6 is flexible enough to handle both benefits and costs in that it requires only the input of the net cash flow. The spreadsheet calculations are identical to those made in Example 11-10.

11.12 Summary

In summary, there are several important points to keep in mind when conducting a replacement or retirement study:

1. The MV of the defender must *not* also be deducted from the purchase price of the challenger *when using the outsider viewpoint* to analyze a replacement problem. This error double-counts the defender's MV and biases the comparison toward the challenger.

2. A sunk cost (i.e., MV − BV < 0) associated with keeping the defender must *not* be added to the purchase price of the best available challenger. This error results in an incorrect penalty that biases the analysis in favor of retaining the defender.

3. The income tax effects of replacement decisions cannot be ignored. The foregone income tax credits associated with keeping the defender may swing the economic preference away from the defender, thus making the challenger the better choice.

4. In Section 11.7 we observed that the economic life of the defender is often one year, which is generally true if annual expenses are high relative to the defender's investment cost when using the outsider viewpoint. Hence, the marginal cost of the defender should be compared with the EUAC at the *economic life* of the challenger to answer the fundamental question "Should the defender be kept for one or more years or disposed of now?" The typical pattern of the EUAC for a defender and a challenger is illustrated in Figure 11-7.

5. The *best available* challengers must be determined. Failure to do so represents unacceptable engineering practice.

	A	B	C	D	E	F	G	H
1	MARR	12%						
2								
3			Actual				Equivalent	
4			Depreciation		Net	Total (Marginal)		
5	End of	MV at End	During	Cost of	Cash Flow	Cash Flow	Annual Worth	
6	Year	of Year	Year	Capital	for Year	for Year	Through Year	
7	0	$ 50,000			(R - E)	(R - E - CR)		
8	1	$ 40,000	$ 10,000	$ 6,000	$ 10,000	$ (6,000)	$ (6,000)	
9	2	$ 32,000	$ 8,000	$ 4,800	$ 15,000	$ 2,200	$ (2,132)	
10	3	$ 25,000	$ 7,000	$ 3,840	$ 18,000	$ 7,160	$ 622	
11	4	$ 21,000	$ 4,000	$ 3,000	$ 13,000	$ 6,000	$ 1,747	
12	5	$ 18,000	$ 3,000	$ 2,520	$ 9,000	$ 3,480	$ 2,020	
13	6	$ 17,000	$ 1,000	$ 2,160	$ 6,000	$ 2,840	$ 2,121	Abandon
14	7	$ 15,000	$ 2,000	$ 2,040	$ 5,000	$ 960	$ 2,006	

FIGURE 11-6   Spreadsheet for Determining Abandonment Time in Example 11-11

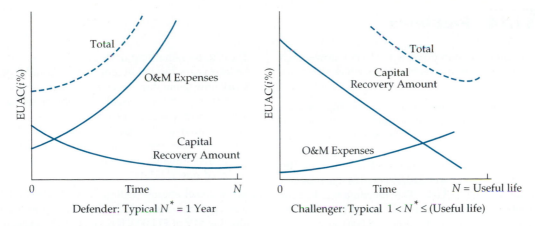

FIGURE 11-7   Typical Pattern of the EUAC for a Defender and a Challenger

6. Any excess capacity, reliability, flexibility, safety, and so on of the challenger may have value to the owner and should be claimed as a dollar benefit if a dollar estimate can be placed on it. Otherwise, this value would be treated as a *nonmonetary* benefit.

11.13   References

Barish, N. N., and S. Kaplan. *Economic Analysis for Engineering and Managerial Decision Making.* New York: McGraw-Hill Book Co., 1978.

Bernhard, R. H., "Improving the Economic Logic Underlying Replacement Age Decisions for Municipal Garbage Trucks: Case Study," *The Engineering Economist* 35, no. 2 (Winter 1990): 129–147.

Lake, D. H., and A. P. Muhlemann, "An Equipment Replacement Problem," *Journal of the Operational Research Society* 30, no. 5 (1979): 405–411.

Leung, L. C., and J. M. A. Tanchoco, "Multiple Machine Replacement within an Integrated Systems Framework," *The Engineering Economist* 32, no. 2 (1987): 89–114.

Machinery and Allied Products Institute. *MAPI Replacement Manual.* Washington, D.C.: Machinery and Allied Products Institute, 1950.

Matsuo, H., "A Modified Approach to the Replacement of an Existing Asset," *The Engineering Economist* 33, no. 2 (Winter 1988): 109–120.

Morris, W. T. *Engineering Economic Analysis.* Reston, VA: Publishing Co., 1976

Naik, M. D., and K. P. Nair. "Multistage Replacement Strategies." *Journal of the Operations Research Society of America,* vol. 13, no. 2, March–April 1965, pp. 279–290.

Oakford, R. V., J. R. Lohmann, and A. Salazar, "A Dynamic Replacement Economy Decision Model," *IIE Transactions* 16, no. 1 (1984): 65–72.

Park, C. S., and G. P. Sharp-Bette. *Advanced Engineering Economics.* New York: John Wiley & Sons, 1990.

Terborgh, George. *Business Investment Management.* Washington, D.C.: Machinery and Allied Products Institute, 1967.

11.14 Problems

The number in parentheses ( ) that follows each problem refers to the section from which the problem is taken.

11-1. A pipeline contractor is considering the purchase of certain pieces of earth-moving equipment to reduce costs. The alternatives are as follows:

Plan A: Retain a backhoe already in use. The contractor still owes $10,000 on this machine but can sell it for $15,000. This machine will last for another six years, with maintenance, insurance, and labor expenses totaling $28,000 per year.

Plan B: Purchase a trenching machine and a highlift to do the trenching and backfill work and sell the current backhoe. The new machines will cost $55,000, will last for six years with a market value of $15,000, and will reduce annual maintenance, insurance, and labor expenses to $20,000.

If the MARR is 8% per year before taxes, which plan should be selected? Use the PW procedure to evaluate both plans on a before-tax basis. (11.4)

11-2. Suppose that you have an old car, which is a real gas guzzler. It is 10 years old and could be sold to a local dealer for $400 cash. Assume its MV two years from now is zero. The annual maintenance expenses will average $800 into the foreseeable future, and the car averages only 10 miles per gallon. Gasoline costs $1.50 per gallon, and you average 15,000 miles per year. You now have an opportunity to replace the old car with a better one that costs $8,000. If you buy it, you will pay cash. Because of a two-year warranty, the maintenance expenses are expected to be negligible. This car averages 30 miles per gallon. Use the IRR method and specify which alternative you should select. Utilize a two-year comparison period and assume that the new car can be sold for $5,000 at the end of year two. Ignore the effects of income taxes and let your MARR be 15% per year. State any other assumptions you make. (11.4)

11-3. A current asset (defender) is being evaluated for potential replacement. It was purchased four years ago at a cost of $62,000. It has been depreciated as a MACRS (GDS) five-year property class asset. The present market value of the defender is

$12,000. Its remaining useful life is estimated to be four years, but it will require additional repair work now (a one-time $4,000 expense) to provide continuing service equivalent to the challenger. The current effective income tax rate is 39%, and the after-tax MARR = 15% per year. Based on an outsider viewpoint, what is the after-tax initial investment in the defender if it is kept (*not* replaced now)? (11.4)

11-4. A diesel engine (defender) was installed 10 years ago at a cost of $50,000. It has a present realizable MV of $14,000. If kept, it can be expected to last for five more years, have annual expenses of $14,000, and have a market value of $8,000 at the end of the five years. This engine can be replaced with an improved version costing $65,000 and having an expected life of 20 years. The challenger will have estimated annual expenses of $9,000 and an ultimate market value of $13,000. It is thought that an engine will be needed indefinitely and that the results of the economy study would not be affected by the consideration of income taxes. Using a before-tax MARR of 15% per year, make an analysis to determine whether to keep or replace the old engine. (11.4, 11.8)

11-5. A three-year-old asset that was originally purchased for $4,500 is being considered for replacement. Its estimated remaining useful life is two years. The new asset under consideration would cost $6,000. The engineering department has made estimates (shown below) of the annual expenses of the two alternatives. The dealer has agreed to place a $2,000 trade-in value on the old asset if the new one is purchased now. It is estimated that the residual (market) value for either of the assets will be zero at any time in the future. If the before-tax MARR is 12% per year, make an analysis of this situation and recommend which course of action should be taken now. Use the repeatability assumption and give a short written explanation of your answer. (11.4, 11.8)

Year	Old Asset	New Asset
1	−$2,000	−$ 500
2	− 2,400	− 1,500
3	—	− 2,500
4	—	− 3,500
5	—	− 4,500

11-6.

a. The replacement of a planing machine is being considered by the Reardorn Furniture Company. The best challenger will cost $30,000 installed, and will have an estimated useful life of 12 years and a $2,000 MV at that time. It is estimated that annual expenses will average $16,000 per year. The defender has a BV of $6,000 and a present MV of $4,000. Data for the defender for the next three years are as follows:

Year	MV at End of Year	BV at End of Year	Expenses During the Year
1	$3,000	$4,500	−$20,000
2	2,500	3,000	− 25,000
3	2,000	1,500	− 30,000

Using a before-tax interest rate of 15% per year, make a comparison to determine whether it is economical to make the replacement now.

b. If the annual expenses for the present machine had been estimated to be $15,000, $18,000, and $23,000 in years one, two, and three, respectively, what replacement strategy should be recommended? (11.7, 11.8)

11-7. Suppose that it is desired to make an after-tax analysis for the situation posed in Problem 11-4. The defender is being depreciated by the straight-line method over 15 years; an estimated market value of $8,000 is being used for depreciation purposes. Assume that if the replacement is made, the challenger will be depreciated as a MACRS (GDS) five-year property class asset. Also, assume that the effective income tax rate is 40%. Use the AW method to determine if the replacement is justified by earning an after-tax MARR of 10% per year or more. (11.4, 11.8)

11-8. You have a machine that was purchased four years ago and depreciated on a five-year straight-line schedule with no market value. The original cost was $150,000, and the machine can last for 10 years or more in its current application. A new machine is now available at a cost of only $100,000. It can be depreciated with the MACRS (GDS) method (five-year property class). The annual expenses of the challenger are only $5,000, while those of the defender are $20,000. The challenger has a useful life greater than 10 years. You find that $40,000 is the best price you can get if you sell the present machine now. Your best projection for the future is that you will need the service provided by one or the other of the two machines for the next five years. The MV of the defender is estimated at $2,000 in five years, but that of the challenger is estimated at $5,000 in five years. If the after-tax MARR is 10% per year, should you sell the defender and purchase the challenger? You do not need both. Assume that the company is in the 40% income tax bracket. (11.4)

11-9. It has been decided to replace an existing piece of equipment with a newer, more productive one that costs $80,000 and has an estimated MV of $20,000 at the end of its useful life of six years. Installation charges for the new equipment will amount to $3,000; this is not added to the capital investment, but will be an expensed item during the first year of operation. MACRS (GDS) depreciation (five-year property class) will be used. The new equipment will reduce direct costs (labor, maintenance, rework, etc.) by $10,000 in the first year, and this amount is expected to increase by $500 each year thereafter during its six-year life. It is also known that the BV of the fully depreciated old machine is $10,000 but that its present fair MV is $14,000. The MV of the old machine will be zero in six years. The effective income tax rate is 40%. (11.4)

a. Determine the prospective after-tax *incremental* cash flow associated with the new equipment if it is believed that the existing machine could perform satisfactorily for six more years.

b. Assume the after-tax MARR is 12% per year. Based on the ERR method, should you replace the defender with the challenger? Assume ϵ = MARR.

11-10. Use the PW method to select the better of the alternatives shown at the top of the first column on page 470. Assume that the defender was installed five years ago and that its MACRS (GDS) property class is seven years. The after-tax MARR is 10% per year, and the effective income tax rate is 40%.

Annual Expenses	Defender: Alternative A	Challenger: Alternative B
Labor	−$300,000	−$250,000
Material	− 250,000	− 100,000
Insurance and property taxes	4% of initial capital investment	None
Maintenance	−$8,000	None
Rental cost	None	−$100,000

Definition of alternatives:

Alternative A: Retain an already owned machine (defender) in service for eight more years.

Alternative B: Sell the old machine and rent a new one (challenger) for eight years.

Alternative A (additional information):

Cost of defender five years ago = −$500,000.
BV now = $111,550.
Estimated market value eight years from now = $50,000.
Present MV = $150,000.

11-11. A steel pedestrian overpass must either be reinforced or replaced. Reinforcement would cost $22,000 and would make the overpass adequate for an additional five years of service. If the overpass is torn down, the scrap value of the steel would exceed the demolition cost by $14,000. If it is reinforced, it is estimated that its net salvage value would be increased by $16,000 at the time it is retired from service. A new prestressed concrete overpass would cost $140,000 and would meet the foreseeable requirements of the next 40 years. Such a design would have no net scrap or salvage value. It is estimated that the annual expenses of the reinforced overpass would exceed those of the concrete overpass by $3,200. Assume that money costs the state 10% per year and that the state pays no taxes. What would you recommend? (11.4)

11-12. Four years ago the Attaboy Lawn Mower Company purchased a piece of equipment for its assembly line. Because of increasing maintenance costs for this equipment, a new piece of machinery is being considered. The cost characteristics of the defender (present equipment) and the challenger are shown in the table below. Suppose that a $3,200 MV is available now for the defender. Perform an after-tax analysis, using an after-tax MARR of 10% per year to determine which alternative to select. The effective income tax rate is 40%. Straight-line depreciation is applied to the defender over nine years, while the challenger qualifies for MACRS (GDS) depreciation (five-year property class). (11.4)

11-13.

a. Find the economic life of an asset having the following projected cash flows. The MARR is 0% per year. (11.6)

Capital investment = −$5,000

MV = 0 (at all times)

Annual expenses = − $3,000 (EOY 1),
− $4,000 (EOY 2),
− $5,000 (EOY 3),
and − $6,000 (EOY 4).

b. Find the economic life of another asset having these cash flow estimates. The MARR is 12% per year. (11.6)

Capital investment = −$10,000

MV = $10,000 (at all times)

Annual expenses = − $3,000 (EOY 1),
− $4,000 (EOY 2),
− $5,000 (EOY 3),
and − $6,000 (EOY 4).

c. Repeat (b), except that MV = $0 at all times. (11.6)

Defender	Challenger	(Pr. 11-12)
Original cost = −$9,000	Purchase cost = −$13,000	
Maintenance = −$300 in first year of use four years ago, increasing by 10% per year thereafter.	Maintenance = −$100 in year one, increasing by 10% per year thereafter.	
Original estimated useful life = nine years	MACRS property class: five years	
Original estimated market value = 0	MV = $3,000 at the end of year five	

	Year 1	Year 2	Year 3	Year 4	(Pr. 11-15)
Annual expenses	−$ 950	−1,050	−1,100	−1,550	
Trade-in value at end of year	2,250	1,800	1,450	1,160	

11-14. Consider a piece of equipment that initially cost $8,000 and has these estimated annual expenses and market values:

End of Year, k	Annual Expenses	MV at End of Year
1	−$3,000	$4,700
2	− 3,000	3,200
3	− 3,500	2,200
4	− 4,000	1,450
5	− 4,500	950
6	− 5,520	600
7	− 6,250	300
8	− 7,750	0

If the cost of money is 8% per year, before taxes, determine the economical life of this equipment. (11.6)

11-15. Robert Roe has just purchased a four-year-old used car, paying $3,000 for it. A friend has suggested that he determine in advance how long he should keep the car so as to ensure the greatest overall economy. Robert has decided that, because of style changes, he would not want to keep the car longer than four years, and he has estimated the annual expenses and trade-in values for years one through four as summarized in the table above. If Robert's capital is worth 12% per year, at the end of which year should he dispose of the car? (11.6)

11-16. Consider a piece of equipment that initially cost $8,000 and has these estimated annual expenses and market values (same as problem 11-14):

End of Year, k	Annual Expenses	MV at End of Year
1	−$3,000	$4,700
2	− 3,000	3,200
3	− 3,500	2,200
4	− 4,000	1,450
5	− 4,500	950
6	− 5,250	600
7	− 6,250	300
8	− 7,750	0

If the after-tax MARR is 7% per year, determine the most economical time to replace this equipment based on an after-tax analysis. MACRS (GDS) depreciation is being used (five-year property class). The effective income tax rate is 40%. (11.6)

11-17. A present asset (defender) has a current market value of $87,000 ($MV_0$). Based on the used equipment market, the estimated market values at the end of the next three years are $MV_1 = $76,000, $MV_2 = $60,000, $MV_3 = $40,000. The annual expenses are $18,000 in present (year zero) dollars, and these expenses are estimated to increase at 4.1% per year. The MARR is 10% per year. The best challenger available has an economic life of six years, and its EUAC over this period is $44,210. Based on this information, when should you plan to replace the defender with the challenger? (11.7, 11.8)

11-18. The present worth of the after-tax cash flows through year k, PW_k, for a defender (three-year remaining useful life) and a challenger (five-year useful life) are given below. Assume the after-tax MARR is 12% per year. Based on this information:

a. What is the economic life (N^*) and the related minimum equivalent uniform annual cost, EUAC, when $k = N^*_{AT}$, for both the defender and the challenger? (11.7, 11.8)

b. When should the defender be replaced by the challenger (based on the present analysis)? Why? (11.7, 11.8)

c. What assumption(s) have you made in answering Part b?

	PW of ATCF Through Year k, PW_k	
Year	Defender	Challenger
1	−$14,020	−$18,630
2	− 28,100	− 34,575
3	− 43,075	− 48,130
4	—	− 65,320
5	—	− 77,910

	End of Year						(Pr. 11-19)
	0	**1**	**2**	**3**	**4**	**5**	
Annual (after-tax) revenues less expenses	−$7,500[a]	$2,000	$2,000	$2,000	$2,000	$2,000	
Abandonment value of machine[b]	—	$6,200	$5,200	$4,000	$2,200	0	

[a] Capital investment
[b] Estimated market value

11-19. A high-speed commercial centrifuge has the after-tax cash flows and abandonment values over its useful life that appear in the table above. The firm's after-tax MARR is 10% per year. Determine the optimal time for the centrifuge to be abandoned if it is acquired and not used for more than five years. (11.9)

11-20. Ten years ago a corporation built a facility at a cost of $400,000 in an area that since has developed into a major retail location. At the time the facility was constructed, it was estimated to have a depreciable life of 20 years, with no market value, and straight-line depreciation has been used. The corporation now finds it would be more convenient to be in a less congested area and can sell the old facility for $250,000. A new facility, in the desired location, would cost $500,000 and have a MACRS (GDS) property class of 10 years. There would be an annual savings of $4,000 per year in expenses. Taxes and insurance on the old facility have been 5% of the initial capital investment per year, while for the new facility they are estimated to be only 3% per year. The study period is 10 years, and the estimated MV of the new facility at the end of 10 years is $200,000. The corporation has a 40% income tax rate, and capital is worth 12% per year after taxes. What would you recommend on the basis of an after-tax IRR analysis? (11.4, 11.10)

11-21. Five years ago an airline installed a baggage conveyor system in a terminal, knowing that within a few years it would have to be moved. The original cost of the installation was $120,000, and through accelerated depreciation methods the company has been able to write off the entire cost. It now finds that it will cost $40,000 to move and upgrade the conveyor. This cost would be recovered over the next six years (the MACRS alternative depreciation system (ADS) with half-

year convention and a five-year class life would be used), which the airline believes is a good estimate of the remaining useful life of the system if moved. It finds that it can purchase a somewhat more efficient conveyor system for an installed cost of $120,000, and this system would result in an estimated reduction in annual expenses of $6,000 in year zero dollars. Annual expenses are expected to inflate by 6% per year. The new system is in the five-year MACRS (GDS) property class, and its estimated market value six years from now, in year zero dollars, is 50% of its installed cost. This MV is estimated to escalate 3% per year. A small airline company, which will occupy the present space, has offered to buy the old conveyor for $90,000.

Annual property taxes and insurance on the present equipment have been $1,500, but it is estimated that they would increase to $1,800 if the equipment is moved and upgraded. For the new system it is estimated that these would be about $2,750 per year. All other expenses would be about equal for the two alternatives. The company is in the 40% income tax bracket. It wishes to obtain at least 10% per year, after taxes, on any invested capital. What would you recommend? (11.4, 11.10)

11-22. Refer to Example 11-9 in Section 11.10. Assume that the hospital's manager of facilities engineering has agreed that the new 120-kW turbine-powered generator (challenger) should be used in the design of the expanded emergency electrical power system. He or she, however, has asked "what is the maximum we can afford to pay to purchase the new 120-kW generator and break even with the contract lease arrangement?" Assume that it (1) would be a MACRS (GDS) seven-year property class asset, (2) would have an estimated market value at the end of 10 years of 35% of the purchase price, in year zero dollars (estimated to escalate 2% per year), and (3) would have annual expenses as previously estimated in Example 11-9. Based on this information and an after-tax, actual dollar analysis, determine the purchase price for the challenger at which the hospital ought to be indifferent to acquiring it through the contract lease arrangement. (11.10)

11-23. A manufacturing company has some existing semiautomated production equipment that it is considering replacing. This equipment has a present MV of $57,000, and a BV of $30,000. It has five more years of straight-line depreciation available (if kept) of $6,000 per year, at which time its BV will be zero. The estimated MV of the equipment five years from now (in year zero dollars) is $18,500. The MV escalation rate on this type of equipment has been averaging 3.2% per year. The total annual expenses are averaging $27,000 per year.

New automated replacement equipment would be leased. Estimated annual expenses for the new equipment are $12,200 per year. The *annual* leasing costs would be $24,300. The MARR (after taxes) is 9% per year, $t = 40\%$, and the analysis period is five years. (Remember: The owner claims depreciation, and the leasing cost is an operating expense).

Based on an after-tax, actual dollar analysis, should the new equipment be leased? Base your answer on the IRR of the incremental cash flow. (11.4)

11-24. A company is considering replacing a turret lathe (defender) with a single-spindle screw machine (challenger). The turret lathe was purchased six years ago at a cost of $80,000, and depreciation has been figured on a 10-year, straight-line basis, using zero salvage value. It can now be sold for $15,000, and if retained would operate satisfactorily for four more years and have zero market value. The screw machine is estimated to have a useful life of 10 years. MACRS (GDS) depreciation would be used (five-year property class). It would require only 50% attendance of an operator who earns $12.00 per hour. The machines would have equal capacities and would be operated eight hours per day, 250 days per year. Maintenance on the turret lathe has been $3,000 per year; for the screw machine it is estimated to be $1,500 per year. Taxes and

insurance on each machine would be 2% annually of the initial capital investment. If capital is worth 10% per year to the company after taxes, and the company has a 40% income tax rate, what is the maximum price it can afford to pay for the screw machine? Assume a four-year analysis period and an imputed market value (Chapter 5) for the challenger at the end of four years. (11.4, 11.10)

11-25 A company has some existing semiautomated production equipment it is considering replacing. This equipment has a present MV of $30,000. It has five more years of straight-line depreciation left (if kept) of $6,000 per year, at which time its BV would be zero. The estimated MV of the equipment five years from now (in year zero dollars) is $18,500. The MV escalation rate on this type of equipment has been averaging 3.2% per year. The actual annual operation and maintenance and other related expenses are $19.30 per operating hour. New automated replacement equipment would be leased. Estimated annual operation and maintenance and other related expenses for the new equipment are $8.70 per operating hour. The annual leasing cost would be $24,300. MARR = 9% per year; $t = 38\%$; and the analysis period is five years. (Remember: the owner gets depreciation, and the annual leasing cost is an operating expense. The user pays annual operation and maintenance expenses (based on problem 11-23). (11.10)

a. What is the breakeven point, in terms of operating hours per year, between keeping the existing equipment (Alternative A) and leasing the new automated production equipment (Alternative B)? Use after-tax, actual dollar analysis and the PW method as applicable.

b. Graph the PW of the net ATCF for each alternative as a function of operating hours per year. Show the breakeven point and indicate the operating range where each alternative is preferred.

11-26. *Brain Teaser*
There are two customers requiring three-phase electrical service, one existing at location A and a new customer at location B. The load at location A is known to be 110 KVA, and at location B it is contracted to be 280 KVA. Both loads are expected to remain constant indefinitely into the future. Already in service at A are three 100-KVA

	Existing and New Transformers		(Pr. 11-26)
	Three 37 1/2 KVA	**Three 100 KVA**	
Capital Investment:			
Equipment	−$900	−$2,100	
Installation	−$340	−$ 475	
Property tax	−2% of capital investment	−2% of capital investment	
Removal cost	−$100	−$ 110	
Salvage value	$100	$ 110	
Useful life (years)	30	30	

transformers that were installed some years ago when the load was much greater. Thus, the alternatives are as follows:

a. Install three 100-KVA transformers (new) at B now and replace those at A with three $37\frac{1}{2}$-KVA transformers only when the existing ones must be retired.

b. Remove the three 100-KVA transformers now at A and relocate them at B. Then install three $37\frac{1}{2}$-KVA transformers (new) at A.

Data for both alternatives are provided in the table above. The existing transformers have 10 years of life remaining. Suppose that the before-tax MARR = 8% per year. Recommend which action to follow after calculating an appropriate criterion for comparing these alternatives. List all assumptions necessary and ignore income taxes. (11.8)

12

Capital Financing and Allocation

*F*or ease of presentation and discussion, we have divided this chapter into two main areas: (1) the long-term sourcing of capital for a firm (capital financing) and (2) the expenditure of capital through development, selection, and implementation of specific projects (capital allocation). Our aim is to give the student an understanding of these basic components of the capital budgeting process so that the important role of the engineer in this complex and strategic function will be made clear.

The following topics are discussed in this chapter:

Capital financing: borrowed funds versus equity

Financing with borrowed capital

Financing with equity capital

Other sources of equity capital

Leasing as a source of capital

Capital allocation among independent projects

Linear programming formulations of capital allocation problems

An overview of corporate capital allocation policy and procedures

12.1 Introduction

Capital financing and allocation are challenging and difficult functions that take place at the highest levels of an organization, such as the office of the chief executive officer or the board of directors. However, information produced at lower levels in the organizational hierarchy directly affects investment proposals that are ultimately considered in the overall capital budget. For example, many alternatives of design and specification will usually have been considered before a major engineering project is placed in top management's capital budget, and the related decisions are an inherent part of the project package (portfolio) recommended to upper management. Appropriate procedures for evaluation of these alternatives should be available and uniformly applied to ensure that the merits of these alternatives are properly considered at all levels of the capital appropriation process.

Capital financing and allocation are normally simultaneous decision processes regarding *how much* and *where* resources will be obtained and expended for future use, particularly in the future production of goods and services. The scope of these activities encompasses

1. How the money is acquired and from what sources
2. How individual capital project alternatives (and combinations of alternatives) are identified and evaluated
3. How minimum requirements of acceptability are set
4. How final project selections are made
5. How postmortem reviews are conducted

12.2 Capital Financing

As discussed in previous chapters, capital plays a significant role in engineering and business projects. Although engineers seldom engage in obtaining the capital for projects, the method by which the capital is to be obtained, and whether it is equity or borrowed capital, may be of great importance to the overall economy of a project.

Most engineering economy studies are concerned with the *total* capital used, without regard for source; this approach, in effect, evaluates the *project* rather than the interests of any particular group of capital suppliers. The illustrations and problems in this book normally evaluate the *project* because in most analyses the choice between alternatives can be made independently of the sources of funds to be used. Hence, up to this point, the firm's overall pool of capital has been regarded as the source of investment funds, and its cost in Chapter 7, for example, was the after-tax MARR. However, the sources of capital may be important in investment analyses in which different sources can be used (e.g., leasing versus borrowing versus equity).

12.2.1 Differences Between Sources of Borrowed Capital and Equity

In previous chapters, various differences in the uses of borrowed (debt) capital and equity capital have been mentioned. These differences can be summarized as follows:

1. When borrowed capital is used, interest must be paid to the suppliers of the capital, and the debt must be repaid at a specified time. The suppliers of debt

capital do not share in the profits resulting from the use of the capital; the interest they receive, of course, comes out of the firm's revenues. In many instances the borrower pledges some type of security to ensure that the money will be repaid. Sometimes the terms of the loan may place limitations on the uses to which the funds may be put, and in some cases restrictions may be put on further borrowing. Interest paid for the use of borrowed funds is a tax-deductible expense for the firm.

2. Equity capital is supplied and used by its owners in the expectation that a *profit* will be earned. There is no assurance that a profit will, in fact, be gained or that the invested capital will be recovered. Similarly, there are no limitations placed on the use of the funds except those imposed by the owners themselves. There is no *explicit* cost for the use of such capital in the ordinary sense of a tax-deductible cost.

12.2.2 Types of Business Organization

A.   Individual Ownership Several types of business organizations have been devised and are employed for obtaining and using capital in business ventures. The simplest and oldest form is individual ownership. An individual uses his or her own capital to establish a business and is the sole owner. Thus, the individual owner controls the enterprise, is entitled to any benefits and profits that accrue, and must assume any losses that occur.

This type of organization has serious limitations relating to continuity and growth. Obviously, the organization ceases upon the death of the owner. Because life is uncertain, the life of the organization is uncertain. Equity funds for expansion must, in general, come from the after-tax profits of the enterprise. At the same time, it is difficult for such an organization to obtain borrowed capital, particularly in large amounts and for an extended period of time, owing to the uncertainty of the life of the organization.

B.   The Partnership One obvious solution to the limited amount of capital that ordinarily can be raised through individual ownership is for two or more persons to become partners and pool their resources so that the required capital will be obtained. This method was widely used in the United States during the nineteenth century.

The partnership has a number of advantages. It is bound by few legal requirements as to its accounts, procedures, tax forms, and other items of operation. Dissolution of the partnership may take place at any time by mere agreement of the partners with practically no consideration of outside persons. It provides an easy method whereby two persons of differing talents may enter into business, each carrying those burdens that he or she can best handle; an example is where one partner is a technician and the other a salesperson.

The partnership, however, has four disadvantages. First, the amount of capital that can be accumulated is definitely limited. Second, the life of the partnership is determined by the lives of the individual partners. When *any* partner dies, the partnership automatically ends. Third, there may be serious disagreement among the partners. Fourth, each member of the partnership is liable for all the debts of the partnership. This particular disadvantage is one of the most serious.

C. The Corporation The corporation is a form of organization that was originated to avoid many of the disadvantages of the individual and partnership forms of ownership of business enterprises. A corporation is a fictitious being, recognized by law, that can pursue almost any type of business transaction in which a real person can engage. It operates under a charter that is granted by a state and is endowed by this charter with certain rights and privileges, such as perpetual life without regard to any change in the persons of its owners, the stockholders. In payment for these privileges and the enjoyment of being a legal entity, the corporation is subject to certain restrictions. It is limited in its field of action by the provisions of its charter. In order to enter new fields of enterprise, it must apply for a revision of its charter or obtain a new one. Special taxes are also assessed against it.

The capital of a corporation is acquired through the sale of stock. The purchasers of the stock are part owners, usually called *stockholders,* of the corporation and its assets. In this manner ownership may be spread throughout the world, and as a result enormous sums of capital can be accumulated. With few exceptions, the stockholders of a corporation, although they are the owners and are entitled to share in the profits, are not liable for the debts of the corporation. *They are thus never compelled to suffer any loss beyond the value of their stock.* Because the life of a corporation is continuous or indefinite, long-term investments can be made and the future faced with some degree of certainty, which makes debt capital (particularly long-term) easier to obtain, and generally at a lower interest cost, for corporations than for individual and partnership types of business organizations.

There are many types of stock, but two are of primary importance. These are *common stock,* which represents ordinary ownership without special guarantees of return on investment; and *preferred stock,* which has certain privileges and restrictions not available for common stock. For instance, dividends on common stock are *not* paid until the fixed percentage return on preferred stock has been paid.

One factor that does not favor the corporate form of business organization is that, except in limited circumstances, the profits of a corporation are subject to double taxation. That is, after the corporation income tax is paid, any remaining profits that are distributed as dividends to stockholders are again taxed as income to those stockholders. Thus, if the corporation's income tax rate is 40% and all remaining profits are distributed to stockholders paying an average of 25% income taxes, $1.00 earned by the corporation becomes $1.00(1 − 0.40) = $0.60 distributed to the stockholder, which becomes $0.60(1 − 0.25) = $0.45 net after taxes. Hence, the actual total income tax rate is ($1.00 − $0.45)/$1.00 = 0.55 = 55%.

12.3 Financing with Debt Capital

There are many situations in which the use of borrowed capital is preferable to the use of equity capital. Expansion through the use of equity capital requires either the existing owners to supply more capital or the sale of additional stock to others, which results in decreasing the percentage ownership of the existing stockholders. If the additional capital is needed for a fairly definite period of time and there is considerable assurance that the existing or future cash flow can readily pay the costs and provide for the repayment of borrowed capital, it may be to the advantage of the existing owners to obtain the needed capital by borrowing. It is common for up to 30% of the total capital of private, competitive corporations to be from debt sources.

If additional debt capital is needed for only a short period of time, usually less than five years and more frequently less than two years, it may be borrowed from a bank or other lending agency by signing a short-term note. Such a note is merely a promise to repay the amount borrowed, with interest, at a fixed future date or dates. The lending agency may require something of tangible value as security for the loan, or at least it will make certain that the financial position of the borrowing organization is such that there is minimal risk involved. Because long commitments of capital are required in most projects, corporations usually resort to bond issues for obtaining long-term debt capital.

12.3.1   Long-Term Bonds

A *bond* is essentially a long-term note given to the lender by the borrower, stipulating the terms of repayment and other conditions. In return for the money loaned, the corporation promises to repay the loan and interest upon it at a specified rate. Because the bond merely represents corporate indebtedness, the bondholder has no voice in the affairs of the business, at least for as long as the interest is paid, and of course is not entitled to any share of the profits.

Bonds usually are issued in units of from $1,000 to $5,000 each, which is known as the *face value*, or *par value*, of the bond. This amount is to be repaid to the lender at the end of a specified period of time. When the face value has been repaid, the bond is said to have been *retired* or *redeemed*. The interest rate quoted on the bond is called the *bond rate*, and the periodic interest payment due is computed as the face value times the bond interest rate per period.

A description of what happens during the normal life cycle of a bond is presented in Figure 12-1.

12.3.2   Bond Retirement

Because stock represents ownership, a corporation does not need to make absolute provision for payments (dividends) to the stockholders. If profits remain after the operating expenses are paid, part or all of these can be divided among the stockholders. Bonds, on the other hand, represent debt, and the interest on them is a cost of doing business. In addition to this periodic cost, the corporation must look forward to the day when the bonds become due and the principal must be repaid to the bondholders.

When it is desired to repay long-term loans and thus reduce corporate indebtedness, a systematic program frequently is adopted for repayment of a bond issue when it becomes due. Some such provision, planned in advance, gives assurance

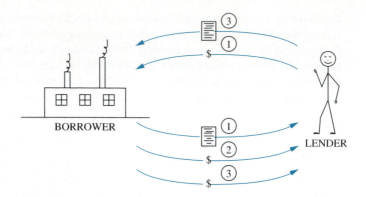

Description of Step	Supplementary Comments
① Borrower sells a bond to the lender. Lender gets bond certificate.	Bonds are issued in even denominations (face values) such as $1,000, $10,000, etc., but amount paid by lender is determined by market supply and demand. The transaction is usually done through a broker.
② Periodic bond interest payments are made to lender.	The amount of each bond interest payment is computed as face value times bond interest rate.
③ Borrower redeems bond by paying principal and getting bond certificate back.	Usually done at end of the stated bond life, the amount paid back is usually the face value.

FIGURE 12-1   Life Cycle of Financing a Bond

to the bondholders and makes the bonds more attractive to the investing public; it may also allow the bonds to be issued at a lower rate of interest.

In many cases the corporation periodically sets aside definite sums that, with the interest they earn, will accumulate to the amount needed to retire the bonds at the time they are due. Because it is convenient to have these periodic deposits equal in amount, the retirement procedure becomes a *sinking fund*. This is one of the most common uses of a sinking fund. By its use, the bondholders know that adequate provision is being made to safeguard their investment. The corporation knows in advance what the annual cost for bond retirement will be.

If a bond issue of $100,000 in 10-year bonds, in $1,000 units, paying 10% nominal interest in semiannual payments must be retired by the use of a sinking fund that earns 8% compounded semiannually, the semiannual cost for retirement will be as follows:

$$A = F(A/F, i\%, N)$$
$$F = \$100,000$$
$$i = 8\%/2 = 4\% \text{ per period}$$
$$N = 2 \times 10 = 20 \text{ periods}$$

Thus,

$$A = \$100,000(0.0336) = \$3,358$$

In addition, the semiannual interest on the bonds must be paid, which would be calculated as follows:

$$\text{Interest} = \$100{,}000 \times \frac{0.10}{2} = \$5{,}000$$

$$\text{Total semiannual cost} = \$3{,}358 + \$5{,}000 = \$8{,}358$$

$$\text{Annual cost} = \$8{,}358 \times 2 = \$16{,}716$$

The total cost for interest and retirement of the entire bond issue over 20 periods (10 years) will be

$$\$8{,}358 \times 20 = \$167{,}160$$

12.4 Financing with Equity Capital

When 100% equity capital is available to finance a project, engineering economic evaluations can be made without explicitly assigning the cost of capital as an expense. That is, interest charges and the business risk associated with debt capital can be down-played so that the analysis is simplified.

12.4.1 Common Stock

The issuance of common stock is an important source of new capital utilized to finance new, modernized, and/or expanded operations of a corporation. Other sources of equity capital include preferred stock, retained earnings, and depreciation reserves.

Establishing value for a share of common stock is not as straightforward as placing value on a bond (refer to Chapter 4). Valuation of common stock is actually a controversial subject because of numerous assumptions regarding the future of dividend growth rates, future stock prices, perceived riskiness of the investment, projected after-tax earnings, and so on.* The value of common stock must be a measure of the earnings that will be received through ownership of the stock, and it is dependent upon several factors, which can probably all be summed up under two headings—dividends and market price. The market price will be affected by dividends as well as by general economic conditions, future prospects of the corporation, the money market, and the preferences of the investing public. In addition, the market price may be drastically altered by speculation, which may bring about a price that is not a true measure of the actual worth of the stock.

In this section, we present a very simple approach for the valuation of common stock and the estimation of per share rate of return expected by the investor. The approach is termed the *dividend valuation model*. Other approaches, such as the earnings model and the investment-opportunities model, are discussed in any good finance textbook.

*See, for example, Franco Modigliani and Merton H. Miller, "The Cost of Capital, Corporation Finance, and the Theory of Investment," *American Economic Review*, Vol. XLVIII, No. 3, June 1958, pp. 261–297; and D. Durand, "The Cost of Capital in an Imperfect Market: a Reply to Modigliani and Miller," *American Economic Review*, Vol. XLIX, No. 4, September 1959, pp. 639–655.

The owner of a share of common stock in a corporation is entitled to receive cash dividends declared by the company, as well as the price of the stock at the time it is sold. If we let the after-tax value of cash dividends received during year k equal Div_k, the current value of a share of common stock in the dividend valuation model can be approximated by the PW of future cash receipts during an N-year ownership period:

$$P_0 \simeq \frac{\text{Div}_1}{(1 + e_a)} + \frac{\text{Div}_2}{(1 + e_a)^2} + \cdots + \frac{\text{Div}_N}{(1 + e_a)^N} + \frac{P_N}{(1 + e_a)^N} \qquad \textbf{(12-1)}$$

where e_a = rate of return per year (%) required by common stockholders (This is the after-tax cost of equity to the corporation.)

P_0 = current value of a share of common stock

P_N = selling price of a share of common stock at the end of N years

The value of e_a must be sufficient to compensate the shareholder for his or her time value of money and for the risk that is believed to be associated with the investment. How we estimate the value of P_N is a further complication in determining P_0.

The dividend valuation model incorporates the two conservative assumptions that dividends are *constant* over the indefinitely long life of a corporation and that $P_0 = P_N$. In this case, the current price of a share of common stock equals the PW of an infinite series of dividend receipts that remain constant in amount:

$$P_0 = \text{Div} \, (P/A, e_a, \infty) = \frac{\text{Div}}{e_a} \qquad \textbf{(12-2)}$$

Thus, if the current selling price of a share of common stock is known and the annual dividend for the past year is also known, the return to equity (common stock) is conservatively estimated to be

$$e_a = \frac{\text{Div}}{P_0} \qquad \textbf{(12-3)}$$

When the future price of the security is assumed to grow at a rate of g (expressed as a decimal) each year, the cost of equity can be approximated by Equation 12-4:

$$e_a = \frac{\text{Div}}{P_0} + g \qquad \textbf{(12-4)}$$

Suppose that a share of common stock is priced at $100 and a dividend of $8 is currently paid annually. The expected annual growth in price is 4% per year. If an investor is willing to purchase this security based on the assumption that dividends remain constant and the price grows at 4% annually, the expected return is about $8/$100 + 0.04 = 0.12, or 12% per year. A second, less risky security being considered may sell for $100 and pay a dividend of $10 annually, with $g = 0$. In this case, $e_1 = 10\%$. If the investor is indifferent between the two securities, an additional expected return of 2% is required to compensate for the extra risk associated with the first investment.

The determination of the cost of all types of equity is difficult in practice. For the purposes of this book, the opportunity cost principle and Equations 12-3 and 12-4 provide a basic, though oversimplified, point of departure for approximating this quantity.

EXAMPLE 12-1

The CMX Corporation is expected to generate perpetual after-tax net earnings of $1,600,000 per year with its existing assets. This firm produces a stable product and has been in business for 75 years. Furthermore, it is 100% equity financed, with 200,000 shares of common stock outstanding, and has a long-standing policy of declaring an annual dividend that is 50% of its after-tax earnings. The remaining 50% of earnings is retained for cash reserves, equipment replacement, and so on.

(a) If investors require a 15% return on their investment, how much would they be willing to pay for a share of CMX common stock if dividends remain constant?

(b) An optimistic investor who owns 1,000 shares of CMX stock believes that its net earnings will grow at a rate of 5% per year in the future. What is the rate of return on CMX stock expected by this investor?

SOLUTION

(a) From Equation 12-2 the current selling price of a share of CMX common stock should be [$1,600,000(0.5)/200,000 shares]/0.15 = $26.67.

(b) The return to equity based on Equation 12-4 would be approximately ($4.000/$26.67) + 0.05 = 0.20, or 20% per year.

12.4.2 Preferred Stock

Preferred stock also represents ownership, but the owner has certain additional privileges and restrictions not assigned to the holder of common stock. Preferred stockholders are guaranteed a dividend on their stock, usually a percentage of its par value, before the holders of the common stock may receive any return. In case of dissolution of the corporation, the assets must be used to satisfy the claims of the preferred stockholders before those of the holders of common stock. Preferred stockholders usually, but not always, have voting rights. Occasionally they are granted certain privileges, such as the election of special representatives on the board of directors, if their preferred dividends are not paid for a specified period.

Because the dividend rate is fixed, preferred stock is a more conservative investment than common stock and has many of the features of long-term bonds. For this reason, the market value of such stock is less likely to fluctuate.

12.4.3 Retained Earnings

Another important source of internal capital for expansion of existing enterprises is retained profits that are reinvested in the business instead of being paid as dividends to the owners. Although this method of financing is used by most companies, there are three factors that tend to limit its use.

Probably the greatest deterrent is the fact that the owners (the stockholders in the case of a corporation) usually expect and demand that they receive some profits from their investment. Therefore, it *usually* is necessary for a large portion (maybe 50% or more) of the profits to be paid to the owners in the form of dividends. Retention of the remaining profits reduces the immediate amount of dividends per share of stock, increases the book value of the stock, and results in greater future dividends and/or market resale value for the stock. Many investors prefer to have some of the profits retained and reinvested to help increase the value of their stock.

The "cost" of this type of capital (retained earnings) is normally assumed to be identical to the rate of return expected by common stockholders. The reason is that retained earnings are reinvested within the corporation, and their opportunity cost to the owners of the firm (common stockholders) should be at least e_a, or else these funds ought to be distributed as dividends.

12.4.4 Depreciation Funds

Funds that are set aside out of revenue as an allowance for depreciation are usually retained and used in a business. These funds are available for reinvestment and are an important internal source of capital for financing new projects. In effect, the depreciation funds provide a revolving investment fund that may be used to the best possible advantage. The funds are thus an important source of capital for financing new ventures within an existing enterprise. Obviously, depreciation funds must be managed so that required capital is available for replacing essential equipment when the time for replacement arrives.

12.5 Leasing as a Source of Capital

As mentioned at the beginning of this chapter, management must address several fundamental issues when obtaining and spending money for long-term investments: (1) How much money is needed for capital expenditures?, (2) How much money can be made available for this purpose?, and (3) What priorities should be used for allocating funds to projects? Normally, the first two questions are answered separately, but in the case of leasing they are resolved simultaneously. As a result, leases are unique in the way their economic advantages and disadvantages are analyzed. The purpose of this section is to demonstrate correct evaluation procedures for simple lease versus purchase situations.

The decision to lease or purchase an asset represents a situation in which the source of capital may affect which alternative is eventually chosen. Leasing is a source of capital generally regarded as a long-term liability similar to a mortgage, whereas the purchase of an asset typically uses funds from the firm's overall pool of capital (much of which is equity). Before considering examples of lease-buy problems, we provide some background information about leases.

For corporations the rent paid on leased assets used in their trade or business is generally deductible as a business expense. To make lease payments deductible as rent, the contract must be for a true leasing arrangement instead of a conditional sales agreement. In a true lease the corporation using the property (lessee) does not acquire ownership in or title to the asset, whereas a conditional sales contract will transfer to the lessee an equity interest in or title to the asset being leased. Hence, the test of whether or not lease payments qualify as business expenses lies in distinguishing between a true lease and a conditional sale.*

For our purposes, it is assumed that a true lease exists and that an asset may be acquired through *leasing* or *purchasing*. Before the decision is made regarding leasing or purchasing, however, the services of the asset in question must have been shown to be economically justifiable. That is, from an investment standpoint the equipment will pay its own way, and the decision regarding how best to finance it constitutes the lease versus purchase decision.

A number of studies have shown that there is no real income tax advantage in leasing. This is particularly true since accelerated methods (e.g., MACRS) have been permitted for depreciation. Assuming a given purchase price, the firm offering a lease contract (lessor) can charge no more for depreciation than can the owner of assets. If assets are leased, the annual lease payments are deducted in computing income taxes; if the assets are purchased, annual depreciation is deducted. Most companies have now come to realize that leasing does not offer major tax advantages.

There may or may not be savings in maintenance expenses through leasing. Any savings will depend on the actual circumstances, which should be carefully evaluated in each case. There is no doubt that leasing usually does simplify maintenance problems, which may be an important factor. Also, many indirect costs, which frequently are difficult to determine, are usually associated with ownership.

These same studies have concluded that a true advantage of leasing lies in allowing a firm to obtain modern equipment that is subject to rapid technological change. Further, leasing for this purpose typically provides an effective hedge against obsolescence and inflation.

The following two examples illustrate correct methods of handling a lease versus purchase study on an after-tax basis; they use the tabular format presented in Chapter 7 (Figure 7-5).

EXAMPLE 12-2

A company is considering building a small office complex at a cost of $100,000 on land that would cost $20,000. It is estimated that the land could be sold at its cost at the end of 50 years and that the building, although fully depreciated on a straight-line basis over the 50 years, probably would have a market value of $20,000. Annual expenses for maintenance, property taxes, insurance, and so on are estimated at $5,250. An alternative is to lease a building for $16,500 per year for a 50-year period. The company ordinarily obtains about 20% on its capital before

---

*For further information, see *Tax Guide for Small Business*, U.S. Internal Revenue Service Publication 334, published annually.

TABLE 12-1   ATCFs for Example 12-2

Year	(A) BTCF	(B) Depreciation	(C) = (A) − (B) Taxable Income	D = −(C) × 40% Cash Flow for Income Taxes	(E) = (A) + (D) ATCF
Purchase Building					
0	−$120,000				−$120,000
1–50	− 5,250	$2,000	−$ 7,250	+$2,900	− 2,350
50	+ 40,000		+ 20,000ᵃ	− 8,000	+ 32,000
Lease Building					
1–50	− 16,500		− 16,500	+ 6,600	− 9,900

ᵃCapital gain on building

taxes and, being in the 40% income tax bracket, about 12% after taxes. What should it do?

SOLUTION

(Using the IRR Method) Table 12-1 will facilitate analysis of the two alternatives. The IRR on the *extra* capital required to purchase rather than to lease the building is the $i'\%$ at which

$$-\$120,000 + (-\$2,350 + \$9,990)(P/A, i'\%, 50)$$
$$+ \$32,000(P/F, i'\%, 50) = 0$$

The $i'\%$ can be found to be approximately 6%. Because 6% < 12%, it would be better to lease the building. Of course, any of the other theoretically correct methods would result in the same recommendation.

EXAMPLE 12-3

An industrial forklift truck can be purchased for $30,000 or leased for a fixed amount of $9,200 per year payable at the *beginning* of each year. The lease contract provides that maintenance expenses are borne by the lessor. Regardless of whether the truck is purchased or leased, the study period is six years. If purchased, annual maintenance expenses are expected to be $1,000 in year zero purchasing power, and they will inflate at 5% per year over the study period. The MV of the truck is expected to be negligible after six years of normal use. Depreciation is determined with the MACRS (GDS) method using a five-year recovery period (deductions occur over six years). The effective income tax rate is 40% and the after-tax MARR, which includes an allowance for general price inflation, is 15%.

Use the AW method and determine whether the forklift truck should be purchased or leased. This firm is profitable in its overall business activity.

SOLUTION

The effects of general price inflation and income taxes on the ATCFs of both alternatives are shown in Table 12-2. The purchase alternative is more costly than the lease alternative (−$6,439 < −$6,348) and would probably not be selected.

TABLE 12-2   ATCFs for Example 12-3

Year	(A) BTCF	(B) Depreciation[a]	(C) = (A) − (B) Taxable Income	D = −(C) × 40% Cash Flow for Income Taxes	(E) = (A) + (D) ATCF
Purchase Truck (A $ Study)[b]					
0	−$30,000				−$30,000
1	− 1,050	$6,000	−$ 7,050	$2,820	1,770
2	− 1,102	9,600	− 10,702	4,281	3,179
3	− 1,158	5,760	− 6,918	2,767	1,609
4	− 1,216	3,456	− 4,672	1,869	653
5	− 1,276	3,456	− 4,732	1,893	617
6	− 1,340	1,728	− 3,058	1,227	−113
Lease Truck (A $ Study)[c]					
0	−$ 9,200		−$ 9,200	$3,680	−$ 5,520
1–5	− 9,200		− 9,200	3,680	− 5,520
6	0	0	0	0	0

[a]MACRS rates are given in Table 7-3.
[b]The AW at MARR = 15% is −$6,439.
[c]The AW at MARR = 15% is −$6,348.

Moreover, if capital is not readily available, the firm may elect to lease the forklift truck because the difference in annual equivalent worths is small. Furthermore, if estimates of maintenance expenses and general price inflation are believed to be inaccurate, the firm would tend to favor leasing as a hedge against an uncertain technological future.

Rather than make use of tabular procedures illustrated in Examples 12-2 and 12-3, models may be developed that yield the same equivalent worths (e.g., PWs) for the lease and purchase alternatives. These will now be summarized.

12.5.1   Cost of the Lease Alternative

The after-tax cost of a lease during year k is given by

$$l_k = L_k(1 - t)$$

where l_k = after-tax lease expense during year k
L_k = before-tax lease expense during year k
t = effective income tax rate

If i, the interest rate the firm expects from the use of its money, is known and fixed, the PW of the after-tax *cost* of the lease during its life of N years is given by

$$\text{PW}_{\text{Lease}}(i\%) = \sum_{k=1}^{N} \frac{L_k(1 - t)}{(1 + i)^k} \tag{12-5}$$

It should be noted that annual maintenance expenses are not included in Equation 12-5 because they are assumed to be borne by the equipment supplier and are included in the annual lease cost L_k. Furthermore, end-of-year cash flows are assumed.

12.5.2 Cost of the Purchase Alternative

The after-tax cost of equipment when it is purchased is a function of the expected annual expenses during the life of the equipment, as well as the purchase price and expected market value. The PW of the after-tax *cost* of purchased equipment is given by

$$PW_{Buy}(i\%) = I - \frac{MV}{(1+i)^N} + \sum_{k=1}^{N} \frac{O\&M_k(1-t) - d_k(t)}{(1+i)^k} \qquad \textbf{(12-6)}$$

where
$$\begin{aligned}
I &= \text{capital investment} \\
MV &= \text{expected market value at end of year } N \\
i &= \text{interest rate per year} \\
N &= \text{life of equipment in years} \\
O\&M_k &= \text{operating and maintenance expense during year } k \\
t &= \text{effective income tax rate} \\
d_k &= \text{depreciation during year } k
\end{aligned}$$

It should be noted that the market value and depreciation credit terms of Equation 12-6 are negative because they reduce costs. Again, the end-of-year cash flow convention is assumed.

12.6 Capital Allocation

To this point we have been discussing capital financing topics that deal with (1) how a company obtains capital (and from what sources) and (2) how much capital and at what cost the company must invest to maintain a successful business enterprise in the years ahead.

The remainder of this chapter examines the capital expenditure decision-making process, also referred to as *capital allocation*. This process involves the planning, evaluation, and management of capital projects. In fact, much of this book has dealt with concepts and techniques required to make correct capital expenditure decisions involving engineering projects. Now our task is to place them in the broader context of upper management's responsibility for proper planning, measurement, and control of the firm's overall portfolio of capital investments.

An outstanding phenomenon of present-day industrialized civilizations is the extent to which engineers and managers, by using capital (money and property), are able to create wealth through activities that transform various types of resources into goods and services. In this regard, Figure 12-2 indicates that the largest industrialized nations in the world consume a significant portion of their gross national product each year by investing in wealth-creating equipment and machinery (so-called intermediate goods of production).

Capital allocation takes place as part of the *investment* function, and it is here that engineering economy plays a significant role in initiating the capital expenditure approval process. As shown in Figure 12-3, engineering economy is heavily

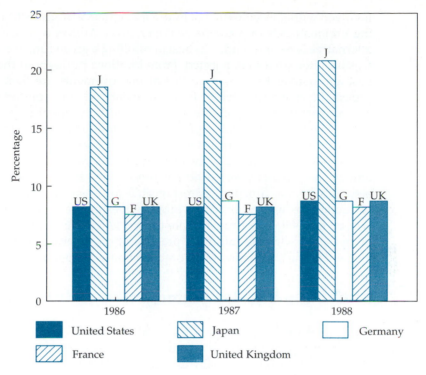

FIGURE 12-2 Fixed Investment in Machinery and Equipment as a Percentage of Gross National Product Source: Office of Technology Assessment, February 1990.

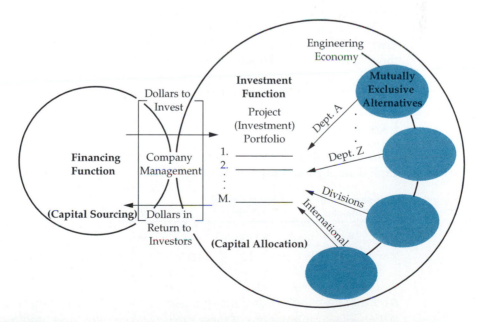

FIGURE 12-3 Overview of Capital Allocation Activities in a Typical Organization

involved with the evaluation of mutually exclusive project alternatives for improving various facets of a company's operations. *Mutually exclusive* means that one alternative is recommended as best in fulfilling a given function and the remaining feasible alternatives are rejected. From locations throughout the organization, the best alternatives for making different improvements are then aggregated into a project (investment) portfolio for a particular budgeting period (e.g., a fiscal year). This project portfolio consists of *independent* investment opportunities, meaning that a *subset* of them will be approved for funding because available capital in any organization is almost always less than the demand for it.

As shown in Figure 12-3, the investment function allocates the available capital among the projects selected on a companywide basis. Management, through its activities in this function, is responsible for ensuring that a reasonable return (in dollars) is earned on these project investments so that providers of capital will be motivated to furnish more capital when the need arises. Thus, it should be apparent why the informed practice of engineering economy is an essential element in the foundation of an organization's competitive culture.

12.7 Allocating Capital Among Independent Projects

Companies are constantly presented with independent opportunities in which they can invest capital across the organization. These opportunities usually represent a collection of the best mutually exclusive alternatives for improving operations in all areas of the company (e.g., manufacturing, research and development, etc.). In most cases the amount of available capital is limited, and additional capital can be obtained only at increasing incremental cost. Thus, companies have a problem of budgeting, or allocating, available capital to numerous possible uses.

One popular approach to capital budgeting uses the PW criterion and was discussed in Chapter 5. If project risks are about equal, the procedure is to compute the PW for each investment opportunity and then to determine the combination of projects that maximizes PW, subject to various constraints on the availability of capital. The following example provides a general review of this procedure.

EXAMPLE 12-4

Consider these five independent projects and determine the best allocation of capital among them if no more than $300,000 is available to invest.

Independent Project	Initial Capital Outlay	PW
A	−$100,000	$25,000
B	− 125,000	30,000
C	− 150,000	35,000
D	− 75,000	40,000

SOLUTION

All possible combinations of these projects taken two, three, and four at a time are shown in Table 12-3, together with the total PW and initial capital outlay of each.

TABLE 12-3   Project Combinations for Example 12-4

Combination	Total PW ($\times 10^3$)	Total Capital Outlay ($\times 10^3$)
AB	$ 55	$225
AC	60	250
AD	65	175
BC	65	275
BD	70	200
CD	75	225
ABC	90	375
ACD	100	325
BCD	105	350
ABD	95	300* **Best**
ABCD	130	450

After eliminating those combinations that violate the $300,000 funds constraint, the proper selection of projects would be ABD and the maximum PW is $95,000. The process of enumerating combinations of projects having nearly identical risks is best accomplished with a computer when large numbers of projects are being evaluated.

Methods for determining which possible projects should be allocated available funds seem to require the exercise of judgment in most realistic capital budgeting problems. Example 12-5 illustrates such a problem and possible methods of solution.

EXAMPLE 12-5

Assume that a firm has five investment opportunities (projects) available, which require the indicated amounts of capital and which have economic lives and prospective after-tax IRRs as shown in Table 12-4. Further, assume that the five ventures are independent of each other; investment in one does not prevent investment in any other, and none is dependent upon the undertaking of another.

Now suppose that the company has unlimited funds available, or at least sufficient funds to finance all these projects, and that capital funds cost the company 6% after taxes. For these conditions the company probably would decide to undertake all projects that offered a return of at least 6%, and thus projects A, B, C, and D would be financed. However, such a conclusion would assume that the risks associated with each project are reasonable in light of the prospective IRR or are no greater than those encountered in the normal projects of the company.

Unfortunately, in most cases the amount of capital is limited, either by absolute amount or by increasing cost. If the total of capital funds available is $60,000, the decision becomes more difficult. Here it would be helpful to list the projects in order of *decreasing* profitability in Table 12-5 (omitting the undesirable project E). Here it is clear that a complication exists. We naturally would wish to undertake

TABLE 12-4	Prospective Projects for a Firm[a]		
Project	**Capital Investment**	**Life (years)**	**Rate of Return (%)**
A	$40,000	5	7
B	15,000	5	10
C	20,000	10	8
D	25,000	15	6
E	10,000	4	5

[a]Here we assume that the indicated rates of return for these projects can be repeated indefinitely by subsequent "replacements."

TABLE 12-5	Prospective Projects of Table 12-4 Ordered by IRR		
Project	**Capital Investment**	**Life (years)**	**Rate of Return (%)**
B	$15,000	5	10
C	20,000	10	8
A	40,000	5	7
D	25,000	15	6

TABLE 12-6	Prospective Projects of Table 12-5 Ordered According to Overall Desirability			
Project	**Capital Investment**	**Life (years)**	**Rate of Return (%)**	**Risk Rating**
C	$20,000	10	8	Lower
A	40,000	5	7	Average
B	15,000	5	10	Higher
D	25,000	15	6	Average

those ventures that have the greatest profit potential. However, if projects B and C are undertaken, there will not be sufficient capital for financing project A, which offers the next greatest rate of return. Projects B, C, and D could be undertaken and would provide an annual return of $4,600 (= $15,000 × 10% + $20,000 × 8% + $25,000 × 6%). If project A were undertaken, together with either B or C, the total annual return would not exceed $4,600.* A further complicating factor is the fact that project D involves a longer life than the others. It is thus apparent that we might *not* always decide to adopt the alternative that offers the greatest profit potential.

The problem of allocating limited capital becomes even more complex when the risks associated with the various available projects are not the same. Assume that the risks associated with project B are determined to be higher than the average

*This return amount is given assuming that the leftover capital could earn no more than 6% per year.

risk associated with projects undertaken by the firm and that those associated with project C are lower than average. The company thus might rank the projects according to their overall desirability, as in Table 12-6. Under these conditions the company might decide to finance projects C and A, thus avoiding one project with a higher than average risk and another having the lowest prospective return and longest life of the group.

▼ 12.8 Linear Programming Formulations of Capital Allocation Problems

For large numbers of independent and/or interrelated investments, the "brute force" enumeration and evaluation of all combinations of projects, as illustrated in Example 12-4, is impractical. This section describes a mathematical procedure for efficiently determining the optimal *portfolio* of projects in industrial capital allocation problems characteristic of Figure 12-3. Only formulations of these problems will be presented in this section; their solution is beyond the scope of the book.

Suppose that the goal of a firm is to maximize its net PW by adopting a capital budget that includes a large number of mutually exclusive combinations of projects. When the number of possible combinations becomes fairly large, manual methods for determining the optimal investment plan tend to become complicated and time consuming, and it is worthwhile to consider linear programming as a solution procedure. The remainder of this section describes how simple capital allocation problems can be formulated as linear programming problems. Linear programming is a mathematical procedure for maximizing (or minimizing) a linear objective function, subject to one or more linear constraint equations. Hopefully, the reader will obtain some feeling for how more involved problems might also be modeled.

Linear programming is a useful technique for solving certain types of multi-period *capital allocation problems* when a firm is not able to implement all projects that may increase its PW. For example, constraints often exist on how much investment capital can be committed during each fiscal year, and interdependencies among projects may affect the extent to which projects can be successfully carried out during the planning period.

The *objective function* of the capital allocation problem can be written as follows:

$$\text{Maximize net PW} = \sum_{j=1}^{m} B_j^* X_j$$

where B_j^* = net PW of investment opportunity j during the planning period being considered

        X_j = fraction of opportunity j that is implemented during the planning period (*Note:* In most problems of interest, X_j will be either 0 or 1; the X_j values are the decision variables)

        m = number of mutually exclusive combinations of alternatives under consideration

In computing the net PW of each mutually exclusive combination, an MARR must be specified.

The following notation is used in writing the constraints for a linear programming model:

c_{ij} = cash outlay (e.g., initial capital investment or annual operating budget) required for opportunity j in time period k

C_k = maximum cash outlay that is permissible in time period k

Typically, two types of constraints are present in capital budgeting problems:

1. Limitations on cash outlays for period k of the planning horizon:

$$\sum_{j=1}^{m} C_{k_j} X_j \leq C_k$$

2. Interrelationships among investment opportunities. The following are examples:

   a. If projects p, q, and r are mutually exclusive, then

   $$X_p + X_q + X_r \leq 1$$

   b. If project r can be undertaken only if project s has already been selected, then

   $$X_r \leq X_s \quad \text{or} \quad X_r - X_s \leq 0$$

   c. If projects u and v are mutually exclusive and project r is dependent (contingent) on the acceptance of u or v, then

   $$X_u + X_v \leq 1$$
   $$\text{and} \qquad X_r \leq X_u + X_v$$

To illustrate the formulation of linear programming models for capital allocation problems, Example 12-6 and Example 12-7 are presented.

EXAMPLE 12-6

Five capital investment projects are being considered for the upcoming budget period (i.e., the year zero capital requirements). Interrelationships and cash flows of the projects are summarized here.

Project	Cash Flow ($000s) for End of Year k					PW at MARR = 10%
	0	1	2	3	4	
B1	−50	20	20	20	20	13.4
B2	−30	12	12	12	12	8.0
C1	−14	4	4	4	4	−1.3
C2	−15	5	5	5	5	0.9
D	−10	6	6	6	6	9.0

Projects B1 and B2 are mutually exclusive. Projects C1 and C2 are mutually exclusive *and* dependent on the acceptance of B2. Finally, project D is dependent on the acceptance of C1.

Using the PW method and MARR $= 10\%$ per year, determine which combination (portfolio) of projects is best if the availability of capital is limited to $48,000.

SOLUTION

The objective function and constraints for this problem are written as follows. Maximize

$$13.4X_{B1} + 8.0X_{B2} - 1.3X_{C1} + 0.9X_{C2} + 9.0X_D$$

subject to

$50X_{B1} + 30X_{B2} + 14X_{C1} + 15X_{C2} + 10X_D \leq 48$
(constraint on investment funds)
$X_{B1} + X_{B2} \leq 1$
($B1$ and $B2$ are mutually exclusive)
$X_{C1} + X_{C2} \leq X_{B2}$
($C1$ or $C2$ is contingent on $B2$)
$X_D \leq X_{C1}$
(D is contingent on $C1$)
$X_j = 0$ or 1
(no fractional projects are allowed)

A problem such as this could be solved readily by using the simplex method of linear programming if the last constraint ($X_j = 0$ or 1) were not present. With that constraint included, the problem is classified as a linear *integer* programming problem. (Many computer programs are available for solving large linear integer programming problems.)

EXAMPLE 12-7

Consider a three-period capital allocation problem having the cash flow estimates shown at the top of the next page. The MARR is 12% and the ceiling on investment funds available is $1,200,000. In addition, there is a constraint on operating funds for support of the alternative selected—$400,000 in year one. From these constraints on funds outlays and the interrelationships among opportunities indicated, we shall formulate this situation in terms of a linear integer programming problem.

SOLUTION

First, the net PW of each investment opportunity at 12% is calculated. The objective function then becomes

$$\text{Maximize net PW} = 135.3X_{A1} + 146.0X_{A2} + 119.3X_{A3} + 164.1X_{B1}$$
$$+ 151.9X_{B2} + 8.7X_{C1} - 13.1X_{C2} + 2.3X_{C3}$$

The budget constraints are the following.
Investment funds constraint:

$$225X_{A1} + 290X_{A2} + 370X_{A3} + 600X_{B1} + 1,200X_{B2}$$
$$+ 160X_{C1} + 200X_{C2} + 225X_{C3} \leq 1,200$$

Investment Opportunity		Net Cash Flow ($000s), End of Year[a]				Net PW ($000s) at 12%[b]
		0	1	2	3	
A1		− 225	150 (60)	150 (70)	150 (70)	+135.3
A2	mutually exclusive	− 290	200 (180)	180 (80)	160 (80)	+146.0
A3		− 370	210 (290)	200 (170)	200 (170)	+119.3
B1	independent	− 600	100 (100)	400 (200)	500 (300)	+164.1
B2		−1,200	500 (250)	600 (400)	600 (400)	+151.9
C1	mutually exclusive and	− 160	70 (80)	70 (50)	70 (50)	+ 8.1
C2	dependent on	− 200	90 (65)	80 (65)	60 (65)	− 13.1
C3	acceptance of A1 or A2	− 225	90 (100)	95 (60)	100 (70)	+ 2.3

[a]Estimates in parentheses are annual operating expenses (which have already been subtracted in the determination of net cash flows).
[b]For example, net PW for $A1 = -\$225{,}000 + \$150{,}000(P/A, 12\%, 3) = +\$135{,}300$.

First year's operating cost constraint:

$$60X_{A1} + 180X_{A2} + 290X_{A3} + 100X_{B1} + 250X_{B2}$$
$$+ 80X_{C1} + 65X_{C2} + 100X_{C3} \leq 400$$

Interrelationships among the investment opportunities give rise to these constraints on the problem:

$$X_{A1} + X_{A2} + X_{A3} \leq 1 \qquad \text{A1, A2, A3 are mutually exclusive}$$

$$\left. \begin{array}{c} X_{B1} \leq 1 \\ X_{B2} \leq 1 \end{array} \right\} \qquad \text{B1, B2 are independent}$$

$$X_{C1} + X_{C2} + X_{C3} \leq X_{A1} + X_{A2} \qquad \begin{array}{l} \text{accounts for dependence of} \\ \text{C1, C2, C3 on A1 or A2} \end{array}$$

Finally, if all decision variables are required to be either 0 (not in the optimal solution) or 1 (included in the optimal solution), the last constraint on the problem would be written:

$$X_j = 0, 1 \quad \text{for } j = A1, A2, A3, B1, B2, C1, C2, C3$$

As can be seen, a fairly simple problem such as this would require a large amount of time to solve by listing and evaluating all mutually exclusive combinations, as suggested in Chapter 5. Consequently, it is recommended that a suitable computer program be used to obtain solutions for all but the most simple capital allocation problems.

12.9 An Overview of Corporate Capital Allocation Policy and Procedures

There is always the possibility that a student of engineering economy has been immersed in such a maze of details by this point that he or she has lost sight of the "enterprise context" in which various types of calculations are performed to evaluate proposed capital expenditures. Therefore, our objective in the remainder of this chapter is to focus on how the results of an engineering economic analysis are used in the corporate capital allocation process. The student should pay close attention to how measures of financial merit, such as present worth and internal rate of return, are used in the corporate capital appropriation process.

Typical corporate capital allocation policy and procedure consist of several sequential steps:

1. Preliminary planning and screening
2. Annual capital expenditure budget
3. Cost of enterprise capital
4. Capital expenditure policies and evaluation procedures
5. Project implementation and postaudit review
6. Communication

12.9.1 Preliminary Planning and Screening

A considerable amount of planning must be accomplished before capital expenditure decisions can be made. The main purpose of capital expenditure planning is to make sure that long-term goals of the organization can be attained. These long-term goals and strategic plans directly tie profit plans to capital budgets. Although budget periods normally range from three to ten years, most large- and medium-sized firms use a five-year period and small firms use a three- to five-year period.

A list of proposed capital project expenditures is prepared by plant and functional managers and forwarded to top management. As these project proposals go up the ranks of the organizational hierarchy, some are eliminated and others are added. To assist management in completing the capital budgeting process, the proposed projects must be classified in some manner. No matter what size the firm is, the two most common methods of classifying proposed projects are by operating division (project type and purpose) and by project size in dollars.

Once the proposed projects are classified, it is necessary to rank them within the portfolio according to various selection criteria. The profitability of invested capital and adherence to the long-term strategy and goals of the business are often ranked as the two top criteria. Two methods frequently used by companies to measure economic merit in the planning stages of a project are the payback period and IRR methods. Projects with long payback periods and/or low IRRs are dropped from further consideration unless there are extenuating circumstances to retain them in the investment portfolio (e.g., projects that must be funded to ensure compliance with legal requirements).

12.9.2   Annual Capital Expenditure Budget

A normal procedure for developing an annual capital expenditure budget within a firm is for division and functional level managers to develop the list of proposed projects, which is then submitted to higher-level management for review and approval. Large projects are normally listed individually, but smaller projects may be combined into summary groups or categories. As might be expected, especially in large companies, top management and the board of directors will usually approve the overall capital budget; division and functional managers will decide on the allocation of capital to most of the individual projects.

Every year a company will have some projects that can be called *noneconomic*. A noneconomic project is one that requires capital investment but provides little or no monetary return. Most companies separate economic and noneconomic projects when requesting funding, and some firms further divide noneconomic projects into various categories such as sustaining, regulatory and environmental, safety and health, and administrative.

For various reasons, not all profitable projects are accepted. A project may be turned down at two stages of the capital budgeting process, the first being at the planning and selection stage and the second at the implementation stage. Although productivity of capital is important, the two major reasons for rejecting a proposed project at either stage are incompatibility with company goals and objectives and unavailability of capital.

12.9.3   Cost of Enterprise Capital

In the long-range planning of a business enterprise, a company decides how big it wants to be, how fast it wants to grow, how much capital it needs, and how it will acquire the capital funds needed. As discussed earlier, the acquisition of these funds from either internal or external sources determines the cost of capital. The most common approach to determining the cost of capital is the after-tax weighted average. Companies using an after-tax cost of capital normally use the current market rate as a basis of cost for each source, and the weighting is based on either the planned debt/equity ratio or the BV of securities.

Some firms use the acquisition cost of capital as the MARR for capital expenditure planning, but others use this as a starting point in developing the MARR for each division. The latter is more prominent in medium-size firms, although most firms tend to use one companywide rate. A revision of a firm's cost of capital is usually made annually.

12.9.4   Capital Expenditure Policies and Evaluation Procedures

Capital expenditure policies and evaluation procedures can be subdivided into two broad parts: (1) management approval levels for projects of different sizes and (2) management control over specific capital expenditures.

Three typical plans for delegating management responsibility for project approvals are as follows:

1. Whenever proposed projects are clearly good in terms of economic desirability according to operating division analysis, the division is given approval power as long as control can be maintained over the total amount invested by each division and as long as the division analyses are considered reliable.
2. Whenever projects represent the execution of policies already established by headquarters, such as routine replacements, the division is given the power to commit funds within the limits of appropriate controls.
3. Whenever a project requires a total commitment of more than a certain amount, this request is sent to higher levels within the organization. The request is often coupled with a budget limitation regarding the maximum total investment that a division may undertake in a budget period.

To illustrate the idea of a larger investment requiring higher administrative approval, the limitations for a particular firm might be as follows:

If the Total Capital Investment Is...

More Than	But Less Than or Equal to	Then Approval Is Required Through
$ 5,000	$ 100,000	Plant Manager
100,000	1,000,000	Division Vice-President
1,000,000	2,500,000	President
2,500,000	—	Board of Directors

The purpose of these policies is to streamline the capital expenditure planning and control process by delegating authority to various management levels to approve projects that can be handled effectively at these levels. This streamling permits top management to concentrate on the most significant capital demands.

Establishing capital expenditure policies is a major responsibility of top management, but responsibility for developing sound economic selection criteria tends to vary by organization. However, regardless of which group develops these criteria, they are applied when a project is proposed and again when it is ready for implementation. When it is initially proposed or later when serious funding is considered, it is customary practice to develop cash flow estimates for a particular project. This information gives some indication of the project size, its profitability, and its overall impact on the firm. It also plays a key role in the development of long-range cash flow forecasts for the company.

12.9.5 Project Implementation and Postaudit Review

The implementation time for a project may be either short or very long, and responsibility for project implementation customarily rests with division management and the project sponsor. Approximately two to six months before implementation, an appropriations request (AR) must be submitted and approved. During the time that a project is being implemented, a periodic progress report is usually submitted to the proper levels of management. This report is used to ensure that the project is on schedule and that management is aware of any problems that may have arisen.

Often a project will have a cost overrun due to the difficulty in estimating future cash flows. Most companies allow a 10% overrun without requiring a new AR to be submitted.

In most firms, division management is responsible for conducting a postaudit review after a project has attained operational status (see Step 7 of the engineering economic analysis procedure discussed in Section 1.4). This review is usually a constructive learning experience that includes a review of project operations and financial performance. The primary objectives of the postaudit appraisal are (1) to determine whether project objectives have been achieved, (2) to discover the degree of conformance to the plan and ascertain where variance occurred, (3) to encourage more careful estimates in the original proposal, and (4) to learn from the results, to identify problems, and to promote better estimates in the future. The postaudit appraisal varies from three months to two years after start-up but is normally done after one year of operation.

12.9.6 Communication

If project proposals are to be transmitted from one organizational unit to another for review and approval, there must be effective means for communication that may range from standard forms to personal appearances. In communicating proposed projects to higher levels in the management structure, it is desirable to use a format that is as standardized as possible to help ensure uniformity and completeness of information and evaluation. In general, the technical and marketing aspects of each proposed project should be completely described in the manner most appropriate to each individual case. However, financial summaries of all proposals should be standardized so that they may be consistently and fairly evaluated.

12.10 Summary

This chapter has provided a broad overview of capital financing (sourcing) and capital allocation (expenditure). Our discussion of capital financing has dealt with where companies get money to continue to grow and prosper and how much it costs to obtain this capital. In this regard, differences between borrowed capital and owner's (equity) capital were made clear. Leasing as a source of capital was also briefly described, and two lease-versus-purchase examples were analyzed.

Our treatment of capital allocation among independent investment opportunities has been built on two important observations. First, the primary concern in capital expenditure activity is to ensure the survival of the company by implementing ideas to maximize future shareholder wealth, which is equivalent to maximization of shareholder PW. Second, engineering economic analysis plays a vital role in deciding which alternative from a large mutually exclusive set is recommended for funding approval and included in a company's overall capital investment portfolio.

12.11   References

BAUMOL, W. J., and R. E. QUANDT. "Investment and Discount Rates Under Capital Rationing—A Programming Approach," *Economic Journal,* vol. 75, no. 298, June 1965, pp. 317–329.

BERNARD, R. H. "Mathematical Programming Models for Capital Budgeting—A Survey, Generalization, and Critique," *Journal of Financial and Quantitative Analysis,* vol. 4, no. 2, 1969, pp. 111–158.

LEVY, H., and M. SARNAT. *Capital Investment and Financial Decisions,* 2nd ed. Englewood Cliffs, N.J.: Prentice-Hall, 1983.

WEINGARTNER, H. M. *Mathematical Programming and the Analysis of Capital Budgeting Problems.* Englewood Cliffs, N.J.: Prentice-Hall, 1963.

12.12   Problems

The number in parentheses () that follows each problem refers to the section from which the problem is taken.

12-1. Describe how an organization's capital financing activities affect the practice of engineering economy. (12.2)

12-2. Why do most engineering economic analyses normally assume that an investor's pool of capital is being used to sponsor a capital expenditure instead of a specific source of capital (e.g., equity versus borrowed funds)? (12.2)

12-3. List five possible sources of funds to a corporation for sponsoring large, capital-intensive projects. (12.4)

12-4. Briefly describe the six basic steps associated with a company's capital allocation procedure. (12.9)

12-5. Contrast the limits on liability for debts of a partnership versus those of a corporation. (12.2)

12-6.
a. What is equity capital, and how is it different from debt capital?
b. Why do bondholders, on the average, receive a lower rate of return than do holders of common stock in the same corporation?
c. Why might a fast-growing company *not* desire to use large amounts of debt capital in its operation? (12.2)

12-7.
a. List at least four characteristics of a corporation.
b. Give an example of how the profits of a corporation are doubly taxed, and discuss how this may retard investment in a firm's stock. (12.2)

12-8. A corporation sold an issue of 20-year bonds, having a total face value of $5,000,000, for $4,750,000. The bonds bear interest at 10%, payable semiannually. The company wishes to establish a sinking fund for retiring the bond issue and will make semiannual deposits that will earn 8%, compounded semiannually. Compute the semiannual cost for interest and redemption of these bonds. (12.3)

12-9. The Yog Manufacturing Company's common stock is presently selling for $32 per share, and annual dividends have been constant at $2.40 per share. If an investor believes that the price of a share of common stock will grow at 5% per year into the foreseeable future, what is the approximate cost of equity to Yog? What assumptions did you make? (12.4)

12-10. A small corporation having a capitalization of $200,000, represented by 2,000 shares of common stock, has been in operation for five years. During this time it has paid no dividends in order to be able to finance its growth through retained profits. It now needs $100,000 in additional capital to finance expansion. It is considering three methods of obtaining the capital: (1) attempting to issue $100,000 in new common stock; (b) borrowing from a bank at 8% interest; and (c) selling five-year bonds bearing interest at 7%, with the restriction that no further indebtedness can be incurred during the life of the bond issue. Discuss briefly the advantages and disadvantages of each method of financing. (12.3, 12.4)

12-11. During the first five years of its life, a small corporation, which has a capitalization of $2,000,000, represented by 2,000 shares of common stock, has paid no dividends in order to finance its expansion out of retained profits. It now needs $1,000,000 in additional capital to finance and stock two new warehouses. Discuss briefly the advantages and disadvantages of obtaining the required capital from (a) selling additional common stock, (b) making a five-year bank loan at 10% interest, and (3) issuing 8%, ten-year bonds that would contain a provision that the corporation could not incur further indebtedness until the bond issue was retired. (12.3, 12.4)

12-12. Refer to Example 12-3. If annual maintenance expenses can range from $800 to $1,300 per year and inflation can vary from 3% to 8% per year, determine whether the forklift truck should be purchased or leased for each combination of extreme values. (12.5)

Annual Maintenance	Annual Inflation Rate (%)	Recommendation
$ 800	3	?
800	8	?
1,300	3	?
1,300	8	?

12-13. An existing piece of equipment has been performing poorly and needs replacing. More modern equipment can be *purchased* for cash using retained earnings (equity funds), or it can be *leased* from a reputable firm. If purchased, the equipment will cost $20,000 and have a depreciable life of five years with no market value. Straight-line depreciation is used by the firm. Because of improved operating characteristics of the equipment, raw materials savings of $5,000 per year are expected to result relative to continued use of the present equipment. However, labor expenses for the new equipment will most likely increase by $2,000 per year and maintenance will go up by $1,000 per year. To lease the new equipment requires a refundable deposit of $2,000, and the end-of-year leasing fee is $6,000. Annual materials savings and extra labor expenses will be the same when purchasing or leasing the equipment, but the company will provide maintenance for its equipment as part

of the leasing fee. The desired after-tax rate of return (IRR) is 15% and the effective income tax rate is 50%. If purchased, it is believed that the equipment can be sold at the end of five years for $1,500 even though $0 was used in calculating depreciation. Determine whether the company should buy or lease the new equipment, assuming that it has been decided to replace the present equipment. (12.5)

12-14. Determine the more economical means of acquiring a business machine if you may either (a) purchase the machine for $5,000 with a probable resale value of $1,000 at the end of five years or (b) lease the machine at an annual rate of $900 per year for five years with an initial deposit of $500, refundable upon returning the machine in good condition. If you own the machine, you will depreciate it for tax purposes at an annual rate of $800. All lease charges are deductible for income tax purposes. As the owner or lessee, you will pay all expenses associated with the operation of the machine.

a. Compare these alternatives by using the AW method. The after-tax MARR is 10% and the effective income tax rate is 50%.

b. How high could the annual leasing fees be such that leasing remains the more desirable alternative? (12.5)

12-15. A firm is considering the development of several new products. The products under consideration are listed in the table below. Products in each group are mutually exclusive. At most one product from each group will be selected. The firm has an MARR of 10% and a budget limitation on development costs of $2,100,000. The life of all products is assumed to be ten years, with no salvage value. Formulate this capital allocation problem as a linear integer programming model. (12.7, 12.8)

Group	Product	Development Cost	Annual Net Cash Income
A	A1	$ 500,000	$ 90,000
	A2	650,000	110,000
	A3	700,000	115,000
B	B1	600,000	105,000
	B2	675,000	112,000
C	C1	800,000	150,000
	C2	1,000,000	175,000

		Proposal			(Pr.12-16)
End of Year	**A**	**B**	**C**	**D**	
0	−$100,000	−$20,000	−$120,000	−$30,000	
1	40,000	6,000	25,000	6,000	
2	40,000	10,000	50,000	10,000	
3	60,000	10,000	85,000	19,000	

12-16. Four proposals are under consideration by your company. Proposals A and C are mutually exclusive; proposals B and D are mutually exclusive and cannot be implemented unless proposal A *or* C has been selected. No more than $140,000 can be spent at time zero. The before-tax MARR is 15%. The estimated cash flows are shown in the table above. Form all mutually exclusive combinations in view of the specified contingencies, and formulate this problem as an integer linear programming model. (12.7, 12.8)

12-17. Three alternatives are being considered for a proposal AS/RS application. Their cash flow estimates are shown in the table in the next column. A and B are mutually exclusive, and C is a high-speed palletizer that represents an optional add-on feature to alternative A. Investment funds are limited to $5,000,000. Another constraint on this project is the engineering personnel needed to design and implement the solution. No more than 10,000 person-hours of engineering time can be committed to the project. Set up a linear integer programming formulation of this capital allocation problem. (12.8)

	Alternative		
	A	**B**	**C**
Initial investment (10^6)	4.0	4.5	1.0
Personnel requirement (hours)	7,000	9,000	3,000
After-tax annual savings, years one through four (10^6)	1.3	2.2	0.9
PW at 10%	0.12	2.47	1.85

12-18. Four proposals are under consideration by your company. Proposals A and C are mutually exclusive; proposals B and D are mutually exclusive and cannot be implemented unless proposal A *or* C has been selected. No more than $140,000 can be spent at time zero. The before-tax MARR is 15%. Formulate this situation in terms of a linear integer programming problem, assuming the variables are either 0 or 1. Relevant data are provided below. (12.8)

	Estimated Cash Flow for Proposals				(Pr.12-18)
End of Year	**A**	**B**	**C**	**D**	
0	−50,000	−20,000	−120,000	−30,000	
1	0	10,000	55,000	15,000	
2	0	10,000	55,000	15,000	
3	83,000	10,000	55,000	15,000	
PW(15%)	4,574	2,832	5,577	4,248	
IRR	18.4%	23.4%	17.8%	23.4%	

Engineering Economy Studies in Investor-Owned Utilities

*O*ur objective in this chapter is to present an economic evaluation technique, called the revenue requirement method, that is widely used by utility companies to select from among proposed projects. Because utilities are expected to minimize the revenues required from customers paying for services, the project having the smallest revenue requirement while providing an acceptable level of service will normally be chosen.

The following topics are discussed in this chapter:

General characteristics of investor-owned utilities

Development of the revenue requirement method

Assumptions of the revenue requirement method

Utility rate regulation

Illustration of the revenue requirement method

Immediate versus deferred investment

Revenue requirement analysis under conditions of inflation

13.1 Background

Investor-owned utilities provide services such as gas, electric power, water, radio and telephone communications, environmental protection, and transportation. Because public utilities are monopolistic, their financing and management have customarily been a government responsibility. However, the last quarter of this century has been characterized by a strong privatization movement that has included power generation and transportation. For example, in the United Kingdom, the British Electric Board has been completely sold to private investors. In the United States, the government has urged and helped private investors to enter the public utilities sector via the Private Power Producers Act. Similarly, most airline companies are privatized in the United States, Japan, and Western Europe.

Most utilities are highly capital intensive and require large amounts of debt (borrowed capital) to finance their initial plant construction and equipment acquisition costs (a 70% ratio of fixed cost to total unit cost is typical). It is, therefore, imperative to ensure a high usage level of the acquired plant and equipment to benefit both the private investor and the public consumer (i.e., society cannot justify having two electric power companies competing on both sides of the same street). Consequently, a compromise is necessary to attain full and efficient utilization of expensive utility resources:

1. The utility is granted a monopolistic franchise in its field of service within a prespecified territory.
2. In return, the community to be served expects the utility to

- Provide capacity to satisfy customer demand
- Maintain acceptable service quality and reliability
- Conform to environmental standards
- Charge a unit price allowing only for a fair profit beyond the unit cost
- Have the capability to attract enough capital to finance new projects

Because city, state, and federal governments, acting on behalf of the public, grant a monopolistic position to the utility, they retain the right to regulate and control the utility's policies. Therefore, a regulatory body, generally in the form of a *public utilities commission,* is established by the public to exercise the desired regulation and control functions. Although public utilities commissions were originally established to prevent utilities from discriminating among customers with respect to services provided and prices charged, their functions have been expanded. They encompass the setting of rates, so that excessive profits are eliminated, and the establishment and maintenance of standards of service. For instance, during the 1970s, energy issues became critical and the major piece of legislation addressing the issue of efficient energy use was the 1979 Public Utilities Regulatory Policies Act (PURPA). This act required utilities to purchase power from available industrial sources and to pay as much for it per kilowatt-hour as they would have if they themselves had generated it.

Thus the regulated, privately owned utility is neither a free-market, for-profit organization nor a fully government-owned, public utility. Its distinctive business objectives and operational characteristics dictate the need for an economic evaluation technique that takes a different viewpoint from those used in free-market enterprises or government-subsidized public utilities.

▼
13.2 General Characteristics of Investor-Owned Utilities

Because of the nature of the services they render, their monopolistic position, and the regulation to which they are subject, investor-owned utilities have a number of unique economic characteristics that must be taken into account in making engineering economy studies. Some of these characteristics are discussed in the following paragraphs.

1. The capital investment per worker and the ratio of fixed costs to variable costs are very high. This means that careful attention must be given to investment problems and to ensuring an adequate flow of capital for expansion.

2. Utilities *must* render whatever service is demanded by customers within established rate schedules. Subject to regulatory safeguards, a utility must expand to meet the growth of the community it serves.

3. A utility is *required* to keep abreast of technical developments in its field that would permit reduction in the cost of its services and improve the quality of the service. This must be done, even though not demanded immediately by customers, to maintain public goodwill and to protect the utility's monopolistic position.

4. The rates charged for a utility's services are based on total costs, including a fair return after income taxes, on the rate-base value of its property. The so-called rate base is roughly equal to the book value of a utility's plant and equipment in service.

5. A basic concept in setting rates for utility services is that the companies must be able to earn enough profits to pay sufficient dividends to attract the capital necessary for rendering the service. If an adequate rate of profit is not obtained, the necessary capital will not be forthcoming from investors, and, as a result, the public will not be able to have the desired utility service. On the other hand, the public has a right to expect to pay no more for the services than will permit an efficiently operated utility to earn the minimum required rate of profit to ensure continuity of service at the level of quality desired.

6. The earnings of a utility are limited by the rate base. As a result, profit on sales is of very little significance. If sales income increases as a result of decreased operating costs—for example, by more effective energy conservation—it may not produce any long-term profit increase. Profits for the current year might be realized, but if the increase were to result in a return that was judged by the regulatory commission to be greater than necessary, a rate reduction would be ordered. Thus the benefits of the improved operation in terms of financial gain to the company would be eliminated.

7. Utilities have much greater stability of income than other companies. The upper limit of earnings, after income taxes, is not usually permitted to exceed about 12% to 16% on equity capital. It should be noted that even though a maximum limit is put on earnings, there is no guarantee of any profit, and

there is no assurance against loss. However, if the utility can show that it is operating efficiently, it can usually obtain permission to increase its rates when needed to produce a fair profit, and may thus be able to attract needed capital. For the well-managed utility company there is a stability of earnings that does not exist for the nonutility company, and this is significant in lowering the cost of capital (debt and equity) that utilities must pay.

8. Because of the stable nature of their business and earnings, utilities commonly finance their capital expenditures with a higher percentage of *borrowed* capital than do nonutility companies. Most nonutilities seldom use more than about 30% debt capital, but many utilities use 50 to 60% borrowed capital as a percentage of their total capitalization.

9. The assets of utilities, on the average, involve longer write-off periods than those of nonutilities. This is caused by the physical nature of the assets and by the fact that the monopolistic situation results in less functional depreciation.

10. Utilities must rely on a larger proportion of new capital for expansion than do other companies, because earnings are so closely regulated that they are only sufficient to pay for the cost of capital. Therefore, if earnings are just high enough to meet the payment of dividends demanded by investors, then very few profits can be retained as surplus to provide for expansion. For example, one utility having equity capital of over $1,375,000,000 has only $104,000,000 in earned surplus, accumulated over more than 40 years.

11. Utilities are much less limited in terms of the availability of capital than are nonutility companies, due to their greater stability of revenues and earnings and to the fact that regulatory agencies recognize that utilities must be permitted to earn a return that will ensure an adequate flow of capital.

13.3 General Concepts of Utility Economy Studies

There are several concepts that are usually inherent to engineering economy studies in regulated utility companies. These are as follows:

1. Economy studies of regulated utilities usually reflect the interests of the customer, whereas those of nonutility companies generally reflect the viewpoint of the owner.

2. Investor-owned utility economy studies usually involve alternative ways of, or alternative programs for, *doing* something. Because a utility is obligated to provide the service demanded by its customers, studies are seldom made of the economy of doing versus not doing. Instead, it is more often a matter of how to do something most economically.

3. Administrative and general supervision expenses frequently are not included. Because these expenses will be about the same for each alternative, they usually may be omitted.

4. The costs of money, depreciation, income taxes, and property taxes are usually expressed in terms of the capital invested.

13.4 Methods of Engineering Economy for Investor-Owned Utility Projects

The economic evaluation method most widely used by privately owned, regulated utilities is the *minimum revenue requirement method.* This method provides a basis for comparing mutually exclusive alternatives. It can be applied to a wide spectrum of regulated businesses that have the characteristics discussed in the previous sections.

> In essence, this method calculates the revenues that a given project must provide just to meet all the costs associated with it, including a fair return to investors.

The relationship between a project's revenue requirements and its costs is shown in Figure 13-1. Because regulatory commissions act on behalf of the consumers of a utility's services, investment project selection should be made in such a way that revenue requirements are minimized.

The payback method, discussed in Chapter 4, is another evaluation method used by utility companies. However, this method is often considered a screening technique, permitting the reduction of the number of candidate alternatives to a manageable size, rather than a final selection method. In electric power utilities the payback method is specifically used to evaluate small discretionary investments such as spare parts or retrofit activities.

The following sections develop and illustrate the revenue requirement method. Examples will be used to explain various facets of this method as it applies to privately owned, regulated utilities.

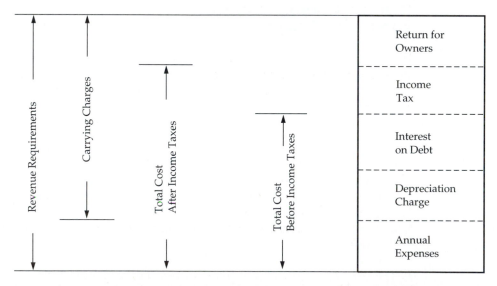

FIGURE 13-1   Relationship of Revenue Requirements and Costs for an Investor-Owned Utility

13.5 Development of the Revenue Requirement Method*

As shown in Figure 13-1, the minimum revenue requirement consists of carrying charges resulting from capital investments that must be recovered, plus all associated expenses that occur periodically (i.e., fuel, O&M expenses, property taxes, and insurance). Carrying charges are also called *total fixed charges.* They include the following:

- Interest on bonds used to partially finance the project
- Equity return requirements for the stockholders
- Income taxes to be paid to state and local governments
- Depreciation charges on the investment

The concept of a fixed-charge rate is widely used in the utility industry. The fixed-charge rate is defined as the annual owning cost of an investment (carrying charges) expressed as a percentage of the investment.

The following equation is used to find the annual carrying charges in year k, CC_k:

$$CC_k = D_{B_k} + [(1 - \lambda)e_a + \lambda i_b] \cdot UI_k + T_k \tag{13-1}$$

where D_{B_k} = book depreciation taken in year k, $1 \leq k \leq N$
  λ = fraction of borrowed money in a utility's total capitalization
  e_a = return to equity capital (as a decimal)
  i_b = cost of borrowed capital (as a decimal)
  UI_k = unrecovered investment at the beginning of year k
  $UI_k = \begin{cases} I \text{ (initial investment)}, k = 1 \\ UI_{k-1} - D_{B_{k-1}}, 2 \leq k \leq N \end{cases}$
  T_k = income taxes paid in year k

Since depreciation claimed for income tax purposes and interest paid on debt are tax deductible, the income tax in any given year is determined by the following equation:

$$T_k = t\left(CC_k - \lambda \cdot i_b \cdot UI_k - D_{T_k}\right) \tag{13-2}$$

where D_{T_k} is the tax depreciation for income tax purposes in year k and t is the effective income tax rate.

Note that carrying charges (CC_k) are a function of income taxes (T_k) in Equation 13-1 and that income taxes (T_k) are a function of carrying charges (CC_k) in Equation 13-2. This can be seen clearly in Figure 13-1. The revenue requirement

---

*The notation in Chapter 13 differs from that used in the rest of the book because various costs of debt and equity capital and resultant costs of total capital are required here. Many of these same concepts are utilized in other chapters but are not developed and employed as rigorously as in Chapter 13.

can be determined if the income taxes are known and, conversely, the income taxes can be computed if the revenue requirement is known. There are two equations and two unknowns (i.e., CC_k and T_k). Solving for T_k we find

$$T_k = [t/(1-t)][(1-\lambda)e_a \cdot UI_k + D_{B_k} - D_{T_k}]$$

(13-3)

The revenue requirement in year k, (RR_k), is

$$RR_k = CC_k + C_k$$

(13-4)

where C_k represents all recurring annual expenses in year k.

13.6 Assumptions of the Revenue Requirement Method

The following assumptions are common when using the revenue requirement method:

1. The total investment in an asset during any year is equal to its beginning-of-year book value.
2. The amount of debt capital invested in an asset during any year is a constant fraction of its book value during that year, and this fraction remains constant throughout the asset's life.
3. Equity and debt capital involve constant rates of return throughout the life of the project.
4. Book depreciation charges are used to retire capital stock and bonds each year in proportion to the debt-equity mix of financing employed.
5. The effective income tax rate is constant over the life of the project.

13.7 Utility Rate Regulation

Utility rates are established during a regulatory rate proceeding. When changes in a utility's cost or income occur because of a change in the company's physical plant, a regulatory rate proceeding takes place to consider whether a new rate is warranted. First, an acceptable return on investors' equity is determined based on factors such as what is required to instill financial confidence in the utility, what other utilities are allowed when operating in the same business risk environment, and what is fair and reasonable. Revenues are then calculated to yield the required return on equity.

H. G. Stoll, of the General Electric Company, distinguishes these two criteria for electric rate regulation: the return on equity and the return on the rate base (Stoll, 1987). The return on common equity criterion calculates the ratio of net income available for common stock (from the utility's income statement) to the average or end-of-year common equity (from the utility's balance sheet). The revenue requirement is then increased or decreased so that the return-on-equity target is achieved. Alternatively, return on the rate base is a more traditional rate regulation criterion. The rate base is defined as follows:

$$
\begin{aligned}
\text{Rate base} = \ & \text{total plant in service} \\
& - \text{accumulated depreciation reserve} \\
& + \text{materials and supplies (optional)} \\
& + \text{fossil fuel inventory (optional)} \\
& + \text{working capital allowance (optional)} \\
& - \text{deferred income taxes (optional)} \\
& - \text{deferred investment tax credit (optional)} \\
& + \text{construction work in progress (optional)}
\end{aligned}
$$

Because of the role and importance of a utility's cost of capital and capitalization structure to the minimum revenue requirement, we discuss selected aspects of an investor-owned utility's financing operation at this point. First, the interest paid on borrowed capital (debt) is tax deductible. Therefore, the after-tax cost of debt, i'_a, is

$$
\begin{aligned}
i'_a &= i'_b - t i'_b \\
&= (1 - t)\left[(1 + i_b)(1 + \bar{f}) - 1\right]
\end{aligned} \tag{13-5}
$$

where
i'_b = inflation-adjusted cost of borrowed capital
$$= \left[(1 + i_b)(1 + \bar{f}) - 1\right]$$
t = effective income tax rate
$\bar{f}$ = average annual inflation rate

Second, the firm's cost of capital depends on the proportion and cost of both debt and equity capital. The *after-tax cost of capital*, K'_a, including an adjustment for inflation, is

$$
\begin{aligned}
K'_a &= \lambda i'_a + (1 - \lambda)e'_a \\
&= \lambda(1 - t)i'_b + (1 - \lambda)e'_a
\end{aligned} \tag{13-6}
$$

where
λ = fraction of borrowed money in the utility's total capitalization
$(1 - \lambda)$ = fraction of equity capital in total capitalization
e'_a = inflation-adjusted equity rate = $\left[(1 + e_a)(1 + \bar{f}) - 1\right]$

The real (inflation-free) after-tax cost of capital, K_a, is

$$
\begin{aligned}
K_a &= \frac{1 + K'_a}{1 + \bar{f}} - 1 \\
&= \frac{\lambda(1 - t)i_b + (1 - \lambda)e_a - \lambda t \bar{f}}{1 + \bar{f}}
\end{aligned} \tag{13-7}
$$

where e_a is the real equity rate.

▼ 13.8 Flow-Through and Normalized Accounting

The revenue requirement method of project comparison presented in Section 13.9 uses the *flow-through* accounting method. Flow-through accounting requires income tax savings (credits) resulting from (1) accelerated depreciation, (2) investment credits (when applicable), and (3) interest paid on funds used during construction to be passed on to a utility's customers in the year that they occur. For instance, straight-line depreciation for rate-setting purposes and accelerated depreciation for determining federal income taxes owed typically combine to *reduce* the revenue requirements for a project when using flow-through accounting methods. However, this accounting method is also frequently employed to compare the relative economics of competing projects, and it produces revenue requirements that are equivalent to those of the after-tax discounted cash flow methods illustrated in Chapter 7.

On the other hand, normalized accounting requires the income tax savings listed previously to be amortized (depreciated) over the life of a project. Normalized accounting is utilized by most investor-owned utilities as a way of protecting the company against unforeseen changes in future income tax rates and state/federal laws that govern their operation. In addition, normalized accounting is almost exclusively used for setting rates on services provided to customers. This accounting method produces revenue requirements that are frequently higher than those resulting from the flow-through method. Because of the additional details associated with normalized accounting, we have elected in this chapter to discuss only the flow-through method for determining revenue requirements. Information concerning normalized accounting can be found in the book by G. T. Stevens (1979, pp. 118–132).

▼ 13.9 Illustration of the Revenue Requirement Method: A Tabular Procedure

Using a tabular form to calculate the annual revenue requirements for a utility project offers an easy-to-manipulate and understandable computational format. The analyst can use tabular columns, as required by a given problem, to account for the various components of revenue requirement shown in Figure 13-1.

EXAMPLE 13-1

This example, evaluating a single investment project, utilizes the following project data and the column-by-column operations shown in Table 13-1:

Project life, N = book life = 4 years
Initial capital investment, I = $7,500
Market value, MV = $1,500
Annual O&M expenses, C = $500

TABLE 13-1   Annual Revenue Requirements for Example 13-1

Year, k	(1) Unrecovered Invest., UI_k	(2) Book Depr., D_{B_k}	(3) Tax Depr., D_{T_k}	(4) Debt Return, $\lambda i_b UI_k$	(5) Equity Return, $(1-\lambda)e_a UI_k$	(6) Income Tax, T_k	(7) Annual Expenses, C_k	(8) RR_k = Cols. 2 + 4 + 5 + 6 + 7
1	$7,500	$1,500	$1,500	$113	$844	$844	$500	$3,780
2	6,000	1,500	1,500	90	675	675	500	3,440
3	4,500	1,500	1,500	68	506	506	500	3,080
4	3,000	1,500	1,500	45	337	337	500	2,720

Real (inflation-free) cost of borrowed money, $i_b = 5\%$ per year
Real (inflation-free) return on equity, $e_a = 16.07\%$ per year
Debt ratio, $\lambda = 0.3$
Effective income tax rate, $t = 50\%$
Book depreciation method = straight line
Tax depreciation method = straight line
Average annual inflation rate, $\bar{f} = 0\%$

For each operating year, k, $1 \leq k \leq 4$, the revenue requirement in year k, RR_k, is calculated by using Equation 13-4. One column is reserved for each term of the carrying charges (see Equation 13-1), and an additional column is used for the recurring annual expenses associated with the project.

For instance, RR_2 is calculated as follows:

Column 1: $UI_2 = UI_1 - D_{B_1}$
$$= \$7,500 - \$1,500 = \$6,000$$

Columns 2 and 3: $D = (I - MV)/N$
$$= (\$7,500 - \$1,500)/4 = \$1,500$$

Column 4: $\lambda i_b \cdot UI_2 = 0.3(0.05)(\$6,000) = \$90$

Column 5: $(1 - \lambda)e_a \cdot UI_2 = 0.7(0.1607)(\$6,000) = \$674.94$

Column 6: $T_2 = [t/(1 - t)][(1 - \lambda)e_a \cdot UI_2 + D_{B_2} - D_{T_2}]$
$$= [0.5/(1 - 0.5)] \cdot [0.7 \cdot 0.1607 \cdot \$6,000 + \$1,500 - \$1,500]$$
$$= \$674.94$$

Column 7: $C_2 = \$500$

Column 8: $RR_2 = \$1,500 + \$500 + \$90 + \$674.94 + \$674.94$
$$= \$3,439.88$$

Calculations for the remaining years are performed similarly. Table 13-1 provides a summary of the results in Example 13-1. Note that no unrecovered investment is left after the end of year four.

It is customary to express the yearly revenue requirements (column 8) as a single measure of worth for the project under study.

Cumulative present worth, equivalent annual worth (also called *levelized revenue requirement, $\overline{RR}$*), and capitalized worth are the three measures most often used by utilities to report the economic merit of a given project. To calculate these quantities, a discounting factor is needed to account for the time value of money. The utility's real after-tax cost of capital, K_a, is usually the interest rate employed for such calculations.

In the preceding example, K_a is determined using Equation 13-7 with an inflation rate of $\bar{f} = 0$:

$$K_a = 0.3 \cdot (1 - 0.5) \cdot 0.05 + (1 - 0.3) \cdot 0.1607 - 0.3 \cdot 0.5 \cdot 0/(1 + 0)$$
$$= 0.12$$

Hence, the present worth of RR as a function of K_a is:

$$PWRR(K_a) = \sum_{k=1}^{N} RR_k \cdot (P/F, K_a\%, k)$$
$$= [\$3{,}799.86(P/F, 12\%, 1) + \$3{,}439.88(P/F, 12\%, 2)$$
$$+ \$3{,}079.92(P/F, 12\%, 3) + \$2{,}719.94(P/F, 12\%, 4)]$$
$$= \$10{,}055.59$$

The levelized revenue requirement is then

$$\overline{RR}(K_a) = PWRR(K_a) \cdot (A/P, K_a\%, N)$$
$$= \$10{,}055.59 \cdot (A/P, 12\%, 4)$$
$$= \$3{,}310.70$$

Finally, the capitalized revenue requirement is

$$CRR(K_a) = \overline{RR}(K_a) \div K_a$$
$$= \$3{,}310.70 \div 0.12$$
$$= \$27{,}589.17$$

In selecting among alternative investment projects, the three quantitative measures above are equivalent. The alternative that minimizes the selected revenue requirement measure represents the most economical choice. Because the utility is obligated to provide services to the public, a request for a rate increase can be presented to the regulatory commission if the investors feel that revenues generated from a project are unsatisfactory.

EXAMPLE 13-2

A public utility must extend electric power service to a small shopping center. A decision must be made as to whether a pole line or an underground system should be used. The pole-line system would cost only $158,000 to install, but because of numerous changes that are anticipated in the development and use of the shopping center, it is estimated that annual maintenance expenses would be

$29,000. An underground system would cost $315,000 to install, but the annual maintenance expenses would not exceed $5,500. Annual property taxes are 1.5% of the capital investment. The company operates with 33% borrowed capital, on which it pays an interest rate of 8% per year. Capital should earn about 11% per year after taxes. For this problem, the after-tax return of 11% is interpreted as the value of K_a. A 20-year study period is to be used, and the effects of inflation on cash flows are to be ignored. The straight-line method is to be used for both book and tax depreciation purposes. Finally, the effective income tax rate is 39.94%.

Before column 5 (equity return) in the tabular procedure can be computed, the value of the return on equity, e_a, has to be determined. By using Equation 13-7, one can write

$$e_a = \left\{ K_a - \lambda \cdot \left[(1 - t) \cdot i_b - t \cdot \bar{f} \cdot (1 - \bar{f}) \right] \right\} / (1 - \lambda)$$
$$= \left\{ 0.11 - 0.33 \cdot \left[(1 - 0.3994) \cdot 0.08 - 0.3994 \cdot 0 \cdot (1 - 0) \right] \right\} / (1 - 0.33)$$
$$= 0.1405$$

Tables 13-2 and 13-3 display the yearly RR results for the pole-line and underground systems, respectively. The values of $\overline{RR}$ are given for each alternative at the bottom of the corresponding RR table. Accordingly, the pole-line system shows a lower $\overline{RR}$ and is, therefore, the system to be chosen based on financial considerations only.

13.10 Immediate Versus Deferred Investment

Because utilities must always be prepared to meet the demands for service placed on them, many engineering economy studies in utility companies involve immediate versus deferred investment to meet future demands. The following is an example.

EXAMPLE 13-3

A water company must decide whether to install a new pumping plant now and abandon a gravity-feed system, which has been fully depreciated, or wait 5 years to install the new plant because of deteriorated piping in the gravity-feed system. Annual O&M expenses and taxes for the gravity system are $45,000. The pumping plant will cost $375,000 to install, and it is estimated that it would have a market value of 5% of its capital investment at the time of removal from service 20 years hence, when a new and larger system will be installed. Annual O&M expenses and property taxes for the proposed plant would be $30,000. The gravity-feed system has no market value now or later.

If the pumping plant is installed now, it would have a useful life of 20 years. If installed 5 years hence, its useful life would be only 15 years, but its market value would still be 5% of its capital investment. Using the revenue requirement method, determine which alternative is better. Straight-line depreciation is assumed for both book and tax purposes. The company operates with 50% borrowed capital,

TABLE 13-2 Pole-Line System Calculations for Example 13-2

Year, k	(1) Unrecovered Investment	(2) Book Deprec.	(3) Tax Deprec.	(4) Debt Return	(5) Equity Return	(6) Income Tax	(7) Annual Expenses	(8) RR_k = Cols. 2 + 4 + 5 + 6 + 7
1	$158,000	$7,900	$7,900	$4,171	$14,875	$9,892	$31,370	$68,208
2	150,100	7,900	7,900	3,963	14,131	9,397	31,370	66,761
3	142,200	7,900	7,900	3,754	13,387	8,902	31,370	65,313
4	134,300	7,900	7,900	3,546	12,644	8,408	31,370	63,867
5	126,400	7,900	7,900	3,337	11,900	7,914	31,370	62,420
6	118,500	7,900	7,900	3,128	11,156	7,419	31,370	60,974
7	110,600	7,900	7,900	2,920	10,412	6,924	31,370	59,526
8	102,700	7,900	7,900	2,711	9,669	6,430	31,370	58,080
9	94,800	7,900	7,900	2,503	8,925	5,935	31,370	56,632
10	86,900	7,900	7,900	2,294	8,181	5,440	31,370	55,185
11	79,000	7,900	7,900	2,086	7,437	4,946	31,370	53,739
12	71,000	7,900	7,900	1,877	6,694	4,452	31,370	52,292
13	63,200	7,900	7,900	1,668	5,950	3,957	31,370	50,845
14	55,300	7,900	7,900	1,460	5,206	3,462	31,370	49,399
15	47,400	7,900	7,900	1,251	4,462	2,967	31,370	47,951
16	39,500	7,900	7,900	1,043	3,719	2,473	31,370	46,504
17	31,600	7,900	7,900	834	2,975	1,978	31,370	45,057
18	23,700	7,900	7,900	626	2,231	1,484	31,370	43,611
19	15,800	7,900	7,900	417	1,487	989	31,370	42,163
20	7,900	7,900	7,900	209	744	495	31,370	40,717

$$\overline{RR} = \$59,496.77$$

TABLE 13-3 Underground System Calculations for Example 13-2

Year, k	(1) Unrecovered Investment	(2) Book Deprec.	(3) Tax Deprec.	(4) Debt Return	(5) Equity Return	(6) Income Tax	(7) Annual Expenses	(8) RR_k = Cols. 2 + 4 + 5 + 6 + 7
1	$315,000	$15,750	$15,750	$8,316	$29,655	$19,721	$10,225	$83,667
2	299,250	15,750	15,750	7,900	28,173	18,735	10,225	80,783
3	283,500	15,750	15,750	7,484	26,690	17,749	10,225	77,899
4	267,750	15,750	15,750	7,069	25,207	16,763	10,225	75,014
5	252,000	15,750	15,750	6,653	23,724	15,777	10,225	72,129
6	236,250	15,750	15,750	6,237	22,242	14,791	10,225	69,245
7	220,500	15,750	15,750	5,821	20,759	13,805	10,225	66,360
8	204,750	15,750	15,750	5,405	19,276	12,819	10,225	63,476
9	189,000	15,750	15,750	4,990	17,793	11,832	10,225	60,590
10	173,250	15,750	15,750	4,574	16,310	10,846	10,225	57,705
11	157,500	15,750	15,750	4,158	14,828	9,861	10,225	54,822
12	141,750	15,750	15,750	3,742	13,345	8,874	10,225	51,936
13	126,000	15,750	15,750	3,326	11,862	7,888	10,225	49,052
14	110,250	15,750	15,750	2,911	10,379	6,902	10,225	46,167
15	94,500	15,750	15,750	2,495	8,897	5,917	10,225	43,283
16	78,750	15,750	15,750	2,079	7,414	4,930	10,225	40,398
17	63,000	15,750	15,750	1,663	5,931	3,944	10,225	37,513
18	47,250	15,750	15,750	1,247	4,448	2,958	10,225	34,629
19	31,500	15,750	15,750	832	2,966	1,972	10,225	31,744
20	15,750	15,750	15,750	416	1,483	986	10,225	28,859

$$\overline{RR} = \$66,304.74$$

on which it pays an interest rate of 7% per year. The equity rate is expected to be about 14% per year, and the company pays an effective income tax rate of 50%.

SOLUTION

First, we determine K_a from Equation 13-7 to be $0.5[(1 - 0.5)(0.07)] + 0.5(0.14) = 0.0875$. Next, from Table 13-4, the levelized revenue requirement of the new pumping plant using the tax-adjusted cost of capital is found to be

$$\overline{RR}(8.75\%) = \$92{,}135$$

From Table 13-5, the levelized revenue requirement of the deferred installation is

$$\overline{RR}(8.75\%) = \$74{,}876$$

Finally, a comparison of the levelized revenue requirements for both alternatives shows that it is more economical to defer the new pumping plant for 5 years.

13.11 Revenue Requirement Analysis Under Conditions of Inflation*

As discussed in Chapter 9, considerable confusion frequently arises when we consider inflation in engineering economy studies because of depreciation and other fixed actual dollar annuities that are not sensitive to inflation. This same difficulty also arises in connection with the revenue requirement method. Examples 13-4 and 13-5 illustrate the correct treatment of inflation in revenue requirement studies.

EXAMPLE 13-4

We now reevaluate Example 13-1 when annual expenses inflate at 10% per year and when the cost of borrowed money and return to equity also increase due to this rate of inflation. It is assumed further that the market value is *not* responsive to inflation. Furthermore, amounts estimated for annual expenses are expressed in year zero purchasing power.

The revenue requirement table is obtained by using the equations employed in Example 13-1, *except* that all year zero (real) quantities are replaced by their equivalent inflation-adjusted values and MACRS (GDS property class is three years) depreciation is included in column 3. In particular, the inflation-adjusted cost of borrowed capital is calculated as

$$i'_b = \left(1 + i_b\right) \cdot \left(1 + \bar{f}\right) - 1$$
$$= (1 + 0.05) \cdot (1 + 0.1) - 1$$
$$= 0.155$$

---

*Remaining examples include MACRS depreciation under the "Tax Depreciation" column to illustrate the situation where book depreciation and tax depreciation are different.

TABLE 13-4 Install New Pumping Plant Now for Example 13-3

Year, k	(1) Unrecovered Investment	(2) Book Deprec.	(3) Tax Deprec.	(4) Debt Return	(5) Equity Return	(6) Income Tax	(7) Annual Expenses	(8) RR = Cols. 2 + 4 + 5 + 6 + 7
1	$375,000.00	$17,812.50	$17,812.50	$13,125.00	$26,250.00	$26,250	$30,000	$113,438
2	357,187.50	17,812.50	17,812.50	12,501.56	25,003.13	25,003	30,000	110,320
3	339,375.00	17,812.50	17,812.50	11,878.13	23,756.25	23,756	30,000	107,203
4	321,562.50	17,812.50	17,812.50	11,254.69	22,509.38	22,509	30,000	104,086
5	303,750.00	17,812.50	17,812.50	10,631.25	21,262.50	21,263	30,000	100,970
6	285,937.50	17,812.50	17,812.50	10,007.81	20,015.62	20,016	30,000	97,852
7	268,125.00	17,812.50	17,812.50	9,384.38	18,768.76	18,769	30,000	94,735
8	250,312.50	17,812.50	17,812.50	8,760.94	17,521.88	17,522	30,000	91,618
9	232,500.00	17,812.50	17,812.50	8,137.50	16,275.00	16,275	30,000	88,501
10	214,687.50	17,812.50	17,812.50	7,514.06	15,028.12	15,028	30,000	85,383
11	196,875.00	17,812.50	17,812.50	6,890.63	13,781.26	13,781	30,000	82,266
12	179,062.50	17,812.50	17,812.50	6,267.19	12,534.38	12,534	30,000	79,148
13	161,250.00	17,812.50	17,812.50	5,643.75	11,287.50	11,288	30,000	76,033
14	143,437.50	17,812.50	17,812.50	5,020.31	10,040.62	10,041	30,000	72,915
15	125,625.00	17,812.50	17,812.50	4,396.88	8,793.76	8,794	30,000	69,798
16	107,812.50	17,812.50	17,812.50	3,773.44	7,546.88	7,547	30,000	66,680
17	90,000.00	17,812.50	17,812.50	3,150.00	6,300.00	6,300	30,000	63,563
18	72,187.50	17,812.50	17,812.50	2,526.56	5,053.12	5,053	30,000	60,446
19	54,375.00	17,812.50	17,812.50	1,903.31	3,806.26	3,806	30,000	57,328
20	36,562.50	17,812.50	17,812.50	1,279.69	2,559.38	2,559	30,000	54,210

$$\overline{RR} = \$92,135$$

TABLE 13-5 Defer Installation of New Pump for 5 Years for Example 13-3

Year, k	(1) Unrecovered Investment	(2) Book Deprec.	(3) Tax Deprec.	(4) Debt Return	(5) Equity Return	(6) Income Tax	(7) Annual Expenses	(8) RR$_k$ = Cols. 2 + 4 + 5 + 6 + 7
1	0	0	0	0	0	0	$45,000	$ 45,000
2	0	0	0	0	0	0	45,000	45,000
3	0	0	0	0	0	0	45,000	45,000
4	0	0	0	0	0	0	45,000	45,000
5	0	0	0	0	0	0	45,000	45,000
6	$375,000	$23,750	$23,750	$13,125.00	$26,250	$26,250	$30,000	$119,375
7	351,250	23,750	23,750	12,293.75	24,588	24,588	30,000	115,220
8	327,500	23,750	23,750	11,462.50	22,925	22,925	30,000	111,063
9	303,750	23,750	23,750	10,631.25	21,263	21,263	30,000	106,907
10	280,000	23,750	23,750	9,800.00	19,600	19,600	30,000	102,750
11	256,250	23,750	23,750	8,968.75	17,938	17,938	30,000	98,595
12	232,500	23,750	23,750	8,137.50	16,275	16,275	30,000	94,438
13	208,750	23,750	23,750	7,306.25	14,673	14,673	30,000	90,402
14	185,000	23,750	23,750	6,475.00	12,950	12,950	30,000	86,125
15	161,050	23,750	23,750	5,643.75	11,288	11,288	30,000	81,970
16	137,500	23,750	23,750	4,812.50	9,625	9,625	30,000	77,813
17	113,750	23,750	23,750	3,981.25	7,963	7,963	30,000	73,657
18	90,000	23,750	23,750	3,150.00	6,300	6,300	30,000	69,500
19	66,250	23,750	23,750	2,318.75	4,638	4,638	30,000	65,345
20	42,500	23,750	23,750	1,487.50	2,975	2,975	30,000	61,188

$$\overline{RR} = \$74,876$$

TABLE 13-6 Solution of Example 13-1 with $\bar{f} = 10\%$

Year, k	(1) Unre-covered Investment	(2) Book Deprec.	(3) Tax Deprec.	(4) Debt Return	(5) Equity Return	(6) Income Tax	(7) Annual Expenses	(8) RR_k = Cols. 2 + 4 + 5 + 6 + 7
1	$7,500	$1,500	$2,500	$349	$1,453	$ 453	$550	$4,305
2	6,000	1,500	3,334	279	1,162	− 671	605	2,875
3	4,500	1,500	1,111	209	872	1,261	666	4,508
4	3,000	1,500	556	140	581	1,525	733	4,479

and the inflation-adjusted equity rate is

$$e'_a = (1 + e_a) \cdot (1 + \bar{f}) - 1$$

$$= (1 + 0.1607) \cdot (1 + 0.1) - 1$$

$$= 0.27677$$

Similarly, annual expenses in year k are

$$C_k = \$500 \cdot (1 + \bar{f})^k, \quad 1 \le k \le 4$$

The results of the revenue requirement analysis are summarized in Table 13-6, and representative calculations are the following:

Column 4: Debt return in year k $= \lambda i'_b \cdot UI_k$
Debt return in year 1 $= (0.3)(0.155)(\$7,500)$
$= \$348.75$

Column 5: Equity return in year $k = (1 - \lambda)e'_a \cdot UI_k$
Equity return in year 1 $= (1 - 0.3)(0.27677)(\$7,500)$
$= \$1,453.04$

Column 6: Income tax $T_k = [t/(1 - t)][(1 - \lambda)e'_a \cdot UI_k + D_{B_k} - D_{T_k}]$
Income tax in year 1 $= [0.5/(1 - 0.5)][(1 - 0.3)(0.27677)(\$7,500)$
$+ \$1,500 - \$2,500] = \$453.$

The inflation-adjusted after-tax cost of capital calculated from Equation 13-6 is

$$K'_a = \lambda(1 - t)i'_b + (1 - \lambda)e'_a$$

$$= 0.3(1 - 0.5)(0.155) + (1 - 0.3)(0.27677)$$

$$= 0.216989$$

$$\cong 21.7\%$$

The levelized revenue requirement of the project under conditions of inflation is then

$$\overline{RR}(K'_a) = [\$4,305.08 \cdot (P/F, 21.7\%, 1) + \$2,875.11 \cdot (P/F, 21.7\%, 2)$$

$$= + \$4,507.66 \cdot (P/F, 21.7\%, 3) + \$4,478.69 \cdot (P/F, 21.7\%, 4)]$$

$$= \$3,996.43$$

EXAMPLE 13-5

The following example illustrates the use of the revenue requirement method under conditions of inflation where different escalation rates for costs are involved.

The nuclear production department of a large electric utility is considering whether to build its own facility for processing liquid radwaste material or to contract this operation to an external vendor. The capital investment required to construct the entire facility is estimated to be $106 million. The facility's useful life is 20 years, at the end of which it will have no market value. Yearly operating expenses are estimated at $3.46 million in 1996 dollars. (The expected initial date of commercial operation is January 1997.)

If the service is contracted to a vendor, it is still necessary for the utility to build a resin recovery system costing $17 million. This system will cost the company $2.1 million in annual operating expenses and $3.0 million in maintenance at the beginning of every five-year service period. The recovery system will last for 20 years and will have no market value at the end of its useful life. In addition, the utility will experience a $3.5 million estimated loss each year due to unplanned plant shutdowns. The vendor's annual fees to the utility are estimated at $5.3 million and will escalate at the rate of 15% per year. The company's O&M expenses for the resin recovery system are expected to escalate at the rate of 10% per year.

Assume that the company operates with 50% of borrowed capital at a real interest rate of 7% per year, and that a real equity rate of 14% per year is to be used. Also assume that the utility has a 50% effective income tax rate and that MACRS depreciation (15-year GDS property class) is to be used. Which alternative should be chosen?

SOLUTION

Because the cost of capital to the utility escalates at 10% per year, inflation-adjusted interest rates must be used.

The inflation-adjusted interest rate on borrowed capital is

$$i'_b = (1 + 0.07) \cdot (1 + 0.1) - 1$$
$$= 0.177$$

The inflation-adjusted return on equity is

$$e'_a = (1 + 0.14) \cdot (1 + 0.1) - 1$$
$$= 0.254$$

Equation 13-6 gives the value of the after-tax, inflation-adjusted cost of capital to the utility:

$$K'_a = 0.5 \cdot (1 - 0.5) \cdot 0.177 + (1 - 0.5) \cdot 0.254$$
$$= 0.17125$$

Tables 13-7 and 13-8 provide the revenue requirement results for the two mutually exclusive investment alternatives being considered by the utility.

Finally, we compute the levelized revenue requirement at 17.125% to be $37.7810 million for the radwaste facility construction alternative and $29.9106

TABLE 13-7 Build the Radwaste Facility (in millions of dollars) (Example 13-5)

Year, k	(1) Unrecovered Investment	(2) Book Deprec.	(3) Tax Deprec.	(4) Debt Return	(5) Equity Return	(6) Income Tax	(7) Operating Expenses	(8) RR_k = Cols. $2 + 4 + 5 + 6 + 7$
1	108.0	5.4	5.4	9.558	13.7160	13.7160	3.46	45.8500
2	102.6	5.4	10.26	9.0801	13.0302	8.1702	3.81	39.4905
3	97.2	5.4	9.234	8.6022	12.3444	8.5104	4.19	39.0470
4	91.8	5.4	8.316	8.1243	11.6586	8.7426	4.61	38.5355
5	86.4	5.4	7.4844	7.6464	10.9728	8.8884	5.07	37.9776
6	81.0	5.4	6.7284	7.1685	10.2870	8.9586	5.57	37.3841
7	75.6	5.4	6.372	6.6906	9.6012	8.6292	6.13	36.4513
8	70.2	5.4	6.372	6.2127	8.9154	7.9434	6.74	35.2115
9	64.8	5.4	6.3828	5.7348	8.2296	7.2468	7.42	34.0312
10	59.4	5.4	6.372	5.2569	7.5438	6.5718	8.16	32.9325
11	54.0	5.4	6.3828	4.7790	6.8580	5.8752	8.97	31.8822
12	48.6	5.4	6.372	4.3011	6.1722	5.2002	9.87	30.9435
13	43.2	5.4	6.3828	3.8232	5.4864	4.5036	10.86	30.0732
14	37.8	5.4	6.372	3.3453	4.8006	3.8286	11.94	29.3145
15	32.4	5.4	6.3828	2.8674	4.1148	3.1320	13.14	28.6542
16	27.0	5.4	3.186	2.3895	3.4290	5.6430	14.45	31.3115
17	21.6	5.4	0	1.9116	2.7432	8.1432	15.90	34.0980
18	16.2	5.4	0	1.4337	2.0574	7.4574	17.49	33.8385
19	10.8	5.4	0	0.9558	1.3716	6.7716	19.24	33.7390
20	5.4	5.4	0	0.4779	0.6858	6.0858	21.16	33.8095

$\overline{RR}$ = $37,781,000

TABLE 13-8 Use Vendor Contract Services (in millions of dollars) (Example 13-5)

Year, k	(1) Unrecovered Investment	(2) Book Deprec.	(3) Tax Deprec.	(4) Debt Return	(5) Equity Return	(6) Income Tax	(7) Lost Capacity	(8) Operating Expenses	(9) Repairs	(10) Vendor Costs	(11) Total Annual Costs	(12) RR_k = Cols. 2 + 4 + 5 + 6 + 7
1	17.00	0.85	0.85	1.5045	2.159	2.159	3.50	2.10	3.00	5.30	13.9	20.5725
2	16.15	0.85	1.615	1.4293	2.0511	1.28605	3.85	2.31	0	6.10	12.26	17.8764
3	15.30	0.85	1.4535	1.3541	1.9431	1.3396	4.235	2.541	0	7.01	13.786	19.2728
4	14.45	0.85	1.309	1.27889	1.8352	1.37615	4.6585	2.7951	0	8.06	15.5136	20.8537
5	13.60	0.85	1.1781	1.2036	1.7272	1.3991	5.12435	3.07461	0	9.27	17.469	22.6489
6	12.75	0.85	1.0591	1.1284	1.6193	1.41015	5.636785	3.38207	4.83153	10.66	24.5104	29.5191
7	11.90	0.85	1.003	1.0532	1.5113	1.3583	6.200464	3.72028	0	12.26	22.1808	26.9535
8	11.05	0.85	1.003	0.9779	1.4034	1.25035	6.8205099	4.09231	0	14.10	25.0128	29.4944
9	10.20	0.85	1.0047	0.9027	1.2954	1.1407	7.502560	4.50154	0	16.21	28.2141	32.4029
10	9.35	0.85	1.003	0.8275	1.1875	1.03445	8.252817	4.95169	0	18.64	31.8445	35.7439
11	8.50	0.85	1.0047	0.7523	1.0795	0.9248	9.07810	5.44686	7.78123	21.44	43.7462	47.3528
12	7.65	0.85	1.003	0.6770	0.97155	0.88855	9.98591	5.99155	0	24.66	40.6375	43.9547
13	6.80	0.85	1.0047	0.6018	0.8636	0.7089	10.9845	6.59070	0	28.36	45.9352	48.9595
14	5.95	0.85	1.003	0.5266	0.75565	0.60265	12.08295	7.24977	0	32.61	51.9428	54.6778
15	5.10	0.85	1.0047	0.4514	0.6477	0.493	13.291244	7.974747	0	37.50	58.766	61.2081
16	4.25	0.85	0.5015	0.3761	0.53975	0.88825	14.6203686	8.77222	12.53175	43.13	79.0544	81.7086
17	3.40	0.85	0	0.3009	0.4318	1.2818	16.082406	9.64944	0	49.60	75.3318	78.1963
18	2.55	0.85	0	0.2257	0.32385	1.17385	17.690646	10.614388	0	57.03	85.3351	87.9086
19	1.70	0.85	0	0.1505	0.2159	1.0659	19.459711	11.675826	0	65.59	96.7255	99.0078
20	0.85	0.85	0	0.0752	0.10795	0.95795	21.405682	12.84341	0	75.43	109.6791	111.6702

$$\overline{RR} = \$29{,}910{,}600$$

million for the service contracting alternative. Therefore, based on economic considerations alone, the company should contract with an external vendor to process its liquid radwaste.

13.12 Summary

Because of the special characteristics of utilities, privately owned utilities are granted a monopolistic franchise by the public. In return, these utilities are expected to satisfy their customers' expressed demands under the control of a regulatory body acting on behalf of the public.

The revenue requirement method was introduced as an appropriate economic evaluation technique for public utility projects. A fundamental principle underlying utility rate regulation is the fact that revenues should be generated so as just to cover the utility service expenses and provide a fair return on equity to investors.

> Recommended choices with the revenue requirement method are the same as those made when using conventional PW and AW methods at a discounting rate equal to the utility's after-tax cost of capital.

The revenue requirement method is equivalent to an after-tax PW or AW analysis of competing alternatives. Only the perspective is different. That is, PW and AW methods evaluate the project from the shareholder's viewpoint, while the revenue requirement method uses the utility customer's viewpoint because rates are regulated by representatives of the public.

13.13 References

COMMONWEALTH EDISON COMPANY. *Engineering Economics*. Chicago: Commonwealth Edison Company, 1975.

JEYNES, P. H. *Profitability and Economic Choice*. Ames: Iowa State University Press, 1968.

MAYER, R. R. "Finding Your Minimum Revenue Requirements." *Industrial Engineering,* vol. 9, no. 4, April 1977, pp. 16–22.

STEVENS, G. T. *Economic and Financial Analysis of Capital Investments*. New York: Wiley, 1979.

STOLL, H. G. *Least-Cost Electric Utility Planning*. New York: Wiley, 1987.

WARD, T. L., and W. G. SULLIVAN. "Equivalence of the Present Worth and Revenue Requirements Method of Capital Investment Analysis," *AIIE Transactions,* vol. 13, no. 1, pp. 29–40.

13.14 Problems

The number in parentheses ( ) that follows each problem refers to the section from which the problem is taken.

13-1.

a. Describe the types of regulation to which investor-owned utilities, but not private industries, are subject. Why is regulation necessary? (13.1)

b. How would economy studies differ in a government-owned utility as opposed to an investor-owned utility? (13.2)

13-2.

a. What advantages to the public result from utility companies? (13.1)

b. What disadvantages might there be to utilities? (13.1)

c. How is regulation of utilities that operate solely within an individual state achieved as opposed to those that provide services to many states (e.g., telephone companies and gas pipelines)? (13.2)

13-3. Briefly summarize the basic characteristics that distinguish investor-owned utilities from nonregulated industries such as steel, automobile, and chemical manufacturing. (13.3)

13-4. Why are most utilities heavily financed with borrowed capital? What characteristics of this industry make it possible to attract large amounts of borrowed capital, and what advantages (disadvantages) are associated with the use of borrowed money? (13.2)

13-5. Explain why it may be in the best interest of the consuming public for a regulatory agency to permit a utility to charge sufficiently high rates to allow it to earn an adequate return on its capital. (13.7)

13-6.

a. In a certain state a member of the Public Utilities Commission said, "I will oppose all rate increases. I am interested only in the rates the customers have to pay today." Comment on the results that could follow if all members of the Commission rigidly followed this philosophy. (13.7)

b. Comment on this statement: "No company that provides an exclusive and required service, such as electric power, should be permitted to make a profit." (13.7)

13-7. Is there justification for a privately owned, regulated utility being permitted to include in its rates the cost of advertising (to encourage the public to increase utilization of its service)? (13.7)

Note: Solve the remaining problems by using the after-tax cost of capital, K_a (or K'_a).

13-8. A telephone company can provide certain facilities having a 10-year life and zero market value by either of two alternatives. Alternative A requires a capital investment of $70,000 and

$3,000 per year for maintenance. Alternative B will have a capital investment of $48,000 and will require $6,000 annually for maintenance. Property taxes and insurance would be 4% of the capital investment per year for either alternative. The after-tax cost of capital is 10%, with 30% being borrowed at a 6% interest rate. The effective income tax rate is 50%. Which alternative will provide the lower annual equivalent revenue requirement? The MACRS (GDS) recovery period is 5 years, and book depreciation is computed with the straight-line method over 10 years. (13.5, 13.8)

13-9. A gas company must decide whether to build a new meter-repair and testing facility now or wait 3 years before doing so. It estimates that until the new facility is built, its annual expenses for these functions will be $90,000 greater than when the new facility is completed. The new facility would cost $900,000 and would not be needed after 20 years. The ultimate market value would be $200,000 at that time. The company uses 40% borrowed capital, paying 8% per year interest (before taxes) for it, and the regulatory body permits it to earn 13.8% per year on its equity capital. Assuming that the company has a 46% income tax rate, determine the equivalent annual revenue requirement for both options and recommend which is better. Assume that book (and tax) depreciation over 20 years is computed with the straight line method. (13.5, 13.8)

13-10. A utility company can construct a modern power plant that can generate power at 24 mils ($0.024) per kilowatt-hour at a 70% load factor. The 24 mils covers all expenses, including profit on capital and also income taxes. A large industrially owned power plant will soon make wholesale power available. To take advantage of the wholesale power, it will cost $180 per kilowatt of capacity to build the necessary transmission line, which will experience a 70% load factor. Annual maintenance expenses for this line will be $0.90 per kilowatt of capacity, and it is to be fully depreciated for book purposes over a 30-year period. A 15-year MACRS recovery period (GDS) will be used for calculating depreciation for income tax purposes. The cost of money for the company is 12% per year, with 40% borrowed capital at an interest rate of 7% per year. The income tax rate is 50%. At what price must the company be able to purchase the power in order for it to be as economical as if it were generated with the new modern plant? (13.5, 13.8)

13-11. Determine the annual revenue requirements for the following proposed 280 KVA transformer bank. (13.5)

Installed cost	= $240,000
Property taxes and insurance/yr	= 2% of installed cost
Market value	= 0
Tax life = book life	= 4 years
Depreciation method (for book purposes)	= Straight line
Depreciation method (for tax purposes)	= MACRS (GDS, 3-year recovery period)
Effective income tax rate	= 0.40
Cost of equity capital	= 20% per year, $(1 - \lambda)$
	= 0.60
Cost of borrowed funds	= 12% per year, $\lambda = 0.40$

Fill in the table below in completing this problem.

13-12. A telephone company must provide a direct current battery unit to a new service region in 1997. The expected useful life of the equipment is seven years. Alternative A requires a capital investment of $75,000 and has O&M expenses of $8,000 per year. The salvage value for tax purposes is zero, and this is also the expected realizable market value. A tax life of five years will be claimed, and MACRS (GDS, 5-year property) depreciation will be used for tax purposes. However, straight-line depreciation over seven years will be taken for rate-setting purposes (i.e., book depreciation).

The after-tax cost of capital (K'_a) is 12% per year, with 40% being borrowed at 8% per year. The effective income tax rate is 40%, and the general

inflation rate is 6% per year. Only O&M expenses are affected by inflation, and the costs of capital given previously include an allowance for anticipated inflationary pressures in the economy.

For Alternative A, answer the following questions. Be sure to state any assumptions you feel are appropriate and necessary. (13.10)

a. What is the actual dollar ATCF in year five of this alternative's useful life?

b. What is the *income tax* entry in an RR table for year five?

13-13. In 1997 the installed cost of a new transformer at the OPEC Utility Company is $50,000. Annual maintenance expenses, which are expected to escalate by 5% each year, are $1,500 in today's dollars. A five-year MACRS (GDS) recovery period is to be used for tax depreciation purposes, and the expected life of the transformer is 8 years. The terminal MV is negligible. Straight-line depreciation is used for determining BV for rate-setting purposes. Borrowed capital represents 40% of the company's capitalization, and it costs 10% per year before taxes. The return to equity is approximately 15% per year. (13.10)

a. If the firm's effective income tax rate is 40%, calculate the RR in year three.

b. If the firm's effective income tax rate is 50%, by how much does the RR in year three increase?

13-14. An electric utility company has an opportunity to build a small hydroelectric generating plant, of 20,000-kW capacity, on a mountain stream where the flow is seasonal. As a consequence, the annual output of energy would be only 40,000,000 kWh. The capital investment would be $2,000,000, and it is estimated that the annual operation and maintenance expenses would be $32,000 during its estimated 30-year economic life. It is believed that the property would have a market value of $200,000 at the end

EOY, k	Unrecovered Investment	Depreciation: Book	Tax	Recurring Annual Expenses	Return to Debt	Return to Equity	Income Tax	Revenue Requirement	(Pr. 13-11)
1									
2									
3									
4									

of the 30-year period. An alternative is to build a geothermal generating plant, which would have the same annual capacity, at a cost of $1,600,000. Because the company would have to pay the owners of the property for the geothermal steam, the estimated annual expense for the steam and operation and maintenance is $120,000. A 30-year contract can be obtained on the steam supply, and it is believed that this period is realistic for the economic life of the plant but that the market value at that time would be little more than zero. Property taxes and insurance on either plant would be 2% of the capital investment per year. The company employs 40% borrowed capital, for which it pays 8.5% interest per year. It earns 13% per year after taxes on total capital, and it has a 50% effective income tax rate. Which development should be undertaken? State all assumptions that you make. (13.8)

13-15. Use the RR method to compare alternatives A and B in Problem 13-8 when the average inflation rate on maintenance is 6% per year. Assume that property taxes do not respond to inflation, and adjust the cost of capital to account for inflation. (13.10)

13-16. A natural gas pipeline company is considering two plans to provide service required by present demand and the forecasted growth of demand for the coming 18 years. Alternative A requires an immediate investment of $700,000 in property that has an estimated life of 18 years, with 10% of the capital investment as the terminal market value. Annual expenses will be $25,000. Annual property taxes will be 2% of the capital investment. Alternative B requires an immediate investment of $400,000 in property that has an estimated life of 18 years, with 20% of the capital investment as the terminal market value. Annual O&M expenses during the first 8 years will be $42,000. After 8 years, an additional investment of $450,000 will be required in property having an estimated life of 10 years with 50% of the additional investment as the terminal market value. After this additional property is installed, annual O&M expenses (for years 9 to

18) of the combined properties will be $72,000. Annual property taxes will be 2% of the initial capital investment of the property in service at any time. The regulatory commission is allowing a 10% per year fair return on depreciated BV to cover the cost of money (K_a) to the utility. Assume that this rate of return will continue throughout the 18 years. The utility company's effective tax rate is 50%. Straight-line depreciation is to be used for book purposes in setting rates, and MACRS (GDS) depreciation is to be used for income tax purposes. The MACRS recovery period is 7 years for all depreciable assets. Half of the utility's financing is by debt with interest at 8% per year. Determine which plan minimizes equivalent annual revenue requirements after property and income taxes have been taken into account. (13.9)

13-17. In making its forecast of requirements in a certain area for the next 30 years, a telephone company has determined that a 600-pair cable is required immediately and a total of 1,000 pairs will be required by the end of 15 years. An underground conduit of sufficient size to handle the cable needs is being installed now at a cost of $10,000. If a 1,000-pair cable is installed now, it will cost $30,000. As an alternative, it can install a 600-pair cable immediately at a cost of $20,000 and install an additional 400-pair cable at the end of 15 years at an estimated cost of $16,000. Because of technical obsolescence, it is company policy to consider the economic life of either installation to be 30 years from the present time. Annual property taxes on either alternative would be 2% of the installed cost, and the market value of all cable and conduit at the end of the 30-year period is estimated to be 10% of the installed cost. The company uses 40% borrowed capital, for which it pays 8% per year. It earns 12% per year after taxes on total capital and has a 50% effective income tax rate. Which alternative would you recommend? Assume that depreciation for tax and book purposes is computed with the straight-line method over 15 years for both alternatives. (13.9)

14

Probabilistic Risk Analysis

The objectives of this chapter are to (1) introduce the use of statistical and probability concepts in decision situations involving risk and uncertainty, (2) illustrate how they can be applied in engineering economic analysis, and (3) discuss the considerations and limitations relative to their application.

The following topics are discussed in this chapter:

The definition of a random variable

The basic characteristics of probability distributions

Evaluation of projects with discrete random variables

Probability trees

Evaluation of projects with continuous random variables

A description of Monte Carlo simulation

Accomplishing Monte Carlo simulation with a computer

Decision tree analysis

14.1 Introduction

In this chapter, selected statistical and probability concepts are used to analyze the economic consequences of some decision situations involving risk and uncertainty and requiring engineering knowledge and input. The probability that a cost, revenue, useful life, or other economic factor value will occur, or that a particular equivalent worth or rate of return value for a cash flow will occur, is usually considered to be the long-run relative frequency with which the event (value) occurs or the subjectively estimated likelihood that it will occur. Factors such as these, having probabilistic outcomes, are called *random variables*.

As discussed in Chapter 1, a decision situation such as a design task, a new venture, an improvement project, or any similar effort requiring engineering knowledge, has two or more alternatives associated with it. The cash flow amounts for each alternative often result from the sum, difference, product, or quotient of random variables such as initial capital investments, operating expenses, revenues, changes in working capital, and other economic factors. Under these circumstances, the measures of economic merit (e.g., equivalent worth and rate of return values) of the cash flows will also be random variables.

> The information about these random variables that is particularly helpful in decision making are their expected values and variances, especially for the economic measures of merit of the alternatives. These derived quantities for the random variables are used to make the uncertainty associated with each alternative more explicit, including any probability of loss. Thus, when uncertainty is considered, the variability in the economic measures of merit and the probability of loss associated with the alternatives are normally used in the decision-making process.

14.2 The Distribution of Random Variables

Capital letters such as $X, Y,$ and Z are usually used to represent random variables and lowercase letters (x, y, z) to denote the particular values that these variables take on in the sample space (i.e., in the set of all possible outcomes for each variable). When a random variable X is considered to follow some *discrete* probability distribution, its *probability mass function* is usually indicated by $p(x)$ and its *cumulative distribution function* by $P(x)$. When a random variable is considered to follow a *continuous* probability distribution, its *probability density function* and its cumulative distribution function are usually indicated by $f(x)$ and $F(x)$, respectively.

14.2.1 Discrete Random Variables

A random variable X is said to be *discrete* if it can take on at most a countable (finite) number of values $(x_1, x_2, \ldots x_L)$. The probability that a discrete random variable X takes on the value x_i is given by

$$\Pr\{X = x_i\} = p(x_i) \text{ for } i = 1, 2, \ldots, L \text{ (i is a *sequential index* of}$$
$$\text{the discrete values, } x_i, \text{ that the variable takes on)}$$

where $p(x_i) \geq 0$ and $\sum_i p(x_i) = 1$.

The probability of events about a discrete random variable can be computed from its probability mass function $p(x)$. For example, the probability of the event that the value of X is contained in the closed interval $[a, b]$ is given by (where the colon is read "such that")

$$\Pr\{a \le X \le b\} = \sum_{i:a \le X_i \le b} p(x_i) \tag{14-1}$$

The probability that the value of X is less than or equal to $x = h$, the cumulative distribution function $P(x)$ for a discrete case, is given by

$$\Pr\{X \le h\} = P(h) = \sum_{i:X_i \le h} p(x_i) \tag{14-2}$$

In most practical applications, discrete random variables represent *countable* data such as the useful life of an asset in years, number of maintenance jobs per week, or number of employees as positive integers.

14.2.2 Continuous Random Variables

A random variable X is said to be continuous if there exists a nonnegative function $f(x)$ such that for any set of real numbers $[c, d]$, where $c < d$, the probability of the event that the value of X is contained in the set is given by

$$\Pr\{c \le X \le d\} = \int_c^d f(x)dx \tag{14-3}$$

and

$$\int_{-\infty}^{\infty} f(x)dx = 1$$

Thus, the probability of events about the continuous random variable X can be computed from its probability density function, and the probability that X assumes exactly any one of its values is 0. Also, the probability that the value of X is less than or equal to a value $x = k$, the cumulative distribution function $F(x)$ for a continuous case, is given by

$$\Pr\{X \le k\} = F(k) = \int_{-\infty}^{k} f(x)dx \tag{14-4}$$

Also, for a continuous case,

$$\Pr\{c \le X \le d\} = \int_c^d f(x)dx = F(d) - F(c) \tag{14-5}$$

In most practical applications, continuous random variables represent *measured* data such as time, cost, and revenue on a continuous scale. Depending upon the situation, the analyst decides to model random variables in engineering economic analysis as either discrete or continuous.

14.2.3 Mathematical Expectation and Selected Statistical Moments

The expected value of a single random variable X, $E(X)$, is a weighted average of the distributed values x that it takes on and is a measure of the central location of the distribution (central tendency of the random variable). The $E(X)$ is the first moment of the random variable about the origin and is called the *mean* (central moment) of the distribution. The expected value is

$$E(X) = \begin{cases} \displaystyle\sum_i x_i p(x_i) & \text{for } x \text{ discrete and } i = 1, 2, \ldots, L \\[2em] \displaystyle\int_{-\infty}^{\infty} x f(x) dx & \text{for } x \text{ continuous} \end{cases} \tag{14-6}$$

While $E(X)$ provides a measure of central tendency, it does not measure how the distributed values x cluster around the mean. The *variance*, $V(X)$, which is non-negative, of a single random variable X is a measure of the dispersion of the values it takes on around the mean. It is the expected value of the square of the difference between the values x and the mean, which is the second moment of the random variable about its mean.

$$E\{[X - E(X)]^2\} = V(X) = \begin{cases} \displaystyle\sum_i [x_i - E(X)]^2 p(x_i) & \text{for } x \text{ discrete} \\[2em] \displaystyle\int_{-\infty}^{\infty} [x - E(X)]^2 f(x) dx & \text{for } x \text{ continuous} \end{cases} \tag{14-7}$$

From the binomial expansion of $[X - E(X)]^2$, it can be easily shown that $V(X) = E(X^2) - [E(X)]^2$. That is, $V(X)$ equals the second moment of the random variable around the origin, which is the expected value of X^2, minus the square of its mean. This is the form often used for calculating the variance of a random variable X.

$$V(X) = \begin{cases} \displaystyle\sum_i x_i^2 p(x_i) - [E(X)]^2 & \text{for } x \text{ discrete} \\[2em] \displaystyle\int_{-\infty}^{\infty} x_i^2 f(x) dx - [E(X)]^2 & \text{for } x \text{ continuous} \end{cases} \tag{14-8}$$

The *standard deviation* of a random variable, $SD(X)$, is the positive square root of the variance; that is, $SD(X) = [V(X)]^{1/2}$.

14.2.4 Multiplication of a Random Variable by a Constant

A common operation performed on a random variable is to multiply it by a constant, for example, the estimated maintenance labor expense for a time period, $Y = cX$, when the number of labor hours per period (X) is a random variable, and the cost per labor hour (c) is a constant. Another example is the PW calculation for a project when the net cash flow amounts, F_k, are random variables, and then each F_k is multiplied by a constant $(P/F, i\%, k)$ to obtain the PW value.

When a random variable, X, is multiplied by a constant, c, the expected value, $E(cX)$, and the variance, $V(cX)$, are

$$E(cX) = cE(X) = \begin{cases} \sum_i cx_i p(x_i) & \text{for } x \text{ discrete} \\ \int_{-\infty}^{\infty} cxf(x)dx & \text{for } x \text{ continuous} \end{cases} \qquad (14\text{-}9)$$

$$\begin{aligned} V(cX) &= E\{[cX - E(cX)]^2\} \\ &= E\{c^2X^2 - 2c^2X \cdot E(X) + c^2[E(X)]^2\} \\ &= c^2E\{[X - E(X)]^2\} \\ &= c^2V(X) \end{aligned} \qquad (14\text{-}10)$$

14.2.5 Multiplication of Two Independent Random Variables

A cash flow random variable, say Z, may result from the product of two other random variables, $Z = XY$. Sometimes, X and Y can be treated as statistically independent random variables. For example, consider the estimated annual expenses, $Z = XY$, for a repair part repetitively procured during the year on a competitive basis, when the unit price (X) and the number of units used per year (Y) are modeled as independent random variables.

When a random variable, Z, is a product of two independent random variables, X and Y, the expected value, $E(Z)$, and the variance, $V(Z)$, are:

$$Z = XY$$
$$E(Z) = E(X)E(Y) \qquad (14\text{-}11)$$
$$\begin{aligned} V(Z) &= E[XY - E(XY)]^2 \\ &= E\{X^2Y^2 - 2XYE(XY) + [E(XY)]^2\} \\ &= EX^2EY^2 - [E(X)E(Y)]^2 \end{aligned}$$

But the variance of a random variable, $V(RV)$, is:

$$V(RV) = E[(RV)^2] - [E(RV)]^2$$
$$E[(RV)^2] = V(RV) + [E(RV)]^2$$

Then

$$V(Z) = \{V(X) + [E(X)]^2\}\{V(Y) + [E(Y)]^2\} - [E(X)]^2[E(Y)]^2$$

or

$$V(Z) = V(X)[E(Y)]^2 + V(Y)[E(X)]^2 + V(X)V(Y) \qquad (14\text{-}12)$$

14.3 Evaluation of Projects with Discrete Random Variables

Expected value and variance concepts apply theoretically to long-run conditions in which it is assumed that the event is going to occur repeatedly. However, application of these concepts is often useful even when investments are not going to be made repeatedly over the long run. In this section, several examples are used to illustrate these concepts with selected economic factors modeled as discrete random variables.

EXAMPLE 14-1

We now apply the expected value and variance concepts to the premixed-concrete plant discussed in Example 10-5. Suppose that the estimated probabilities of attaining various capacity utilizations are as follows:

Capacity (%)	Probability
50	0.10
65	0.30
75	0.50
90	0.10

It is desired to determine the expected value and variance of *annual revenue*. Subsequently, the expected value and variance of AW for the project can be computed. By evaluating both $E(AW)$ and $V(AW)$ for the concrete plant, indications of the venture's average profitability and its uncertainty are obtained. The calculations are shown in Tables 14-1 and 14-2.

SOLUTION

Expected value of annual revenue: $\sum (A \times B) = \$575,100$

Variance of annual revenue: $\sum (A \times C) - (575,100)^2 = 6,360 \times 10^6 (\$)^2$

TABLE 14-1   Solution for Annual Revenue (Example 14-1)

i	Capacity (%)	(B) Probability, $p(x_i)$	(A) Revenue[a] x_i	(A) × (B) Expected Revenue	(C) = (B)² x_i^2	(A) × (C)
1	50	0.10	$405,000	$ 40,500	1.64×10^{11}	0.164×10^{11}
2	65	0.30	526,500	157,950	2.77×10^{11}	0.831×10^{11}
3	75	0.50	607,500	303,750	3.69×10^{11}	1.845×10^{11}
4	90	0.10	729,000	72,900	5.31×10^{11}	0.531×10^{11}
				$575,100		$3.371 \times 10^{11}(\$)^2$

[a]From Table 10-1 with revenue for Capacity = 75% added

TABLE 14-2   Solution for AW (Example 14-1)

i	Capacity (%)	(A) $p(x_i)$	(B) AW,[a] x_i	(A) × (B) Expected AW	(C) = (B)^2 (AW)^2	(A) × (C)
1	50	0.10	−$25,093	−$2,509	0.63×10^9	0.063×10^9
2	65	0.30	22,136	6,641	0.49×10^9	0.147×10^9
3	75	0.50	53,622	26,811	2.88×10^9	1.440×10^9
4	90	0.10	100,850	10,085	10.17×10^9	1.017×10^9
				$41,028		$2.667 \times 10^9 (\$)^2$

[a]From Table 10-1 with AW for Capacity = 75% added

Expected value of AW: $\sum(A \times B) = \$41,028$

Variance of AW: $\sum(A \times C) - (41,028)^2 = 9,837 \times 10^5 (\$)^2$

Standard deviaton of AW: $31,364

The standard deviation of AW, SD (AW), is less than the expected AW, E(AW), and only the 50% capacity utilization situation results in a negative AW. Consequently, with this additional information, the investors in this undertaking may well judge the venture to be an acceptable one.

There are situations, such as flood control projects, in which future losses due to natural or human-made risks can be decreased by increasing the amount of capital that is invested. Drainage channels or dams, built to control floodwaters, may be constructed in different sizes, costing different amounts. If they are correctly designed and used, the larger the size, the smaller will be the resulting damage loss when a flood occurs. As we might expect, the most economical size would provide satisfactory protection against most floods, although it could be anticipated that some overloading and damage might occur at infrequent periods.

EXAMPLE 14-2

A drainage channel in a mountain community where flash floods are experienced has a capacity sufficient to carry 700 cubic feet per second. Engineering studies produce the following data regarding the probability that a given water flow in any one year will be exceeded and the cost of enlarging the channel:

Water Flow (ft³/sec)	Probability of a Greater Flow Occurring in Any One Year	Capital Investment to Enlarge Channel to Carry This Flow
700	0.20	—
1,000	0.10	−$20,000
1,300	0.05	−30,000
1,600	0.02	−44,000
1,900	0.01	−60,000

TABLE 14-3   Expected Equivalent Annual Cost (Example 14-2)

Water Flow (ft³/sec)	Capital Recovery Amount	Expected Annual Property Damage[a]	Total Expected Equivalent Annual Cost
700		$-\$20,000(0.20) = -\$4,000$	$-\$4,000$
1,000	$-\$20,000(0.0839) = -\$1,678$	$-20,000(0.10) = -2,000$	$-3,678$
1,300	$-30,000(0.0839) = -2,517$	$-20,000(0.05) = -1,000$	$-3,517$
1,600	$-44,000(0.0839) = -3,692$	$-20,000(0.02) = -400$	$-4,092$
1,900	$-60,000(0.0839) = -5,034$	$-20,000(0.01) = -200$	$-5,234$

[a]These amounts are obtained by multiplying $20,000 by the probability of greater water flow occurring.

Records indicate that the average property damage amounts to $20,000 when serious overflow occurs. It is believed that this would be the average damage whenever the storm flow is *greater* than the capacity of the channel. Reconstruction of the channel would be financed by 40-year bonds bearing 8% interest per year. It is thus computed that the capital recovery amount for debt repayment (principal of the bond plus interest) would be 8.39% of the capital investment, because $(A/P, 8\%, 40) = 0.0839$. It is desired to determine the most economical channel size (water flow capacity).

SOLUTION

The total expected equivalent annual cost for the structure and property damage for all alternative channel sizes would be as shown in Table 14-3. These calculations show that the minimum expected annual cost would be achieved by enlarging the channel so that it would carry 1,300 cubic feet per second, with the expectation that a greater flood might occur in 1 year out of 20 on the average and cause property damage of $20,000.

Note that when loss of life or limb might result, as in Example 14-2, there usually is considerable pressure to disregard pure economy and build such projects in recognition of the nonmonetary values associated with human safety.

The following example illustrates the same principles as in Example 14-2, except that it applies to safety alternatives involving electrical circuits.

EXAMPLE 14-3

Three alternatives are being evaluated for the protection of electrical circuits, with the following required investments and probabilities of failure:

Alternative	Capital Investment	Probability of Loss in Any Year
A	$-\$ 90,000$	0.40
B	$-100,000$	0.10
C	$-160,000$	0.01

TABLE 14-4   Expected Equivalent Annual Cost (Example 14-3)

Alter- native	Capital Recovery Amount = Capital Investment × $(A/P, 12\%, 8)$	Annual Maintenance = Capital Investment × (0.10)	Expected Annual Cost of Failure	Total Expected Equivalent Annual Cost
A	−$ 90,000(0.2013) = −$18,117	−$ 9,000	−$94,000(0.40) = −$37,600	−$64,717
B	−100,000(0.2013) = −20,130	−10,000	−94,000(0.10) = −9,400	−39,530
C	−160,000(0.2013) = −32,208	−16,000	−94,000(0.01) = −940	−49,148

If a loss does occur, it will cost $80,000 with a probability of 0.65, and $120,000 with a probability of 0.35. The probabilities of loss in any year are independent of the probabilities associated with the resultant cost of a loss if one does occur. Each alternative has a useful life of eight years and no market value at that time. The MARR is 12% per year, and annual maintenance expenses are expected to be 10% of the capital investment. It is desired to determine which alternative is best based on expected total annual costs (Table 14-4).

SOLUTION

The expected cost of a loss, if it occurs, can be calculated as follows:

$$-\$80{,}000(0.65) - \$120{,}000(0.35) = -\$94{,}000$$

Thus, Alternative B is the best based on total expected equivalent annual cost, which is a long-run average cost. However, one might rationally choose Alternative C to reduce significantly the chance of an $80,000 or $120,000 loss occurring in any year in return for a 24.3% increase in the total expected equivalent annual cost.

In Examples 14-1 through 14-3, a revenue or cost factor was modeled as a discrete random variable, with project life assumed certain. A second type of situation involves the cash flow estimates being certain but the project life being modeled as a random variable. This is illustrated in Example 14-4, where project life is modeled as a discrete random variable.

EXAMPLE 14-4

The heating, ventilating, and air-conditioning (HVAC) system in a commercial building has become unreliable and inefficient. Rental income is being hurt, and the annual expenses of the system continue to increase. Your engineering firm has been hired by the owners to (1) perform a technical analysis of the system, (2) develop a preliminary design for rebuilding the system, and (3) accomplish an engineering economic analysis to assist the owners in making a decision. The estimated capital investment cost and annual savings in O&M expenses, based on the preliminary design, are shown in the following table. The estimated annual increase in rental income with a modern HVAC system has been developed by the owner's marketing staff and is also provided in the following table. These estimates are considered reliable because of the extensive information available. The useful life of the rebuilt system, however, is quite uncertain. The estimated probabilities of various useful lives are provided. Assume that the MARR = 12% per year and the market value of the rebuilt system at the end of its useful life is zero. Based on this information, what is the E(PW), V(PW), and SD(PW) of the project's cash flows? Also, what is the probability of the PW ≥ 0? What decision would you make regarding the project, and how would you justify your decision using the available information?

Economic Factor	Estimate	Useful Life, Year (N)	$p(N)$
Capital investment	−$521,000	12	0.1
Annual savings	48,600	13	0.2
Increased annual revenue	31,000	14	0.3
		15	0.2
		16	0.1
		17	0.05
		18	0.05

SOLUTION

The PW of the project's cash flows, as a function of project life (N), is

$$PW(12\%)_N = -\$521,000 + \$79,600(P/A, 12\%, N)$$

The calculation of the value of E(PW) = $9,984, and the value of $E[(PW)^2] = 577.527 \times 10^6(\$)^2$, are shown in Table 14-5. Then, by using Equation 14-8, the variance of the PW is

$$V(PW) = E[(PW)^2] - [E(PW)]^2$$
$$= 577.527 \times 10^6 - (\$9,984)^2$$
$$= 477.847 \times 10^6(\$)^2$$

TABLE 14-5   Calculation of E(PW) and E[(PW)²] (Example 14-4)

(1) Useful Life (N)	(2) PW(N)	(3) $p(N)$	(4) = (2) × (3) $E[PW(N)]$	(5) = (2)² $[PW(N)]^2$	(6) = (3) × (5) $p(N)[PW(N)]^2$
12	−$27,926	0.1	−$2,793	779.86×10^6	77.986×10^6
13	− 9,689	0.2	− 1,938	93.88×10^6	18.776×10^6
14	6,605	0.3	1,982	43.63×10^6	13.089×10^6
15	21,148	0.2	4,230	447.24×10^6	89.448×10^6
16	34,130	0.1	3,413	$1,164.86 \times 10^6$	116.486×10^6
17	45,720	0.05	2,286	$2,090.32 \times 10^6$	104.516×10^6
18	56,076	0.05	2,804	$3,144.52 \times 10^6$	157.226×10^6
			E(PW) = $9,984		$E[(PW)^2] = 577.527 \times 10^6(\$)^2$

The SD(PW) is equal to the positive square root of the variance, $V(PW)$:

$$SD(PW) = [V(PW)]^{1/2} = (477.847 \times 10^6)^{1/2}$$
$$= \$21,859$$

Based on the PW of the project as a function of N (column 2), and the probability of each PW(N) value occurring (column 3), the probability of the PW being ≥ 0 is

$$Pr\{PW \geq 0\} = 1 - (0.1 + 0.2) = 0.7$$

The results of the engineering economic analysis indicate that the project is a reasonable business action. The $E(PW)$ of the project is positive ($\$9,984$), and the probability of the PW being greater than zero is favorable (0.7). The weakest indicator is the SD(PW) value, which is over two times the $E(PW)$ value.

14.3.1   Probability Trees

The discrete distribution of cash flows sometimes occurs in each time period. A *probability tree diagram* is useful in describing the prospective cash flows, and the probability of each value occurring, for this situation. Example 14-5 is a problem of this type.

EXAMPLE 14-5

The uncertain cash flows for a small improvement project are described by the probability tree diagram in Figure 14-1 (note that the probabilities emanating from each node sum to one). The analysis period is two years, and the MARR $= 12\%$ per year. Based on this information, (a) what are the $E(PW)$, $V(PW)$, and SD(PW) of the project, (b) what is the probability that PW ≤ 0, and (c) which analysis result(s) favor approval of the project and which ones appear unfavorable?

SOLUTION
(a) The calculation of the values for $E(PW)$ and $E[(PW)^2]$ is shown in Table 14-6. In column 2, PW_j, is the PW of branch j in the tree diagram. The probability of each branch occurring, $p(j)$, is shown in column 3. For example, proceeding from the right node for each cash flow in Figure 14-1 to the left node, we have $p(1) = (0.3)(0.2) = 0.06$, and $p(9) = (0.5)(0.3) = 0.15$.

$$E(PW) = \sum_j (PW_j)p(j) = \$39.56$$

Then

$$V(PW) = E[(PW)^2] - [E(PW)]^2$$
$$= 15,227 - (\$39.56)^2$$
$$= 13,662(\$)^2$$

and

$$SD(PW) = [V(PW)]^{1/2} = (13,662)^{1/2} = \$116.88$$

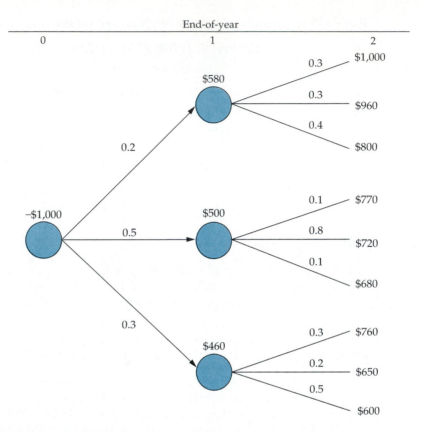

FIGURE 14-1   Probability Tree Diagram for Example 14-5

| **TABLE 14-6** | **Calculation of $E(PW)$ and $E[(PW)^2]$ (Example 14-5)** |

	(1) Net Cash Flow			(2)	(3)	(4) = (2) × (3)	(5) = (2)²	(6) = (3) × (5)
EOY j	0	1	2	PW$_j$	p(j)	$E(PW_j)$	$(PW_j)^2$	$E[(PW_j)^2]$
1	−$1,000	$580	$1,000	$315	0.06	$18.90	$²99,225	$²5,953
2	− 1,000	580	960	283	0.06	16.99	80,089	4,805
3	− 1,000	580	800	156	0.08	12.45	24,336	1,947
4	− 1,000	500	770	60	0.05	3.04	3,600	180
5	− 1,000	500	720	20	0.40	8.17	400	160
6	− 1,000	500	680	− 11	0.05	− 0.57	121	6
7	− 1,000	460	760	17	0.09	1.49	289	26
8	− 1,000	460	650	− 71	0.06	− 4.27	5,044	302
9	− 1,000	460	600	−111	0.15	−16.64	12,321	1,848
						$E(PW)$ = $39.56		$E[(PW)^2]$ = 15,227$²

(b) Based on the entries in column 2, PW_j, and column 3, $p(j)$, we have

$$Pr\{PW \leq 0\} = p(6) + p(8) + p(9)$$
$$= 0.05 + 0.06 + 0.15$$
$$= 0.26$$

(c) The analysis results that favor approval of the project are $E(PW) = \$39.56$, which is greater than zero only by a small amount, and $Pr\{PW > 0\} = 1 - 0.26 = 0.74$. However, the $SD(PW) = \$116.92$ is approximately three times the $E(PW)$. This indicates a relatively high variability in the measure of economic merit, (PW) of the project and is unfavorable.

14.3.2 An Application Perspective

One of the major problems in computing expected values is the determination of the probabilities. In most situations there is no precedent for the particular venture being considered. Therefore, probabilities seldom can be based on historical data and rigorous statistical procedures. In most cases the analyst, or person making the decision, must make a judgment based on all available information in estimating the probabilities. This fact makes some people hesitate to use the expected-value concept, because they cannot see the value of applying such a technique to improve the evaluation of uncertainty when so much apparent subjectivity is present.

Although this argument has merit, the fact is that engineering economy studies always deal with future events and there must be an extensive amount of estimating. Furthermore, even if the probabilities could be based accurately on past history, there rarely is any assurance that the future will repeat the past. Thus, structured methods for assessing subjective probabilities are often used in practice.* Also, even if we must estimate the probabilities, the very process of doing so requires us to make explicit the uncertainty that is inherent in all estimates going into the analysis. Such structured thinking is likely to produce better results than little or no thinking about such matters.

14.4 Evaluation of Projects with Continuous Random Variables

In Section 14.1, we discussed using the variance of a random variable, in addition to its expected value, in decision making. Thus, the uncertainty associated with an alternative can be represented more realistically. This was demonstrated in Examples 14-1, 14-4, and 14-5, when a revenue factor, a cost factor, and the project life, respectively, were modeled as *discrete* random variables. In each of these examples, the expected value and the variance of the project's equivalent worth

---

*For further information, see W. G. Sullivan and W. W. Claycombe, *Fundamentals of Forecasting* (Reston, Va.: Reston Publishing Co., 1977), Chapter 6.

were determined and used in the evaluation. Also, in the latter two examples, the probability of the PW being greater than or less than zero was computed.

In this section, we continue to use the "closed-form" analysis procedure to compute mathematically the expected values and the variances of probabilistic factors, but we model the selected probabilistic factors as *continuous* random variables. In each example, simplifying assumptions are made about the distribution of the random variable and the statistical relationship among the values it takes on. When the situation is more complicated, such as a problem that involves probabilistic cash flows and probabilistic project lives, a second general procedure that utilizes Monte Carlo simulation is normally used. This is the subject of Section 14.5.

> Two frequently used assumptions about uncertain cash flow amounts are that they are distributed according to a normal distribution* and are statistically independent. Underlying these assumptions is a general characteristic of many cash flows in that they result from a number of different and independent factors.

The advantage of using statistical independence as a simplifying assumption, when appropriate, is that no correlation between the cash flow amounts (e.g., the net annual cash flow amounts for an alternative) is being assumed. Consequently, if we have a linear combination of two or more independent cash flow amounts, say $PW = c_0F_0 + \cdots + c_NF_N$, where the c_k values are coefficients and the F_k values are periodic net cash flows, the expression for the $V(PW)$, based on Equation 14-10, reduces to

$$V(PW) = \sum_{k=0}^{N} c_k^2 V(F_k) \qquad \text{(14-13)}$$

And, based on Equation 14-9, we have

$$E(PW) = \sum_{k=0}^{N} c_k E(F_k) \qquad \text{(14-14)}$$

EXAMPLE 14-6

For the following project cash flow estimates, find $E(PW)$, $V(PW)$, and $SD(PW)$. Assume that the annual net cash flow amounts are normally distributed and statistically independent, and that the MARR = 15% per year.

End of Year, k	Expected Value of Net Cash Flow, F_k	SD of Net Cash Flow, F_k
0	−$7,000	0
1	3,500	$600
2	3,000	500
3	2,800	400

---

*This frequently encountered continuous probability function is discussed in any good statistics book, such as R. E. Walpole and R. H. Myers, *Probability and Statistics for Engineers and Scientists* (New York: Macmillan Publishing Co., 1989), pp. 139–154.

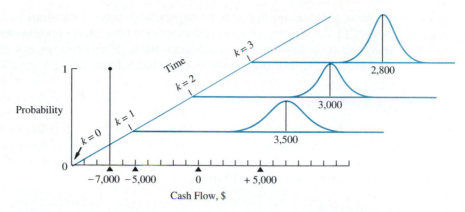

FIGURE 14-2   Probabilistic Cash Flows over Time (Example 14-6)

A graphical portrayal of these normally distributed cash flows is shown in Figure 14-2.

SOLUTION

The expected PW, based on Equation 14-14, is calculated as follows where $E(F_k)$ is the expected net cash flow in year k ($0 \leq k \leq N$) and c_k is the single-payment PW factor $(P/F, 15\%, k)$:

$$E(\text{PW}) = \sum_{k=0}^{3} (P/F, 15\%, k)E(F_k)$$

$$= -\$7{,}000 + \$3{,}500(P/F, 15\%, 1) + \$3{,}000(P/F, 15\%, 2)$$

$$+ \$2{,}800(P/F, 15\%, 3)$$

$$= \$153$$

To determine $V(\text{PW})$, we use the relationship in Equation 14-13. Thus,

$$V(\text{PW}) = \sum_{k=0}^{3} (P/F, 15\%, k)^2 V(F_k)$$

$$= 0^2 1^2 + 600^2 (P/F, 15\%, 1)^2 + 500^2 (P/F, 15\%, 2)^2$$

$$+ 400^2 (P/F, 15\%, 3)^2$$

$$= 484{,}324\$^2$$

and

$$\text{SD(PW)} = [V(\text{PW})]^{1/2} = \$696$$

When we can assume that a random variable, say the PW of a project's cash flow, is normally distributed with a mean, $E(\text{PW})$, and a variance, $V(\text{PW})$, we can compute the probability of events about the random variable occurring. This assumption can be made, for example, when we have some knowledge of the shape of the distribution of the random variable, and when it is appropriate to do so. Also, this assumption may be supportable when a random variable, such as a project's PW value, is a linear combination of other independent random variables (say, the cash flow amounts, F_k), regardless of whether or not the form of the probability distribution(s) of these variables is known.*

EXAMPLE 14-7

Refer to Example 14-6. For this problem, what is the probability that the IRR of the cash flow estimates is less than the MARR, Pr{IRR < MARR}? Assume that the PW is a normally distributed random variable, with its mean and variance equal to the values calculated in Example 14-6.

SOLUTION

For a decreasing PW(i) function having a unique IRR, the probability that the IRR is less than the MARR is the same as the probability that PW is less than zero. Consequently, by using the standardized normal distribution in Appendix E, we can determine the probability that PW is less than zero.†

$$Z = \frac{PW - E(PW)}{SD(PW)} = \frac{0 - 153}{696} = -0.22$$

$$Pr\{PW \leq 0\} = Pr\{Z \leq -0.22\}$$

From Appendix E we find that Pr{Z ≤ −0.22} = 0.4129.

EXAMPLE 14-8

The estimated cash flow data for a project are shown in the following table for the five-year study period being used. Each annual net cash flow amount, F_k, is a linear combination of two statistically independent random variables, X_k and Y_k, where X_k is a revenue factor and Y_k is an expense factor. The X_k cash flow amounts are statistically independent of each other, and the same is true of the Y_k amounts. Both X_k and Y_k are continuous random variables, but the form of their probability distributions is not known. The MARR = 20% per year. Based on this information, (a) what are the E(PW), V(PW), and SD(PW) values of the project's cash flows, and (b) what is the probability that the PW is less than zero, that is, Pr{PW ≤ 0}, and is the project economically attractive?

---

*The theoretical basis of this assumption is the central limit theorem of statistics. For a summary discussion of the supportability of this assumption under different conditions, see C. S. Park and G. P. Sharpe-Bette, *Advanced Engineering Economics* (New York: John Wiley and Sons, 1990), pp. 420–421.

†A random variable, X, is normally distributed with mean μ and standard deviation σ in accordance with the following equation:

$$f(X) = \frac{1}{\sigma \sqrt{2\pi}} \exp\left\{ -\left[\frac{(X - \mu)^2}{2\sigma^2} \right] \right\}$$

The standardized normal distribution, $f(Z)$, of the variable $Z = (X - \mu)/\sigma$ has a mean of 0 and a standard deviation of 1.

End of Year k	Net Cash Flow $F_k = a_k X_k - b_k Y_k$	Expected Values X_k	Expected Values Y_k	Standard Deviations (SD) X_k	Standard Deviations (SD) Y_k
0	$F_0 = X_0 + Y_0$	$0	-$100,000	$0	$10,000
1	$F_1 = X_1 + Y_1$	60,000	-20,000	4,500	2,000
2	$F_2 = X_2 + 2Y_2$	65,000	-15,000	8,000	1,200
3	$F_3 = 2X_3 + 3Y_3$	40,000	-9,000	3,000	1,000
4	$F_4 = X_4 + 2Y_4$	70,000	-20,000	4,000	2,000
5	$F_5 = 2X_5 + 2Y_5$	55,000	-18,000	4,000	2,300

SOLUTION

(a) The calculation of the $E(F_k)$ and $V(F_k)$ values of the project's annual net cash flows are shown in Table 14-7. The $E(PW)$ is calculated using Equation 14-14 as follows:

$$E(PW) = \sum_{k=0}^{5} (P/F, 20\%, k)E(F_k)$$

$$= -\$100,000 + \$40,000(P/F, 20\%, 1) + \cdots$$

$$+ \$74,000(P/F, 20\%, 5)$$

$$= \$32,517$$

Then the $V(PW)$ is calculated using Equation 14-13 as follows:

$$V(PW) = \sum_{k=0}^{5} (P/F, 20\%, k)^2 V(F_k)$$

$$= 100.0 \times 10^6 + (24.25 \times 10^6)(P/F, 20\%, 1)^2 + \cdots$$

$$+ (85.16 \times 10^6)(P/F, 20\%, 5)^2$$

$$= 186.75 \times 10^6 (\$)^2$$

Finally

$$SD(PW) = [V(PW)]^{1/2}$$

$$= [186.75 \times 10^6]^{1/2}$$

$$= \$13,666$$

TABLE 14-7 Calculation of $E(F_k)$ and $V(F_k)$ (Example 14-8)

End of Year k	F_k	$E(F_k) = a_k E(X_k) + b_k E(Y_k)$		$V(F_k) = a_k^2 V(X_k) + (-b_k)^2 V(Y_k)$	
0	F_0	$0 - \$100,000 =	-$100,000	$0 + (1)^2(10,000)^2 =$	$100.0 \times 10^6 \$^2$
1	F_1	$60,000 - 20,000 =$	40,000	$(4,500)^2 + (1)^2 (2,000)^2 =$	24.25×10^6
2	F_2	$65,000 - 2(15,000) =$	35,000	$(8,000)^2 + (2)^2 (1,200)^2 =$	69.76×10^6
3	F_3	$2(40,000) - 3(9,000) =$	53,000	$(2)^2(3,000)^2 + (3)^2 (1,000)^2 =$	45.0×10^6
4	F_4	$70,000 - 2(20,000) =$	30,000	$(4,000)^2 + (2)^2 (2,000)^2 =$	32.0×10^6
5	F_5	$2(55,000) - 2(18,000) =$	74,000	$(2)^2(4,000)^2 + (2)^2 (2,300)^2 =$	85.16×10^6

(b) The PW of the project's net cash flow is a linear combination of the annual net cash flow amounts, F_k, that are independent random variables. Each of these random variables, in turn, is a linear combination of the independent random variables X_k and Y_k. We can also observe in Table 14-7 that the V(PW) calculation does not include any dominant $V(F_k)$ value. Therefore, we have a reasonable basis on which to assume that the PW of the project's net cash flow is approximately normally distributed, with E(PW) = \$32,517 and SD(PW) = \$13,666.

Based on this assumption, we have

$$Z = \frac{PW - E(PW)}{SD(PW)} = \frac{0 - \$32,517}{\$13,666} = -2.3794$$

$$Pr\{PW \le 0\} = Pr\{Z \le -2.3794\}$$

From Appendix E we find that $Pr\{Z \le -2.3794\} = 0.0087$. Therefore, the probability of loss on this project is negligible. Based on this result, the E(PW) > 0, and the SD(PW) $= 0.42[E(PW)]$, the project is economically attractive and not too risky.

14.5 Evaluation of Uncertainty Using Monte Carlo Simulation*

The development of computers has resulted in the increased use of Monte Carlo simulation as an important tool for analysis of project uncertainties. For complicated problems, Monte Carlo simulation generates random outcomes for probabilistic factors so as to imitate the randomness inherent in the original problem. In this manner, a solution to a rather complex problem can be inferred from the behavior of these random outcomes.

To perform a simulation analysis, the first step is to construct an analytical model that represents the actual decision situation. This may be as simple as developing an equation for the PW of a proposed industrial robot in an assembly line or as complex as examining the effects of proposed environmental regulations on typical petroleum refinery operations. The second step is develop a probability distribution from subjective or historical data for each uncertain factor in the model. Sample outcomes are randomly generated by using the probability distribution for each uncertain quantity and are then used to determine a *trial* outcome for the model. Repeating this sampling process a large number of times leads to a frequency distribution of trial outcomes for a desired measure of merit, such as PW or AW. The resulting frequency distribution can then be used to make probabilistic statements about the original problem.

---

*Adapted from W. G. Sullivan and R. Gordon Orr, "Monte Carlo Simulation Analyzes Alternatives in Uncertain Economy," *Industrial Engineering* Vol. 14, No. 11 (November 1982). Reprinted with permission from *Industrial Engineering* magazine. Copyright ©Institute of Industrial Engineers, Inc., 25 Technology Park/Atlanta, Norcross, Georgia.

TABLE 14-8   Probability Distribution for Useful Life

Number of Years, N	Pr(N)
3 ⎤	0.20 ⎤
5 ⎱ possible	0.40 ⎱
7 ⎰ values	0.25 ⎰ $\sum \Pr(N) = 1.00$
10 ⎦	0.15 ⎦

To illustrate the Monte Carlo simulation procedure, suppose that the probability distribution for the useful life of a piece of machinery has been estimated as shown in Table 14-8. The useful life can be simulated by assigning random numbers to each value such that they are proportional to the respective probabilities. (A random number is selected in a manner such that each number has an equal probability of occurrence.) Because two-digit probabilities are given in Table 14-8, random numbers can be assigned to each outcome, as shown in Table 14-9. Next, a single outcome is simulated by choosing a number at random from a table of random numbers.* For example, if any random number between and including 00 and 19 is selected, the useful life is three years. As a further example, the random number 74 corresponds to a life of seven years.

If the probability distribution that describes a random variable is *normal*, a slightly different approach is followed. Here the simulated outcome is based on the mean and standard deviation of the probability distribution and on a random normal deviate, which is a random number of standard deviations above or below the mean of a standardized normal distribution. An abbreviated listing of typical random normal deviates is shown in Table 14-10. For normally distributed random

TABLE 14-9   Assignment of Random Numbers

Number of Years, N	Random Numbers
3	00–19
5	20–59
7	60–84
10	85–99

TABLE 14-10   Random Normal Deviates (RNDs)

−1.565	0.690	−1.724	0.705	0.090
0.062	−0.072	0.778	−1.431	0.240
0.183	−1.012	−0.844	−0.227	−0.448
−0.506	2.105	0.983	0.008	0.295
1.613	−0.225	0.111	−0.642	−0.292

_____

*The last two digits of randomly chosen telephone numbers in a telephone directory are usually quite close to being random numbers.

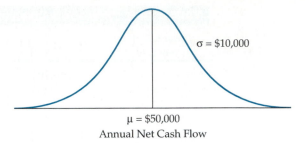

FIGURE 14-3
Normally Distributed Annual Cash
Flow

TABLE 14-11	Example of the Use of RNDs	
Year	RND	Annual Net Cash Flow [$50,000 + RND ($10,000)]
1	0.090	$50,900
2	0.240	52,400
3	−0.448	45,520
4	0.295	52,950
5	−0.292	47,080

variables, the simulated outcome is based on Equation 14-15:

$$\text{Outcome value} = \text{mean} + [\text{random normal deviate} \times \text{standard deviation}] \qquad \text{(14-15)}$$

For example, suppose that an *annual* net cash flow is assumed to be normally distributed, with a mean of $50,000 and a standard deviation of $10,000, as shown in Figure 14-3.

Simulated cash flows for a period of five years are listed in Table 14-11. Notice that the average annual net cash flow is $248,850/5, which equals $49,770. This approximates the known mean of $50,000 with an error of 0.46%.

If the probability distribution that describes a random event is *uniform* and continuous, with a minimum value of A and a maximum value of B, another procedure should be followed to determine the simulated outcome. Here the outcome can be computed with Equation 14-16:

$$\text{Simulation outcome} = A + \frac{RN}{RN_m}[B - A] \qquad \text{(14-16)}$$

where RN_m is the maximum possible random number (9 if one digit is used, 99 if two are used, etc.) and RN is the random number actually selected. This equation should be used when the minimum outcome, A, and the maximum outcome, B, are known.

For example, suppose that the market value in year N is assumed to be uniformly and continuously distributed between $8,000 and $12,000. A value of this random variable would be generated as follows with a random number of 74:

$$\text{Simulation outcome} = \$8,000 + \frac{74}{99} (\$12,000 - \$8,000) = \$10,990$$

Proper use of these procedures, coupled with an accurate model, will result in an approximation of the actual outcome. But how many simulation trials are necessary for an *accurate* approximation of, for example, the average outcome? In general, the greater the number of trials, the more accurate the approximation of the mean and standard deviation will be. One method of determining whether or not a sufficient number of trials has been conducted is to keep a running average of results. At first, this average will vary considerably from trial to trial. The amount of change between successive averages should decrease as the number of simulation trials increases. Eventually, this running (cumulative) average should level off at an accurate approximation.

EXAMPLE 14-9

Monte Carlo simulation can also simplify the analysis of a more complex problem. The following estimates relate to a capital investment opportunity being considered by a large manufacturer of air-conditioning equipment. Subjective probability functions have been estimated for the four independent uncertain factors as follows.

CAPITAL INVESTMENT   Normally distributed with a mean of −$50,000 and a standard deviation of $1,000.

USEFUL LIFE   Uniformly and continuously distributed with a minimum life of 10 years and a maximum life of 14 years.

ANNUAL REVENUE

$35,000 with a probability of 0.4

$40,000 with a probability of 0.5

$45,000 with a probability of 0.1

ANNUAL EXPENSE   Normally distributed, with a mean of −$30,000 and a standard deviation of $2,000.

The management of this company wishes to determine whether or not the capital investment will be a profitable one. The interest rate is 10% per year. In order to answer this question, the PW of the venture will be simulated.

SOLUTION

To illustrate the Monte Carlo simulation procedure, five trial outcomes are computed manually in Table 14-12. The estimate of the average present worth based on this very small sample is $19,010/5 = $3,802. For more accurate results, hundreds or even thousands of repetitions would be required.

The applications of Monte Carlo simulation for investigating uncertainty are many and varied. However, remember that the results can be no more accurate than the model and the probability estimates used. In all cases the procedure and rules are the same: careful study of the problem and development of the model;

TABLE 14-12   Monte Carlo Simulation of PW Involving Four Independent Factors (Example 14-9)

Trial Number	Random Normal Deviate (RND_1)	Capital Investment, I $[-\$50,000 + RND_1\ (\$1,000)]$	Three-Digit RNs	Project Life, N $[10 + \frac{RN}{999}(14 - 10)]$	Project Life, N (Nearest Integer)
1	+1.003	−$48,997	807	13.23	13
2	+0.358	−49,642	657	12.63	13
3	−1.294	−51,294	488	11.95	12
4	+0.019	−49,981	282	11.13	11
5	−0.147	−50,147	504	12.02	12

	One-Digit RN	Annual Revenue, R $35,000 for 0–3 40,000 for 4–8 45,000 for 9	RND_2	Annual Expense, E $[-\$30,000 + RND_2\ (\$2,000)]$	$PW = I$ $+(R + E)(P/A, 10\%, N)$
1	2	$35,000	+0.036	−$29,928	−$12,969
2	0	35,000	−0.605	−31,210	− 22,720
3	4	40,000	−1.470	−32,940	− 3,189
4	9	45,000	−1.864	−33,728	+ 23,232
5	8	40,000	+1.223	−27,554	+ 34,656
				Total	+$19,010

accurate assessment of the probabilities involved; true randomization of outcomes as required by the Monte Carlo simulation procedure; and calculation and analysis of the results. Furthermore, a sufficiently large number of Monte Carlo trials should always be used to reduce the estimation error to an acceptable level.

14.6 Performing Monte Carlo Simulation with a Computer

It is apparent from the preceding section that a Monte Carlo simulation of a complex project requiring several thousand trials can be accomplished only with the help of a computer. Indeed, there are numerous simulation programs that can be obtained from software companies and universities. To illustrate the computational features and output of a typical simulation program, Example 14-9 has been evaluated with a computer program developed by R. Gordon Orr, a former industrial engineering graduate student at the University of Tennessee. The computer queries and user responses (in boxes) are shown in Figure 14-4. Simulation results for 3,160 trials are shown in Figure 14-5. (This number of trials was needed for the cumulative average PW to stabilize to a variation of ±0.5%.)

The average PW is $7,759.60, which is larger than the $3,801 obtained from Table 14-12. This underscores the importance of having a sufficient number of simulation trials to ensure reasonable accuracy in Monte Carlo analyses.

The histogram in Figure 14-5 indicates that the *median PW* of this investment is $6,700 and that the dispersion of PW trial outcomes is considerable. The standard deviation of simulated trial outcomes is one way to measure this dispersion. Based

```
THE FOLLOWING PROGRAM USES MONTE CARLO SIMULATION
TECHNIQUES AS APPLIED TO RISK ANALYSIS PROBLEMS OF
ENGINEERING ECONOMY.

WILL YOU BE USING A REMOTE PRINTER FOR OUTPUT ? (Y OR
N)  Y

INPUT A RANDOM NUMBER BETWEEN 1 AND 1000.  199

MAXIMUM NUMBER OF ITERATIONS YOU WISH TO RUN ?  1000

WHAT INTEREST RATE (PERCENT) IS TO BE USED ?  10

THE DATA FOR EACH RANDOM VARIABLE INVOLVED MAY BE
FORMULATED AS FOLLOWS:

1.      SINGLE VALUE OR ANNUITY
2.      SINGLE VALUE WITH UNIFORM GRADIENT
3.      SINGLE VALUE WITH GEOMETRIC GRADIENT
4.      DISCRETE DISTRIBUTION
5.      UNIFORM DISTRIBUTION
6.      NORMAL DISTRIBUTION
7.      A SERIES OF YEARLY CASH FLOWS
8.      SALVAGE VALUE DEPENDENT ON PROJECT LIFE
9.      TRIANGULAR DISTRIBUTION

INFORMATION FOR INITIAL CASH FLOW:

    DISTRIBUTION IDENTIFICATION NUMBER =    6

    MEAN VALUE =    -50000

    STANDARD DEVIATION =    1000

INFORMATION FOR YEARLY CASH FLOW:

    THIS CASH FLOW MAY CONSIST OF A NUMBER OF
    DIFFERENT ELEMENTS WHICH MAY FOLLOW DIFFERENT
    DISTRIBUTIONS.
    PLEASE INPUT THE DATA ONE ELEMENT AT A TIME AND
    YOU WILL BE PROMPTED FOR ADDITIONAL INFORMATION.
    DISTRIBUTION IDENTIFICATION NUMBER =    4

    NUMBER OF VALUES =    3
```

(continued)

FIGURE 14-4   Sample Monte Carlo Simulation—Computer Queries and User Responses

```
        INPUT VALUES IN ASCENDING ORDER:

     VALUE 1 =    35000
           WITH PROBABILITY     0.4

     VALUE 2 =    40000
           WITH PROBABILITY     0.5

     VALUE 3 =    45000
           WITH PROBABILITY     0.1

     IS THERE ADDITIONAL ANNUAL CASH FLOW DATA? (Y OR N)    Y

     DISTRIBUTION IDENTIFICATION NUMBER =    6

     MEAN VALUE =    -30000

     STANDARD DEVIATION =    2000

     IS THERE ADDITIONAL ANNUAL CASH FLOW DATA? (Y OR N)    N

  INFORMATION FOR SALVAGE VALUE:

     DISTRIBUTION IDENTIFICATION NUMBER =    1

     CASH VALUE =   0
     INFORMATION FOR PROJECT LIFE:
           DISTRIBUTION IDENTIFICATION NUMBER =   5

     MINIMUM VALUE =    10

     MAXIMUM VALUE =    14

     EXPECTED VALUE OF PRESENT WORTH =          7759.60
     VARIANCE OF PRESENT WORTH =          680623960.00
     STANDARD DEVIATION OF PRESENT WORTH =      26088.77
     PROBABILITY THAT PRESENT WORTH IS GREATER THAN
     ZERO =                              0.595

     EXPECTED VALUE OF ANNUAL WORTH =          1114.15
     VARIANCE OF ANNUAL WORTH =           14611587.00
     STANDARD DEVIATION OF ANNUAL WORTH =       3822.51
     PROBABILITY THAT ANNUAL WORTH IS GREATER THAN
     ZERO =                              0.595
```

FIGURE 14-4   (*continued*)

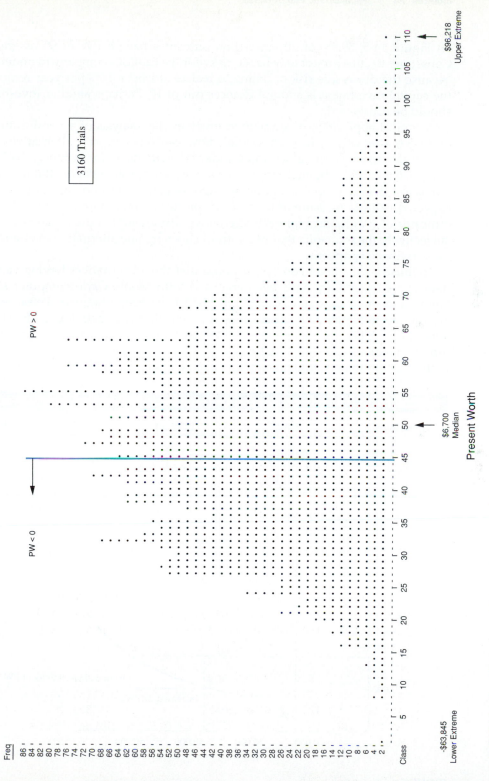

FIGURE 14-5 Histogram of PW for Example 14-9

on Figure 14-5, 59.5% of all simulation outcomes have a PW of $0 or greater. Consequently, this project may be too risky for the cautious company to undertake because the down-side risk of failing to realize at least a 10% per year return on the capital investment is about 4 chances out of 10. Perhaps another investment should be considered.

A typical application of simulation involves the analysis of several mutually exclusive alternatives. In such studies, how can one compare alternatives that have different expected values and standard deviations of, for instance, PW? One approach is to select the alternative that *minimizes* the probability of attaining a PW that is less than zero. Another popular response to this question utilizes a graph of expected value (a measure of the reward) plotted against standard deviation (an indicator of uncertainty) for each alternative. An attempt is then made to assess subjectively the trade-offs that result from choosing one alternative over another in pairwise comparisons.

To illustrate the latter concept, suppose that three alternatives having varying degrees of uncertainty have been analyzed with Monte Carlo computer simulation, and the results shown in Table 14-13 have been obtained. These results are plotted in Figure 14-6, where it is apparent that Alternative C is inferior to Alternative A because of its lower $E(PW)$ and larger standard deviation. Therefore, C offers a smaller PW that has a greater amount of uncertainty associated with it! Unfortunately, the choice of B versus A is not as clear because the increased

TABLE 14-13	**Simulation Results for Three Mutually Exclusive Alternatives**		
Alternative	E(PW)	SD(PW)	E(PW) ÷ SD(PW)
A	$37,382	$1,999	18.70
B	49,117	2,842	17.28
C	21,816	4,784	4.56

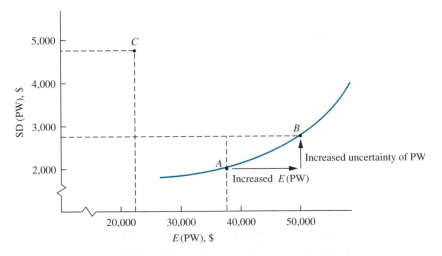

FIGURE 14-6   Graphical Summary of Computer Simulation Results

expected PW of B has to be balanced against the increased uncertainty of B. This trade-off *may or may not* favor B, depending on management's attitude toward accepting the additional uncertainty associated with a larger expected reward. The comparison also presumes that Alternative A is acceptable to the decision maker. One simple procedure for choosing between A and B is to rank alternatives based on the ratio of E(PW) to SD(PW). In this case, Alternative A would be chosen because it has the more favorable (larger) ratio.

▼ 14.7   Decision Trees*

> Decision trees, also commonly called *decision flow networks* and *decision diagrams*, are powerful means of depicting and facilitating the analysis of important problems, especially those that involve sequential decisions and variable outcomes over time. Decision trees have great usefulness in practice because they make it possible to break down a large, complicated problem into a series of smaller simple problems, and they enable objective analysis and decision making that includes explicit consideration of the risk and effect of the future.

The name *decision tree* is appropriate, because it shows branches for each possible alternative for a given decision and branches for each possible outcome (event) that can result from each alternative. Such networks reduce abstract thinking to a logical visual pattern of cause and effect. When costs and benefits are associated with each branch and probabilities are estimated for each possible outcome, analysis of the decision flow network can clarify choices and risks.

14.7.1   Deterministic Example

The most basic form of decision tree occurs when each alternative can be assumed to result in a single outcome—that is, when certainty is assumed. The replacement problem in Figure 14-7 illustrates this. The problem as shown reflects that the decision about whether or not to replace the defender (old machine) with the new machine (challenger) is not just a one-time decision, but rather one that recurs periodically. That is, if the decision is made to keep the old machine at decision point one, then later, at decision point two, a choice again has to be made. Similarly, if the old machine is chosen at decision point two, then a choice again has to be made at decision point three. For each alternative, the cash inflow and duration of the project are shown above the arrow and the capital investment is shown below the arrow.

For this problem, the question initially seems to be which alternative to choose at decision point one. But an intelligent choice at decision point one should take into account the later alternatives and decisions that stem from it. Hence, the correct procedure in analyzing this type of problem is to start at the most distant decision

---

*Adapted (except Section 14.7.3) from John R. Canada/William G. Sullivan, *Economic and Multiattribute Evaluation of Advanced Manufacturing Systems,* © 1989, pp. 341–343, 347. Reprinted by permission of Prentice Hall, Englewood Cliffs, New Jersey.

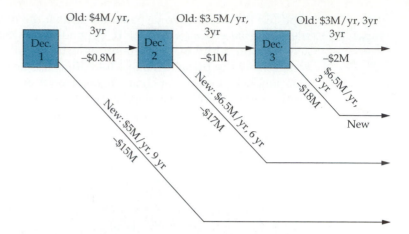

FIGURE 14-7

Deterministic Replacement
Example

point, determine the best alternative and quantitative result of that alternative, and then "roll back" to each successive decision point, repeating the procedure until finally the choice at the initial or present decision point is determined. By this procedure, one can make a present decision that directly takes into account the alternatives and expected decisions of the future.

For simplicity in this example, timing of the monetary outcomes will first be neglected, which means that a dollar has the same value regardless of the year in which it occurs. Table 14-14 shows the necessary computations and decisions using a nine-year study period. Note that the monetary outcome of the best alternative at decision point three ($7.0M for the *old*) becomes part of the outcome for the old alternative at decision point two. Similarly, the best alternative at decision point two ($22.0M for the *new*) becomes part of the outcome for the defender alternative at decision point one.

The computations in Table 14-14 show that the answer is to keep the old alternative now and plan to replace it with the new one at the end of three years (at decision point two). But this does not mean that the old machine should necessarily

TABLE 14-14   Monetary Outcomes and Decisions at Each Point—Deterministic Replacement Example of Figure 14-7[a]

Decision Point	Alternative	Monetary Outcome		Choice
3	Old	$3M(3) − $2M	= $ 7.0M	Old
	New	$6.5M(3) − $18M	= $ 1.5M	
2	Old	$7M + $3.5M(3) − $1M	= $16.5M	
	New	$6.5M(6) − $17M	= $22.0M	New
1	Old	$22.0M + $4M(3) − $0.8M	= $33.2M	Old
	New	$5M(9) − $15M	= $30.0M	

[a]Interest = 0% per year, that is, ignoring timing of cash flows.

TABLE 14-15	Decision at Each Point with Interest = 25% Per Year for Deterministic Replacement Example of Figure 14-7			
Decision Point	**Alternative**	**PW of Monetary Outcome**		**Choice**
3	Old	$3M(P/A, 3) - $2M$ $3M(1.95) - $2M$	= $3.85M	Old
	New	$6.5M(P/A, 3) - $18M$ $6.5M(1.95) - $18M$	= -$5.33M	
2	Old	$3.85(P/F, 3) + $3.5M(P/A, 3) - $1M$ $3.85(0.512) + $3.5M(1.95) - $1M$	= $7.79M	Old
	New	$6.5M(P/A, 6) - $17M$ $6.5M(2.95) - $17M$	= $2.18M	
1	Old	$7.79M(P/F, 3) + $4(P/A, 3) - $0.8M$ $7.79M(0.512) + $4M(1.95) - $0.8M$	= $10.99M	Old
	New	$5.0M(P/A, 9) - $15M$ $5.0M(3.46) - $15M$	= $2.30M	

be kept for a full three years and a new machine bought without question at the end of that period. Conditions may change at any time, necessitating a fresh analysis—probably a decision tree analysis—based on estimates that are reasonable in light of conditions at that time.

14.7.1.1 Deterministic Example Considering Timing

For decision tree analyses, which involve working from the most distant to the nearest decision point, the easiest way to take into account the timing of money is to use the PW approach and thus *discount all monetary outcomes to the decision points in question*. To demonstrate, Table 14-15 shows computations for the same replacement problem of Figure 14-7 using an interest rate of 25% per year.

Note from Table 14-15 that when taking into account the effect of timing by calculating PWs at each decision point, the indicated choice is not only to keep the old alternative at decision point one, but also to keep the old alternative at decision points two and three as well. This result is not surprising since the high interest rate tends to favor the alternatives with lower capital investments, and it also tends to place less weight on long-term returns (benefits).

14.7.2 General Principles of Diagramming

The proper diagramming of a decision problem is, in itself, generally very useful to the understanding of the problem, and it is essential to correct subsequent analysis.

The placement of decision points (nodes) and chance outcome nodes from the initial decision point to the base of any later decision point should give an accurate representation of the information that will and will not be available when the choice represented by the decision point in question actually has to be made. The decision tree diagram should show the following (normally a square symbol is used to depict a decision node, and a circle symbol is used to depict a chance outcome node):

1. All initial or immediate alternatives among which the decision maker wishes to choose.
2. All uncertain outcomes and future alternatives that the decision maker wishes to consider because they may directly affect the consequences of initial alternatives.
3. All uncertain outcomes that the decision maker wishes to consider because they may provide information that can affect his or her future choices among alternatives and hence indirectly affect the consequences of initial alternatives.

Note that the alternatives at any decision point and the outcomes at any chance outcome node must be:

1. Mutually exclusive; that is, no more than one can be chosen.
2. Collectively exhaustive; that is, one event must be chosen or something must occur if the decision point or outcome node is reached.

14.7.3 Decision Trees with Random Outcomes

The deterministic replacement problem discussed in Section 14.7.1 introduced the concept of sequential decisions using assumed certainty for alternative outcomes. However, an engineering problem requiring sequential decisions often includes random outcomes, and decision trees are very useful in structuring this type of situation. The decision tree diagram helps to make the problem explicit and assists in its analysis. This is illustrated in Examples 14-10 through 14-12.

EXAMPLE 14-10

The Ajax Corporation manufactures compressors for commercial air conditioning systems. A new compressor design is being evaluated as a potential replacement for the most frequently used unit. The new design involves major changes that have the expected advantage of better operating efficiency. From the perspective of a typical user, the new compressor (as an assembled component in an air conditioning system) would have an increased investment of $8,600 relative to the present unit, and an annual expense saving dependent upon the extent to which the design goal is met in actual operations.

Estimates by the multidisciplinary design team of the new compressor achieving four levels (percentages) of the efficiency design goal, and the probability and annual expense saving at each level are as follows:

Level (Percentage) of Design Goal Met (%)	Probability $p(L)$	Annual Expense Saving
90	0.25	$3,470
70	0.40	2,920
50	0.25	2,310
30	0.10	1,560

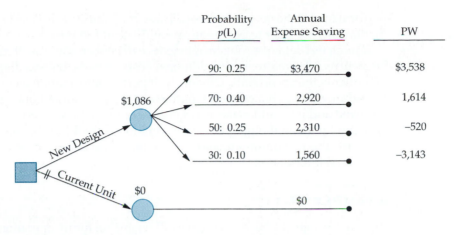

Probability $p(L)$	Annual Expense Saving	PW
90: 0.25	$3,470	$3,538
70: 0.40	2,920	1,614
50: 0.25	2,310	−520
30: 0.10	1,560	−3,143

FIGURE 14-8   Single-State Decision Tree (Example 14-10)

Based on a before-tax analysis (MARR = 18% per year, analysis period = 6 years, and salvage value = 0) and $E(PW)$ as the decision criterion, is the new compressor design economically preferable to the current unit?

SOLUTION

The single-stage decision tree diagram for the design alternatives is shown in Figure 14-8. The PWs associated with each of the efficiency design goal levels being met are as follows:

$$PW(18\%)_{90} = -\$8,600 + \$3,470(P/A, 18\%, 6) = \$3,538$$
$$PW(18\%)_{70} = -\$8,600 + \$2,920(P/A, 18\%, 6) = \$1,614$$
$$PW(18\%)_{50} = -\$8,600 + \$2,310(P/A, 18\%, 6) = -\$520$$
$$PW(18\%)_{30} = -\$8,600 + \$1,560(P/A, 18\%, 6) = -\$3,143$$

Based on these values, the $E(PW)$ of each installed unit of the new compressor is as follows:

$$E(PW) = 0.25(\$3,538) + 0.40(\$1,614) + 0.25(-\$520) + 0.10(-\$3,143)$$
$$= \$1,086$$

The $E(PW)$ of the current unit is zero since the cash flow estimates for the new design are incremental amounts relative to the present design. Therefore, the analysis indicates that the new design is economically preferable to the present design (the parallel lines across the current unit path on the diagram indicate that it was not selected).

14.7.3.1   The Expected Value of Perfect Information (EVPI)   The probability estimates of achieving the efficiency design goal levels, $p(L)$, developed by the design team in Example 14-10 reflect the uncertainty about the future operational

performance of the new compressor. These probabilities are based on present information and are prior to obtaining any additional test data.

If more test data were obtained to reduce the uncertainty, additional costs would be incurred. Therefore, these additional costs must be balanced against the value of reducing the uncertainty. Obviously, if perfect information were available about the future operational efficiency of the new compressor, all uncertainty would be removed and we could make an optimal decision between the current unit and the new design. Even though perfect information is unobtainable, its expected value indicates the maximum amount (upper limit) we should consider spending for additional information.

EXAMPLE 14-11

Refer to Example 14-10. What is the EVPI regarding future operational performance of the new compressor to the typical user of an air conditioning system?

SOLUTION

We can calculate the EVPI by comparing the optimal decision based on perfect information with the original decision in Example 14-10 to select the new compressor. This comparison is shown in Table 14-16. Based on this comparison, the EVPI to the typical user of a unit is

$$\text{EVPI} = \$1,530 - \$1,086 = \$444$$

14.7.3.2   The Use of Additional Information to Reduce Uncertainty The solution to Example 14-11 shows that there is some potential value to be obtained from additional test information about the operational performance of the new compressor. From the viewpoint of the typical user of a new unit versus the current unit, the maximum estimated value of additional information is $444.

The members of the management team of Ajax Corporation are strongly customer focused and want to exceed customers' expectations about the performance of their products. Thus, they asked the design team to estimate the value of data that could be obtained from an additional comprehensive test of prototypes of the new compressor. The information from the test will not be perfect, because

TABLE 14-16   Expected Value of Perfect Information (Example 14-11)

Level (%) of Design Goal Met	Probability $p(L)$	Decision with Perfect Information		Prior Decision (New Design)
		Decision	Outcome	
90	0.25	New	$3,538	$3,538
70	0.40	New	1,614	1,614
50	0.25	Current	0	−520
30	0.10	Current	0	−3,143
		Expected value: $1,530		$1,086

the test cannot determine exactly the long-term operational performance of the new design under diverse customer applications. However, the imperfect test information may reduce uncertainty and merit the additional cost of obtaining it.

We can evaluate the value of additional information before obtaining it only if we can estimate the reliability of the experiment being used. Therefore, the design team addressed the reliability of additional test information in predicting the future operational performance of the new compressor. The estimates developed by the design team, the calculated *revised probabilities* of achieving the new efficiency design goal levels, and the value of additional test information from a user's perspective are discussed in Example 14-12.

EXAMPLE 14-12

Refer to Examples 14-10 and 14-11. The members of the design team are confident that data from the additional comprehensive test of prototype compressors will show whether future operational performance will be favorable (60% or more of the design goal met) or not favorable (60% of the design goal not met). Based on obtaining these results from the test, and by using current engineering data within Ajax Corporation, the design team developed the *conditional probability* estimates below.

Comprehensive Test Outcome	Conditional Probabilities of Test Outcome Given the Level (%) of the Design Goal Met			
	90	70	50	30
Favorable (F)	0.95	0.85	0.30	0.05
Not favorable (NF)	0.05	0.15	0.70	0.95
Sum:	1.00	1.00	1.00	1.00

For example, given that the future operational performance of the new compressor is 90% of the design goal, the conditional probability of the comprehensive test results being favorable is estimated to be 0.95, and the conditional probability of the test results being not favorable is estimated to be 0.05. That is to say, $p(F/90) = 0.95$, and $p(NF/90) = 0.05$, where the slash mark means "given."

Based on these conditional probabilities (reliability estimates of test results) and the choice of either doing the test or not doing it, (a) calculate the revised probabilities of the four efficiency design goal levels being met and (b) estimate the value of doing the comprehensive test to the typical user of a new compressor unit.

SOLUTION

(a) The two-stage decision tree diagram, which includes an initial decision on whether or not to do the test, is shown in Figure 14-9. In order to calculate the revised probabilities, we need to determine the joint probabilities of each level of the efficiency design goal being met and each test outcome occurring, as well as the marginal probability of each test outcome. These probabilities are shown in

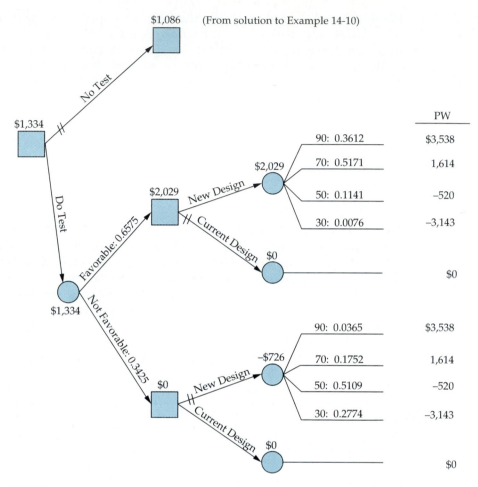

FIGURE 14-9   Two-Stage Decision Tree (Example 14-12)

Table 14-17. As an example, the joint probabilities of the efficiency design goal being met at the 90% level and each test outcome occurring are calculated as follows:

$$p(90, F) = p(F/90) \cdot p(90) = (0.95)(0.25) = 0.2375$$
$$p(90, NF) = p(NF/90) \cdot p(90) = (0.05)(0.25) = 0.0125$$

The remaining six joint probabilities are determined in a similar manner. The sum of the joint probabilities over the four design goal levels gives the marginal probability of each test outcome occurring. That is to say, $p(F) = 0.6575$ and $p(NF) = 0.3425$. Similarly, the sums of the joint probabilities over the test outcomes are the marginal probabilities of achieving the new efficiency design goal levels [and are the same as the prior probabilities, $p(L)$, in Example 14-10]. The revised [e.g., $p(90) = p(90, F)/p(F) = 0.2375/0.6575 = 0.3612$] and marginal probabilities based on Table 14-17 are also plotted in Figure 14-9. (b) The estimated value (to the typical user of a new compressor unit) of doing

TABLE 14-17   Joint and Marginal Probabilities (Example 14-12)

Level (%) of Design Goal Met	Joint Probabilities		Marginal Probabilities, p(L)
	Favorable (F)	**Not Favorable (NF)**	
90	0.2375	0.0125	0.25
70	0.3400	0.0600	0.40
50	0.0750	0.1750	0.25
30	0.0050	0.0950	0.10
Marginal probabilities of test outcome:	0.6575	0.3425	1.00 (Sum)

the additional comprehensive test can be determined by using the data shown in Figure 14-9. By starting on the right side of Figure 14-9 and working toward the left side, we can calculate the E(PW) of the new design for both a favorable test outcome ($2,029) and an unfavorable test outcome (−$726). Based on these results, the choices at the two decision nodes are the new design alternative for a favorable test outcome and the current design alternative for a not favorable test outcome, respectively.

Based on these alternative design selections at the two decision nodes, the E(PW) at the chance node for the "do test" option is $1,334. The expected value of the comprehensive test, before the additional cost is considered, is $1,334 − $1,086 = $248, where $1,086 is the E(PW) of the new design without additional test information (Example 14-10).

The management team at Ajax Corporation used this information to help make a final decision about the additional comprehensive test of the new compressor design. Since the estimated total cost of the test was less than the expected value to the typical user of a unit ($248) times the estimated number of units to be sold in one year, and because of their strong customer focus, management decided to proceed with the additional test.

14.8 Spreadsheet Applications

Earlier in this chapter we learned how Monte Carlo simulation can simplify the analysis of relatively complex problems. To reduce the estimation error it is recommended that a large number of trials (as many as several thousand) be used. This is a formidable task if you were to perform this analysis using hand calculations. In this section, a spreadsheet model for Monte Carlo simulation is presented.

At the heart of Monte Carlo simulation is the generation of random numbers. Most spreadsheet packages include a RAND( ) function that returns a random number between zero and one. Other advanced statistical functions, such as NORMSINV( ), will return the inverse of a cumulative distribution function (the standard normal distribution in this case). This function can be used to generate random normal deviates. The spreadsheet model shown in Figure 14-10 makes use of these functions in performing a Monte Carlo simulation.

	A	B	C	D	E	F	G	H	I	J
1	MARR	10%								
2										
3			Capital Investment	Mean	Std. Dev.		Annual		Cumulative	
4			Annual Expenses	-$50,000	$1,000		Revenues	Prob.	Probability	
5				-$30,000	$2,000		$35,000	0.4	0.4	
6				Minimum	Maximum		$40,000	0.5	0.9	
7			Project Life	10	14		$45,000	0.1	1	
8										
9									Annual	
10	Trial	RND$_1$	Capital Investment	Uniform RN [0,1]	Project Life	Uniform RN [0,1]	Annual Revenues	RND$_2$	Annual Expenses	PW
11	1	-0.346	-50346	0.912	14	0.315	35,000	-0.268	-30536	$ (17,461)
12	2	+0.453	-49547	0.213	11	0.413	40,000	+1.397	-27206	$ 33,551
13	3	+1.092	48908	0.992	14	0.146	35,000	+0.382	-29236	$ (6,446)
14	4	-0.064	-50064	0.688	13	0.898	40,000	+0.807	-28386	$ 32,434
15	5	+0.983	-49017	0.638	13	0.212	35,000	+0.158	-29684	$ (11,256)
16	6	+1.274	-48726	0.477	12	0.895	40,000	+0.189	-29622	$ 21,986
17	7	-1.083	-51083	0.157	11	0.766	40,000	-0.003	-30006	$ 13,829
18	8	+0.535	-49465	0.771	13	0.239	35,000	-0.513	-31026	$ (21,236)
19	9	+0.167	-49833	0.488	12	0.470	40,000	-1.168	-32336	$ 2,387
20	10	-0.499	-50499	0.073	10	0.982	45,000	+1.061	-27878	$ 54,708
21										
22									Average	$ 6,164

FIGURE 14-10 Spreadsheet Model for Monte Carlo Simulation

564

The project evaluated in Example 14-9 is used to demonstrate the use of a spreadsheet for Monte Carlo simulation. The probabilistic functions for the four independent uncertain factors are included in the spreadsheet model. The required capital investment and annual expenses are normally distributed with the mean and standard deviations shown. The project life is expected to be uniformly distributed between 10 and 14 years. A discrete probability distribution has been compiled for annual revenues. The associated cumulative probability distribution (shown in column I, rows 4–6) is computed by the spreadsheet model.

Random normal deviates are generated in column B and column H to compute each trial's capital investment and annual expenses values. A uniform random number is generated in column D for the purposes of obtaining the project life. The trial value of project life makes use of the function ROUND() to return integer values. A second uniform random number is generated to obtain the annual revenues. The random number is compared to the cumulative distribution function for annual revenues and the appropriate value is placed in column G. The present worth of each trial is computed in column J.

Figure 14-10 shows only 10 trials. The average present worth over these trials was computed to be $6,164. More trials can be created simply by copying blocks of cells. The average present worth found using this spreadsheet model over 1000 trials was $7,949 (which is very close to the expected value of the present worth). The formulas for the highlighted cells are given in the following table.

Cell	Contents
I5	= I4 + H5
B11	= NORMSINV(RAND())
C11	= ($D$3 + $E$3 * B11)
D11	= RAND()
E11	= ROUND($D$7 + RAND() * ($E$7 − $D$7))
F11	= RAND()
G11	= IF(F11 <= I4,G4,IF(F11 <= I5,G5,G6))
H11	= NORMSINV(RAND())
I11	= ($D$4 + $E$4 * H11)
J11	= C11 − PV($B$1, E11, G11 + I11)
J22	= AVERAGE(J11 : J20)

14.9 Summary

Engineering economy involves decision making among competing uses of scarce capital resources. The consequences of resultant decisions usually extend far into the future. In this chapter we have presented various statistical and probability concepts that address the fact that the consequences (cash flows, project lives, etc.) of engineering alternatives can never be known with certainty, including Monte Carlo computer simulation techniques and decision tree analysis. Cash inflow and cash outflow factors, as well as project life, were modeled as discrete and continuous random variables. The resulting impact of uncertainty on the economic measures of merit for an alternative was analyzed. Included in the discussion were several considerations and limitations relative to the use of these methods in application.

Regrettably, there is no quick and easy answer to the question "How should uncertainty best be considered in an engineering economic evaluation?" Generally, simple procedures (e.g., breakeven analysis and sensitivity analysis, discussed in Chapter 10) allow modest discrimination among alternatives to be made on the basis of uncertainties present, and they are relatively inexpensive to apply. Additional discrimination among alternatives is possible with more complex procedures that utilize probabilistic concepts. These procedures, however, are more difficult to apply and require additional time and expense.

14.10 References

BONINI, C. P. "Risk Evaluation of Investment Projects," *OMEGA*, vol. 3, no. 6, 1975, pp. 735–750.

HERTZ, D. B., and H. THOMAS. *Risk Analysis and Its Applications*. New York: John Wiley & Sons, 1983.

HILLIER, F. S. *The Evaluation of Risky Interrelated Investments*. Amsterdam: North-Holland, 1969.

HULL, J. C. *The Evaluation of Risk in Business Investment*. New York: Pergamon Press, 1980.

MAGEE, J. F. "Decision Trees for Decision Making," *Harvard Business Review*, vol. 42, no. 4, July-August 1964, pp. 126–138.

PARK, C. S., and G. SHARPE-BETTE. *Advanced Engineering Economics*. New York: John Wiley & Sons, 1990.

RIGGS, J. L., and T. M. WEST. *Engineering Economics*, 3rd ed. New York: McGraw-Hill, 1987.

ROSE, L. M. *Engineering Investment Decisions: Planning Under Uncertainty*. Amsterdam: Elsevier, 1976.

WAGLE, B. "A Statistical Analysis of Risk in Capital Investment Projects," *Operational Research Quarterly*, vol. 18, 1967, pp. 13–33.

WALPOLE, R. E., and R. H. MEYERS. *Probability and Statistics for Engineers*, 4th ed. New York: Macmillan, 1989.

14.11 Problems

The number in parentheses () that follows each problem refers to the section from which the problem is taken.

14-1. Assume that the estimated net annual benefits for a project during each year of its life have the following probabilities:

Net Annual Benefits (NAB)	p(NAB)
$2,000	0.40
3,000	0.50
4,000	0.10

The life is three years for certain and the capital investment is −$7,000, with negligible salvage value. The MARR is 15% per year. Determine $E(PW)$ and the probability that PW is greater than zero [i.e., $\Pr(PW \geq 0)$]. (14.3)

14-2. A bridge is to be constructed now as part of a new road. Engineers have determined that traffic density on the new road will justify a two-lane bridge at the present time. Because of uncertainty regarding future use of the road, the time at which an extra two lanes will be required is currently being studied. The estimated probabilities of having to widen the bridge to four lanes at various times in the future are as follows:

Widen Bridge In	Probability
3 years	0.1
4 years	0.2
5 years	0.3
6 years	0.4

The two-lane bridge will cost $2,000,000 and the four-lane bridge, if built at one time, will cost $3,500,000. The cost of widening a two-lane bridge will be an extra $2,000,000 plus $250,000 for every year that construction is delayed. If money can earn 12% per year, what would you recommend? (14.3)

14-3. In Problem 14-2, perform an analysis to determine how sensitive the choice of a four-lane bridge built at one time versus a four-lane bridge that is constructed in two stages is to the interest rate. Will an interest rate of 15% per year reverse the initial decision? At what interest rate(s) would the two-lane bridge be preferred? (14.3)

14-4. In a building project, the amount of concrete to be poured during the next week is uncertain. The foreman has estimated the following probabilities:

Amount (cubic yards)	Probability
1,000	0.1
1,200	0.3
1,300	0.3
1,500	0.2
2,000	0.1

Determine the expected value (amount) of concrete to be poured next week. Also compute the variance and standard deviation of concrete to be poured. (14.3)

14-5. Consider the following two random variables, P and Q.

Price, P	$p(P)$	Quantity Sold, Q	$p(Q)$
$6	$\frac{1}{3}$	10	$\frac{1}{3}$
5	$\frac{1}{3}$	15	$\frac{1}{3}$
4	$\frac{1}{3}$	20	$\frac{1}{3}$

Assume that P and Q are independent. What are the mean, variance, and standard deviation of the probability distribution for revenue? (14.3)

14-6. A small dam is being planned for a river tributary that is subject to frequent flooding. From past experience, the probabilities that water flow will exceed the design capacity of the dam during a year, plus relevant cost information, are as follows:

Design of Dam	Probability of Greater Flow During a Year	Capital Investment
A	0.100	−$180,000
B	0.050	− 195,000
C	0.025	− 208,000
D	0.015	− 214,000
E	0.006	− 224,000

Estimated annual damages that occur if water flows exceed design capacity are $150,000, $160,000, $175,000, $190,000, and $210,000 for design A, B, C, D, and E, respectively. The life of the dam is expected to be 50 years, with negligible salvage value. For an interest rate of 8% per year, determine which design should be implemented. What nonmonetary considerations might be important to the selection? (14.3)

14-7. A diesel generator is needed to provide auxiliary power in the event that the primary source of power is interrupted. Various generator designs are available, and more expensive generators tend to have higher reliabilities should they be called on to produce power. Estimates of reliabilities, capital investment costs, O & M expenses, and damages resulting from a complete power failure (i.e., the standby generator fails to operate) are given in the table at the top of page 568 for three alternatives. If the life of each generator is 10 years and MARR = 10% per year, which generator should be chosen if you assume one main power failure per year? Does your choice change if you assume two main power failures per year (O & M expenses remain the same)? (14.3)

14-8. The owner of a ski resort is considering installing a new ski lift, which will cost $900,000. Expenses for operating and maintaining the lift are estimated to be $1,500 per day when operating. The U.S. Weather Service estimates that

Alternative	Capital Investment	O&M Expenses/Year	Reliability	Cost of Power Failure	Market Value	(Pr. 14-7)
R	−$200,000	−$5,000	0.96	−$400,000	$40,000	
S	− 170,000	− 7,000	0.95	− 400,000	25,000	
T	− 214,000	− 4,000	0.98	− 400,000	38,000	

there is a 60% probability of 80 days of skiing weather per year, a 30% probability of 100 days per year, and a 10% probability of 120 days per year. The operators of the resort estimate that during the first 80 days of adequate snow in a season, an average of 500 people will use the lift each day, at a fee of $10 each. If 20 additional days are available, the lift will be used by only 400 people per day during the extra period, and if 20 more days of skiing are available, only 300 people per day will use the lift during those days. The owners wish to recover any invested capital within five years and want at least a 25% per year rate of return before taxes. Based on a before-tax analysis, should the lift be installed? (14.3)

14-9. Refer to Problem 14-8. Assume the following changes: the study period is eight years; the ski lift will be depreciated using the MACRS Alternative Depreciation System (ADS); the ADS recovery period is seven years; the MARR = 15% per year (after-tax); and the effective income tax rate (t) is 40%. Based on this information, what is the $E(PW)$ and $SD(PW)$ of the ATCF? Interpret the analysis results and make a recommendation on installing the ski lift. (14.3)

14-10. Every time an automatic welding machine fails, the cost in idle labor and repairs is $1,000. The outage time averages four working hours, and the plant works 2,000 hours a year. Suppose that the machine could fail a maximum of 250 times in a given year. The probabilities of failure during a year are assessed by a certain trade association to be as follows: no failures, 0.050; 1, 0.113; 2, 0.209; 3, 0.333; 4, 0.201; 5, 0.080; 6, 0.009; 7, 0.003; 8, 0.001; 9, 0.0008; 10, 0.00008. A standby machine with an expected life of 10 years can be procured for $10,000, with a $2,000 market value. The annual expenses to keep it ready to run are $500. The probability of its breaking down during a standby run are nil. The MARR is 15% per year. Based on this information, (a) what are the $E(ACF)$, $V(ACF)$, and $SD(ACF)$ for the present welding machine? (b) If you assume that the

discrete probability distribution of the expected ACF (present machine) can be approximated by a normal distribution, what is the probability that the $E(ACF)$ is less than the AC of the standby machine? What recommendation would you make regarding purchase of the standby machine? AC = annual cost, and ACF = annual cost of failure. (14.3, 14.4)

14-11. The purchase of a new piece of electronic measuring equipment for use in a continuous metal forming process is being considered. If this equipment were purchased, the capital cost would be $418,000, and the estimated savings are $148,000 per year. The useful life of the equipment in this application is uncertain. The estimated probabilities of different useful lives occurring are shown in the following table. Assume that the MARR = 15% per year before taxes, and the market value at the end of its useful life is equal to zero. Based on a before-tax analysis, (a) what are the $E(PW)$, $V(PW)$, and $SD(PW)$ associated with the purchase of the equipment, and (b) what is the probability that the PW is less than zero? Make a recommendation and give your supporting logic based on the analysis results. (14.3)

Useful Life, Years (N)	$p(N)$
3	0.1
4	0.1
5	0.2
6	0.3
7	0.2
8	0.1

14-12. The probability tree diagram at the top of page 569 describes the uncertain cash flows for an engineering project. The analysis period is two years, and the $MARR = 15\%$ per year. Based on this information, (a) what are the $E(PW)$, $V(PW)$, and $SD(PW)$ of the project, and (b) what is the probability that PW ≥ 0? (14.3)

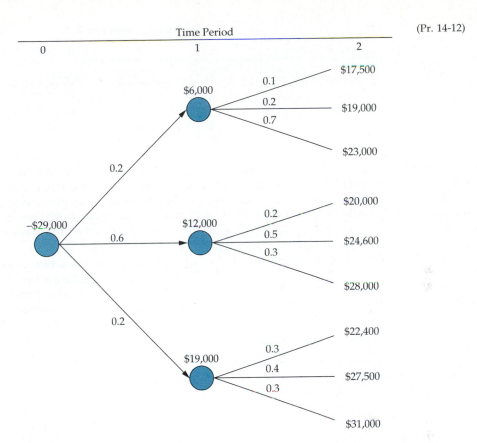

(Pr. 14-12)

14-13. A potential project has an initial capital investment of $100,000. Net annual revenues minus expenses are estimated to be $40,000 (A$) in the first year and to increase at the rate of 6.48% per year. However, the useful life of the primary equipment is uncertain, as shown in the following table. (Assume that i_c = MARR = 15% per year and f = 4% per year.) Based on this information, (a) what are the $E(PW)$ and $SD(PW)$ for this project, (b) what is the $\Pr\{PW \geq 0\}$, and (c) what is the $E(AW)$ in R$? Do you consider the project economically acceptable, questionable, or not acceptable, and why? (14.3)

Useful Life, Years (N)	$p(N)$
1	0.03
2	0.10
3	0.30
4	0.30
5	0.17
6	0.10

14-14. A proposed venture has an initial capital investment of $80,000, annual revenues minus expenses of $30,000, and an uncertain useful life, N, as follows:

N	Probability of N
1	0.05
2	0.15
3	0.20
4	0.30
5	0.20
6	0.05
7	0.05

Determine the $E(PW)$ and $SD(PW)$ of this investment when the MARR is 20% per year. Also, what is the $\Pr\{PW \leq 0\}$? (14.3)

14-15. Suppose that a random variable (e.g., market value for a piece of equipment) is normally distributed, with mean = −$175 and variance = 25^2. What is the probability that the actual market value is *at least* $171? (14.4)

| End of Year, k | Net Cash Flow | Three-Point Estimates for X_k | | | (Pr. 14-18) |
		L	**M**	**H**	
0	$F_0 = X_0$	$-\$38,000$	$-\$41,000$	$-\$45,000$	
1	$F_1 = 2X_1$	$-\ 1,900$	$-\ 2,200$	$-\ 2,550$	
2	$F_2 = X_2$	$9,800$	$10,600$	$11,400$	
3	$F_3 = 4X_3$	$5,600$	$6,100$	$6,400$	
4	$F_4 = 5X_4$	$4,600$	$4,800$	$5,100$	
5	$F_5 = X_5$	$16,500$	$17,300$	$18,300$	

14-16. The AW of project R-2 is normally distributed, with a mean of $1,500 and a variance of $810,000(\$)^2$. Determine the probability that this project's AW is less than $1,700. (14.4)

14-17. For the following cash flow estimates, determine the $E(PW)$ and $V(PW)$. Also find the probability that the PW will exceed $0. The cash flows are statistically independent and normally distributed, and the MARR is 12% per year. (14.4)

End of Year, k	Expected Value of Cash Flow	Standard Deviation of Cash Flow
0	$-\$14,000$	0
1	6,000	$ 800
2	4,000	400
3	4,000	400
4	8,000	1,000

14-18. The use of three estimates (defined here as H=high, L=low, and M=most likely) for random variables is a practical technique for modeling uncertainty in some engineering economic analyses. Assume that the mean and variance of the random variable, X_k, in this situation can be estimated by $E(X_k) = (^1/_6)(H + 4M + L)$ and $V(X_k) = [(H - L)/6]^2$. The estimated net cash flow data for one alternative associated with a project are shown in the table above.

The random variables, X_k, are assumed to be statistically independent, and the applicable

MARR = 15% per year. Based on this information, (a) what are the mean and variance of the PW, (b) what is the probability that PW ≥ 0 (state any assumptions that you make), and (c) is this the same as the probability that the IRR is acceptable? Explain. (14.4)

14-19. Consider Problem 14-8 when, in addition to uncertainty regarding number of skiing days per year, the useful life of the venture is *also* uncertain as follows:

Useful Life, Years (N)	$p(N)$
4	0.2
5	0.6
6	0.2

Finally, the market value (MV) of the ski lift is a function of the venture's life:

$$MV = \$10,000(7 - N)$$

a. Set up a table and use Monte Carlo simulation to determine five trial outcomes of the venture's before-tax AW. Recall that the MARR is 25% per year.

b. Based on your simulation outcomes, should the lift be installed? State any assumptions you make. (14.5, 14.6)

14-20. Consider the estimates for a new piece of manufacturing equipment shown below: (14.5, 14.6)

Factor	Expected Value	Type of Probability Distribution	(Pr. 14-20)
Capital investment	$-\$150,000$	Known with certainty	
Market value	$ 2,000 (13 - N)$	Normal, $\sigma = \$500$	
Annual savings	$ 70,000$	Normal, $\sigma = \$4,000$	
Annual expenses	$-\$ 43,000$	Normal, $\sigma = \$2,000$	
Useful life, N	13 years	Uniform in $[8, 18]$	
MARR	8%/year	Known with certainty	

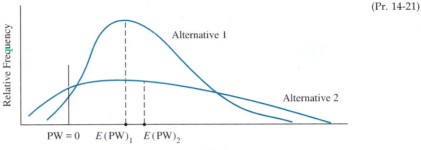

(Pr. 14-21)

Present Worth at 12% per year

a. Set up a table and simulate five trials of the equipment's PW.

b. Compute the mean of the five trials and recommend whether or not the equipment should be purchased.

14-21. Simulation results are available for two mutually exclusive alternatives. A large number of trials have been run with a computer, with the results shown at the top of this page.

Discuss the issues that may arise when attempting to decide between these two alternatives. (14.5, 14.6)

14-22. A large mudslide caused by heavy rains will cost Sabino County $1,000,000 per occurrence in lost property tax revenues. In any given year there is 1 chance in 100 that a major mudslide will occur.

A civil engineer has proposed constructing a culvert on a mountain where mudslides are likely. The culvert will reduce the likelihood of a mudslide to almost zero. The capital investment would be $50,000, and annual maintenance expenses would be $2,000 in the first year, increasing by 5% per year thereafter.

If the life of the culvert is expected to be 20 years and the cost of capital to Sabino County is 7% per year, should the culvert be built? (14.3)

14-23. There is a 50% chance that Dynamics General Corporation will require a conversion from *baseband* Ethernet cabling in a certain manufacturing plant to *broadband* Ethernet cabling at a rate of 1,000 feet per year (there is a 50% chance that the conversion will occur at a rate of 500 feet per year). The recabling will be required if a new government contract is negotiated with them now, which is 10 years prior to expiration of the present contract.

The vice president of manufacturing feels that if the company *waits* for the present contract to expire in 10 years, there is a 25% chance that the new contract will require 5,000 feet per year of conversion from baseband to broadband Ethernet cable and a 75% chance of conversion at the rate of 1,000 feet per year. Each 1,000 feet of Ethernet cable conversion costs $100,000.

Property tax for either type of cable is the same and should not be considered. All new cabling will be broadband Ethernet, and at an MARR of 20% per year, it can be considered to have a perpetual life. The terms of any new contract will call for a continuing annual program of converting existing baseband cable to broadband cable until the total anticipated cabling requirement of 20,000 feet has been completed.

Draw a decision tree for this problem, including the appropriate cash flows (before taxes) and probabilities. State any assumptions you make. Carefully label all alternatives. (14.7)

14-24. If $p(Y)$ is four times as great as $p(X)$, and if the MARR is 0% per year, which alternative in the tree diagram at the top of page 572 (i.e., "buy" or "don't buy") would you recommend? The salvage value is zero. (14.7)

14-25. If the interest rate is 8% per year, what decision would you make based on the decision tree diagram on page 572? (14.7)

14-26. The vice president of operations at a plant that manufactures components for hydraulic systems is considering an improvement to present production capability. The decision situation has been reduced to three alternatives. The first alternative would result in significant changes in present operations, including increased automation. The second alternative would involve fewer

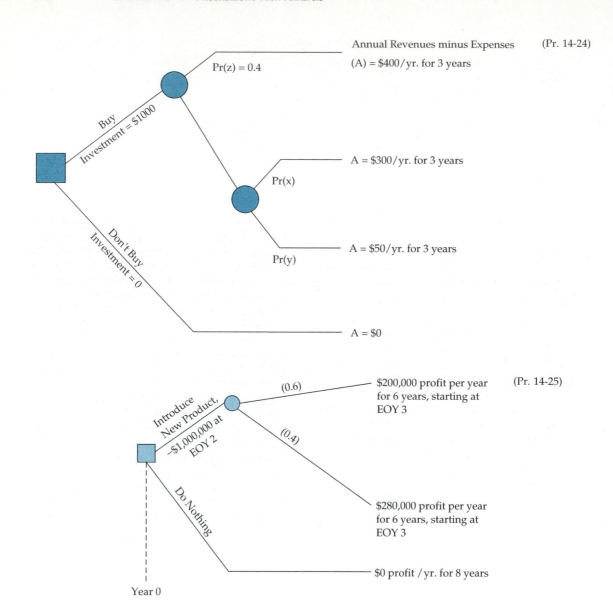

Annual Revenues minus Expenses   (Pr. 14-24)
(A) = $400/yr. for 3 years

A = $300/yr. for 3 years

A = $50/yr. for 3 years

A = $0

Pr(z) = 0.4

Buy
Investment = $1000

Don't Buy
Investment = 0

Pr(x)

Pr(y)

(Pr. 14-25)

Introduce
New Product,
−$1,000,000 at
EOY 2

(0.6) $200,000 profit per year
for 6 years, starting at
EOY 3

(0.4) $280,000 profit per year
for 6 years, starting at
EOY 3

Do Nothing $0 profit / yr. for 8 years

Year 0

Alternative	Capital Investment	Future Sales	Annual Revenue	(Pr. 14-26)
1	−$300,000	Good	$142,000	
		Average	119,000	
		Poor	50,000	
2	− 85,000	Good	66,000	
		Average	46,000	
		Poor	17,000	

changes in present operations and not include any new automation. The third alternative is to make no changes (do nothing).

The incremental capital investment and incremental annual revenue for the first two alternatives, relative to present operations, are shown at the bottom of page 572. The annual revenue estimates are based on future sales of the components. The sales department estimates the probability of good, average, and poor future sales as 0.30, 0.60, and 0.10, respectively.

Draw a single-stage decision tree for this situation. Then, based on a before-tax analysis (MARR = 20% per year, analysis period equal to five years, and zero market value for all alternatives) and the E(PW) as a decision criterion, determine which alternative is preferred. What would be the expected value of perfect information (EVPI) about future sales in this case? (14.7)

14-27. Refer to Problem 14-26. At the end of the analysis of the single-stage decision tree diagram by the vice president of operations, the plant management team realized that additional information about future sales of the hydraulic components should reduce the uncertainty involved. Therefore, they asked the sales department to survey their customers and improve the information about future sales conditions. The management team's estimates of the *conditional probabilities* of survey outcomes for each potential sales condition are shown above.

Based on this information, develop a two-stage decision tree for this situation. Calculate the revised probabilities of the three future sales conditions occurring. What is the estimated value to the plant of conducting the sales survey (before considering any additional cost involved)? (14.7)

Appendixes

Accounting and Its Relationship to Engineering Economy

A.1 Introduction

Engineering economy studies are made for the purpose of determining whether capital should be invested in a project or whether it should be used differently than it presently is being used. These studies always deal, at least for one of the alternatives being considered, with something that currently is not being done. Such studies provide information upon which investment and managerial decisions about future operations can be based. Thus, the engineer doing an economic analysis might be termed an *alternatives fortune teller*.

After a decision to invest capital in a project has been made and the capital has been invested, those who supply and manage the capital want to know the financial results. Therefore, procedures are established so that financial events relating to the investment can be recorded and summarized and financial productivity determined. At the same time, through the use of proper financial information, controls can be established and used to help guide the venture toward the desired financial goals. Financial accounting and cost accounting are the procedures that provide these necessary services in a business organization. Accounting studies thus are concerned with *past* and *current* financial events. Thus, the accountant might be termed a *financial historian*.

The accountant is somewhat like a data recorder in a scientific experiment. Such a recorder reads the pertinent gauges and meters and records all the essential data during the course of an experiment. From these it is possible to determine the results of the experiment and to prepare a report. Similarly, the accountant records all significant financial events connected with an investment, and from these data he or she can determine what the results have been and can prepare financial reports. By taking cognizance of what is happening during the course of an experiment and making suitable corrections—thereby gaining more information and better results from the experiment—engineers and managers must rely on accounting reports to make corrective decisions in order to improve the current and future financial performance of the business.

Accounting is generally a source of much of the past financial data needed to make estimates of future financial conditions. Accounting is also a prime source of data for *postmortem*, or after-the-fact, analyses that might be made regarding how well an investment project has turned out compared to the results that were predicted in the engineering economy study.

A proper understanding of the origins and meaning of accounting data is needed in order to use or not use those data properly in making projections into the future and in comparing actual versus predicted results.

A.2 Accounting Fundamentals

Accounting is often referred to as the language of business. Engineers should make serious efforts to learn about a firm's accounting practice so that they can better communicate with top management. This section contains an extremely brief and simplified exposition of the elements of financial accounting in recording and summarizing transactions affecting the finances of the enterprise. These fundamentals apply to any entity (such as an individual or a corporation) called here a *firm*.

All accounting is based on the *fundamental accounting equation*, which is

$$\text{Assets} = \text{liabilities} + \text{owners' equity} \qquad \text{(A-1)}$$

where *assets* are those things of monetary value that the firm possesses, *liabilities* are those things of monetary value that the firm owes, and *owners' equity* is the worth of what the firm owes to its stockholders (also referred to as *equities, net worth*, etc.). For example, typical accounts in each term of Equation A-1 are as follows:

Asset Accounts	=	Liability Accounts	+	Owner's Equity Accounts
Cash		Short-term debt		
Receivables		Payables		Capital stock
Inventories		Long-term debt		
Equipment				Retained earnings
Buildings				(income retained
Land				in the firm)

The fundamental accounting equation defines the format of the *balance sheet*, which is one of the two most common accounting statements and which shows the financial position of the firm at any given point in time.

Another important, and rather obvious, accounting relationship is

$$\text{Revenues} - \text{expenses} = \text{profit (or loss)} \qquad \text{(A-2)}$$

This relationship defines the format of the *income statement* (also commonly known as a *profit-and-loss statement*), which summarizes the revenue and expense results of operations *over a period of time*. Equation A-1 can be expanded to take account of profit defined in Equation A-2:

$$\text{Assets} = \text{liabilities} + (\text{beginning owners' equity} + \text{revenue} - \text{expenses}) \qquad \text{(A-3)}$$

Profit is the increase in money value (not to be confused with cash) that results from a firm's operations and is available for distribution to stockholders. It therefore represents the return on owners' invested capital.

A useful analogy is that a balance sheet is like a snapshot of the firm at an instant in time, whereas an income statement is a summarized moving picture of the firm over an interval of time. It is also useful to note that revenue serves to increase owners' interests in a firm, but an expense serves to decrease the owners' equity amount for a firm.

To illustrate the workings of accounts in reflecting the decisions and actions of a firm, suppose that an individual decides to undertake an investment opportunity and the following sequence of events occurs over a period of one year:

1. Organize XYZ firm and invest $3,000 cash as capital.
2. Purchase equipment for a total cost of $2,000 by paying cash.
3. Borrow $1,500 through a note to the bank.
4. Manufacture year's supply of inventory through the following:
   a. Pay $1,200 cash for labor.
   b. Incur $400 accounts payable for material.
   c. Recognize the partial loss in value (depreciation) of the equipment amounting to $500.
5. Sell on credit all goods produced for year, 1,000 units at $3 each. Recognize that the accounting cost of these goods is $2,100, resulting in an increase in equity (through profits) of $900.
6. Collect $2,200 of accounts receivable.
7. Pay $300 of accounts payable and $1,000 of bank note.

A simplified version of the accounting entries recording the same information in a format that reflects the effects on the fundamental accounting equation (with a "+" denoting an increase and a "−" denoting a decrease) is shown in Figure A-1. A summary of results is shown in Figure A-2.

It should be noted that the profit for a period serves to increase the value of the owners' equity in the firm by that amount. Also, it is significant that the net cash flow from operation of $700 (= $2,200 − $1,200 − $300) is not the same as profit. This amount was recognized in transaction 4c, in which capital consumption (depreciation) for equipment of $500 was declared. Depreciation serves to convert part of an asset into an expense, which is then reflected in a firm's profits, as

Transaction

	Account	1	2	3	4	5	6	7	Balances at End of Year
Assets	Cash	+$3,000	−$2,000	+$1,500	−$1,200	+$3,000	+$2,200	−$1,300	+$2,200
	Accounts receivable						− 2,200		+ 800
	Inventory		+ 2,000		+ 2,100 − 500	− 2,100			0
	Equipment			+ 1,500					+ 1,500
equals									$4,500
Liabilities	Accounts payable				+ 400			− 300	+ 100
	Bank note			+ 1,500				− 1,100	+ 500
plus									
Owners' equity	Equity	+ 3,000				+ 900			+ 3,900
									$4,500

FIGURE A-1 Accounting Effects of Transactions: XYZ Firm

XYZ Firm Balance Sheet
As of Dec. 31, 1997

Assets		Liabilities and Owners' Equity	
Cash	$2,200	Bank note	$ 500
Accounts receivable	800	Accounts payable	100
Equipment	1,500	Equity	3,900
Total	$4,500	Total	$4,500

XYZ Firm Income Statement
for Year Ending Dec. 31, 1997

			Cash Flow
Operating revenues (Sales)		$3,000	$2,200
Operating costs (Inventory depleted)			
Labor	$1,200		−1,200
Material	400		− 300
Depreciation	500		0
		$2,100	
Net income (profits)		$ 900	$ 700

FIGURE A-2   Balance Sheet and Income Statement Resulting from Transactions Shown in Figure A-1

seen in Equation A-2. Thus, the profit was $900, or $200 more than the net cash flow. For our purposes, revenue is recognized when it is earned, and expenses are recognized when they are incurred.

One important and potentially misleading indicator of after-the-fact financial performance that can be obtained from Figure A-2 is "annual rate of return." If the invested capital is taken to be the owners' (equity) investment, the annual rate of return at the end of this particular year is $900/$3,900 = 23%.

Financial statements are usually most meaningful if figures are shown for two or more years (or other reporting periods such as quarters or months) or for two or more individuals or firms. Such comparative figures can be used to reflect trends or financial indications that are useful in enabling investors and management to determine the effectiveness of investments *after* they have been made.

A.3 Cost Accounting

Cost accounting, or management accounting, is a phase of accounting that is of particular importance in engineering economic analysis because it is concerned principally with decision making and control in a firm. Consequently, cost accounting is the source of much of the cost data needed in making engineering economy studies. Modern cost accounting may satisfy any or all of the following objectives:

1. Determination of the actual cost of products or services
2. Provision of a rational basis for pricing goods or services
3. Provision of a means for allocating and controlling expenditures
4. Provision of information on which operating decisions may be based and by means of which operating decisions may be evaluated

Although the basic objectives of cost accounting are simple, the exact determination of costs usually is not. As a result, some of the procedures used are arbitrary devices that make it possible to obtain reasonably accurate answers for most cases but that may contain a considerable percentage of error in other cases, particularly with respect to the actual cash flow involved.

A.4 The Elements of Cost

One of the first problems in cost accounting is that of determining the elements of cost that arise in the production of an article or the rendering of a service. A study of how these costs occur gives an indication of the accounting procedure that must be established to give satisfactory cost information. Also, an understanding of the procedure that is used to account for these costs makes it possible to use them more intelligently.

From an engineering and managerial viewpoint in manufacturing enterprises, it is common to consider the general elements of cost to be *direct materials, direct labor,* and *overhead.* Such terms as *burden* and *indirect costs* are often used synonymously with *overhead,* and overhead costs are often divided into several subcategories.

Ordinarily, the materials that can be conveniently and economically charged directly to the cost of the product are called *direct materials.* Several guiding principles are used when we decide whether a material is classified as a direct material. In general, direct materials should be readily measurable, be of the same quantity in identical products, and be used in economically significant amounts. Those materials that do not meet these criteria are classified as *indirect materials* and are a part of the charges for overhead. For example, the exact amount of glue and sandpaper used in making a chair would be difficult to determine. Still more difficult would be the measurement of the exact amount of coal that was used to produce the steam that generated the electricity that was used to heat the glue. Some reasonable line must be drawn beyond which no attempt is made to measure directly the material that is used for each unit of production.

Labor costs also are ordinarily divided into *direct* and *indirect* categories. Direct labor costs are those that can be conveniently and easily charged to the product or service in question. Other labor costs, such as for supervisors, material handlers, and design engineers, are charged as indirect labor and are thus included as part of overhead costs. It is often imperative to know what is included in direct labor and direct material cost data before attempting to use them in engineering economy studies.

In addition to indirect materials and indirect labor, there are numerous other cost items that must be incurred in the production of products or the rendering of services. Property taxes must be paid; accounting and personnel departments must be maintained; buildings and equipment must be purchased and maintained; supervision must be provided. It is essential that these necessary *overhead* costs be attached to each unit produced in proper proportion to the benefits received. Proper allocation of these overhead costs is not easy, and some factual, yet reasonably simple, method of allocation must be used.

As might be expected where solutions attempt to meet conflicting requirements such as exist in overhead-cost allocation, the resulting procedures are empirical approximations that are accurate in some cases and less accurate in others.*

There are many methods of allocating overhead costs among the products or services produced. The most commonly used methods involve allocation in proportion to direct labor cost, direct labor hours, direct materials cost, sum of direct labor and direct materials cost, or machine hours. In these methods, it is necessary to estimate what the total overhead costs will be if standard costs are being determined. Accordingly, *total overhead costs are customarily associated with a certain level of production*, which is an important condition that should always be remembered when dealing with unit-cost data. These costs can be correct only for the conditions for which they were determined.

To illustrate one method of allocation of overhead costs, consider the method that assumes that overhead is incurred in direct proportion to the cost of direct labor used. With this method the overhead rate (overhead per dollar of direct labor) and the overhead cost per unit would be

$$\text{Overhead rate} = \frac{\text{total overhead in dollars for period}}{\text{direct labor in dollars for period}}$$

$$\text{Overhead cost/unit} = \text{overhead rate} \times \text{direct labor cost/unit} \qquad \textbf{(A-4)}$$

Suppose that for a future period (say, a quarter) the total overhead cost is expected to be $100,000 and the total direct labor cost is expected to be $50,000. From this, the overhead rate = $100,000/$50,000 = $2 per dollar of direct labor cost. Suppose further that for a given unit of production (or job) the direct labor cost is expected to be $60. From Equation A-4, the overhead cost for the unit of production would be 60 × $2 = $120.

This method obviously is simple and easy to apply. In many cases it gives quite satisfactory results. However, in many other instances, it gives only very approximate results because some items of overhead, such as depreciation and taxes, have very little relationship to labor costs. Quite different total costs may be obtained for the same product when different procedures are used for the allocation of overhead costs. The magnitude of the difference will depend on the extent to which each method produces or fails to produce results that realistically capture the facts.

_____

*Section A.7 discusses a relatively recent methodology, called *activity-based cost management*, for avoiding badly distorted cost estimates caused by traditional overhead allocations.

▼

A.5 Cost Accounting Example

This relatively simple example involves a job order system in which costs are assigned to work by job number. Schematically, this process is illustrated in the following diagram.

Costs are assigned to jobs in this manner:

1. Raw materials attach to jobs via material requisitions.
2. Direct labor attaches to jobs via direct labor tickets.
3. Overhead cannot be attached to jobs directly but must have an allocation procedure that relates it to one of the resource factors, such as direct labor, which is already accumulated by the job.

Consider how an order for 100 tennis rackets accumulates costs at the Bowling Sporting Good Company:

Job #161	100 tennis rackets
Labor rate	$7 per hour
Leather	50 yards at $2 per yard
Gut	300 yards at $0.50 per yard
Graphite	180 pounds at $3 per pound
Labor hours for the job	200 hours
Total annual factory overhead costs	$600,000
Total annual direct labor hours	200,000 hours

The three major costs are now attached to the job. Direct labor and material expenses are straightforward:

Job #161		
Direct labor	200 × $7 =	$1,400
Direct material	leather: 50 × $2 =	100
	gut: 300 × $0.5 =	150
	graphite: 180 × $3 =	540
Prime costs (direct labor + direct materials)		$2,190

Notice that this cost is not the total cost. We must somehow find a way to attach (allocate) factory costs that cannot be directly identified to the job but are nevertheless involved in producing the 100 rackets. Costs such as the power to run the graphite molding machine, the depreciation on this machine, the depreciation of the factory building, and the supervisor's salary constitute overhead for this company. These overhead costs are part of the cost structure of the 100 rackets but cannot be directly traced to the job. For instance, do we really know how much machine obsolescence is attributable to the 100 rackets? Probably not. Therefore, we must allocate these overhead costs to the 100 rackets using the overhead rate determined as follows:

$$\text{Overhead rate} = \frac{\$600,000}{200,000} = \$3 \text{ per direct labor hour}$$

This means that $600 ($3 × 200) of the total annual overhead cost of $600,000 would be allocated to Job #161. Thus, the total cost of Job #161 would be

Direct labor	$1,400
Direct materials	790
Factory overhead	600
	$2,790

The cost of manufacturing each racket is thus $27.90. If selling expenses and administrative expenses are allocated as 40% of the cost of goods sold, the total expense of a tennis racket becomes 1.4($27.90) = $39.06.

A.6 The Use of Accounting Costs in Engineering Economy Studies

When we recognize that accounting costs are linked to a definite set of conditions and that they are the result of certain arbitrary decisions concerning the allocation of overhead costs, it is apparent that they should not be used without modification in cases where the conditions are different from those for which they were determined. Engineering economy studies invariably deal with situations that now are *not* being done. Thus, ordinary accounting costs normally cannot be used without modification in these economy studies. However, if we understand how the accounting costs were determined, we should be able to break them down into their component elements, and then we often find that these cost elements will supply much of the cost information that is needed for an engineering economy study. Thus, an understanding of the basic objectives and procedures of cost accounting will enable the engineer doing an economic analysis to make best use of available cost information and to avoid needless work and serious mistakes.

It should not be assumed that the figures contained in accounting reports are absolutely correct and indicative, even though they have been prepared with the utmost care by highly professional accountants, because accounting procedures often must include certain assumptions that are based on subjective judgment or the current tax laws. For example, the years of life on which depreciation expense for a particular asset is based has to be determined or assumed and the estimate may turn out to have caused unrealistic depreciation expenses and book values in accounting reports. Also, there are many accepted practices in accounting that may provide unrealistic information for management control purposes. For example, the net book value of an asset is generally declared in the balance sheet at the original price (cost basis) minus any accumulated depreciation, even though it may be recognized that the true value of the asset at a particular time is far above or below this reported book value.

A.7   Bringing Cost Management Up to Date*

Traditional accounting systems allocate overhead costs using a volume-oriented base, such as direct labor hours or direct material dollars. The cost allocation bases of direct labor and/or production quantity were designed primarily for inventory valuation. As a consequence, traditional cost accounting methods are fully effective only when direct labor (and/or direct materials) is the dominant cause of cost.

Although traditional standard cost systems were effective in the past, changes in manufacturing technologies, such as the just-in-time philosophy, robotics, CAD, and flexible manufacturing systems, have made traditional cost models somewhat obsolete. Rapid technological advancement has resulted in the restructuring of manufacturing cost patterns (e.g., the direct labor and inventory components of total cost are decreasing, while those of technology depreciation, engineering, and data processing are increasing). Due to the changing nature of these cost components, existing cost accounting systems and cost management practices do not adequately support the objectives of advanced manufacturing. In fact, the direct labor component of product cost now accounts for as little as 5% of product cost, and overhead accounts for as much as 500%. In automated environments, volume-based allocations distort product costs because the measures used to allocate overhead do not cause the costs. As a result, product cost distortion occurs due to high overhead rates that are inflated by many costs that should be directly traceable to the product rather than arbitrarily allocated on the basis of volume. Several components of a product's cost that should be traced to the product include hidden overhead costs such as material movement, order processing, process planning, rework, warranty maintenance, production planning and control, and quality assurance.

Assume that a company uses a traditional cost accounting system that applies overhead based on direct labor dollars (Figure A-3). The product cost is computed as $550 with a sales price of $660, resulting in a reported net profit of $110 per unit.

The manufacture of the product involves a significant number of automated processes, however. Automated production requires a heavy investment in depreciation, software, and maintenance support. An analysis of these costs and others traced to the product results in a completely different financial outcome (Figure A-4). The new product cost after the analysis was $925. With a sales price of $660, the company was losing $265 on each unit produced.

---

*Excerpted from J. A. Brimson, "Bringing Cost Management Up to Date," *Manufacturing Engineering,* vol. 102, no. 12, June 1988, pp. 49–51. Reproduced with permission of the Society of Manufacturing Engineers, Dearborn, MI.

Traditional Cost Accounting		
Sales price		$660
Direct labor	$ 50	
Direct material	300	
Overhead	200	
Total production cost		550
Net profit		$110

FIGURE A-3   A Traditional Cost Accounting System Applies Overhead Based on Direct Labor Dollars

New Cost Accounting		
Sales price		$660
Cost		
Directly traceable		
Direct labor	$ 50	
Direct material	300	
Technology	200	
Scrap and Rework	50	
Imputed cost		
Raw materials inventory	20	
WIP inventory	60	
Other direct cost	90	
		770
Nontraceable overhead		155
Total cost		925
Net loss		($265)

FIGURE A-4   New Cost Management Techniques, Which Consider the Ramifications of Automation, Produce Different Financial Results

A primary reason for this distortion is that the overhead rate is inflated by potentially traceable direct costs. The inflated overhead cost is then allocated to the products using a direct labor base. For this allocation method to be correct, there must be a complementary relationship between labor and technology. In other words, reported product cost is often based on a choice of accounting methods that do not mirror the manufacturing process.

A.7.1   Activity-Based Costing Systems

Activity-based cost management systems track hidden overhead costs to the specific activities that cause them and thus provide a more reliable product cost.

Four key concepts differentiate activity-based costing from traditional costing systems, allowing activity-based systems to provide more accurate product cost data:

I. *Activity accounting.* In an activity-based system, the cost of the product is the sum of all costs required to manufacture and deliver the product. The activities a firm pursues consume its resources, and resource availability and usage create costs.

Activity accounting decomposes an organization into an activity structure that provides a cause-and-effect rationale for how fundamental objectives and their associated activities create costs and result in outputs. According to Brimson, an effective activity accounting system uses the following approach:*

  A. Determine the fundamental activities that must be accomplished to satisfy a firm's objectives. Activities permit identification of how a company deploys its resources to achieve its basic aims.

  B. Determine the causal relationships that permit outputs (performance) to be attributed to inputs (resources). A large number of these relationships will be based on non-volume-related measures such as the number of parts in a new design.

  C. Ascertain the output of an activity in terms of a measure of *activity volume* through which costs of a business process vary most directly (e.g., number of machine setups required for a complex design).

  D. Trace activities to products (or other objects) and determine how much of each activity's output is dedicated to them. A cost structure, known as a *bill of activities*, is used to describe each product's pattern of activity consumption.

  E. Determine critical success factors by which activities of the enterprise can be aligned with stated strategic objectives. This step indicates how effectively desired performance is being accomplished through activities undertaken by a firm.

  F. Take action, using a continuous improvement philosophy, on the productivity opportunities identified in Steps A–E. Because activity cost is the ratio of resources consumed by an activity to the measured output of the activity, a means for evaluating effectiveness and efficiency (i.e., productivity) is available to managers. Various alternatives for making desired changes in activity patterns, through investment or organizational means, can now be realistically appraised.

II. *Cost drivers.* A cost driver is an event that affects the cost/performance of a group of related activities. Familiar cost drivers include number of machine setups, number of engineering change notices, and number of purchase orders. Cost drivers reflect the demands placed on activities at both the activity and product levels. By controlling the cost driver, unnecessary costs can be eliminated, resulting in improved product cost.

---

*J. A. Brimson, *Activity Accounting: An Activity-Based Costing Approach* (New York: John Wiley & Sons, 1991).

III. *Direct traceability.* Direct traceability involves attributing costs to those products or processes that consume resources. Many hidden overhead costs can be effectively traced to products, thus providing a more accurate product cost.

IV. *Non-value-added costs.* In manufacturing processes, customers may perceive that certain activities add no value to the product. Through identification of cost drivers, a firm can pinpoint these unnecessary costs. Activity-based cost systems identify and place a cost on the activities performed (value-adding and non-value-adding) so that management can determine desired changes in resource requirements for each activity. In contrast, traditional cost systems accumulate costs by budgetary line items and by functions.

These four basic concepts are embodied in activity-based costing systems and lead to more accurate costing information. In addition, activity-based costing systems provide more flexibility than conventional costing systems because they produce a variety of cost figures useful for technology accounting, product costing, and life cycle analysis. In addition, these cost figures can be applied in making various special decisions, including inventory valuation, budgeting/forecasting, product line analysis, make/buy decisions, and design to cost.

A.7.2   Examples of Activity-Based Costing

The objective of this section is to show with two examples how activity-based costing (ABC) can be used to capture more accurately the estimated cost of manufacturing a product.

In general terms, an ABC system identifies and then classifies the major activities of a facility's production process into four main "base" groups: *unit-level, batch-level, product-level,* and *facility-level* activities. The ABC approach assumes that not all overhead resources are consumed in proportion to the number of units produced. ABC introduces these hierarchical levels to ensure that the final estimate of product cost mirrors the manufacturing process as closely as possible.

Unit-level costs are those costs that can be directly apportioned to volume. These costs may include such activities as direct labor hour costs, directly attributable material costs, and costs per machine operating hour.

Batch-level costs are those costs that can be directly apportioned to the particular batch run. For this type of cost, certain activities are consumed in direct proportion to the number of batch runs for each product. These batch-level costs may include setup, ordering, material handling, and transportation costs.

Product-level costs are those costs that can be directly apportioned to the product, which assumes that certain activities are consumed to develop or permit production of different products. These product-level costs may include such activities as research and development (R&D), parts and material acquisition and inventory costs, technical administration, and specialized preproduction safety and manufacturing training.

Facility-level costs cause problems in an ABC environment because these costs are associated with the sustainment of a general manufacturing process. These facility-level costs may include such activities as travel costs, directors' fees, and general administration and can include a large segment of the estimated product cost.

EXAMPLE A-1

To illustrate the key concepts listed above, we consider the XYZ Company, which makes pagers for the telecommunications industry. Currently, the company is making two models of pagers with equal annual production volume of 50,000 each. Ever since its inception, the company has been using a volume-based absorption (traditional) costing system for product costing. Thus, the company uses direct labor hours as the allocation base for all its manufacturing overhead costs. The total manufacturing overhead costs of the company as well as the total direct material and direct labor costs of both products for the current year are shown in Figure A-5.

Recently, the company installed an activity-based costing system and subsequently identified six main activities to be responsible for most of the manufacturing overhead costs. The overhead costs were distributed to these six activities through activity accounting. Also, six corresponding cost drivers and their budget consequences for the current year were established. The activities, their cost drivers, resultant overhead costs, and budgeted rates are provided in Figure A-6.

Category	
	Model A
Direct material cost:	$ 2,800,600
Direct labor cost:	$ 350,000
Direct labor hours:	35,000
	Model B
Direct material cost:	$ 1,500,000
Direct labor cost:	$ 250,000
Direct labor hours:	25,000
Total manufacturing overhead costs:	$12,000,000

FIGURE A-5   Manufacturing Costs for Model A and Model B (Current Year)

Activity	Costs	Cost Driver	Budgeted Rate
Production:	$8,000,000	# of machine hours	200,000 hours
Engineering:	$1,000,000	# of engineering change orders (ECO)	40,000 orders
Material handling	$1,000,000	# of material moves	60,000 moves
Receiving:	$ 800,000	# of batches	500 batches
Quality assurance:	$ 800,000	# of inspections	20,000 inspections
Packing & shipping:	$ 400,000	# of products	100,000 products

FIGURE A-6   Activity-Based Costing Data

Activity	Model A	Model B
Production:	50,000 hours	150,000 hours
Engineering:	15,000 orders	25,000 orders
Material handling	20,000 moves	40,000 moves
Receiving:	150 batches	350 batches
Quality assurance:	6,000 inspections	14,000 inspections
Pack & ship:	50,000 products	50,000 products

FIGURE A-7   Cost Driver Rates for Models A and B

Besides establishing the data in Figure A-6, the activity-based costing system also measured the levels of each activity (or cost driver rate) required by the two products. These data are given in Figure A-7.

For post-audit purposes, XYZ Company ran a cost comparison between the two costing systems. To arrive at product costs for Model A and Model B, the following two sections show the computations involved in the traditional (volume-based absorption) costing system and the activity-based costing system, respectively.

Cost Computations Using the Traditional Costing Method

1. Obtain the overall manufacturing overhead cost rate:

   Cost Rate = (Total Overhead Costs)/(Total Direct Labor Hours)

   = $12,000,000/60,000 hours

   = $200 per direct labor hour

2. Compute total cost per unit for each model as shown below:

	Model A	Model B
Direct material cost:	$2,800,600	$1,500,000
Direct labor cost:	$350,000	$250,000
Allocated overhead:	(35,000 hours × $200) = $7,000,000	(25,000 hours × $200) = $5,000,000
Total costs:	$10,150,600	$6,750,000
Cost per unit (based on 50,000 units):	$203	$135

Cost Computations Using Activity-Based Costing

1. Establish an applied activity rate of the six activities:

   Applied activity rate = (Activity Cost in Figure A-6)

   ÷ (Budgeted Cost Driver Rate in Figure A-6)

2. Determine the activity cost to be traced to each pager model:

Cost traced to each pager model = Cost Driver Rate (from Figure A-7)

$\times$ Applied Activity Rate (from step #1.)

3. Compute total cost per unit for each model as shown below:

	Applied Activity Rate	Model A	Model B
Production:	$40/mach. hr*	$2,000,000†	$6,000,000
Engineering:	$25/ECO	$375,000	$625,000
Material handling:	$16.67/move	$333,400	$666,600
Receiving:	$1,600/batch	$240,000	$560,000
Quality assurance:	$40/inspection	$240,000	$560,000
Pack & ship:	$4/product	$200,000	$200,000
Total overhead		$3,388,400	$8,611,600
Costs:			$12,000,000
Direct material cost:		$2,800,600	$1,500,000
Direct labor cost:		$350,000	$250,000
Total costs:		$6,539,000	$10,361,600
Cost per unit:		$130.78	$207.23

*$40/machine-hour = $8,000,000/200,000 machine-hours.
† $2,000,000 = $40/machine-hour $\times$ 50,000 hours.

In summary, the two costing systems produced different cost estimates, as summarized in Figure A-8. Using volume-based absorption (traditional) costing, Model A costs more to make than Model B. However, based on activity-based costing, Model A is *less expensive* to make than Model B. Traditional costing bases its cost estimates on the assumption that Model A is responsible for more overhead costs than Model B, using direct labor hours as an allocation base. However, because of differences in the number of transactions that affect indirect (overhead) costs, the reality is that Model B actually incurs more overhead costs than Model A, which is shown by the analysis carried out with activity-based costing. Consequently, this example demonstrates how a traditional costing method generates distorted product costs that do not accurately reflect the actual amount of overhead costs incurred by the product. Companies making decisions based on distorted costs may unknowingly price some of their products out of

	Model A	Model B
Traditional costing	$203.00	$135.00
Activity-based costing	$130.78	$207.24

FIGURE A-8　　Product Cost Comparison (Traditional vs. ABC)

the market while selling others at a loss. Or they may make key design decisions, based on competitor target costs, that do not reflect the true costs of manufacturing their new or revised products (see Chapter 8 for target cost discussion).

EXAMPLE A-2

This second example is constructed such that the attributes of ABC are clearly exhibited. Perhaps more important than representing the "advantages" of ABC, it is an illustration of the differences between ABC and traditional volume-based absorption costing (VBC).

The attributes that make ABC most productive and valuable are essentially diversity of products and high support costs and overheads. Certainly these characteristics are common to most markets today, and the trend is increasingly moving in that direction. Consider this simplified yet realistic example of ABC versus VBC:

Exhibit A-1. This exhibit shows the basic business scenario. It gives the detailed production budgets, volumes, and costs. The bottom half of the page shows the

	Production: Scheduled Data for Year				
	Baseball	**Glove**	**Bat**	**Pitching Machine**	**Totals**
Budget units of production	20,000	10,000	5,000	200	35,200
Material cost per unit	$0.45	$5.00	$0.75	$2,000.00	n/a
Direct labor hours per unit	0.05	2.00	0.1	50.0	n/a
Direct labor cost per hour	$5.00	$5.00	$5.00	$5.00	$5.00
Machine hours per unit	0.1	0.1	0.2	100.0	24,000
Parts required per unit	3	4	1	250	n/a
Production orders (total budgeted)	500	25	100	100	725
Production set-ups (total budgeted)	1,000	100	100	100	1,300
Number of shipments	400	250	25	100	775

			Activity Conversion Costs		
				Cost–Driver	
	Cost Driver		**Costs**	**Total Units**	**Rate**
Material handling	# of moves		$50,000	155,000	$0.32
Production planning dept.	# of prod. orders		40,000	725	$55.17
Set-up indirect labor	# of set-ups		25,000	1,300	$19.23
Depreciation on machinery	Machine hours		725,000	24,000	$30.21
Quality & finishing	Dir. labor hours		150,000	31,500	$4.76
Shipping department	# of shipments		100,000	775	$129.03
	Total indirect costs		**$1,090,000**		
	Indirect costs per direct labor hour		**$34.60** *{Traditional Volume-Based Costing}*		

EXHIBIT A-1   Business Scenario for a Small Firm

ABC "cost driver" calculations (there are six) and the VBC traditional overhead allocation of $34.60 per direct labor hour.

Note that the model shows a capital intensive business. Depreciation represents about 67% of total indirect costs.

The model has only four products. More diversity would better highlight the ABC advantages but would add unwanted complexity to the model. The products are very different in terms of production and market. The pitching machine requires much machine time, whereas the other products use hardly any. Direct labor is an insignificant factor for the ball and bat but very significant for the glove (the glove requires 40 times the labor of the ball and 20 times the labor of the bat).

Exhibit A-2. Traditional Volume-Based Costing—All indirect costs are allocated based on the overall rate of $34.60 per direct labor hour.

The glove with relatively high direct labor content appears to be a loser. It sells for $57.00 but is assigned costs totaling $84.21—*a loss of $27.21 per unit!* Note that all other products seem to have very healthy margins. *Management would certainly have to consider dropping the glove from production and marketing.*

Exhibit A-3. Activity-Based Costing—Now we see completely different results in terms of margins and profitability. The glove is now the only profitable product with the others being losers!

| | Traditional Volume-Based Product Costs | | | | |
	Baseball	Glove	Bat	Pitching Machine	Totals
Units produced	20,000	10,000	5,000	200	35,200
Direct material costs	$9,000	$50,000	$3,750	$400,000	$462,750
Direct labor costs	5,000	100,000	2,500	50,000	157,500
Overhead allocations (based on D.L. @$34.60/hr.)	34,603	692,063	17,302	346,032	1,090,000
Total product costs	**$48,603**	**$842,063**	**$23,552**	**$796,032**	**$1,710,250**
Per unit costs					
Direct costs	$0.70	$15.00	$1.25	$2,250.00	
Overhead	1.73	69.21	3.46	1,730.16	
Total cost per unit	*$2.43*	*$84.21*	*$4.71*	*$3,980.16*	
Selling price	$4.45	$57.00	$10.00	$5,000.00	
Volume-based margin	$2.02	$(27.21)	$5.29	$1,019.84	
	45%	−48% LOSS	53%	20%	

EXHIBIT A-2   Traditional Volume-Based Costing for Four Products

		Baseball	Glove	Bat	Pitching Machine	Totals
				Activity-Based Product Costs		
Units produced		20,000	10,000	5,000	200	35,200
Direct material costs		$9,000	$50,000	$3,750	$400,000	$462,750
Direct labor costs		5,000	100,000	2,500	50,000	157,500
Overhead costs:	*ABC rate*					
Material handling	$0.32	19,355	12,903	1,613	16,129	50,000
Production planning dept.	$55.17	27,586	1,379	5,517	5,517	40,000
Set-up indirect labor	$19.23	19,231	1,923	1,923	1,923	25,000
Depreciation on machinery	$30.21	60,417	30,208	30,208	604,167	725,000
Quality & finishing	$4.76	4,762	95,238	2,381	47,619	150,000
Shipping department	$129.03	51,613	32,258	3,226	12,903	100,000
Total product costs		$196,964	$323,910	$51,118	$1,138,258	$1,710,250
Per unit costs:						
Direct costs		$0.70	$15.00	$1.25	$2,250.00	
Overhead		9.15	17.39	8.97	3,441.29	
Total cost per unit		$9.85	$32.39	$10.22	$5,691.29	
Selling price		$4.45	$57.00	$10.00	$5,000.00	
Activity-based margin		$(5.40)	$24.61	$(0.22)	$(691.29)	
		−121% LOSS	43%	−2% LOSS	−14% LOSS	

EXHIBIT A-3   Activity-Based Costing for Four Products

The four products in Exhibits A-1 through A-3 were designed to show differences between ABC and VBC.

Out of total costs of $1,710,250 direct labor represents only $157,500, less than ten percent. In this case, direct labor used in production varied significantly in proportion to total product costs. The glove had the highest proportional costs. Not surprisingly, using VBC allocation methods on labor loaded up the indirect costs on the glove, which looked like a terrible product. However, production statistics in Exhibit A-1 show that the glove hardly uses any indirect resources, less than all the others. Setting up a glove manufacturing shop outside the company would require little indirect support and would probably be a profitable business.

Conversely, the pitching machine uses about 83% of the machinery's capacity, and the machinery is costly. Using direct labor to allocate indirect costs results in allocation of only $1,730 of indirect costs to the glove—hardly near the 83% of costs on just machinery depreciation alone! Allocation based on machine hours of usage doubles the indirect cost assignment to the pitching machine, which, again, is logical because most of the indirect costs are machinery and the pitching machine seems to use most of the machinery.

When the products are costed on a resource usage basis (ABC), only the glove seems to be profitable, which is a much more accurate reflection of the real costs of doing business.

This information is excellent feedback for both production and marketing. The pitching machine is only slightly unprofitable. Perhaps it should command a better price. Is it unique and a better product than the competition? Is it perhaps priced too low, based on the prior cost information (VBC) that indicated a 20% margin? Can it be redesigned to save some costs?

The bat is only a 2% margin loser. Redesign of production should be considered. A slight change in any production factor will make the bat profitable.

The baseball uses 1,000 set-ups. Set-ups should be re-examined. Perhaps we will end up with more set-ups but with reduced costs. Serious redesign is needed here; otherwise this product just may not be good for the business.

ABC offers product profitability insights in areas that might otherwise cloud product costing. For strategic purposes the following two factors need to be considered:

1. Many of the costs are fixed. Elimination of a product line may not reduce costs in total. In the short term it may be best to keep some products even if they show fully costed losses. For long-term investment decisions the full (ABC) costs are more relevant.

2. As indicated with the discussion of set-ups, ABC costs may give improper direction if followed "blindly." High set-up costs assigned by a cost driver just means you have high set-up costs. The way to reduce the set-up costs may not be to reduce the number of set-ups, but it may be to redesign the set-up process. (Perhaps you end up increasing the number of set-ups?)

ABC helps in identifying potential areas to pursue; it does not necessarily give the cost reduction answers; it just gives the direction to take in pursuing these difficult questions.

The graphs presented in Figure A-9 show the differing effects that ABC and VBC have on unit costs as overhead levels are varied. Overhead costs were varied by using the following levels: $0, $100,000, $500,000, $725,000, $1,000,000, and $2,000,000. All other factors remained exactly the same.

Note that at low overheads (around $100,000) the pitching machine and glove are indifferent to costing method.

The most significant point of Figure A-9 is that as overhead increases, the cost methodology becomes increasingly important. Absolute differences in unit product costs increase with higher overhead. Importantly, the rate of increase in unit cost differences (differences in slopes) increases in all cases. The rate of increase in slope differences was least for the baseball. Proportionately, VBC costs (based

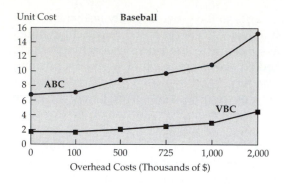

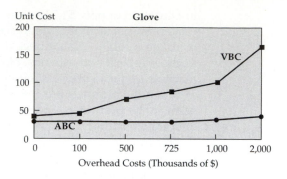

Note: Overhead Cost Axis Not to Scale

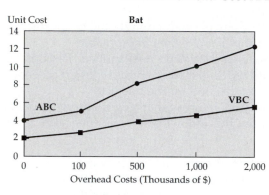

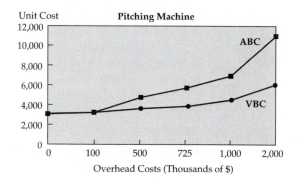

Notes: —All Products except the Glove show higher overhead costs for ABC than VBC.

—The glove has relatively constant assignment of overhead costs under ABC.

—The Pitching Machine actually has a transition where at zero overhead ABC has slightly lower unit cost, then at $100,000 of overhead ABC is slightly higher. Differences increase at increasing rate thereafter.

—The Glove, the Bat (and the Pitching Machine) all have increasing differences in slopes of ABC and VBC cost. The Baseball slope differences increase at a much lower rate.

FIGURE A-9   Unit Cost Charts

on direct labor) for the baseball was closest to ABC cost assignments as the levels of overhead increased.

Also, note that as overhead goes from $0 to $2,000,000 the overhead assignment from ABC remains roughly constant for the glove. Under VBC the overhead assignment increases significantly.

B

Abbreviations and Notation*

CHAPTER 2

C_F	total fixed cost
C_V	total variable cost
c_v	variable cost per unit
C_T	total cost
C_U	average unit cost
D	demand for a product or service in units that maximizes profit
D^*	optimal demand or production volume
D'	breakeven point
$\hat{D}$	demand or production volume that will produce maximum revenue

CHAPTER 3

A	equal and uniform end-of-period cash flows (or equivalent end-of-period values)
APR	annual percentage rate (nominal interest)
A_1	end of period 1 cash flow in a geometric sequence of cash flows
$\overline{A}$	an amount of money flowing uniformly and continuously over a specified period of time
BOY	beginning of year
EOY	end of year
F	a future equivalent sum of money
$\bar{f}$	a geometric change from one time period to the next in cash flows or equivalent values

---

*Listed by the chapter in which they first appear.

G	an arithmetic (i.e., uniform) change from one period to the next in cash flows or equivalent values
I	total interest earned or paid (simple interest)
i	effective interest rate per interest period
i_{CR}	an interest rate called the *convenience rate*
k	an index for time periods
P	principal amount of a loan; a present equivalent sum of money
M	number of compounding periods per year
N	number of interest periods
r	nominal interest rate per period (usually a year)
$\underline{r}$	a nominal interest rate that is continuously compounded

CHAPTER 4

AW($i\%$)	equivalent uniform annual worth, computed at $i\%$ interest, of one or more cash flows
B/C	benefit/cost ratio
$\underline{B}$	equivalent annual worth of benefits
CR($i\%$)	equivalent annual cost of capital recovery, computed at $i\%$ interest
$\underline{E}$	equivalent annual expenses
ERR	external rate of return
FW($i\%$)	future equivalent worth, calculated at $i\%$ interest, of one or more cash flows
I	initial investment for a project
IRR	internal rate of return, also designated $i'\%$
MARR	minimum attractive rate of return
N	the length of the study period (usually years)
O&M	equivalent annual operating and maintenance expenses
PW($i\%$)	present equivalent worth, computed at $i\%$ interest, of one or more cash flows
$\underline{R}$	equivalent annual revenues (or savings)
S	salvage (residual) value at the end of the study period
Θ	payback period
Θ'	discounted payback period
V_N	value (price) of a bond N periods prior to redemption
Z	face value of a bond

CHAPTER 5

A $\rightarrow$ B	increment (or, incremental net cash flow) between Alternative A and Alternative B (read: A to B)
Δ(B $-$ A)	incremental net cash flow (difference) calculated from the cash flow of Alternative B minus the cash flow of Alternative A (read: delta B minus A)

CHAPTER 6

$\underline{B}$	equivalent uniform annual benefits of a proposed project
$\overline{B}/C$	benefit/cost ratio
CR	capital recovery amount (a cost)
I	initial investment
O&M	equivalent uniform annual operating and maintenance expenses

CHAPTER 7

ACRS	accelerated cost recovery system
ADS	alternative depreciation system
B	cost basis
BV	book value of an asset
d	depreciation deduction
d^*	cumulative depreciation over a specified period of time
GDS	general depreciation system
MACRS	modified accelerated cost recovery system
MV	market value of an asset; the price that a buyer will pay for a particular type of property
N	useful life of an asset (ADR life)
p	property classification for MACRS depreciation
R	ratio of depreciation in a particular year to book value at the beginning of the same year
r_k	ACRS or MACRS depreciation rate (a decimal)
ATCF	after-tax cash flow
BTCF	before-tax cash flow
Int	interest expense
E	annual expense
NIAT	net income after taxes
NIBT	net income before taxes
R	gross annual revenue
SV_N	salvage value of an asset at the end of useful life
T	income taxes
t	effective income tax rate

CHAPTER 8

$\bar{I}_n$	an unweighted or a weighted index number dependent on the calculation
K	the number of input resource units needed to produce the first output unit
u	the output unit number
Z_u	the number of input resource units needed to produce output unit number u

CHAPTER 9

A$	actual (current) dollars
b	base time period index
e_j	total price escalation (or deescalation) rate for good or service j
e_j'	differential price inflation (or deflation) rate for good or service j
f_e	annual devaluation rate (rate of annual change in the exchange rate) between the currency of a foreign country and the U.S. dollar
f	general inflation rate
i_c	combined (nominal) interest rate; also called the *market interest rate*
i_r	real interest rate
i_{us}	rate of return in terms of a combined (market) interest rate relative to U.S. dollars
i_{fc}	rate of return in terms of a combined (market) interest rate relative to the currency of a foreign country
R$	real (constant) dollars

CHAPTER 11

EUAC	equivalent uniform annual cost
TC_k	total (marginal) cost for year k

CHAPTER 12

B^*	present worth of an investment opportunity in a specified budgeting period
c	cash outlay for expenses in capital allocation problems
Div	cash dividends (after taxes)
g	annual growth rate for the value of common stock and other equity interests
e_a	annual rate of return to owners of a firm (stockholders)
L	before-tax lease expense
l	after-tax lease expense
m	number of mutually exclusive projects being considered
X	binary decision variable (= 0 or 1) in capital allocation problems

CHAPTER 13

C	recurring annual expenses
CC	carrying charges
D_B	book depreciation
D_T	depreciation taken for income tax purposes
e_a	return on equity (no inflation)
e_a'	return on equity (inflation-adjusted)

i_b	cost of borrowed capital (no inflation)
i_b'	cost of borrowed capital (inflation-adjusted)
K_a	weighted after-tax cost of capital (no inflation)
K_a'	weighted after-tax cost of capital (inflation-adjusted)
λ	fraction of total capitalization represented by borrowed money, also called the debt ratio
RR	annual revenue requirement
$\overline{RR}$	levelized revenue requirement
T	income taxes paid
t	effective income tax rate
UI	unrecovered investment

CHAPTER 14

$E(X)$	mean of a random variable
EVPI	expected value of perfect information
$f(x)$	probability density function of a continuous random variable
$F(x)$	cumulative distribution function of a continuous random variable
$p(x)$	probability mass function of a discrete random variable
$p(x_i)$	probability that a discrete random variable takes on the value x_i
$P(x)$	cumulative distribution function of a discrete random variable
$\Pr\{\cdots\}$	probability of the described event occurring
SD(X)	standard deviation of a random variable
$V(X)$	variance of a random variable

Interest and Annuity Tables for Discrete Compounding

(For various values of i from $\frac{1}{4}$% to 25%)

i = effective interest rate per period (usually one year)
N = number of compounding periods

$$(F/P, i\%, N) = (1 + i)^N \qquad\qquad (A/F, i\%, N) = \frac{i}{(1 + i)^N - 1}$$

$$(P/F, i\%, N) = \frac{1}{(1 + i)^N} \qquad\qquad (A/P, i\%, N) = \frac{i(1 + i)^N}{(1 + i)^N - 1}$$

$$(F/A, i\%, N) = \frac{(1 + i)^N - 1}{i} \qquad\qquad (P/G, i\%, N) = \frac{1}{i}\left[\frac{(1 + i)^N - 1}{i(1 + i)^N} - \frac{N}{(1 + i)^N}\right]$$

$$(P/A, i\%, N) = \frac{(1 + i)^N - 1}{i(1 + i)^N} \qquad\qquad (A/G, i\%, N) = \frac{1}{i} - \frac{N}{(1 + i)^N - 1}$$

TABLE C-1 Discrete Compounding; $i = \frac{1}{4}\%$

| | Single Payment | | Uniform Series | | | | Uniform Gradient | | |
| | Compound Amount Factor | Present Worth Factor | Compound Amount Factor | Present Worth Factor | Sinking Fund Factor | Capital Recovery Factor | Gradient Present Worth Factor | Gradient Uniform Series Factor | |
N	To Find F Given P F/P	To Find P Given F P/F	To Find F Given A F/A	To Find P Given A P/A	To Find A Given F A/F	To Find A Given P A/P	To Find P Given G P/G	To Find A Given G A/G	N
1	1.0025	0.9975	1.0000	0.9975	1.0000	1.0025	0.000	0.0000	1
2	1.0050	0.9950	2.0025	1.9925	0.4994	0.5019	0.995	0.4994	2
3	1.0075	0.9925	3.0075	2.9851	0.3325	0.3350	2.980	0.9983	3
4	1.0100	0.9901	4.0150	3.9751	0.2491	0.2516	5.950	1.4969	4
5	1.0126	0.9876	5.0251	4.9627	0.1990	0.2015	9.901	1.9950	5
6	1.0151	0.9851	6.0376	5.9478	0.1656	0.1681	14.826	2.4927	6
7	1.0176	0.9827	7.0527	6.9305	0.1418	0.1443	20.722	2.9900	7
8	1.0202	0.9802	8.0704	7.9107	0.1239	0.1264	27.584	3.4869	8
9	1.0227	0.9778	9.0905	8.8885	0.1100	0.1125	35.406	3.9834	9
10	1.0253	0.9753	10.1133	9.8639	0.0989	0.1014	44.184	4.4794	10
11	1.0278	0.9729	11.1385	10.8368	0.0898	0.0923	53.913	4.9750	11
12	1.0304	0.9705	12.1664	11.8073	0.0822	0.0847	64.589	5.4702	12
13	1.0330	0.9681	13.1968	12.7753	0.0758	0.0783	76.205	5.9650	13
14	1.0356	0.9656	14.2298	13.7410	0.0703	0.0728	88.759	6.4594	14
15	1.0382	0.9632	15.2654	14.7042	0.0655	0.0680	102.244	6.9534	15
16	1.0408	0.9608	16.3035	15.6650	0.0613	0.0638	116.657	7.4469	16
17	1.0434	0.9584	17.3443	16.6235	0.0577	0.0602	131.992	7.9401	17
18	1.0460	0.9561	18.3876	17.5795	0.0544	0.0569	148.245	8.4328	18
19	1.0486	0.9537	19.4336	18.5332	0.0515	0.0540	165.411	8.9251	19
20	1.0512	0.9513	20.4822	19.4845	0.0488	0.0513	183.485	9.4170	20
21	1.0538	0.9489	21.5334	20.4334	0.0464	0.0489	202.463	9.9085	21
22	1.0565	0.9466	22.5872	21.3800	0.0443	0.0468	222.341	10.3995	22
23	1.0591	0.9442	23.6437	22.3241	0.0423	0.0448	243.113	10.8901	23
24	1.0618	0.9418	24.7028	23.2660	0.0405	0.0430	264.775	11.3804	24
25	1.0644	0.9395	25.7646	24.2055	0.0388	0.0413	287.323	11.8702	25
30	1.0778	0.9278	31.1133	28.8679	0.0321	0.0346	413.185	14.3130	30
36	1.0941	0.9140	37.6206	34.3865	0.0266	0.0291	592.499	17.2306	36
40	1.1050	0.9050	42.0132	38.0199	0.0238	0.0263	728.740	19.1673	40
48	1.1273	0.8871	50.9312	45.1787	0.0196	0.0221	1040.055	23.0209	48
60	1.1616	0.8609	64.6467	55.6524	0.0155	0.0180	1600.085	28.7514	60
72	1.1969	0.8355	78.7794	65.8169	0.0127	0.0152	2265.557	34.4221	72
84	1.2334	0.8108	93.3419	75.6813	0.0107	0.0132	3029.759	40.0331	84
100	1.2836	0.7790	113.4500	88.3825	0.0088	0.0113	4191.242	47.4216	100
∞				400.0000		0.0025			∞

TABLE C-2 Discrete Compounding; $i = \frac{1}{2}\%$

	Single Payment		Uniform Series				Uniform Gradient		
	Compound Amount Factor	Present Worth Factor	Compound Amount Factor	Present Worth Factor	Sinking Fund Factor	Capital Recovery Factor	Gradient Present Worth Factor	Gradient Uniform Series Factor	
	To Find F Given P	To Find P Given F	To Find F Given A	To Find P Given A	To Find A Given F	To Find A Given P	To Find P Given G	To Find A Given G	
N	F/P	P/F	F/A	P/A	A/F	A/P	P/G	A/G	N
1	1.0050	0.9950	1.0000	0.9950	1.0000	1.0050	0.000	0.0000	1
2	1.0100	0.9901	2.0050	1.9851	0.4988	0.5038	0.990	0.4988	2
3	1.0151	0.9851	3.0150	2.9702	0.3317	0.3367	2.960	0.9967	3
4	1.0202	0.9802	4.0301	3.9505	0.2481	0.2531	5.901	1.4938	4
5	1.0253	0.9754	5.0503	4.9259	0.1980	0.2030	9.803	1.9900	5
6	1.0304	0.9705	6.0755	5.8964	0.1646	0.1696	14.655	2.4855	6
7	1.0355	0.9657	7.1059	6.8621	0.1407	0.1457	20.449	2.9801	7
8	1.0407	0.9609	8.1414	7.8230	0.1228	0.1278	27.176	3.4738	8
9	1.0459	0.9561	9.1821	8.7791	0.1089	0.1139	34.824	3.9668	9
10	1.0511	0.9513	10.2280	9.7304	0.0978	0.1028	43.387	4.4589	10
11	1.0564	0.9466	11.2792	10.6770	0.0887	0.0937	52.853	4.9501	11
12	1.0617	0.9419	12.3356	11.6189	0.0811	0.0861	63.214	5.4406	12
13	1.0670	0.9372	13.3972	12.5562	0.0746	0.0796	74.460	5.9302	13
14	1.0723	0.9326	14.4642	13.4887	0.0691	0.0741	86.584	6.4190	14
15	1.0777	0.9279	15.5365	14.4166	0.0644	0.0694	99.574	6.9069	15
16	1.0831	0.9233	16.6142	15.3399	0.0602	0.0652	113.424	7.3940	16
17	1.0885	0.9187	17.6973	16.2586	0.0565	0.0615	128.123	7.8803	17
18	1.0939	0.9141	18.7858	17.1728	0.0532	0.0582	143.663	8.3658	18
19	1.0994	0.9096	19.8797	18.0824	0.0503	0.0553	160.036	8.8504	19
20	1.1049	0.9051	20.9791	18.9874	0.0477	0.0527	177.232	9.3342	20
21	1.1104	0.9006	22.0840	19.8880	0.0453	0.0503	195.243	9.8172	21
22	1.1160	0.8961	23.1944	20.7841	0.0431	0.0481	214.061	10.2993	22
23	1.1216	0.8916	24.3104	21.6757	0.0411	0.0461	233.677	10.7806	23
24	1.1272	0.8872	25.4320	22.5629	0.0393	0.0443	254.082	11.2611	24
25	1.1328	0.8828	26.5591	23.4456	0.0377	0.0427	275.269	11.7407	25
30	1.1614	0.8610	32.2800	27.7941	0.0310	0.0360	392.632	14.1265	30
36	1.1967	0.8356	39.3361	32.8710	0.0254	0.0304	557.560	16.9621	36
40	1.2208	0.8191	44.1588	36.1722	0.0226	0.0276	681.335	18.8359	40
48	1.2705	0.7871	54.0978	42.5803	0.0185	0.0235	959.919	22.5437	48
60	1.3489	0.7414	69.7700	51.7256	0.0143	0.0193	1448.646	28.0064	60
72	1.4320	0.6983	86.4089	60.3395	0.0116	0.0166	2012.348	33.3504	72
84	1.5204	0.6577	104.0739	68.4530	0.0096	0.0146	2640.664	38.5763	84
100	1.6467	0.6073	129.3337	78.5426	0.0077	0.0127	3562.793	45.3613	100
∞				200.0000		0.0050			∞

TABLE C-3 Discrete Compounding; $i = \frac{3}{4}\%$

	Single Payment		Uniform Series				Uniform Gradient		
	Compound Amount Factor	Present Worth Factor	Compound Amount Factor	Present Worth Factor	Sinking Fund Factor	Capital Recovery Factor	Gradient Present Worth Factor	Gradient Uniform Series Factor	
N	To Find F Given P F/P	To Find P Given F P/F	To Find F Given A F/A	To Find P Given A P/A	To Find A Given F A/F	To Find A Given P A/P	To Find P Given G P/G	To Find A Given G A/G	N
1	1.0075	0.9926	1.0000	0.9926	1.0000	1.0075	0.000	0.0000	1
2	1.0151	0.9852	2.0075	1.9777	0.4981	0.5056	0.985	0.4981	2
3	1.0227	0.9778	3.0226	2.9556	0.3308	0.3383	2.941	0.9950	3
4	1.0303	0.9706	4.0452	3.9261	0.2472	0.2547	5.853	1.4907	4
5	1.0381	0.9633	5.0756	4.8894	0.1970	0.2045	9.706	1.9851	5
6	1.0459	0.9562	6.1136	5.8456	0.1636	0.1711	14.487	2.4782	6
7	1.0537	0.9490	7.1595	6.7946	0.1397	0.1472	20.181	2.9701	7
8	1.0616	0.9420	8.2132	7.7366	0.1218	0.1293	26.775	3.4608	8
9	1.0696	0.9350	9.2748	8.6716	0.1078	0.1153	34.254	3.9502	9
10	1.0776	0.9280	10.3443	9.5996	0.0967	0.1042	42.606	4.4384	10
11	1.0857	0.9211	11.4219	10.5207	0.0876	0.0951	51.817	4.9253	11
12	1.0938	0.9142	12.5076	11.4349	0.0800	0.0875	61.874	5.4110	12
13	1.1020	0.9074	13.6014	12.3423	0.0735	0.0810	72.763	5.8954	13
14	1.1103	0.9007	14.7034	13.2430	0.0680	0.0755	84.472	6.3786	14
15	1.1186	0.8940	15.8137	14.1370	0.0632	0.0707	96.988	6.8606	15
16	1.1270	0.8873	16.9323	15.0243	0.0591	0.0666	110.297	7.3413	16
17	1.1354	0.8807	18.0593	15.9050	0.0554	0.0629	124.389	7.8207	17
18	1.1440	0.8742	19.1947	16.7792	0.0521	0.0596	139.249	8.2989	18
19	1.1525	0.8676	20.3387	17.6468	0.0492	0.0567	154.867	8.7759	19
20	1.1612	0.8612	21.4912	18.5080	0.0465	0.0540	171.230	9.2516	20
21	1.1699	0.8548	22.6524	19.3628	0.0441	0.0516	188.325	9.7261	21
22	1.1787	0.8484	23.8223	20.2112	0.0420	0.0495	206.142	10.1994	22
23	1.1875	0.8421	25.0010	21.0533	0.0400	0.0475	224.668	10.6714	23
24	1.1964	0.8358	26.1885	21.8891	0.0382	0.0457	243.892	11.1422	24
25	1.2054	0.8296	27.3849	22.7188	0.0365	0.0440	263.803	11.6117	25
30	1.2513	0.7992	33.5029	26.7751	0.0298	0.0373	373.263	13.9407	30
36	1.3086	0.7641	41.1527	34.4468	0.0243	0.0318	524.992	16.6946	36
40	1.3483	0.7416	46.4464	34.4469	0.0215	0.0290	637.469	18.5058	40
48	1.4314	0.6986	57.5207	40.1848	0.0174	0.0249	886.840	22.0691	48
60	1.5657	0.6387	75.4241	48.1734	0.0133	0.0208	1313.519	27.2665	60
72	1.7126	0.5839	95.0070	55.4768	0.0105	0.0180	1791.246	32.2882	72
84	1.8732	0.5338	116.4269	62.1540	0.0086	0.0161	2308.128	37.1357	84
100	2.1111	0.4737	148.1445	70.1746	0.0068	0.0143	3040.745	43.3311	100
∞				133.3333		0.0075			∞

TABLE C-4 Discrete Compounding; $i = 1\%$

	Single Payment		Uniform Series					Uniform Gradient		
	Compound Amount Factor	Present Worth Factor	Compound Amount Factor	Present Worth Factor	Sinking Fund Factor	Capital Recovery Factor		Gradient Present Worth Factor	Gradient Uniform Series Factor	
	To Find F Given P F/P	To Find P Given F P/F	To Find F Given A F/A	To Find P Given A P/A	To Find A Given F A/F	To Find A Given P A/P		To Find P Given G P/G	To Find A Given G A/G	
N										N
1	1.0100	0.9901	1.0000	0.9901	1.0000	1.0100		0.000	0.0000	1
2	1.0201	0.9803	2.0100	1.9704	0.4975	0.5075		0.980	0.4975	2
3	1.0303	0.9706	3.0301	2.9410	0.3300	0.3400		2.922	0.9934	3
4	1.0406	0.9610	4.0604	3.9020	0.2463	0.2563		5.804	1.4876	4
5	1.0510	0.9515	5.1010	4.8534	0.1960	0.2060		9.610	1.9801	5
6	1.0615	0.9420	6.1520	5.7955	0.1625	0.1725		14.321	2.4710	6
7	1.0721	0.9327	7.2135	6.7282	0.1386	0.1486		19.917	2.9602	7
8	1.0829	0.9235	8.2857	7.6517	0.1207	0.1307		26.381	3.4478	8
9	1.0937	0.9143	9.3685	8.5660	0.1067	0.1167		33.696	3.9337	9
10	1.1046	0.9053	10.4622	9.4713	0.0956	0.1056		41.844	4.4179	10
11	1.1157	0.8963	11.5668	10.3676	0.0865	0.0965		50.807	4.9005	11
12	1.1268	0.8874	12.6825	11.2551	0.0788	0.0888		60.569	5.3815	12
13	1.1381	0.8787	13.8093	12.1337	0.0724	0.0824		71.113	5.8607	13
14	1.1495	0.8700	14.9474	13.0037	0.0669	0.0769		82.422	6.3384	14
15	1.1610	0.8613	16.0969	13.8651	0.0621	0.0721		94.481	6.8143	15
16	1.1726	0.8528	17.2579	14.7179	0.0579	0.0679		107.273	7.2886	16
17	1.1843	0.8444	18.4304	15.5623	0.0543	0.0643		120.783	7.7613	17
18	1.1961	0.8360	19.6147	16.3983	0.0510	0.0610		134.996	8.2323	18
19	1.2081	0.8277	20.8109	17.2260	0.0481	0.0581		149.895	8.7017	19
20	1.2202	0.8195	22.0190	18.0456	0.0454	0.0554		165.466	9.1694	20
21	1.2324	0.8114	23.2392	18.8570	0.0430	0.0530		181.695	9.6354	21
22	1.2447	0.8034	24.4716	19.6604	0.0409	0.0509		198.566	10.0998	22
23	1.2572	0.7954	25.7163	20.4558	0.0389	0.0489		216.066	10.5626	23
24	1.2697	0.7876	26.9734	21.2434	0.0371	0.0471		234.180	11.0237	24
25	1.2824	0.7798	28.2432	22.0232	0.0354	0.0454		252.895	11.4831	25
30	1.3478	0.7419	34.7849	25.8077	0.0287	0.0387		355.002	13.7557	30
36	1.4308	0.6989	43.0769	30.1075	0.0232	0.0332		494.621	16.4285	36
40	1.4889	0.6717	48.8863	32.8346	0.0205	0.0305		596.856	18.1776	40
48	1.6122	0.6203	61.2226	37.9740	0.0163	0.0263		820.146	21.5976	48
60	1.8167	0.5504	81.6697	44.9550	0.0122	0.0222		1192.806	26.5333	60
72	2.0471	0.4885	104.7099	51.1504	0.0096	0.0196		1597.867	31.2386	72
84	2.3067	0.4335	130.6723	56.6485	0.0077	0.0177		2023.315	35.7170	84
100	2.7048	0.3697	170.4814	63.0289	0.0059	0.0159		2605.776	41.3426	100
∞				100.0000		0.0100				∞

TABLE C-5　Discrete Compounding; $i = 2\%$

	Single Payment		Uniform Series				Uniform Gradient	
	Compound Amount Factor	Present Worth Factor	Compound Amount Factor	Present Worth Factor	Sinking Fund Factor	Capital Recovery Factor	Gradient Present Worth Factor	Gradient Uniform Series Factor
	To Find F Given P	To Find P Given F	To Find F Given A	To Find P Given A	To Find A Given F	To Find A Given P	To Find P Given G	To Find A Given G
N	F/P	P/F	F/A	P/A	A/F	A/P	P/G	A/G	N
---	---	---	---	---	---	---	---	---	---
1	1.0200	0.9804	1.0000	0.9804	1.0000	1.0200	0.000	0.0000	1
2	1.0404	0.9612	2.0200	1.9416	0.4950	0.5150	0.961	0.4950	2
3	1.0612	0.9423	3.0604	2.8839	0.3268	0.3468	2.846	0.9868	3
4	1.0824	0.9238	4.1216	3.8077	0.2426	0.2626	5.617	1.4752	4
5	1.1041	0.9057	5.2040	4.7135	0.1922	0.2122	9.240	1.9604	5
6	1.1262	0.8880	6.3081	5.6014	0.1585	0.1785	13.680	2.4423	6
7	1.1487	0.8706	7.4343	6.4720	0.1345	0.1545	18.904	2.9208	7
8	1.1717	0.8535	8.5830	7.3255	0.1165	0.1365	24.878	3.3961	8
9	1.1951	0.8368	9.7546	8.1622	0.1025	0.1225	31.572	3.8681	9
10	1.2190	0.8203	10.9497	8.9826	0.0913	0.1113	38.955	4.3367	10
11	1.2434	0.8043	12.1687	9.7868	0.0822	0.1022	46.998	4.8021	11
12	1.2682	0.7885	13.4121	10.5753	0.0746	0.0946	55.671	5.2642	12
13	1.2936	0.7730	14.6803	11.3484	0.0681	0.0881	64.948	5.7231	13
14	1.3195	0.7579	15.9739	12.1062	0.0626	0.0826	74.800	6.1786	14
15	1.3459	0.7430	17.2934	12.8493	0.0578	0.0778	85.202	6.6309	15
16	1.3728	0.7284	18.6393	13.5777	0.0537	0.0737	96.129	7.0799	16
17	1.4002	0.7142	20.0121	14.2919	0.0500	0.0700	107.555	7.5256	17
18	1.4282	0.7002	21.4123	14.9920	0.0467	0.0667	119.458	7.9681	18
19	1.4568	0.6864	22.8406	15.6785	0.0438	0.0638	131.814	8.4073	19
20	1.4859	0.6730	24.2974	16.3514	0.0412	0.0612	144.600	8.8433	20
21	1.5157	0.6598	25.7833	17.0112	0.0388	0.0588	157.796	9.2260	21
22	1.5460	0.6468	27.2990	17.6580	0.0366	0.0566	171.380	9.7055	22
23	1.5769	0.6342	28.8450	18.2922	0.0347	0.0547	185.331	10.1317	23
24	1.6084	0.6217	30.4219	18.9139	0.0329	0.0529	199.631	10.5547	24
25	1.6406	0.6095	32.0303	19.5235	0.0312	0.0512	214.259	10.9745	25
30	1.8114	0.5521	40.5681	22.3965	0.0246	0.0446	291.716	13.0251	30
36	2.0399	0.4902	51.9944	25.4888	0.0192	0.0392	392.041	15.3809	36
40	2.2080	0.4529	60.4020	27.3555	0.0166	0.0366	461.993	16.8885	40
48	2.5871	0.3865	79.3535	30.6731	0.0126	0.0326	605.966	19.7556	48
60	3.2810	0.3048	114.0515	34.7609	0.0088	0.0288	823.698	23.6961	60
72	4.1611	0.2403	158.0570	37.9841	0.0063	0.0263	1034.056	27.2234	72
84	5.2773	0.1895	213.8666	40.5255	0.0047	0.0247	1230.419	30.3616	84
100	7.2446	0.1380	312.2323	43.0984	0.0032	0.0232	1464.753	33.9863	100
∞				50.0000		0.0200			∞

TABLE C-6 Discrete Compounding; $i = 3\%$

	Single Payment		Uniform Series					Uniform Gradient	
	Compound Amount Factor	Present Worth Factor	Compound Amount Factor	Present Worth Factor	Sinking Fund Factor	Capital Recovery Factor	Gradient Present Worth Factor	Gradient Uniform Series Factor	
N	To Find F Given P F/P	To Find P Given F P/F	To Find F Given A F/A	To Find P Given A P/A	To Find A Given F A/F	To Find A Given P A/P	To Find P Given G P/G	To Find A Given G A/G	N
1	1.0300	0.9709	1.0000	0.9709	1.0000	1.0300	0.000	0.0000	1
2	1.0609	0.9426	2.0300	1.9135	0.4926	0.5226	0.943	0.4926	2
3	1.0927	0.9151	3.0909	2.8286	0.3235	0.3535	2.773	0.9803	3
4	1.1255	0.8885	4.1836	3.7171	0.2390	0.2690	5.438	1.4631	4
5	1.1593	0.8626	5.3091	4.5797	0.1884	0.2184	8.889	1.9409	5
6	1.1941	0.8375	6.4684	5.4172	0.1546	0.1846	13.076	2.4138	6
7	1.2299	0.8131	7.6625	6.2303	0.1305	0.1605	17.955	2.8819	7
8	1.2668	0.7894	8.8923	7.0197	0.1125	0.1425	23.481	3.3450	8
9	1.3048	0.7664	10.1591	7.7861	0.0984	0.1284	29.612	3.8032	9
10	1.3439	0.7441	11.4639	8.5302	0.0872	0.1172	36.309	4.2565	10
11	1.3842	0.7224	12.8078	9.2526	0.0781	0.1081	43.533	4.7049	11
12	1.4258	0.7014	14.1920	9.9540	0.0705	0.1005	51.248	5.1485	12
13	1.4685	0.6810	15.6178	10.6350	0.0640	0.0940	59.420	5.5872	13
14	1.5126	0.6611	17.0863	11.2961	0.0585	0.0885	68.014	6.0210	14
15	1.5580	0.6419	18.5989	11.9379	0.0538	0.0838	77.000	6.4500	15
16	1.6047	0.6232	20.1569	12.5611	0.0496	0.0796	86.348	6.8742	16
17	1.6528	0.6050	21.7616	13.1661	0.0460	0.0760	96.028	7.2936	17
18	1.7024	0.5874	23.4144	13.7535	0.0427	0.0727	106.014	7.7081	18
19	1.7535	0.5703	25.1169	14.3238	0.0398	0.0698	116.279	8.1179	19
20	1.8061	0.5537	26.8704	14.8775	0.0372	0.0672	126.799	8.5229	20
21	1.8603	0.5375	28.6765	15.4150	0.0349	0.0649	137.550	8.9231	21
22	1.9161	0.5219	30.5368	15.9369	0.0327	0.0627	148.509	7.3186	22
23	1.9736	0.5067	32.4529	16.4436	0.0308	0.0608	159.657	9.7093	23
24	2.0328	0.4919	34.4265	16.9355	0.0290	0.0590	170.971	10.0954	24
25	2.0938	0.4776	36.4593	17.4131	0.0274	0.0574	182.434	10.4768	25
30	2.4273	0.4120	47.5754	19.6004	0.0210	0.0510	241.361	12.3141	30
35	2.8139	0.3554	60.4621	21.4872	0.0165	0.0465	301.627	14.0375	35
40	3.2620	0.3066	75.4012	23.1148	0.0133	0.0433	361.750	15.6502	40
45	3.7816	0.2644	92.7199	24.5187	0.0108	0.0408	420.633	17.1556	45
50	4.3839	0.2281	112.7969	25.7298	0.0089	0.0389	477.480	18.5575	50
60	5.8916	0.1697	163.0534	27.6756	0.0061	0.0361	583.053	21.0674	60
80	10.6409	0.0940	321.3630	30.2008	0.0031	0.0331	756.087	25.0353	80
100	19.2186	0.0520	607.2877	31.5989	0.0016	0.0316	879.854	27.8444	100
∞				33.3333		0.0300			∞

TABLE C-7 Discrete Compounding; $i = 4\%$

	Single Payment		Uniform Series					Uniform Gradient	
	Compound Amount Factor	Present Worth Factor	Compound Amount Factor	Present Worth Factor	Sinking Fund Factor	Capital Recovery Factor	Gradient Present Worth Factor	Gradient Uniform Series Factor	
	To Find F Given P	To Find P Given F	To Find F Given A	To Find P Given A	To Find A Given F	To Find A Given P	To Find P Given G	To Find A Given G	
N	F/P	P/F	F/A	P/A	A/F	A/P	P/G	A/G	N
1	1.0400	0.9615	1.0000	0.9615	1.0000	1.0400	0.000	0.0000	1
2	1.0816	0.9246	2.0400	1.8861	0.4902	0.5302	0.925	0.4902	2
3	1.1249	0.8890	3.1216	2.7751	0.3203	0.3603	2.703	0.9739	3
4	1.1699	0.8548	4.2465	3.6299	0.2355	0.2755	5.267	1.4510	4
5	1.2167	0.8219	5.4163	4.4518	0.1846	0.2246	8.555	1.9216	5
6	1.2653	0.7903	6.6330	5.2421	0.1508	0.1908	12.506	2.3857	6
7	1.3159	0.7599	7.8983	6.0021	0.1266	0.1666	17.066	2.8433	7
8	1.3686	0.7307	9.2142	6.7327	0.1085	0.1485	22.181	3.2944	8
9	1.4233	0.7026	10.5828	7.4353	0.0945	0.1345	27.801	3.7391	9
10	1.4802	0.6756	12.0061	8.1109	0.0833	0.1233	33.881	4.1773	10
11	1.5395	0.6496	13.4864	8.7605	0.0741	0.1141	40.377	4.6090	11
12	1.6010	0.6246	15.0258	9.3851	0.0666	0.1066	47.248	5.0343	12
13	1.6651	0.6006	16.6268	9.9856	0.0601	0.1001	54.455	5.4533	13
14	1.7317	0.5775	18.2919	10.5631	0.0547	0.0947	61.962	5.8659	14
15	1.8009	0.5553	20.0236	11.1184	0.0499	0.0899	69.736	6.2721	15
16	1.8730	0.5339	21.8245	11.6523	0.0458	0.0858	77.744	6.6720	16
17	1.9479	0.5134	23.6975	12.1657	0.0422	0.0822	85.958	7.0656	17
18	2.0258	0.4936	25.6454	12.6593	0.0390	0.0790	94.350	7.4530	18
19	2.1068	0.4746	27.6712	13.1339	0.0361	0.0761	102.893	7.8342	19
20	2.1911	0.4564	29.7781	13.5903	0.0336	0.0736	111.565	8.2091	20
21	2.2788	0.4388	31.9692	14.0292	0.0313	0.0713	120.341	8.5779	21
22	2.3699	0.4220	34.2480	14.4511	0.0292	0.0692	129.202	8.9407	22
23	2.4647	0.4057	36.6179	14.8568	0.0273	0.0673	138.128	9.2973	23
24	2.5633	0.3901	39.0826	15.2470	0.0256	0.0656	147.101	9.6479	24
25	2.6658	0.3751	41.6459	15.6221	0.0240	0.0640	156.104	9.9925	25
30	3.2434	0.3083	56.0849	17.2920	0.0178	0.0578	201.062	11.6274	30
35	3.9461	0.2534	73.6522	18.6646	0.0136	0.0536	244.877	13.1198	35
40	4.8010	0.2083	95.0255	19.7928	0.0105	0.0505	286.530	14.4765	40
45	5.8412	0.1712	121.0294	20.7200	0.0083	0.0483	325.403	15.7047	45
50	7.1067	0.1407	152.6671	21.4822	0.0066	0.0466	361.164	16.8122	50
60	10.5196	0.0951	237.9907	22.6235	0.0042	0.0442	422.997	18.6972	60
80	23.0498	0.0434	551.2450	23.9154	0.0018	0.0418	511.116	21.3718	80
100	50.5049	0.0198	1237.6237	24.5050	0.0008	0.0408	563.125	22.9800	100
∞				25.0000		0.0400			∞

TABLE C-8 Discrete Compounding; $i = 5\%$

N	Single Payment — Compound Amount Factor — To Find F Given P — F/P	Single Payment — Present Worth Factor — To Find P Given F — P/F	Uniform Series — Compound Amount Factor — To Find F Given A — F/A	Uniform Series — Present Worth Factor — To Find P Given A — P/A	Uniform Series — Sinking Fund Factor — To Find A Given F — A/F	Uniform Series — Capital Recovery Factor — To Find A Given P — A/P	Uniform Gradient — Gradient Present Worth Factor — To Find P Given G — P/G	Uniform Gradient — Gradient Uniform Series Factor — To Find A Given G — A/G	N
1	1.0500	0.9524	1.0000	0.9524	1.0000	1.0500	0.000	0.0000	1
2	1.1025	0.9070	2.0500	1.8594	0.4878	0.5378	0.907	0.4878	2
3	1.1576	0.8638	3.1525	2.7232	0.3172	0.3672	2.635	0.9675	3
4	1.2155	0.8227	4.3101	3.5460	0.2320	0.2820	5.103	1.4391	4
5	1.2763	0.7835	5.5256	4.3295	0.1810	0.2310	8.237	1.9025	5
6	1.3401	0.7462	6.8019	5.0757	0.1470	0.1970	11.968	2.3579	6
7	1.4071	0.7107	8.1420	5.7864	0.1228	0.1728	16.232	2.8052	7
8	1.4775	0.6768	9.5491	6.4632	0.1047	0.1547	20.970	3.2445	8
9	1.5513	0.6446	11.0266	7.1078	0.0907	0.1407	26.127	3.6758	9
10	1.6289	0.6139	12.5779	7.7217	0.0795	0.1295	31.652	4.0991	10
11	1.7103	0.5847	14.2068	8.3064	0.0704	0.1204	37.499	4.5144	11
12	1.7959	0.5568	15.9171	8.8633	0.0628	0.1128	43.624	4.9219	12
13	1.8856	0.5303	17.7130	9.3936	0.0565	0.1065	49.988	5.3215	13
14	1.9799	0.5051	19.5986	9.8986	0.0510	0.1010	56.554	5.7133	14
15	2.0789	0.4810	21.5786	10.3797	0.0463	0.0963	63.288	6.0973	15
16	2.1829	0.4581	23.6575	10.8378	0.0423	0.0923	70.160	6.4736	16
17	2.2920	0.4363	25.8404	11.2741	0.0387	0.0887	77.141	6.8423	17
18	2.4066	0.4155	28.1324	11.6896	0.0355	0.0855	84.204	7.2034	18
19	2.5270	0.3957	30.5390	12.0853	0.0327	0.0827	91.328	7.5569	19
20	2.6533	0.3769	33.0660	12.4622	0.0302	0.0802	98.488	7.9030	20
21	2.7860	0.3589	35.7193	12.8212	0.0280	0.0780	105.667	8.2416	21
22	2.9253	0.3418	38.5052	13.1630	0.0260	0.0760	112.846	8.5730	22
23	3.0715	0.3256	41.4305	13.4886	0.0241	0.0741	120.009	8.8971	23
24	3.2251	0.3101	44.5020	13.7986	0.0225	0.0725	127.140	9.2140	24
25	3.3864	0.2953	47.7271	14.0939	0.0210	0.0710	134.228	9.5238	25
30	4.3219	0.2314	66.4388	15.3725	0.0151	0.0651	168.623	10.9691	30
35	5.5160	0.1813	90.3203	16.3742	0.0111	0.0611	200.581	12.2498	35
40	7.0400	0.1420	120.7998	17.1591	0.0083	0.0583	229.545	13.3775	40
45	8.9850	0.1113	159.7002	17.7741	0.0063	0.0563	255.315	14.3644	45
50	11.4674	0.0872	209.3480	18.2559	0.0048	0.0548	277.915	15.2233	50
60	18.6792	0.0535	353.5837	18.9293	0.0028	0.0528	314.343	16.6062	60
80	49.5614	0.0202	971.2288	19.5965	0.0010	0.0510	359.646	18.3526	80
100	131.5013	0.0076	2610.0252	19.8479	0.0004	0.0504	381.749	19.2337	100
∞				20.0000		0.0500			∞

TABLE C-9 Discrete Compounding; $i = 6\%$

	Single Payment		Uniform Series				Uniform Gradient		
	Compound Amount Factor	Present Worth Factor	Compound Amount Factor	Present Worth Factor	Sinking Fund Factor	Capital Recovery Factor	Gradient Present Worth Factor	Gradient Uniform Series Factor	
	To Find F Given P	To Find P Given F	To Find F Given A	To Find P Given A	To Find A Given F	To Find A Given P	To Find P Given G	To Find A Given G	
N	F/P	P/F	F/A	P/A	A/F	A/P	P/G	A/G	N
1	1.0600	0.9434	1.0000	0.9434	1.0000	1.0600	0.000	0.0000	1
2	1.1236	0.8900	2.0600	1.8334	0.4854	0.5454	0.890	0.4854	2
3	1.1910	0.8396	3.1836	2.6730	0.3141	0.3741	2.569	0.9612	3
4	1.2625	0.7921	4.3746	3.4651	0.2286	0.2886	4.946	1.4272	4
5	1.3382	0.7473	5.6371	4.2124	0.1774	0.2374	7.935	1.8836	5
6	1.4185	0.7050	6.9753	4.9173	0.1434	0.2034	11.459	2.3304	6
7	1.5036	0.6651	8.3938	5.5824	0.1191	0.1791	15.450	2.7676	7
8	1.5938	0.6274	9.8975	6.2098	0.1010	0.1610	19.842	3.1952	8
9	1.6895	0.5919	11.4913	6.8017	0.0870	0.1470	24.577	3.6133	9
10	1.7908	0.5584	13.1808	7.3601	0.0759	0.1359	29.602	4.0220	10
11	1.8983	0.5268	14.9716	7.8869	0.0668	0.1268	34.870	4.4213	11
12	2.0122	0.4970	16.8699	8.3838	0.0593	0.1193	40.337	4.8113	12
13	2.1329	0.4688	18.8821	8.8527	0.0530	0.1130	45.963	5.1920	13
14	2.2609	0.4423	21.0151	9.2950	0.0476	0.1076	51.713	5.5635	14
15	2.3966	0.4173	23.2760	9.7122	0.0430	0.1030	57.555	5.9260	15
16	2.5404	0.3936	25.6725	10.1059	0.0390	0.0990	63.459	6.2794	16
17	2.6928	0.3714	28.2129	10.4773	0.0354	0.0954	69.401	6.6240	17
18	2.8543	0.3503	30.9057	10.8276	0.0324	0.0924	75.357	6.9597	18
19	3.0256	0.3305	33.7600	11.1581	0.0296	0.0896	81.306	7.2867	19
20	3.2071	0.3118	36.7856	11.4699	0.0272	0.0872	87.230	7.6051	20
21	3.3996	0.2942	39.9927	11.7641	0.0250	0.0850	93.114	7.9151	21
22	3.6035	0.2775	43.3923	12.0416	0.0230	0.0830	98.941	8.2166	22
23	3.8197	0.2618	46.9958	12.3034	0.0213	0.0813	104.701	8.5099	23
24	4.0489	0.2470	50.8156	12.5504	0.0197	0.0797	110.381	8.7951	24
25	4.2919	0.2330	54.8645	12.7834	0.0182	0.0782	115.973	9.0722	25
30	5.7435	0.1741	79.0582	13.7648	0.0126	0.0726	142.359	10.3422	30
35	7.6861	0.1301	111.4348	14.4982	0.0090	0.0690	165.743	11.4319	35
40	10.2857	0.0972	154.7620	15.0463	0.0065	0.0665	185.957	12.3590	40
45	13.7646	0.0727	212.7435	15.4558	0.0047	0.0647	203.110	13.1413	45
50	18.4202	0.0543	290.3359	15.7619	0.0034	0.0634	217.457	13.7964	50
60	32.9877	0.0303	533.1282	16.1614	0.0019	0.0619	239.043	14.7909	60
80	105.7960	0.0095	1746.5999	16.5091	0.0006	0.0606	262.549	15.9033	80
100	339.3021	0.0029	5638.3681	16.6175	0.0002	0.0602	272.047	16.3711	100
∞				16.6667		0.0600			∞

TABLE C-10 Discrete Compounding; $i = 7\%$

| | Single Payment | | Uniform Series | | | | | | Uniform Gradient | | |
| | Compound Amount Factor | Present Worth Factor | Compound Amount Factor | Present Worth Factor | Sinking Fund Factor | Capital Recovery Factor | | | Gradient Present Worth Factor | Gradient Uniform Series Factor | |
N	To Find F Given P F/P	To Find P Given F P/F	To Find F Given A F/A	To Find P Given A P/A	To Find A Given F A/F	To Find A Given P A/P			To Find P Given G P/G	To Find A Given G A/G	N
1	1.0700	0.9346	1.0000	0.9346	1.0000	1.0700			0.000	0.0000	1
2	1.1449	0.8734	2.0700	1.8080	0.4831	0.5531			0.873	0.4831	2
3	1.2250	0.8163	3.2149	2.6243	0.3111	0.3811			2.506	0.9549	3
4	1.3108	0.7629	4.4399	3.3872	0.2252	0.2952			4.795	1.4155	4
5	1.4026	0.7130	5.7507	4.1002	0.1739	0.2439			7.647	1.8650	5
6	1.5007	0.6663	7.1533	4.7665	0.1398	0.2098			10.978	2.3032	6
7	1.6058	0.6227	8.6540	5.3893	0.1156	0.1856			14.715	2.7304	7
8	1.7182	0.5820	10.2598	5.9713	0.0975	0.1675			18.789	3.1465	8
9	1.8385	0.5439	11.9780	6.5152	0.0835	0.1535			23.140	3.5517	9
10	1.9672	0.5083	13.8164	7.0236	0.0724	0.1424			27.716	3.9461	10
11	2.1049	0.4751	15.7836	7.4987	0.0634	0.1334			32.467	4.3296	11
12	2.2522	0.4440	17.8885	7.9427	0.0559	0.1259			37.351	4.7025	12
13	2.4098	0.4150	20.1406	8.3577	0.0497	0.1197			42.330	5.0648	13
14	2.5785	0.3878	22.5505	8.7455	0.0443	0.1143			47.372	5.4167	14
15	2.7590	0.3624	25.1290	9.1079	0.0398	0.1098			52.446	5.7583	15
16	2.9522	0.3387	27.8881	9.4466	0.0359	0.1059			57.527	6.0897	16
17	3.1588	0.3166	30.8402	9.7632	0.0324	0.1024			62.592	6.4110	17
18	3.3799	0.2959	33.9990	10.0591	0.0294	0.0994			67.622	6.7225	18
19	3.6165	0.2765	37.3790	10.3356	0.0268	0.0968			72.599	7.0242	19
20	3.8697	0.2584	40.9955	10.5940	0.0244	0.0944			77.509	7.3163	20
21	4.1406	0.2415	44.8652	10.8355	0.0223	0.0923			82.339	7.5990	21
22	4.4304	0.2257	49.0057	11.0612	0.0204	0.0904			87.079	7.8725	22
23	4.7405	0.2109	53.4361	11.2722	0.0187	0.0887			91.720	8.1369	23
24	5.0724	0.1971	58.1767	11.4693	0.0172	0.0872			96.255	8.3923	24
25	5.4274	0.1842	63.2490	11.6536	0.0158	0.0858			100.677	8.6391	25
30	7.6123	0.1314	94.4608	12.4090	0.0106	0.0806			120.972	9.7487	30
35	10.6766	0.0937	138.2369	12.9477	0.0072	0.0772			138.135	10.6687	35
40	14.9745	0.0668	199.6351	13.3317	0.0050	0.0750			152.293	11.4233	40
45	21.0023	0.0476	285.7495	13.6055	0.0035	0.0735			163.756	12.0360	45
50	29.4570	0.0339	406.5289	13.8007	0.0025	0.0725			172.905	12.5287	50
60	57.9464	0.0173	813.5204	14.0392	0.0012	0.0712			185.768	13.2321	60
80	224.2344	0.0045	3189.0627	14.2220	0.0003	0.0703			198.075	13.9273	80
100	867.7163	0.0012	12381.6618	14.2693	0.0001	0.0701			202.200	14.1703	100
∞				14.2857		0.0700					∞

TABLE C-11 Discrete Compounding; $i = 8\%$

	Single Payment		Uniform Series					Uniform Gradient		
	Compound Amount Factor	Present Worth Factor	Compound Amount Factor	Present Worth Factor	Sinking Fund Factor	Capital Recovery Factor	Gradient Present Worth Factor	Gradient Uniform Series Factor	
N	To Find F Given P F/P	To Find P Given F P/F	To Find F Given A F/A	To Find P Given A P/A	To Find A Given F A/F	To Find A Given P A/P	To Find P Given G P/G	To Find A Given G A/G	N
1	1.0800	0.9259	1.0000	0.9259	1.0000	1.0800	0.000	0.0000	1
2	1.1664	0.8573	2.0800	1.7833	0.4808	0.5608	0.857	0.4808	2
3	1.2597	0.7938	3.2464	2.5771	0.3080	0.3880	2.445	0.9487	3
4	1.3605	0.7350	4.5061	3.3121	0.2219	0.3019	4.650	1.4040	4
5	1.4693	0.6806	5.8666	3.9927	0.1705	0.2505	7.372	1.8465	5
6	1.5869	0.6302	7.3359	4.6229	0.1363	0.2163	10.523	2.2763	6
7	1.7138	0.5835	8.9228	5.2064	0.1121	0.1921	14.024	2.6937	7
8	1.8509	0.5403	10.6366	5.7466	0.0940	0.1740	17.806	3.0985	8
9	1.9990	0.5002	12.4876	6.2469	0.0801	0.1601	21.808	3.4910	9
10	2.1589	0.4632	14.4866	6.7101	0.0690	0.1490	25.977	3.8713	10
11	2.3316	0.4289	16.6455	7.1390	0.0601	0.1401	30.266	4.2395	11
12	2.5182	0.3971	18.9771	7.5361	0.0527	0.1327	34.634	4.5957	12
13	2.7196	0.3677	21.4953	7.9038	0.0465	0.1265	39.046	4.9402	13
14	2.9372	0.3405	24.2149	8.2442	0.0413	0.1213	43.472	5.2731	14
15	3.1722	0.3152	27.1521	8.5595	0.0368	0.1168	47.886	5.5945	15
16	3.4259	0.2919	30.3243	8.8514	0.0330	0.1130	52.264	5.9046	16
17	3.7000	0.2703	33.7502	9.1216	0.0296	0.1096	56.588	6.2037	17
18	3.9960	0.2502	37.4502	9.3719	0.0267	0.1067	60.843	6.4920	18
19	4.3157	0.2317	41.4463	9.6036	0.0241	0.1041	65.013	6.7697	19
20	4.6610	0.2145	45.7620	9.8181	0.0219	0.1019	69.090	7.0369	20
21	5.0338	0.1987	50.4229	10.0168	0.0198	0.0998	73.063	7.2940	21
22	5.4365	0.1839	55.4568	10.2007	0.0180	0.0980	76.926	7.5412	22
23	5.8715	0.1703	60.8933	10.3711	0.0164	0.0964	80.673	7.7786	23
24	6.3412	0.1577	66.7648	10.5288	0.0150	0.0950	84.300	8.0066	24
25	6.8485	0.1460	73.1059	10.6748	0.0137	0.0937	87.804	8.2254	25
30	10.0627	0.0994	113.2832	11.2578	0.0088	0.0888	103.456	9.1897	30
35	14.7853	0.0676	172.3168	11.6546	0.0058	0.0858	116.092	9.9611	35
40	21.7245	0.0460	259.0565	11.9246	0.0039	0.0839	126.042	10.5699	40
45	31.9204	0.0313	386.5056	12.1084	0.0026	0.0826	133.733	11.0447	45
50	46.9016	0.0213	573.7702	12.2335	0.0017	0.0817	139.593	11.4107	50
60	101.2571	0.0099	1253.2133	12.3766	0.0008	0.0808	147.300	11.9015	60
80	471.9548	0.0021	5886.9354	12.4735	0.0002	0.0802	153.800	12.3301	80
100	2199.7613	0.0005	27484.5157	12.4943	a	0.0800	155.611	12.4545	100
∞				12.5000		0.0800			∞

aLess than 0.0001.

613

TABLE C-12 Discrete Compounding; $i = 9\%$

	Single Payment		Uniform Series				Uniform Gradient		
	Compound Amount Factor	Present Worth Factor	Compound Amount Factor	Present Worth Factor	Sinking Fund Factor	Capital Recovery Factor	Gradient Present Worth Factor	Gradient Uniform Series Factor	
	To Find F Given P	To Find P Given F	To Find F Given A	To Find P Given A	To Find A Given F	To Find A Given P	To Find P Given G	To Find A Given G	
N	F/P	P/F	F/A	P/A	A/F	A/P	P/G	A/G	N
1	1.0900	0.9174	1.0000	0.9174	1.0000	1.0900	0.000	0.0000	1
2	1.1881	0.8417	2.0900	1.7591	0.4785	0.5685	0.842	0.4785	2
3	1.2950	0.7722	3.2781	2.5313	0.3051	0.3951	2.386	0.9426	3
4	1.4116	0.7084	4.5731	3.2397	0.2187	0.3087	4.511	1.3925	4
5	1.5386	0.6499	5.9847	3.8897	0.1671	0.2571	7.111	1.8282	5
6	1.6771	0.5963	7.5233	4.4859	0.1329	0.2229	10.092	2.2498	6
7	1.8280	0.5470	9.2004	5.0330	0.1087	0.1987	13.375	2.6574	7
8	1.9926	0.5019	11.0285	5.5348	0.0907	0.1807	16.888	3.0512	8
9	2.1719	0.4604	13.0210	5.9952	0.0768	0.1668	20.571	3.4312	9
10	2.3674	0.4224	15.1929	6.4177	0.0658	0.1558	24.373	3.7978	10
11	2.5804	0.3875	17.5603	6.8052	0.0569	0.1469	28.248	4.1510	11
12	2.8127	0.3555	20.1407	7.1607	0.0497	0.1397	32.159	4.4910	12
13	3.0658	0.3262	22.9534	7.4869	0.0436	0.1336	36.073	4.8182	13
14	3.3417	0.2992	26.0192	7.7862	0.0384	0.1284	39.963	5.1326	14
15	3.6425	0.2745	29.3609	8.0607	0.0341	0.1241	43.807	5.4346	15
16	3.9703	0.2519	33.0034	8.3126	0.0303	0.1203	47.585	5.7245	16
17	4.3276	0.2311	36.9737	8.5436	0.0270	0.1170	51.282	6.0024	17
18	4.7171	0.2120	41.3013	8.7556	0.0242	0.1142	54.886	6.2687	18
19	5.1417	0.1945	46.0185	8.9501	0.0217	0.1117	58.387	6.5236	19
20	5.6044	0.1784	51.1601	9.1285	0.0195	0.1095	61.777	6.7674	20
21	6.1088	0.1637	56.7645	9.2922	0.0176	0.1076	65.051	7.0006	21
22	6.6586	0.1502	62.8733	9.4424	0.0159	0.1059	68.205	7.2232	22
23	7.2579	0.1378	69.5319	9.5802	0.0144	0.1044	71.236	7.4357	23
24	7.9111	0.1264	76.7898	9.7066	0.0130	0.1030	74.143	7.6384	24
25	8.6231	0.1160	84.7009	9.8226	0.0118	0.1018	76.927	7.8316	25
30	13.2677	0.0754	136.3075	10.2737	0.0073	0.0973	89.028	8.6657	30
35	20.4140	0.0490	215.7108	10.5668	0.0046	0.0946	98.359	9.3083	35
40	31.4094	0.0318	337.8824	10.7574	0.0030	0.0930	105.376	9.7957	40
45	48.3273	0.0207	525.8587	10.8812	0.0019	0.0919	110.556	10.1603	45
50	74.3575	0.0134	815.0836	10.9617	0.0012	0.0912	114.325	10.4295	50
60	176.0313	0.0057	1944.7921	11.0480	0.0005	0.0905	118.968	10.7683	60
80	986.5517	0.0010	10950.5741	11.0998	0.0001	0.0901	122.431	11.0299	80
100	5529.0408	0.0002	61422.6755	11.1091	a	0.0900	123.234	11.0930	100
∞				11.1111		0.0900			∞

a Less than 0.0001

TABLE C-13   Discrete Compounding; $i = 10\%$

| | Single Payment | | Uniform Series | | | | Uniform Gradient | | |
| | Compound Amount Factor | Present Worth Factor | Compound Amount Factor | Present Worth Factor | Sinking Fund Factor | Capital Recovery Factor | Gradient Present Worth Factor | Gradient Uniform Series Factor | |
N	To Find F Given P F/P	To Find P Given F P/F	To Find F Given A F/A	To Find P Given A P/A	To Find A Given F A/F	To Find A Given P A/P	To Find P Given G P/G	To Find A Given G A/G	N
1	1.1000	0.9091	1.0000	0.9091	1.0000	1.1000	0.000	0.0000	1
2	1.2100	0.8264	2.1000	1.7355	0.4762	0.5762	0.826	0.4762	2
3	1.3310	0.7513	3.3100	2.4869	0.3021	0.4021	2.329	0.9366	3
4	1.4641	0.6830	4.6410	3.1699	0.2155	0.3155	4.378	1.3812	4
5	1.6105	0.6209	6.1051	3.7908	0.1638	0.2638	6.862	1.8101	5
6	1.7716	0.5645	7.7156	4.3553	0.1296	0.2296	9.684	2.2236	6
7	1.9487	0.5132	9.4872	4.8684	0.1054	0.2054	12.763	2.6216	7
8	2.1436	0.4665	11.4359	5.3349	0.0874	0.1874	16.029	3.0045	8
9	2.3579	0.4241	13.5795	5.7590	0.0736	0.1736	19.422	3.3724	9
10	2.5937	0.3855	15.9374	6.1446	0.0627	0.1627	22.891	3.7255	10
11	2.8531	0.3505	18.5312	6.4951	0.0540	0.1540	26.396	4.0641	11
12	3.1384	0.3186	21.3843	6.8137	0.0468	0.1468	29.901	4.3884	12
13	3.4523	0.2897	24.5227	7.1034	0.0408	0.1408	33.377	4.6988	13
14	3.7975	0.2633	27.9750	7.3667	0.0357	0.1357	36.801	4.9955	14
15	4.1772	0.2394	31.7725	7.6061	0.0315	0.1315	40.152	5.2789	15
16	4.5950	0.2176	35.9497	7.8237	0.0278	0.1278	43.416	5.5493	16
17	5.0545	0.1978	40.5447	8.0216	0.0247	0.1247	46.582	5.8071	17
18	5.5599	0.1799	45.5992	8.2014	0.0219	0.1219	49.640	6.0526	18
19	6.1159	0.1635	51.1591	8.3649	0.0195	0.1195	52.583	6.2861	19
20	6.7275	0.1486	57.2750	8.5136	0.0175	0.1175	55.407	6.5081	20
21	7.4002	0.1351	64.0025	8.6487	0.0156	0.1156	58.110	6.7189	21
22	8.1403	0.1228	71.4027	8.7715	0.0140	0.1140	60.689	6.9189	22
23	8.9543	0.1117	79.5430	8.8832	0.0126	0.1126	63.146	7.1085	23
24	9.8497	0.1015	88.4973	8.9847	0.0113	0.1113	65.481	7.2881	24
25	10.8347	0.0923	98.3471	9.0770	0.0102	0.1102	67.696	7.4580	25
30	17.4494	0.0573	164.4940	9.4269	0.0061	0.1061	77.077	8.1762	30
35	28.1024	0.0356	271.0244	9.6442	0.0037	0.1037	83.987	8.7086	35
40	45.2593	0.0221	442.5926	9.7791	0.0023	0.1023	88.953	9.0962	40
45	72.8905	0.0137	718.9048	9.8628	0.0014	0.1014	92.454	9.3740	45
50	117.3909	0.0085	1163.9085	9.9148	0.0009	0.1009	94.889	9.5704	50
60	304.4816	0.0033	3034.8164	9.9672	0.0003	0.1003	97.701	9.8023	60
80	2048.4002	0.0005	20474.0021	9.9951	^{a}	0.1000	99.561	9.9609	80
100	13780.6123	0.0001	137796.1234	9.9993	^{a}	0.1000	99.920	9.9927	100
∞				10.0000		0.1000			∞

^{a} Less than 0.0001

615

TABLE C-14   Discrete Compounding; $i = 12\%$

| | Single Payment | | Uniform Series | | | | Uniform Gradient | | |
| | Compound Amount Factor | Present Worth Factor | Compound Amount Factor | Sinking Fund Factor | Present Worth Factor | Capital Recovery Factor | Gradient Present Worth Factor | Gradient Uniform Series Factor | |
N	To Find F Given P, F/P	To Find P Given F, P/F	To Find F Given A, F/A	To Find A Given F, A/F	To Find P Given A, P/A	To Find A Given P, A/P	To Find P Given G, P/G	To Find A Given G, A/G	N
1	1.1200	0.8929	1.0000	1.0000	0.8929	1.1200	0.000	0.0000	1
2	1.2544	0.7972	2.1200	0.4717	1.6901	0.5917	0.797	0.4717	2
3	1.4049	0.7118	3.3744	0.2963	2.4018	0.4163	2.221	0.9246	3
4	1.5735	0.6355	4.7793	0.2092	3.0373	0.3292	4.127	1.3589	4
5	1.7623	0.5674	6.3528	0.1574	3.6048	0.2774	6.397	1.7746	5
6	1.9738	0.5066	8.1152	0.1232	4.1114	0.2432	8.930	2.1720	6
7	2.2107	0.4523	10.0890	0.0991	4.5638	0.2191	11.644	2.5515	7
8	2.4760	0.4039	12.2997	0.0813	4.9676	0.2013	14.471	2.9131	8
9	2.7731	0.3606	14.7757	0.0677	5.3282	0.1877	17.356	3.2574	9
10	3.1058	0.3220	17.5487	0.0570	5.6502	0.1770	20.254	3.5847	10
11	3.4785	0.2875	20.6546	0.0484	5.9377	0.1684	23.129	3.8953	11
12	3.8960	0.2567	24.1331	0.0414	6.1944	0.1614	25.952	4.1897	12
13	4.3635	0.2292	28.0291	0.0357	6.4235	0.1557	28.702	4.4683	13
14	4.8871	0.2046	32.3926	0.0309	6.6282	0.1509	31.362	4.7317	14
15	5.4736	0.1827	37.2797	0.0268	6.8109	0.1468	33.920	4.9803	15
16	6.1304	0.1631	42.7533	0.0234	6.9740	0.1434	36.367	5.2147	16
17	6.8660	0.1456	48.8837	0.0205	7.1196	0.1405	38.697	5.4353	17
18	7.6900	0.1300	55.7497	0.0179	7.2497	0.1379	40.908	5.6427	18
19	8.6128	0.1161	63.4397	0.0158	7.3658	0.1358	42.998	5.8375	19
20	9.6463	0.1037	72.0524	0.0139	7.4694	0.1339	44.968	6.0202	20
21	10.8038	0.0926	81.6987	0.0122	7.5620	0.1322	46.819	6.1913	21
22	12.1003	0.0826	92.5026	0.0108	7.6446	0.1308	48.554	6.3514	22
23	13.5523	0.0738	104.6029	0.0096	7.7184	0.1296	50.178	6.5010	23
24	15.1786	0.0659	118.1552	0.0085	7.7843	0.1285	51.693	6.6406	24
25	17.0001	0.0588	133.3339	0.0075	7.8431	0.1275	53.105	6.7708	25
30	29.9599	0.0334	241.3327	0.0041	8.0552	0.1241	58.782	7.2974	30
35	52.7996	0.0189	431.6635	0.0023	8.1755	0.1223	62.605	7.6577	35
40	93.0510	0.0107	767.0914	0.0013	8.2438	0.1213	65.116	7.8988	40
45	163.9876	0.0061	1358.2300	0.0007	8.2825	0.1207	66.734	8.0572	45
50	289.0022	0.0035	2400.0182	0.0004	8.3045	0.1204	67.762	8.1597	50
60	897.5969	0.0011	7471.6411	0.0001	8.3240	0.1201	68.810	8.2664	60
80	8658.4831	0.0001	72145.6925	a	8.3324	0.1200	69.359	8.3241	80
100	83522.2657	a	696010.5477	a	8.3332	0.1200	69.434	8.3321	100
∞					8.3333	0.1200			∞

^a Less than 0.0001

TABLE C-15 Discrete Compounding; $i = 15\%$

	Single Payment		Uniform Series				Uniform Gradient		
	Compound Amount Factor	Present Worth Factor	Compound Amount Factor	Present Worth Factor	Sinking Fund Factor	Capital Recovery Factor	Gradient Present Worth Factor	Gradient Uniform Series Factor	
	To Find F Given P	To Find P Given F	To Find F Given A	To Find P Given A	To Find A Given F	To Find A Given P	To Find P Given G	To Find A Given G	
N	F/P	P/F	F/A	P/A	A/F	A/P	P/G	A/G	N
1	1.1500	0.8696	1.0000	0.8696	1.0000	1.1500	0.000	0.0000	1
2	1.3225	0.7561	2.1500	1.6257	0.4651	0.6151	0.756	0.4651	2
3	1.5209	0.6575	3.4725	2.2832	0.2880	0.4380	2.071	0.9071	3
4	1.7490	0.5718	4.9934	2.8550	0.2003	0.3503	3.786	1.3263	4
5	2.0114	0.4972	6.7424	3.3522	0.1483	0.2983	5.775	1.7228	5
6	2.3131	0.4323	8.7537	3.7845	0.1142	0.2642	7.937	2.0972	6
7	2.6600	0.3759	11.0668	4.1604	0.0904	0.2404	10.192	2.4498	7
8	3.0590	0.3269	13.7268	4.4873	0.0729	0.2229	12.481	2.7813	8
9	3.5179	0.2843	16.7858	4.7716	0.0596	0.2096	14.755	3.0922	9
10	4.0456	0.2472	20.3037	5.0188	0.0493	0.1993	16.980	3.3832	10
11	4.6524	0.2149	24.3493	5.2337	0.0411	0.1911	19.129	3.6549	11
12	5.3503	0.1869	29.0017	5.4206	0.0345	0.1845	21.185	3.9082	12
13	6.1528	0.1625	34.3519	5.5831	0.0291	0.1791	23.135	4.1438	13
14	7.0757	0.1413	40.5047	5.7245	0.0247	0.1747	24.973	4.3624	14
15	8.1371	0.1229	47.5804	5.8474	0.0210	0.1710	26.693	4.5650	15
16	9.3576	0.1069	55.7175	5.9542	0.0179	0.1679	28.296	4.7522	16
17	10.7613	0.0929	65.0751	6.0472	0.0154	0.1654	29.783	4.9251	17
18	12.3755	0.0808	75.8364	6.1280	0.0132	0.1632	31.157	5.0843	18
19	14.2318	0.0703	88.2118	6.1982	0.0113	0.1613	32.421	5.2307	19
20	16.3665	0.0611	102.4436	6.2593	0.0098	0.1598	33.582	5.3651	20
21	18.8215	0.0531	118.8101	6.3125	0.0084	0.1584	34.645	5.4883	21
22	21.6447	0.0462	137.6316	6.3587	0.0073	0.1573	35.615	5.6010	22
23	24.8915	0.0402	159.2764	6.3988	0.0063	0.1563	36.499	5.7040	23
24	28.6252	0.0349	184.1678	6.4338	0.0054	0.1554	37.302	5.7979	24
25	32.9190	0.0304	212.7930	6.4641	0.0047	0.1547	38.031	5.8834	25
30	66.2118	0.0151	434.7451	6.5660	0.0023	0.1523	40.753	6.2066	30
35	133.1755	0.0075	881.1702	6.6166	0.0011	0.1511	42.359	6.4019	35
40	267.8635	0.0037	1779.0903	6.6418	0.0006	0.1506	43.283	6.5168	40
45	538.7693	0.0019	3585.1285	6.6543	0.0003	0.1503	43.805	6.5830	45
50	1083.6574	0.0009	7217.7163	6.6605	0.0001	0.1501	44.096	6.6205	50
60	4363.9987	0.0002	29219.9916	6.6651	a	0.1500	44.343	6.6530	60
80	71750.8794	a	478332.5293	6.6666	a	0.1500	44.436	6.6656	80
100	1174313.4507	a	7828749.6713	6.6667	a	0.1500	44.444	6.6666	100
∞				6.6667		0.1500			∞

a Less than 0.0001

TABLE C-16 Discrete Compounding; $i = 18\%$

	Single Payment		Uniform Series					Uniform Gradient	
	Compound Amount Factor	Present Worth Factor	Compound Amount Factor	Present Worth Factor	Sinking Fund Factor	Capital Recovery Factor		Gradient Present Worth Factor	Gradient Uniform Series Factor
	To Find F Given P	To Find P Given F	To Find F Given A	To Find P Given A	To Find A Given F	To Find A Given P		To Find P Given G	To Find A Given G
N	F/P	P/F	F/A	P/A	A/F	A/P		P/G	A/G	N
1	1.1800	0.8475	1.0000	0.8475	1.0000	1.1800		0.000	0.0000	1
2	1.3924	0.7182	2.1800	1.5656	0.4587	0.6387		0.718	0.4587	2
3	1.6430	0.6086	3.5724	2.1743	0.2799	0.4599		1.935	0.8902	3
4	1.9388	0.5158	5.2154	2.6901	0.1917	0.3717		3.483	1.2947	4
5	2.2878	0.4371	7.1542	3.1272	0.1398	0.3198		5.231	1.6728	5
6	2.6996	0.3704	9.4420	3.4976	0.1059	0.2859		7.083	2.0252	6
7	3.1855	0.3139	12.1415	3.8115	0.0824	0.2624		8.967	2.3526	7
8	3.7589	0.2660	15.3270	4.0776	0.0652	0.2452		10.829	2.6558	8
9	4.4355	0.2255	19.0859	4.3030	0.0524	0.2324		12.633	2.9358	9
10	5.2338	0.1911	23.5213	4.4941	0.0425	0.2225		14.353	3.1936	10
11	6.1759	0.1619	28.7551	4.6560	0.0348	0.2148		15.972	3.4303	11
12	7.2876	0.1372	34.9311	4.7932	0.0286	0.2086		17.481	3.6470	12
13	8.5994	0.1163	42.2187	4.9095	0.0237	0.2037		18.877	3.8449	13
14	10.1472	0.0985	50.8180	5.0081	0.0197	0.1997		20.158	4.0250	14
15	11.9737	0.0835	60.9653	5.0916	0.0164	0.1964		21.327	4.1887	15
16	14.1290	0.0708	72.9390	5.1624	0.0137	0.1937		22.389	4.3369	16
17	16.6722	0.0600	87.0680	5.2223	0.0115	0.1915		23.348	4.4708	17
18	19.6733	0.0508	103.7403	5.2732	0.0096	0.1896		24.212	4.5916	18
19	23.2144	0.0431	123.4135	5.3162	0.0081	0.1881		24.988	4.7003	19
20	27.3930	0.0365	146.6280	5.3527	0.0068	0.1868		25.681	4.7978	20
21	32.3238	0.0309	174.0210	5.3837	0.0057	0.1857		26.300	4.8851	21
22	38.1421	0.0262	206.3448	5.4099	0.0048	0.1848		26.851	4.9632	22
23	45.0076	0.0222	244.4868	5.4321	0.0041	0.1841		27.339	5.0329	23
24	53.1090	0.0188	289.4945	5.4509	0.0035	0.1835		27.773	5.0950	24
25	62.6686	0.0160	342.6035	5.4669	0.0029	0.1829		28.156	5.1502	25
30	143.3706	0.0070	790.9480	5.5168	0.0013	0.1813		29.486	5.3448	30
35	327.9973	0.0030	1816.6516	5.5386	0.0006	0.1806		30.177	5.4485	35
40	750.3783	0.0013	4163.2130	5.5482	0.0002	0.1802		30.527	5.5022	40
45	1716.6839	0.0006	9531.5771	5.5523	0.0001	0.1801		30.701	5.5293	45
50	3927.3569	0.0003	21813.0937	5.5541	*a*	0.1800		30.786	5.5428	50
60	20555.1400	*a*	114189.6665	5.5553	*a*	0.1800		30.847	5.5526	60
80	563067.6604	*a*	3128148.1133	5.5555	*a*	0.1800		30.863	5.5554	80
∞				5.5556		0.1800				∞

a Less than 0.0001

TABLE C-17   Discrete Compounding; $i = 20\%$

	Single Payment		Uniform Series					Uniform Gradient	
	Compound Amount Factor	Present Worth Factor	Compound Amount Factor	Present Worth Factor	Sinking Fund Factor	Capital Recovery Factor	Gradient Present Worth Factor	Gradient Uniform Series Factor	
	To Find F Given P	To Find P Given F	To Find F Given A	To Find P Given A	To Find A Given F	To Find A Given P	To Find P Given G	To Find A Given G	
N	F/P	P/F	F/A	P/A	A/F	A/P	P/G	A/G	N
1	1.2000	0.8333	1.0000	0.8333	1.0000	1.2000	0.000	0.0000	1
2	1.4400	0.6944	2.2000	1.5278	0.4545	0.6545	0.694	0.4545	2
3	1.7280	0.5787	3.6400	2.1065	0.2747	0.4747	1.852	0.8791	3
4	2.0736	0.4823	5.3680	2.5887	0.1863	0.3863	3.299	1.2742	4
5	2.4883	0.4019	7.4416	2.9906	0.1344	0.3344	4.906	1.6405	5
6	2.9860	0.3349	9.9299	3.3255	0.1007	0.3007	6.581	1.9788	6
7	3.5832	0.2791	12.9159	3.6046	0.0774	0.2774	8.255	2.2902	7
8	4.2998	0.2326	16.4991	3.8372	0.0606	0.2606	9.883	2.5756	8
9	5.1598	0.1938	20.7989	4.0310	0.0481	0.2481	11.434	2.8364	9
10	6.1917	0.1615	25.9587	4.1925	0.0385	0.2385	12.887	3.0739	10
11	7.4301	0.1346	32.1504	4.3271	0.0311	0.2311	14.233	3.2893	11
12	8.9161	0.1122	39.5805	4.4392	0.0253	0.2253	15.467	3.4841	12
13	10.6993	0.0935	48.4966	4.5327	0.0206	0.2206	16.588	3.6597	13
14	12.8392	0.0779	59.1959	4.6106	0.0169	0.2169	17.601	3.8175	14
15	15.4070	0.0649	72.0351	4.6755	0.0139	0.2139	18.510	3.9588	15
16	18.4884	0.0541	87.4421	4.7296	0.0114	0.2114	19.321	4.0851	16
17	22.1861	0.0451	105.9306	4.7746	0.0094	0.2094	20.042	4.1976	17
18	26.6233	0.0376	128.1167	4.8122	0.0078	0.2078	20.681	4.2975	18
19	31.9480	0.0313	154.7400	4.8435	0.0065	0.2065	21.244	4.3861	19
20	38.3376	0.0261	186.6880	4.8696	0.0054	0.2054	21.740	4.4643	20
21	46.0051	0.0217	225.0256	4.8913	0.0044	0.2044	22.174	4.5334	21
22	55.2061	0.0181	271.0307	4.9094	0.0037	0.2037	22.555	4.5941	22
23	66.2474	0.0151	326.2369	4.9245	0.0031	0.2031	22.887	4.6475	23
24	79.4968	0.0126	392.4842	4.9371	0.0025	0.2025	23.176	4.6943	24
25	95.3962	0.0105	471.9811	4.9476	0.0021	0.2021	23.428	4.7352	25
30	237.3763	0.0042	1181.8816	4.9789	0.0008	0.2008	24.263	4.8731	30
35	590.6682	0.0017	2948.3411	4.9915	0.0003	0.2003	24.661	4.9406	35
40	1469.7716	0.0007	7343.8578	4.9966	0.0001	0.2001	24.847	4.9728	40
45	3657.2620	0.0003	18281.3099	4.9986	0.0001	0.2001	24.932	4.9877	45
50	9100.4382	0.0001	45497.1908	4.9995	^{a}	0.2000	24.970	4.9945	50
60	56347.5144	^{a}	281732.5718	4.9999	^{a}	0.2000	24.994	4.9989	60
80	2160228.4620	^{a}	10801137.3101	5.0000	^{a}	0.2000	25.000	5.0000	80
∞				5.0000		0.2000			∞

^{a} Less than 0.0001

TABLE C-18 Discrete Compounding; $i = 25\%$

	Single Payment		Uniform Series					Uniform Gradient		
	Compound Amount Factor	Present Worth Factor	Compound Amount Factor	Present Worth Factor	Sinking Fund Factor	Capital Recovery Factor		Gradient Present Worth Factor	Gradient Uniform Series Factor	
N	To Find F Given P F/P	To Find P Given F P/F	To Find F Given A F/A	To Find P Given A P/A	To Find A Given F A/F	To Find A Given P A/P		To Find P Given G P/G	To Find A Given G A/G	N
1	1.2500	0.8000	1.0000	0.8000	1.0000	1.2500		0.000	0.0000	1
2	1.5625	0.6400	2.2500	1.4400	0.4444	0.6944		0.640	0.4444	2
3	1.9531	0.5120	3.8125	1.9520	0.2623	0.5123		1.664	0.8525	3
4	2.4414	0.4096	5.7656	2.3616	0.1734	0.4234		2.893	1.2249	4
5	3.0518	0.3277	8.2070	2.6893	0.1218	0.3718		4.204	1.5631	5
6	3.8147	0.2621	11.2588	2.9514	0.0888	0.3388		5.514	1.8683	6
7	4.7684	0.2097	15.0735	3.1611	0.0663	0.3163		6.773	2.1424	7
8	5.9605	0.1678	19.8419	3.3289	0.0504	0.3004		7.947	2.3872	8
9	7.4506	0.1342	25.8023	3.4631	0.0388	0.2888		9.021	2.6048	9
10	9.3132	0.1074	33.2529	3.5705	0.0301	0.2801		9.987	2.7971	10
11	11.6415	0.0859	42.5661	3.6564	0.0235	0.2735		10.846	2.9663	11
12	14.5519	0.0687	54.2077	3.7251	0.0184	0.2684		11.602	3.1145	12
13	18.1899	0.0550	68.7596	3.7801	0.0145	0.2645		12.262	3.2437	13
14	22.7374	0.0440	86.9495	3.8241	0.0115	0.2615		12.833	3.3559	14
15	28.4217	0.0352	109.6868	3.8593	0.0091	0.2591		13.326	3.4530	15
16	35.5271	0.0281	138.1085	3.8874	0.0072	0.2572		13.748	3.5366	16
17	44.4089	0.0225	173.6357	3.9099	0.0058	0.2558		14.109	3.6084	17
18	55.5112	0.0180	218.0446	3.9279	0.0046	0.2546		14.415	3.6698	18
19	69.3889	0.0144	273.5558	3.9424	0.0037	0.2537		14.674	3.7222	19
20	86.7362	0.0115	342.9447	3.9539	0.0029	0.2529		14.893	3.7667	20
21	108.4202	0.0092	429.6809	3.9631	0.0023	0.2523		15.078	3.8045	21
22	135.5253	0.0074	538.1011	3.9705	0.0019	0.2519		15.233	3.8365	22
23	169.4066	0.0059	673.6264	3.9764	0.0015	0.2515		15.363	3.8634	23
24	211.7582	0.0047	843.0329	3.9811	0.0012	0.2512		15.471	3.8861	24
25	264.6978	0.0038	1054.7912	3.9849	0.0009	0.2509		15.562	3.9052	25
30	807.7936	0.0012	3227.1743	3.9950	0.0003	0.2503		15.832	3.9628	30
35	2465.1903	0.0004	9856.7613	3.9984	0.0001	0.2501		15.937	3.9858	35
40	7523.1638	0.0001	30088.6554	3.9995	*a*	0.2500		15.977	3.9947	40
45	22958.8740	*a*	91831.4962	3.9998	*a*	0.2500		15.992	3.9980	45
50	70064.9232	*a*	280255.6929	3.9999	*a*	0.2500		15.997	3.9993	50
60	652530.4468	*a*	2610117.7872	4.0000	*a*	0.2500		16.000	3.9999	60
∞				4.0000		0.2500				∞

a Less than 0.0001

Interest and Annuity Tables for Continuous Compounding

(For various values of r from 5% to 20%)

r = nominal interest rate per period, compounded continuously

N = number of compounding periods

$$(F/P, \underline{r}\%, N) = e^{rN}$$

$$(P/A, \underline{r}\%, N) = \frac{e^{rN} - 1}{e^{rN}(e^r - 1)}$$

$$(P/F, \underline{r}\%, N) = e^{-rN} = \frac{1}{e^{rN}}$$

$$(F/\overline{A}, \underline{r}\%, N) = \frac{e^{rN} - 1}{r}$$

$$(F/A, \underline{r}\%, N) = \frac{e^{rN} - 1}{e^r - 1}$$

$$(P/\overline{A}, \underline{r}\%, N) = \frac{e^{rN} - 1}{re^{rN}}$$

TABLE D-1 Continuous Compounding; $r = 5\%$

	Discrete Flows				Continuous Flows		
	Single Payment		Uniform Series		Uniform Series		
	Compound Amount Factor	Present Worth Factor	Compound Amount Factor	Present Worth Factor	Compound Amount Factor	Present Worth Factor	
N	To Find F Given P F/P	To Find P Given F P/F	To Find F Given A F/A	To Find P Given A P/A	To Find F Given $\overline{A}$ $F/\overline{A}$	To Find P Given $\overline{A}$ $P/\overline{A}$	N
1	1.0513	0.9512	1.0000	0.9512	1.0254	0.9754	1
2	1.1052	0.9048	2.0513	1.8561	2.1034	1.9033	2
3	1.1618	0.8607	3.1564	2.7168	3.2367	2.7858	3
4	1.2214	0.8187	4.3183	3.5355	4.4281	3.6254	4
5	1.2840	0.7788	5.5397	4.3143	5.6805	4.4240	5
6	1.3499	0.7408	6.8237	5.0551	6.9972	5.1836	6
7	1.4191	0.7047	8.1736	5.7598	8.3814	5.9062	7
8	1.4918	0.6703	9.5926	6.4301	9.8365	6.5936	8
9	1.5683	0.6376	11.0845	7.0678	11.3662	7.2474	9
10	1.6487	0.6065	12.6528	7.6743	12.9744	7.8694	10
11	1.7333	0.5769	14.3015	8.2512	14.6651	8.4610	11
12	1.8221	0.5488	16.0347	8.8001	16.4424	9.0238	12
13	1.9155	0.5220	17.8569	9.3221	18.3108	9.5591	13
14	2.0138	0.4966	19.7724	9.8187	20.2751	10.0683	14
15	2.1170	0.4724	21.7862	10.2911	22.3400	10.5527	15
16	2.2255	0.4493	23.9032	10.7404	24.5108	11.0134	16
17	2.3396	0.4274	26.1287	11.1678	26.7929	11.4517	17
18	2.4596	0.4066	28.4683	11.5744	29.1921	11.8686	18
19	2.5857	0.3867	30.9279	11.9611	31.7142	12.2652	19
20	2.7183	0.3679	33.5137	12.3290	34.3656	12.6424	20
21	2.8577	0.3499	36.2319	12.6789	37.1530	13.0012	21
22	3.0042	0.3329	39.0896	13.0118	40.0833	13.3426	22
23	3.1582	0.3166	42.0938	13.3284	43.1639	13.6673	23
24	3.3201	0.3012	45.2519	13.6296	46.4023	13.9761	24
25	3.4903	0.2865	48.5721	13.9161	49.8069	14.2699	25
26	3.6693	0.2725	52.0624	14.1887	53.3859	14.5494	26
27	3.8574	0.2592	55.7317	14.4479	57.1485	14.8152	27
28	4.0552	0.2466	59.5891	14.6945	61.1040	15.0681	28
29	4.2631	0.2346	63.6443	14.9291	65.2623	15.3086	29
30	4.4817	0.2231	67.9074	15.1522	69.6338	15.5374	30
35	5.7546	0.1738	92.7346	16.1149	95.0921	16.5245	35
40	7.3891	0.1353	124.613	16.8646	127.781	17.2933	40
45	9.4877	0.1054	165.546	17.4484	169.755	17.8920	45
50	12.1825	0.0821	218.105	17.9032	223.650	18.3583	50
55	15.6426	0.0639	285.592	18.2573	292.853	18.7214	55
60	20.0855	0.0498	372.247	18.5331	381.711	19.0043	60
65	25.7903	0.0388	483.515	18.7479	495.807	19.2245	65
70	33.1155	0.0302	626.385	18.9152	642.309	19.3961	70
75	42.5211	0.0235	809.834	19.0455	830.422	19.5296	75
80	54.5981	0.0183	1045.39	19.1469	1071.963	19.6337	80
85	70.1054	0.0143	1347.84	19.2260	1382.108	19.7147	85
90	90.0171	0.0111	1736.20	19.2875	1780.342	19.7778	90
95	115.584	0.0087	2234.87	19.3354	2291.686	19.8270	95
100	148.413	0.0067	2875.17	19.3727	2948.263	19.8652	100

TABLE D-2   Continuous Compounding; r = 8%

	Discrete Flows				Continuous Flows		
	Single Payment		Uniform Series		Uniform Series		
	Compound Amount Factor	Present Worth Factor	Compound Amount Factor	Present Worth Factor	Compound Amount Factor	Present Worth Factor	
	To Find F Given P	To Find P Given F	To Find F Given A	To Find P Given A	To Find F Given A	To Find P Given A	
N	F/P	P/F	F/A	P/A	$F/\bar{A}$	$P/\bar{A}$	N
1	1.0833	0.9231	1.0000	0.9231	1.0411	0.9610	1
2	1.1735	0.8521	2.0833	1.7753	2.1689	1.8482	2
3	1.2712	0.7866	3.2568	2.5619	3.3906	2.6672	3
4	1.3771	0.7261	4.5280	3.2880	4.7141	3.4231	4
5	1.4918	0.6703	5.9052	3.9584	6.1478	4.1210	5
6	1.6161	0.6188	7.3970	4.5771	7.7009	4.7652	6
7	1.7507	0.5712	9.0131	5.1483	9.3834	5.3599	7
8	1.8965	0.5273	10.7637	5.6756	11.2060	5.9088	8
9	2.0544	0.4868	12.6602	6.1624	13.1804	6.4156	9
10	2.2255	0.4493	14.7147	6.6117	15.3193	6.8834	10
11	2.4109	0.4148	16.9402	7.0265	17.6362	7.3152	11
12	2.6117	0.3829	19.3511	7.4094	20.1462	7.7138	12
13	2.8292	0.3535	21.9628	7.7629	22.8652	8.0818	13
14	3.0649	0.3263	24.7920	8.0891	25.8107	8.4215	14
15	3.3201	0.3012	27.8569	8.3903	29.0015	8.7351	15
16	3.5966	0.2780	31.1770	8.6684	32.4580	9.0245	16
17	3.8962	0.2567	34.7736	8.9250	36.2024	9.2917	17
18	4.2207	0.2369	38.6698	9.1620	40.2587	9.5384	18
19	4.5722	0.2187	42.8905	9.3807	44.6528	9.7661	19
20	4.9530	0.2019	47.4627	9.5826	49.4129	9.9763	20
21	5.3656	0.1864	52.4158	9.7689	54.5694	10.1703	21
22	5.8124	0.1720	57.7813	9.9410	60.1555	10.3494	22
23	6.2965	0.1588	63.5938	10.0998	66.2067	10.5148	23
24	6.8120	0.1466	69.8903	10.2464	72.7620	10.6674	24
25	7.3891	0.1353	76.7113	10.3817	79.8632	10.8083	25
26	8.0045	0.1249	84.1003	10.5067	87.5559	10.9384	26
27	8.6711	0.1153	92.1048	10.6220	95.8892	11.0584	27
28	9.3933	0.1065	100.776	10.7285	104.917	11.1693	28
29	10.1757	0.0983	110.169	10.8267	114.696	11.2716	29
30	11.0232	0.0907	120.345	10.9174	125.290	11.3660	30
35	16.4446	0.0608	185.439	11.2765	193.058	11.7399	35
40	24.5325	0.0408	282.547	11.5172	294.157	11.9905	40
45	36.5982	0.0273	427.416	11.6786	444.978	12.1585	45
50	54.5982	0.0183	643.535	11.7868	669.977	12.2711	50
55	81.4509	0.0123	965.947	11.8593	1005.64	12.3465	55
60	121.510	0.0082	1446.93	11.9079	1506.38	12.3971	60
65	181.272	0.0055	2164.47	11.9404	2253.40	12.4310	65
70	270.426	0.0037	3234.91	11.9623	3367.83	12.4538	70
75	403.429	0.0025	4831.83	11.9769	5030.36	12.4690	75
80	601.845	0.0017	7214.15	11.9867	7510.56	12.4792	80
85	897.847	0.0011	10768.1	11.9933	11210.6	12.4861	85
90	1339.43	0.0007	16070.1	11.9977	16730.4	12.4907	90
95	1998.20	0.0005	23979.7	12.0007	24964.9	12.4937	95
100	2980.96	0.0003	35779.3	12.0026	37249.5	12.4958	100

TABLE D-3   Continuous Compounding; $r = 10\%$

	Discrete Flows				Continuous Flows		
	Single Payment		Uniform Series		Uniform Series		
	Compound Amount Factor	Present Worth Factor	Compound Amount Factor	Present Worth Factor	Compound Amount Factor	Present Worth Factor	
N	To Find F Given P F/P	To Find P Given F P/F	To Find F Given A F/A	To Find P Given A P/A	To Find F Given $\overline{A}$ $F/\overline{A}$	To Find P Given $\overline{A}$ $P/\overline{A}$	N
1	1.1052	0.9048	1.0000	0.9048	1.0517	0.9516	1
2	1.2214	0.8187	2.1052	1.7236	2.2140	1.8127	2
3	1.3499	0.7408	3.3266	2.4644	3.4986	2.5918	3
4	1.4918	0.6703	4.6764	3.1347	4.9182	3.2968	4
5	1.6487	0.6065	6.1683	3.7412	6.4872	3.9347	5
6	1.8221	0.5488	7.8170	4.2900	8.2212	4.5119	6
7	2.0138	0.4966	9.6391	4.7866	10.1375	5.0341	7
8	2.2255	0.4493	11.6528	5.2360	12.2554	5.5067	8
9	2.4596	0.4066	13.8784	5.6425	14.5960	5.9343	9
10	2.7183	0.3679	16.3380	6.0104	17.1828	6.3212	10
11	3.0042	0.3329	19.0563	6.3433	20.0417	6.6713	11
12	3.3201	0.3012	22.0604	6.6445	23.2012	6.9881	12
13	3.6693	0.2725	25.3806	6.9170	26.6930	7.2747	13
14	4.0552	0.2466	29.0499	7.1636	30.5520	7.5340	14
15	4.4817	0.2231	33.1051	7.3867	34.8169	7.7687	15
16	4.9530	0.2019	37.5867	7.5886	39.5303	7.9810	16
17	5.4739	0.1827	42.5398	7.7713	44.7395	8.1732	17
18	6.0496	0.1653	48.0137	7.9366	50.4965	8.3470	18
19	6.6859	0.1496	54.0634	8.0862	56.8589	8.5043	19
20	7.3891	0.1353	60.7493	8.2215	63.8906	8.6466	20
21	8.1662	0.1225	68.1383	8.3440	71.6617	8.7754	21
22	9.0250	0.1108	76.3045	8.4548	80.2501	8.8920	22
23	9.9742	0.1003	85.3295	8.5550	89.7418	8.9974	23
24	11.0232	0.0907	95.3037	8.6458	100.232	9.0928	24
25	12.1825	0.0821	106.327	8.7278	111.825	9.1791	25
26	13.4637	0.0743	118.509	8.8021	124.637	9.2573	26
27	14.8797	0.0672	131.973	8.8693	138.797	9.3279	27
28	16.4446	0.0608	146.853	8.9301	154.446	9.3919	28
29	18.1741	0.0550	163.298	8.9852	171.741	9.4498	29
30	20.0855	0.0498	181.472	9.0349	190.855	9.5021	30
35	33.1155	0.0302	305.364	9.2212	321.154	9.6980	35
40	54.5981	0.0183	509.629	9.3342	535.982	9.8168	40
45	90.0171	0.0111	846.404	9.4027	890.171	9.8889	45
50	148.413	0.0067	1401.65	9.4443	1474.13	9.9326	50
55	244.692	0.0041	2317.10	9.4695	2436.92	9.9591	55
60	403.429	0.0025	3826.43	9.4848	4024.29	9.9752	60
65	665.142	0.0015	6314.88	9.4940	6641.42	9.9850	65
70	1096.63	0.0009	10417.6	9.4997	10956.3	9.9909	70
75	1808.04	0.0006	17182.0	9.5031	18070.7	9.9945	75
80	2980.96	0.0003	28334.4	9.5051	29799.6	9.9966	80
85	4914.77	0.0002	46721.7	9.5064	49137.7	9.9980	85
90	8103.08	0.0001	77037.3	9.5072	81020.8	9.9988	90
95	13359.7	a	127019	9.5076	133587	9.9993	95
100	22026.5	a	209425	9.5079	220255	9.9995	100

aLess than 0.0001.

624

TABLE D-4 Continuous Compounding; $r = 12\%$

	Discrete Flows				Continuous Flows		
	Single Payment		Uniform Series		Uniform Series		
	Compound Amount Factor	Present Worth Factor	Compound Amount Factor	Present Worth Factor	Compound Amount Factor	Present Worth Factor	
N	To Find F Given P F/P	To Find P Given F P/F	To Find F Given A F/A	To Find P Given A P/A	To Find F Given A F/Ā	To Find P Given A P/Ā	N
1	1.1275	0.8869	1.0000	0.8869	1.0625	0.9423	1
2	1.2712	0.7866	2.1275	1.6735	2.2604	1.7781	2
3	1.4333	0.6977	3.3987	2.3712	3.6111	2.5194	3
4	1.6161	0.6188	4.8321	2.9900	5.1340	3.1768	4
5	1.8221	0.5488	6.4481	3.5388	6.8510	3.7599	5
6	2.0544	0.4868	8.2703	4.0256	8.7869	4.2771	6
7	2.3164	0.4317	10.3247	4.4573	10.9679	4.7357	7
8	2.6117	0.3829	12.6411	4.8402	13.4308	5.1426	8
9	2.9447	0.3396	15.2528	5.1798	16.2057	5.5034	9
10	3.3201	0.3012	18.1974	5.4810	19.3343	5.8234	10
11	3.7434	0.2671	21.5176	5.7481	22.8618	6.1072	11
12	4.2207	0.2369	25.2610	5.9850	26.8391	6.3589	12
13	4.7588	0.2101	29.4817	6.1952	31.3235	6.5822	13
14	5.3656	0.1864	34.2405	6.3815	36.3796	6.7802	14
15	6.0496	0.1653	39.6061	6.5468	42.0804	6.9558	15
16	6.8210	0.1466	45.6557	6.6934	48.5080	7.1116	16
17	7.6906	0.1300	52.4767	6.8235	55.7551	7.2498	17
18	8.6711	0.1153	60.1673	6.9388	63.9261	7.3723	18
19	9.7767	0.1023	68.8384	7.0411	73.1390	7.4810	19
20	11.0232	0.0907	78.6151	7.1318	83.5265	7.5774	20
21	12.4286	0.0805	89.6383	7.2123	95.2383	7.6628	21
22	14.0132	0.0714	102.067	7.2836	108.443	7.7387	22
23	15.7998	0.0633	116.080	7.3469	123.332	7.8059	23
24	17.8143	0.0561	131.880	7.4030	140.119	7.8655	24
25	20.0855	0.0498	149.694	7.4528	159.046	7.9184	25
26	22.6464	0.0442	169.780	7.4970	180.386	7.9654	26
27	25.5337	0.0392	192.426	7.5362	204.448	8.0070	27
28	28.7892	0.0347	217.960	7.5709	231.577	8.0439	28
29	32.4597	0.0308	246.749	7.6017	262.164	8.0766	29
30	36.5982	0.0273	279.209	7.6290	296.652	8.1056	30
35	66.6863	0.0150	515.200	7.7257	547.386	8.2084	35
40	121.510	0.0082	945.203	7.7788	1004.25	8.2648	40
45	221.406	0.0045	1728.72	7.8079	1836.72	8.2957	45
50	403.429	0.0025	3156.38	7.8239	3353.57	8.3127	50
55	735.095	0.0014	5757.75	7.8327	6117.46	8.3220	55
60	1339.43	0.0007	10497.8	7.8375	11153.6	8.3271	60
65	2440.60	0.0004	19134.6	7.8401	20330.0	8.3299	65
70	4447.07	0.0002	34872.0	7.8416	37050.6	8.3315	70
75	8103.08	0.0001	63547.3	7.8424	67517.4	8.3323	75
80	4764.8	^{a}	115797	7.8428	123032	8.3328	80

^{a}Less than 0.0001.

TABLE D-5   Continuous Compounding; r = 15%

	Discrete Flows				Continuous Flows		
	Single Payment		Uniform Series		Uniform Series		
	Compound Amount Factor	Present Worth Factor	Compound Amount Factor	Present Worth Factor	Compound Amount Factor	Present Worth Factor	
N	To Find F Given P F/P	To Find P Given F P/F	To Find F Given A F/A	To Find P Given A P/A	To Find F Given A $F/\overline{A}$	To Find P Given A $P/\overline{A}$	N
1	1.1618	0.8607	1.0000	0.8607	1.0789	0.9286	1
2	1.3499	0.7408	2.1618	1.6015	2.3324	1.7279	2
3	1.5683	0.6376	3.5117	2.2392	3.7887	2.4158	3
4	1.8221	0.5488	5.0800	2.7880	5.4808	3.0079	4
5	2.1170	0.4724	6.9021	3.2603	7.4467	3.5176	5
6	2.4596	0.4066	9.0191	3.6669	9.7307	3.9562	6
7	2.8577	0.3499	11.4787	4.0168	12.3843	4.3337	7
8	3.3201	0.3012	14.3364	4.3180	15.4674	4.6587	8
9	3.8574	0.2592	17.6565	4.5773	19.0495	4.9384	9
10	4.4817	0.2231	21.5139	4.8004	23.2113	5.1791	10
11	5.2070	0.1920	25.9956	4.9925	28.0465	5.3863	11
12	6.0496	0.1653	31.2026	5.1578	33.6643	5.5647	12
13	7.0287	0.1423	37.2522	5.3000	40.1913	5.7182	13
14	8.1662	0.1225	44.2809	5.4225	47.7745	5.8503	14
15	9.4877	0.1054	52.4471	5.5279	56.5849	5.9640	15
16	11.0232	0.0907	61.9348	5.6186	66.8212	6.0619	16
17	12.8071	0.0781	72.9580	5.6967	78.7140	6.1461	17
18	14.8797	0.0672	85.7651	5.7639	92.5315	6.2186	18
19	17.2878	0.0578	100.645	5.8217	108.585	6.2810	19
20	20.0855	0.0498	117.933	5.8715	127.237	6.3348	20
21	23.3361	0.0429	138.018	5.9144	148.907	6.3810	21
22	27.1126	0.0369	161.354	5.9513	174.084	6.4208	22
23	31.5004	0.0317	188.467	5.9830	203.336	6.4550	23
24	36.5982	0.0273	219.967	6.0103	237.322	6.4845	24
25	42.4024	0.0235	256.565	6.0338	276.807	6.5099	25
26	49.5211	0.0202	299.087	6.0541	322.683	6.5317	26
27	57.3975	0.0174	348.489	6.0715	375.983	6.5505	27
28	66.6863	0.0150	405.886	6.0865	437.909	6.5667	28
29	77.4785	0.0129	472.573	6.0994	509.856	6.5806	29
30	90.0171	0.0111	550.051	6.1105	593.448	6.5926	30
35	190.566	0.0052	1171.36	6.1467	1263.78	6.6317	35
40	403.429	0.0025	2486.67	6.1638	2682.86	6.6501	40
45	854.059	0.0012	5271.19	6.1719	5687.06	6.6589	45
50	1808.04	0.0006	11166.0	6.1757	12046.9	6.6630	50
55	3827.63	0.0003	23645.3	6.1775	25510.8	6.6649	55
60	8103.08	0.0001	50064.1	6.1784	54013.9	6.6658	60
65	17154.2	a	105993	6.1788	114355	6.6663	65
70	36315.5	a	224393	6.1790	242097	6.6665	70
75	76879.9	a	475047	6.1791	512526	6.6666	75
80	162755	a	1005680	6.1791	1085030	6.6666	80

a Less than 0.0001.

TABLE D-6 Continuous Compounding; $r = 20\%$

	Discrete Flows				Continuous Flows		
	Single Payment		Uniform Series		Uniform Series		
	Compound Amount Factor	Present Worth Factor	Compound Amount Factor	Present Worth Factor	Compound Amount Factor	Present Worth Factor	
N	To Find F Given P F/P	To Find P Given F P/F	To Find F Given A F/A	To Find P Given A P/A	To Find F Given $\overline{A}$ $F/\overline{A}$	To Find P Given $\overline{A}$ $P/\overline{A}$	N
1	1.2214	0.8187	1.0000	0.8187	1.1070	0.9063	1
2	1.4918	0.6703	2.2214	1.4891	2.4591	1.6484	2
3	1.8221	0.5488	3.7132	2.0379	4.1106	2.2559	3
4	2.2255	0.4493	5.5353	2.4872	6.1277	2.7534	4
5	2.7183	0.3679	7.7609	2.8551	8.5914	3.1606	5
6	3.3201	0.3012	10.4792	3.1563	11.6006	3.4940	6
7	4.0552	0.2466	13.7993	3.4029	15.2760	3.7670	7
8	4.9530	0.2019	17.8545	3.6048	19.7652	3.9905	8
9	6.0496	0.1653	22.8075	3.7701	25.2482	4.1735	9
10	7.3891	0.1353	28.8572	3.9054	31.9453	4.3233	10
11	9.0250	0.1108	36.2462	4.0162	40.1251	4.4460	11
12	11.0232	0.0907	45.2712	4.1069	50.1159	4.5464	12
13	13.4637	0.0743	56.2944	4.1812	62.3187	4.6286	13
14	16.4446	0.0608	69.7581	4.2420	77.2232	4.6959	14
15	20.0855	0.0498	86.2028	4.2918	95.4277	4.7511	15
16	24.5325	0.0408	106.288	4.3325	117.633	4.7962	16
17	29.9641	0.0334	130.821	4.3659	144.820	4.8331	17
18	36.5982	0.0273	160.785	4.3932	177.991	4.8634	18
19	44.7012	0.0224	197.383	4.4156	218.506	4.8881	19
20	54.5981	0.0183	242.084	4.4339	267.991	4.9084	20
21	66.6863	0.0150	296.682	4.4489	328.432	4.9250	21
22	81.4509	0.0123	363.369	4.4612	402.254	4.9386	22
23	99.4843	0.0101	444.820	4.4713	492.422	4.9497	23
24	121.510	0.0082	544.304	4.4795	602.552	4.9589	24
25	148.413	0.0067	665.814	4.4862	737.066	4.9663	25
26	181.272	0.0055	814.227	4.4917	901.361	4.9724	26
27	221.406	0.0045	995.500	4.4963	1102.03	4.9774	27
28	270.426	0.0037	1216.91	4.5000	1347.13	4.9815	28
29	330.299	0.0030	1487.33	4.5030	1646.50	4.9849	29
30	403.429	0.0025	1817.63	4.5055	2012.14	4.9876	30
35	1096.63	0.0009	4948.60	4.5125	5478.17	4.9954	35
40	2980.96	0.0003	13459.4	4.5151	14899.8	4.9983	40
45	8103.08	0.0001	36594.3	4.5161	40510.4	4.9994	45
50	22026.5	a	99481.4	4.5165	110127	4.9998	50
55	59874.1	a	270426	4.5166	299366	4.9999	55
60	162755	a	735103	4.5166	813769	5.0000	60

a Less than 0.0001.

*S*tandardized Normal Distribution Function

Areas Under the Normal Curve[a]

z	0.00	0.01	0.02	0.03	0.04	0.05	0.06	0.07	0.08	0.09
−3.4	0.0003	0.0003	0.0003	0.0003	0.0003	0.0003	0.0003	0.0003	0.0003	0.0002
−3.3	0.0005	0.0005	0.0005	0.0004	0.0004	0.0004	0.0004	0.0004	0.0004	0.0003
−3.2	0.0007	0.0007	0.0006	0.0006	0.0006	0.0006	0.0006	0.0005	0.0005	0.0005
−3.1	0.0010	0.0009	0.0009	0.0009	0.0008	0.0008	0.0008	0.0007	0.0007	0.0007
−3.0	0.0013	0.0013	0.0013	0.0012	0.0012	0.0011	0.0011	0.0011	0.0010	0.0010
−2.9	0.0019	0.0018	0.0017	0.0017	0.0016	0.0016	0.0015	0.0015	0.0014	0.0014
−2.8	0.0026	0.0025	0.0024	0.0023	0.0023	0.0022	0.0021	0.0021	0.0020	0.0019
−2.7	0.0035	0.0034	0.0033	0.0032	0.0031	0.0030	0.0029	0.0028	0.0027	0.0026
−2.6	0.0047	0.0045	0.0044	0.0043	0.0041	0.0040	0.0039	0.0038	0.0037	0.0036
−2.5	0.0062	0.0060	0.0059	0.0057	0.0055	0.0054	0.0052	0.0051	0.0049	0.0048
−2.4	0.0082	0.0080	0.0078	0.0075	0.0073	0.0071	0.0069	0.0068	0.0066	0.0064
−2.3	0.0107	0.0104	0.0103	0.0099	0.0096	0.0094	0.0091	0.0089	0.0087	0.0084
−2.2	0.0139	0.0136	0.0132	0.0129	0.0125	0.0122	0.0119	0.0116	0.0113	0.0110
−2.1	0.0179	0.0174	0.0170	0.0166	0.0162	0.0158	0.0154	0.0150	0.0146	0.0143
−2.0	0.0228	0.0222	0.0217	0.0212	0.0207	0.0202	0.0197	0.0192	0.0118	0.0183
−1.9	0.0287	0.0281	0.0274	0.0268	0.0262	0.0256	0.0250	0.0244	0.0239	0.0233
−1.8	0.0359	0.0352	0.0344	0.0336	0.0329	0.0322	0.0314	0.0307	0.0301	0.0294
−1.7	0.0446	0.0436	0.0427	0.0418	0.0409	0.0401	0.0392	0.0384	0.0375	0.0367
−1.6	0.0548	0.0537	0.0526	0.0516	0.0505	0.0495	0.0485	0.0475	0.0465	0.0455
−1.5	0.0668	0.0655	0.0643	0.0630	0.0618	0.0606	0.0594	0.0582	0.0571	0.0559
−1.4	0.0808	0.0793	0.0778	0.0764	0.0749	0.0735	0.0722	0.0708	0.0694	0.0681
−1.3	0.0968	0.0951	0.0934	0.0918	0.0901	0.0885	0.0869	0.0853	0.0838	0.0823
−1.2	0.1151	0.1131	0.1112	0.1093	0.1075	0.1056	0.1038	0.1020	0.1003	0.0985
−1.1	0.1357	0.1335	0.1314	0.1292	0.1271	0.1251	0.1230	0.1210	0.1190	0.1170
−1.0	0.1587	0.1562	0.1539	0.1515	0.1492	0.1469	0.1446	0.1423	0.1401	0.1379
−0.9	0.1841	0.1841	0.1788	0.1762	0.1736	0.1711	0.1685	0.1660	0.1635	0.1611
−0.8	0.2119	0.2090	0.2061	0.2033	0.2005	0.1977	0.1949	0.1922	0.1894	0.1867
−0.7	0.2420	0.2389	0.2358	0.2327	0.2296	0.2266	0.2236	0.2206	0.2177	0.2148
−0.6	0.2743	0.2709	0.2676	0.2643	0.2611	0.2578	0.2546	0.2514	0.2483	0.2451
−0.5	0.3085	0.3050	0.3015	0.2981	0.2946	0.2912	0.2877	0.2843	0.2810	0.2776
−0.4	0.3446	0.3409	0.3372	0.3336	0.3300	0.3264	0.3228	0.3192	0.3156	0.3121
−0.3	0.3821	0.3783	0.3745	0.3707	0.3669	0.3632	0.3594	0.3557	0.3520	0.3483
−0.2	0.4207	0.4168	0.4129	0.4090	0.4052	0.4013	0.3974	0.3936	0.3897	0.3859
−0.1	0.4602	0.4562	0.4522	0.4483	0.4443	0.4404	0.4364	0.4325	0.4286	0.4247
−0.0	0.5000	0.4960	0.4920	0.4880	0.4840	0.4801	0.4761	0.4721	0.4681	0.4641

[a]From R. E. Walpole and R. H. Myers, *Probability and Statistics for Engineers and Scientists*, 2nd ed. (New York: Macmillan, 1978), p. 513.

(continued)

Areas Under the Normal Curve (cont'd)

z	0.00	0.01	0.02	0.03	0.04	0.05	0.06	0.07	0.08	0.09
0.0	0.5000	0.5040	0.5080	0.5120	0.5160	0.5199	0.5239	0.5279	0.5319	0.5359
0.1	0.5398	0.5438	0.5478	0.5517	0.5557	0.5596	0.5636	0.5675	0.5714	0.5753
0.2	0.5793	0.5832	0.5871	0.5910	0.5948	0.5987	0.6026	0.6064	0.6103	0.6141
0.3	0.6179	0.6217	0.6255	0.6293	0.6331	0.6368	0.6406	0.6443	0.6480	0.6517
0.4	0.6554	0.6591	0.6628	0.6664	0.6700	0.6736	0.6772	0.6808	0.6844	0.6879
0.5	0.6915	0.6950	0.6985	0.7019	0.7054	0.7088	0.7123	0.7157	0.7190	0.7224
0.6	0.7257	0.7291	0.7324	0.7357	0.7389	0.7422	0.7454	0.7486	0.7517	0.7549
0.7	0.7580	0.7611	0.7642	0.7673	0.7704	0.7734	0.7764	0.7794	0.7823	0.7852
0.8	0.7881	0.7910	0.7939	0.7967	0.7995	0.8023	0.8051	0.8078	0.8106	0.8133
0.9	0.8159	0.8186	0.8212	0.8238	0.8264	0.8289	0.8315	0.8340	0.8365	0.8389
1.0	0.8413	0.8438	0.8461	0.8485	0.8508	0.8531	0.8554	0.8577	0.8599	0.8621
1.1	0.8643	0.8665	0.8686	0.8708	0.8729	0.8749	0.8770	0.8790	0.8810	0.8830
1.2	0.8849	0.8869	0.8888	0.8907	0.8925	0.8944	0.8962	0.8980	0.8997	0.9015
1.3	0.9032	0.9049	0.9066	0.9082	0.9099	0.9115	0.9131	0.9147	0.9162	0.9177
1.4	0.9192	0.9207	0.9222	0.9236	0.9251	0.9265	0.9278	0.9292	0.9306	0.9319
1.5	0.9332	0.9345	0.9357	0.9370	0.9382	0.9394	0.9406	0.9418	0.9429	0.9441
1.6	0.9452	0.9463	0.9474	0.9484	0.9495	0.9505	0.9515	0.9525	0.9535	0.9545
1.7	0.9554	0.9564	0.9573	0.9582	0.9591	0.9599	0.9608	0.9616	0.9625	0.9633
1.8	0.9641	0.9649	0.9656	0.9664	0.9671	0.9678	0.9686	0.9693	0.9699	0.9706
1.9	0.9713	0.9719	0.9726	0.9732	0.9738	0.9744	0.9750	0.9756	0.9761	0.9767
2.0	0.9772	0.9778	0.9783	0.9788	0.9793	0.9798	0.9803	0.9808	0.9812	0.9817
2.1	0.9821	0.9826	0.9830	0.9834	0.9838	0.9842	0.9846	0.9850	0.9854	0.9857
2.2	0.9861	0.9864	0.9868	0.9871	0.9875	0.9878	0.9881	0.9884	0.9887	0.9890
2.3	0.9893	0.9896	0.9898	0.9901	0.9904	0.9906	0.9909	0.9911	0.9913	0.9916
2.4	0.9918	0.9920	0.9922	0.9925	0.9927	0.9929	0.9931	0.9932	0.9934	0.9936
2.5	0.9938	0.9940	0.9941	0.9943	0.9945	0.9946	0.9948	0.9949	0.9951	0.9952
2.6	0.9953	0.9955	0.9956	0.9957	0.9959	0.9960	0.9961	0.9962	0.9963	0.9964
2.7	0.9965	0.9966	0.9967	0.9968	0.9969	0.9970	0.9971	0.9972	0.9973	0.9974
2.8	0.9974	0.9975	0.9976	0.9977	0.9977	0.9978	0.9979	0.9979	0.9980	0.9981
2.9	0.9981	0.9982	0.9982	0.9983	0.9984	0.9984	0.9985	0.9985	0.9986	0.9986
3.0	0.9987	0.9987	0.9987	0.9988	0.9988	0.9989	0.9989	0.9989	0.9990	0.9990
3.1	0.9990	0.9991	0.9991	0.9991	0.9992	0.9992	0.9992	0.9992	0.9993	0.9993
3.2	0.9993	0.9993	0.9994	0.9994	0.9994	0.9994	0.9994	0.9995	0.9995	0.9995
3.3	0.9995	0.9995	0.9995	0.9996	0.9996	0.9996	0.9996	0.9996	0.9996	0.9997
3.4	0.9997	0.9997	0.9997	0.9997	0.9997	0.9997	0.9997	0.9997	0.9997	0.9998

Selected References

AMERICAN TELEPHONE AND TELEGRAPH COMPANY, Engineering Department. *Engineering Economy,* 3rd ed. New York: American Telephone and Telegraph Co., 1977.

AU, T., and T. P. AU. *Engineering Economics for Capital Investment Analysis,* 2nd ed. Boston: Allyn and Bacon, 1992.

BARISH, N. N., and S. KAPLAN. *Economic Analysis for Engineering and Managerial Decision Making.* New York: McGraw-Hill Book Company, 1978.

BIERMAN, H., JR., and S. SMIDT. *The Capital Budgeting Decision,* 8th ed. New York: Macmillan, 1993.

BLANK, L. T., and A. J. TARQUIN. *Engineering Economy,* 3rd ed. New York: McGraw-Hill, 1989.

BRIMSON, J. A. *Activity Accounting: An Activity-Based Approach.* New York: John Wiley & Sons, 1991.

BUSSEY, L. E., and T. G. ESCHENBACH. *The Economic Analysis of Industrial Projects,* 2nd ed. Englewood Cliffs, NJ: Prentice-Hall, 1992.

CAMPEN, J. T. *Benefit, Cost, and Beyond.* Cambridge, MA: Ballinger Publishing Company, 1986.

CANADA, J. R., and W. G. SULLIVAN. *Economic and Multiattribute Analysis of Advanced Manufacturing Systems.* Englewood Cliffs, NJ: Prentice-Hall, 1989.

CANADA, J. R., W. G. SULLIVAN, and J. A. WHITE. *Capital Investment Decision Analysis for Engineering and Management.* Englewood Cliffs, NJ: Prentice-Hall, 1996.

CLARK, J. J., T. J. HINDELANG, and R. E. PRITCHARD. *Capital Budgeting: Planning and Control of Capital Expenditures.* Englewood Cliffs, NJ: Prentice-Hall, 1979.

COCHRANE, J. L., and M. ZELENY. *Multiple Criteria Decision Making.* Columbia, SC: University of South Carolina, 1973.

COLLIER, C. A., and W. B. LEDBETTER. *Engineering Cost Analysis,* 2nd ed. New York: Harper & Row, 1987.

Engineering Economist, The. A quarterly journal jointly published by the Engineering Economy Division of the American Society for Engineering Education and the Institute of Industrial Engineers. Published by IIE, Norcross, GA.

Engineering News-Record. Published monthly by McGraw-Hill, New York.

ENGLISH, J. M., ed. *Cost Effectiveness: Economic Evaluation of Engineering Systems.* New York: John Wiley & Sons, 1968.

ESCHENBACH, T. G. *Engineering Economy: Applying Theory to Practice.* Chicago: Richard D. Irwin, 1995.

FABRYCKY, W. J., and G. J. THUESEN. *Economic Decision Analysis,* 3rd ed. Englewood Cliffs, NJ: Prentice-Hall, 1984.

FLEISCHER, G. A. *Risk and Uncertainty: Non-deterministic Decision Making in Engineering Economy.* Norcross, GA: Institute of Industrial Engineers, Publication EE-75-1, 1975.

FLEISCHER, G. A. *Introduction to Engineering Economy.* Boston: PWS Publishing Company, 1994.

GRANT, E. L., W. G. IRESON, and R. S. LEAVENWORTH. *Principles of Engineering Economy,* 8th ed. New York: John Wiley & Sons, 1990.

GOICOECHEA, A., D. R. HANSEN, and L. DUCKSTEIN. *Multiobjective Decision Analysis with Engineering and Business Applications.* New York: John Wiley & Sons, 1982.

HAPPEL, J., and D. JORDAN. *Chemical Process Economics,* 2nd ed. New York: Marcel Dekker, 1975.

Harvard Business Review. Published bimonthly by the Harvard University Press, Boston.

HULL, J. C. *The Evaluation of Risk in Business Investment.* New York: Pergamon Press, 1980.

Industrial Engineering. A monthly magazine published by the Institute of Industrial Engineers, Norcross, GA.

Internal Revenue Service Publication 534. *Depreciation.* U.S. Government Printing Office, revised periodically (December 1995).

JELEN, F. C., and J. H. BLACK. *Cost and Optimization Engineering,* 3rd ed. New York: McGraw-Hill, 1991.

JONES, B. W. *Inflation in Engineering Economic Analysis.* New York: John Wiley & Sons, 1982.

KAHL, A. L., and W. F. RENTZ. *Spreadsheet Applications in Engineering Economics.* St. Paul: West Publishing Company, 1992.

KENNEY, R. L., and H. RAIFFA. *Decisions with Multiple Objectives: Preferences and Value Trade-offs.* New York: John Wiley & Sons, 1976.

KLEINFELD, IRA H. *Engineering and Managerial Economics.* New York: Holt, Rinehart & Winston, 1986.

LASSER, J. K. *Your Income Tax.* New York: Simon & Schuster (see latest edition).

MACHINERY AND ALLIED PRODUCTS INSTITUTE. *MAPI Replacement Manual.* Washington, DC: Machinery and Allied Products Institute, 1950.

MALLIK, A. K. *Engineering Economy with Computer Applications.* Mahomet, IL: Engineering Technology, 1979.

MAO, J. *Quantitative Analysis of Financial Decisions.* New York: Macmillan, 1969.

MATTHEWS, L. M. *Estimating Manufacturing Costs: A Practical Guide for Managers and Estimators.* New York: McGraw-Hill, 1983.

MAYER, R. R. *Capital Expenditure Analysis for Managers and Engineers.* Prospect Heights, IL: Waveland Press, 1978.

MERRETT, A. J., and A. SYKES. *The Finance and Analysis of Capital Projects.* New York: John Wiley & Sons, 1963.

MISHAN, E. J. *Cost-Benefit Analysis.* New York: Praeger Publishers, 1976.

MORRIS, W. T. *Engineering Economic Analysis.* Reston, VA: Reston Publishing, 1976.

MORRIS, W. T. *Decision Analysis.* Columbus, OH: Grid, 1977.

NEWNAN, D. G., and B. JOHNSON. *Engineering Economic Analysis,* 5th ed. San Jose, CA: Engineering Press, 1995.

OAKFORD, R. V. *Capital Budgeting: A Quantitative Evaluation of Investment Alternatives.* New York: John Wiley & Sons, 1970.

OSTWALD, P. F. *Cost Estimating for Engineering and Management,* 3rd ed. Englewood Cliffs, NJ: Prentice-Hall, 1992.

PARK, C. S. *Contemporary Engineering Economics*. Reading, MA: Addison-Wesley Publishing Company, 1993.

PARK, C. S., and G. P. SHARP-BETTE. *Advanced Engineering Economics.* New York: John Wiley & Sons, 1990.

PARK, W. R., and D. E. JACKSON. *Cost Engineering Analysis: A Guide to Economic Evaluation of Engineering Projects,* 2nd ed. New York: John Wiley & Sons, 1984.

PETERS, M. S., and K. D. TIMMERHAUS. *Plant Design and Economics for Chemical Engineers,* 3rd ed. New York: McGraw-Hill, 1980.

PORTER, M. E. *Competitive Strategy: Techniques for Analyzing Industries and Competitors*. New York: The Free Press, 1980.

ROSE, L. M. *Engineering Investment Decisions: Planning Under Uncertainty*. Amsterdam: Elsevier, 1976.

SMITH, G. W. *Engineering Economy: The Analysis of Capital Expenditures,* 4th ed. Ames, IA: Iowa State University Press, 1987.

STEINER, H. M. *Engineering Economic Principles*. New York: McGraw-Hill, 1992.

STERMOLE, F. J., and J. M. STERMOLE. *Economic Evaluation and Investment Decision Methods,* 6th ed. Golden, CO: Investment Evaluations Corp., 1987.

STEVENS, G. T. *Economic and Financial Analysis of Capital Investment*. New York: John Wiley & Sons, 1979.

STEWART, R. D. *Cost Estimating*. New York: John Wiley & Sons, 1982.

STEWART, R. D., and WYSKIDA, R. M., eds. *Cost Estimators' Reference Manual*. New York: John Wiley & Sons, 1987.

SULLIVAN, W. G., and W. W. CLAYCOMBE. *Fundamentals of Forecasting*. Reston, VA: Reston Publishing, 1977.

TAYLOR, G. A. *Managerial and Engineering Economy,* 3rd ed. New York: Van Nostrand Reinhold, 1980.

TERBORGH, G. *Business Investment Management*. Washington, DC: Machinery and Allied Products Institute, 1967.

THUESEN, G. J., and W. J. FABRYCKY. *Engineering Economy,* 8th ed. Englewood Cliffs, NJ: Prentice-Hall, 1993.

VANHORNE, J. C. *Financial Management and Policy,* 5th ed. Englewood Cliffs, NJ: Prentice-Hall, 1980.

WEINGARTNER, H. M. *Mathematical Programming and the Analysis of Capital Budgeting Problems*. Englewood Cliffs, NJ: Prentice-Hall, 1975.

WELLINGTON, A. M. *The Economic Theory of Railway Location*. New York: John Wiley & Sons, 1887.

WHITE, J. A., M. H. AGEE, and K. E. CASE. *Principles of Engineering Economic Analysis,* 3rd ed. New York: John Wiley & Sons, 1989.

WOODS, D. R. *Financial Decision Making in the Process Industry*. Englewood Cliffs, NJ: Prentice-Hall, 1975

YOUNG, D. *Modern Engineering Economy*. New York: John Wiley & Sons, 1993.

ZELENY, M. *Multiple Criteria Decision Making*. New York: McGraw-Hill, 1982.

Answers to Selected Even-Numbered Problems

CHAPTER 2

2-6.
  a. $D^* = 227$ units/year
  b. $\dfrac{d^2 \text{ (profit)}}{dD^2} = -4.4$

2-8. $D^* = 240$ units/month; maximum profit = $4,960/month

2-10.
  a. $D^* = 5,685,000$ board feet/month; maximum profit = $615,691/month
  b. $D'_1 = 2,175,000$ board feet/month; $D'_2 = 9,195,000$ board feet/month

2-12. $D^* = 1,000$ units/month; maximum profit = $0/month

2-14. $D' = 3,112$ pumps/month; reduced costs result in a 22.75% reduction in breakeven point

2-16. Minimum C_u occurs at an output volume of $12,000,000/year

2-18.
  a. $D' = 40,000$ units/year
  b. Profit (90% of capacity) = $2,500,000/yr; Profit (100% of capacity) = $2,800,000/yr

2-20. Harvest crop in 2.5 weeks to receive maximum revenue = $6,125

2-22. Velocity = 10.25 mph

2-26. Select tool steel to minimize overall cost/piece ($0.71/piece)

2-28.
  a. Purchase item (cost = $8.50/item)
  b. Manufacture item (cost = $8.65/item)

2-30. Recommend renting a ditch-digging machine (total cost = $14,464)

2-32. Select the method capable of recovering 62% of the available ore deposit

2-34.
  a. Breakeven point = 2,500 miles
  b. New breakeven point = 1,100 miles
  c. Rent car from Motor Pool

CHAPTER 3

3-2. $I = \$7,560$

3-4. $200 (years 1–4); $100 (years 5–8); total interest paid = $1,200

3-6.
  a. Total interest paid = $2,400
  b. Year 3 principal payment = $2,070.60; total interest paid = $1,660.60

3-8. Alternative A: $P_0 = -\$200,588$; Alternative B: $P_0 = -\$196,061$. Alternative B should be selected to minimize equivalent costs

3-10. $F = \$2,961$

3-12. Total interest = $6,000; total interest paid in 3-11 = $6,380

3-14. $i = 14.0\%$

3-18. $P = \$20,113$

3-20. $A = \$3,693$

3-22.
   a. $N = 8$ years   **b.** $i = 15.11\%$
   c. $P = \$720.96$ **d.** $A = \$277.40$

3-24.
   c. $P_0 = 8.3333A$

3-26. $F_4 = \$14,490$, select E

3-28. $N = 60$ years

3-30. $A = -\$677.81$

3-32. $P = \$32,309.80$

3-34. $P_0 = \$11,090.40$

3-36. $F_{12} = \$3,269.12$

3-38. $A_1 = \$797.47; A_2 = \$617.40$

3-40. $P_0 = \$384.56$

3-42. $W = \$714.25$

3-44. $A = \$1,417.13$

3-46. $F_{12} = \$9,409.09$

3-48. $P_0 = \$471.37; A = \$90.55$

3-50. P_0 (rental income) $= \$7,920.56 < \$8,000 = P_0$ (investment), therefore not a good investment

3-52. $Z = \$1,256.05$

3-54. $P_0 = \$981.30$

3-56. $P_0 = \$100(P/A, 10\%, 4) + \$100(P/G, 10\%, 8)$

3-58. $A = \$127.59$

3-60. $A = \$657.77$

3-62. $Q = \$435.75$

3-64. $B = \$11,050$

3-66. $P_0 = \$3,917.90$

3-68.
   a. $P_0 = \$61,217.76$ **b.** $A = \$12,323.14$
   c. $A_0 = \$9,345.79$

3-70. $P_0 = \$23,853.75$

3-72.
   a. $i_{CR} = 2\%$ **b.** $P_0 = \$36,204.86$
   c. $P_0 = \$29,896 + 34.22G$
   d. $G = \$184.36$

3-74. $F = \$24,653.61$

3-76. $A = \$1,430$; select D

3-78.
   a. $A = \$169.60; i_{eff/year} = 6.17\%$
   b. $i_{eff/qtr} = 1.51\%$

3-80. $P_0 = \$3,153.40$

3-82. $F = \$20,138.25$

3-84. $F_7 = \$374.10$

3-86. $P_4 = \$99,058.32$

3-88. $P_0 = \$11,359$

3-90. $A = \$312$; select C

3-92. $A = \$558.68$

3-94. $P_0 = \$4,653.33$

3-96.
   a. False   **b.** False   **c.** False
   d. True   **e.** False   **f.** True
   g. False   **h.** False   **i.** False

3-98. $Z = \$4,837.88$

3-100. $P_0 = \$601.04$

3-102. $A = \$880.44$

3-104.
   a. $F = \$708,410.83$ **b.** $A = 129,248.46$

3-106. $N = 5$ years

3-108. Difference $= \$634.50$

CHAPTER 4

4-2.
   a. PW $= \$5,717.70$ **b.** IRR $= 15.3\%$

4-4.
   a. PW $= \$2,911.60$; FW $= \$5,855.60$; AW $= \$868.70$
   b. IRR $= 27.2\%$, yes, the IRR is acceptable
   c. ERR $= 21\%$

4-6. PW $= -\$171,452.50$, the company should not invest in the new product line

4-8. $V_N = \$5,762.25$

4-10.
   a. $V_N = \$7,689$ **b.** $A = \$150,893$

4-12. $i = 7.5\%$ every six months

4-14. $\$5,671.40$ was repaid on September 1

4-20. AW $= -\$4,429.61$, the building was not a good investment

4-22.
   a. PW$(\infty) = -\$3,000$ **b.** $\theta' = 6$ years
   c. AW $= -\$166.70$

4-24. A.P.R. $= 22.8\%$ compounded monthly

4-26. IRR $= 33.1\%$, the investment should be made

4-28. Company could pay up to $19,778 for the new brake

4-30. IRR = 21.5%

4-32.
   a. FW = −$1,128.30, project is not acceptable
   b. Since PW(years 0–5) = −$43,614.70, therefore $\theta' > 5$ years and the project is not acceptable

4-34. ERR = 12.3%

4-38. $i_{\text{eff/year}}$ = 40.9%

4-40.
   a. FW = $12,094.14   **b.** PW = $470.44

4-42. $i_{\text{eff/year}}$ = 28.3%

4-44. i' = 1.24% per year

4-46. $i_{\text{eff/year}}$ = 1.55%, select D

4-48.
   b. θ = 4 years

4-50.
   a. IRR = 24.7%, project is profitable
   b. θ = 4 years, project is not acceptable

4-52.
   b. ERR = 25.9%

CHAPTER 5

5-2.
   a. Select design 3 (AW = $141.10)
   b. Select design 3 (FW = $2,886.16)

5-4. Select the apartment house (AW = $32,016)

5-6.
   a. Select product 2 (PW = $12,897)
   b. IRR$_\Delta$ = 10.4%, select product 2

5-8. Select design D3

5-10. Both methods result in the selection of Alternative B

5-12.
   a. Select Alternative C (AW = −$25,781)
   b. Select Alternative C

5-14. Both methods select motor B (PW = −$3,470.54)

5-16. Assume repeatability, select Alternative C (AW = $60.00)

5-18. Recommend boiler A (AW = −$17,584)

5-20. Design 2 is the preferred design (PW = $5,790)

5-22.
   a. Select Alternative 1, AW = $4,552
   b. Select Alternative 1, AW = $4,552
   c. Select Alternative 1, FW = $30,933

5-24.
   a. Select Alternative E1 (AW = −$16,990)
   b. E2: AW = −$19,255

5-26. Select design D1 (CW = −$147,000)

5-28.
   a. CW = $34,591   **b.** N = 80 years

5-30. Select crane B with auger attachment (AW = −$60,366)

5-32. Select projects A1 and C1 for investment

5-34. Select projects X and Y for investment (PW = $17,520)

5-36. Select Alternative S1 (CW = −$150,927)

5-38. A 50-story building should be recommended

5-40.
   a. Select Alternative S
   b. Select Alternative S (AW = −$8,788)

CHAPTER 6

6-6.
   a. Projects B, C, and E are acceptable for funding
   b. B > C > E (> D > A)
   c. Select project B

6-8. Select RS-511

6-10.
   a. Option B should be recommended
   b. Select option B

6-12. If an option must be chosen, build flood-control dam

6-14. The toll bridge should be constructed (B/C = 1.28)

6-16. Construct the levee (ΔB/ΔC = 1.17)

6-18.
   a. Select design 3   **b.** Select design 3

CHAPTER 7

7-8.
   a. d_3^* = $180,550.50, select #5
   b. d_4 = $14,449.50, select #3
   c. BV$_2$ = $43,329, select #4

7-12.
a. $d_3 = \$3,428.57$; $BV_5 = \$42,857.15$
b. $d_3 = \$5,485.71$; $BV_5 = \$32,571.43$
c. $d_3 = \$6,297.38$; $BV_5 = \$27,759.86$
d. $d_3 = \$10,494$; $BV_5 = \$13,386$
e. $d_3 = \$4,285.71$; $BV_5 = \$40,714.30$

7-14. $PW_\Delta = \$26,865$

7-16.
a. $d_3 = \$21,984$ **b.** $BV_4 = \$19,785.60$
c. $d_4^* = \$70,014.96$

7-18.
a. cost depletion $= \$280,000$
b. % depletion allowance $= \$236,250$

7-20. Depletion unit (year 2) $= \$0.371/\text{mcf}$

7-22.
a. Federal income tax paid $= \$578,000$
b. NIAT $= \$1,122,000$
c. ATCF $= \$2,322,000$

7-24. $t = 37.96\%$; $t = 41.92\%$

7-26. Federal income tax with venture $= \$11,250$; Federal income tax without venture $= \$9,250$

7-28. $PW_\Delta = \$2,439$, so the new grinder should be purchased

7-30. BTCF $= \$31,200$

7-32.
a. The new equipment should not be purchased ($PW = -\$767$)
b. The new equipment should not be purchased ($PW = -\$807$)

7-34.
a. MARR (before taxes) $= 25\%$
c. $MV_8 - BV_8 = \$10,000$
e. Do not purchase the new machine ($PW = -\$25,082$)

7-36. $PW = \$19,116$

7-38.
b. AW $= -\$37,115$

7-40.
a. $d_4 = \$741$
b. Income tax owed $= \$800$
c. $ATCF_2 = -\$1,846$

CHAPTER 8

8-6. $C_{1996} = \$262,780$

8-8. $C = \$232,400$

8-10.
a. $Z_8 = 108.0$ hours, $Z_{50} = 94.3$ hours
b. $C_5 = 117.5$ hours/design

8-12. % Reduction in overhead costs $= 31.4\%$ (with respect to present costs)

8-14. Manufacturing cost $= \$221.43/100$ wire cutters; Selling price $= \$248/100$ wire cutters

8-16. Initial estimate of manufacturing cost $= \$2,285/\text{computer}$; Target cost $= 2,000/\text{computer}$

CHAPTER 9

9-2. $N = 9$ years

9-4. Alternative B has least negative PW at time 0 ($PW = -\$369,070$)

9-8. $C_A = \$1,027,299$

9-12. $N = 5$ years

9-14.
a. FW (in A$) $= \$144,105$
b. FW (in R$) $= \$44,932$

9-16. Product A: $\bar{f} = 6.00\%$; Product B: $\bar{f} = 8.33\%$

9-20. $PW = -\$12,234$

9-22.
a. AW (in A$) $= -\$1,859$
a. AW (in R$) $= -\$1,309$

9-24.
a. Investment $= -\$393,790,000$
b. Investment $= -\$393,790,000$

9-26. $PW = -\$359,665$

9-28.
a. Maximum total investment (including new equipment) $= -\$356,557$

9-30.
a. $i_{fc} = 36.08\%$ **b.** $i_{fc} = 18.44\%$

9-32. Yes, the project is acceptable, $PW(i_{US} = 22.7\%) = \$70,583,300$

9-34. The investment appears to be acceptable ($PW = \$10,263$)

CHAPTER 10

10-6.
a. $H = 867$ hours/year
b. Select XYZ brand motor (AW $= -\$17,987$)

10-8. Electrical energy cost $= 1.88$ cents/kW-hr

10-10.
  a. The project is economically attractive (AW = $232,625)
  b. The decision is most sensitive to changes in occupancy rate

10-18.
  a. Optimistic: AW = $23,330; Most likely: AW = $14,325; Pessimistic: AW = −$9,184

10-20. Build the four-lane bridge now (PW = −$350,000)

10-22. Select the ABC brand motor (AW = −$9,831)

CHAPTER 11

11-2. Keep the old car, IRR_Δ = 14.3%

11-4. Keep the old engine

11-6.
  a. Make the replacement now
  b. Wait at least two more years before making the replacement

11-8. Keep the defender (PW = −$69,836)

11-10. Select the challenger, Alternative B (PW = −$1,440,423)

11-12. Keep the defender (PW = −$3,602)

11-14. Economic life = 5 years (EUAC = $5,382)

11-16. Economic life = 6 years (EUAC = $3,571)

11-18.
  a. Defender: N^* = 1 year, Challenger: N^* = 3 years
  b. Replace defender two years from now

11-20. Keep the defender, IRR_Δ = 1.1%

11-22. Breakeven purchase price = −$216,480

11-24. Maximum affordable price = −$108,508

11-26. Relocate existing transformers (Alternative B), PW = −$4,239

CHAPTER 12

12-8. A = $302,500

12-14.
  a. Lease the machine, AW = −$500
  b. Lease the machine as long as the annual lease rate is ≤ $1,410

CHAPTER 13

13-8. Select Alternative B, $\overline{RR}$ = $17,498

13-10. Company must be able to purchase power for 18.25 mils per kilowatt-hour.

13-12.
  a. $ATCF_5$ (in A$) = −$1,641
  b. T_5 (in A$) = $3,543

13-14. Build geothermal generating plant ($\overline{RR}$ = $525,088)

13-16. Alternative A is marginally preferred to Alternative B ($\overline{RR}_A$ = $145,056; $\overline{RR}_B$ = $145,338)

CHAPTER 14

14-2. E(PW: 4-lanes now) = −$3,500,000 and E(PW: 2 now + 2 later) = −$3,839,500, ∴ build the 4-lane bridge now

14-4. E(concrete) = 1,350 cubic yards; V(concrete) = 66,500 (cubic yards)^2; SD(concrete) = 258 cubic yards

14-6. Implement design E; E(PW) = −$239,414

14-8. Do not install lift; E(PW) = −$85,142

14-10.
  a. E(ACF) = −$2,825; V(ACF) = 1,789,175 ($)^2; SD(ACF) = $1,338
  b. Pr(ACF > −$2,394) = 0.37

14-12.
  a. E(PW) = $175; V(PW) = 28.04 × 10^6 ($)^2; SD(PW) = $5,295
  b. Pr(PW ≥ 0) = 0.68

14-14. E(PW) = −$7,599; SD(PW) = $20,118; Pr(PW ≤ 0) = 0.70

14-16. Pr(AW < $1,700) = 0.5871

14-18.
  a. E(PW) = $1,354; V(PW) = 1,639,240 ($)^2
  b. Pr(PW ≥ 0) = 0.8554

14-22. Build the culvert, PW = −$81,448

14-24. Select don't buy (PW_{buy} = −$340)

14-26. Select Alternative 2, E(PW) = $61,839; EVPI = $9,089

Index